Seaweeds of the British Isles

A collaborative project of the British Phycological Society
and the Natural History Museum

Volume 3 Fucophyceae (Phaeophyceae)

Part 1

Robert L Fletcher

Natural History Museum, London

First published by the Natural History Museum,
Cromwell Road, London SW7 5BD

This edition printed and published by Pelagic Publishing, 2011,
in association with the Natural History Museum, London

ISBN 978-1-907807-11-4

This book is a reprint edition of 0-565-00992-3.

A catalogue record for this book is available from the British Library.

Seaweeds of the British Isles

Publishing in six volumes

Volume 1	*Rhodophyta* *
Volume 2	*Chlorophyta*
Volume 3	*Fucophyceae (Phaeophyceae)***
Volume 4	*Tribophyceae*†
Volume 5	*Cyanophyta*
Volume 6	*Prymnesiophyta*

*In three parts – Parts 1 and 2A are available now.
**In two parts.
†Available now.

Contents

Preface

The present book represents the first part of a contributory volume (Vol. 3) to the series *Seaweeds of the British Isles* which is a collaborative project of the British Phycological Society and the British Museum (Natural History). It is concerned with the brown algae or Fucophyceae recorded around the coasts of the British Isles and will be published in two parts. It was at the first Annual General Meeting of the British Phycological Society, held in London in January 1953, that the initial decision to begin a new British marine algal Flora was taken, and within the newly formed 'Flora Committee', Dr Helen Blackler and Dr Mary Parke were given the responsibility for the brown algae. Following the resignation of Dr Blackler from the Committee in 1967, Dr George Russell was asked to prepare the brown algal section of the flora and in 1972 invited Dr Robert Fletcher to assist him in this task. With financial support from the Natural Environment Research Council, Dr Fletcher completed work on 12 families of brown algae, the written account of which forms the content of the present part 1. This part also contains an Introduction (jointly prepared by Dr R. L. Fletcher and Dr G. Russell), provisional generic key and a glossary. Part 2 will treat the remaining families and will contain a full generic key, glossary and index.

Acknowledgements

I would like to express my sincere appreciation to the following: Professor E. B. Gareth Jones, without whose encouragement, advice, continued support and assistance throughout my period at Portsmouth Polytechnic this book would not have been produced; the Natural Environment Research Council for both full-time and part-time financial support between 1978–1984 (NERC grants GR3/3500 and GR3/4690 to Professor E. B. Gareth Jones); Dr G. Russell for supplying specimens and records and offering much useful advice on the brown algae; Dr D. E. G. Irvine for his detailed critical and valuable examination of the manuscript in his capacity as scientific editor of the present volume; Mrs L. Irvine for considerable help and advice with nomenclatural problems; members of the British Phycological Society's Flora Committee and in particular Mr P. James for all his support, patience and editorial assistance during the final stages of manuscript submission; Dr P. Pedersen for reading the manuscript and making many useful comments: Dr W. F. Farnham for providing specimens, particularly from the sublittoral, and for the generous and unselfish use of his library; Dr Y. M. Chamberlain for her advice and many helpful discussions; Mr J. Price and Mr I. Tittley for advice and useful discussions on the brown algae; the staff of the Publications Section at the British Museum (Natural History) for seeing the book through the press; the technical staff at the Marine Laboratory, Portsmouth Polytechnic, especially Mr J. Hepburn, Mr N. Thomas and Mrs A. Davis: Mr A. E. W. Hawton, Mr C. Derrick and Mr K. Purdy for printing some of the photographs; Mrs P. Davies for her valuable and helpful assistance in typing out the numerous drafts of the text.

I am also greatly indebted to the following for the loan or gifts of specimens, records, etc: Dr J. Berryman, Dr M. C. H. Blackler, Professor G. Blunden, Dr C.-F. Boudouresque, Dr E. M. Burrows, Dr Y. M. Chamberlain, Dr J. J. P. Clokie, Dr A. Critchley, Mrs T. Edelstein, Dr W. F. Farnham, Dr P. W. G. Gray, Dr M. D. Guiry, Dr K. Hiscock, Mrs S. Hiscock, Dr R. G. Hooper, Mr S. I. Honey, Miss C. Howson, Dr D. E. G. Irvine, Mrs L. Irvine, Mr N. A. Jephson, Dr D. M. John, Dr W. E. Jones, Dr J. M. Kain, Dr C. A. Maggs, Dr J. McLachlan, Mr O. Morton, Dr I. Munda, Professor T. A. Norton, Dr P. Pedersen, Mr M. J. Picken, Mr H. T. Powell, Mr J. Price, Dr W. Prud'homme van Reine, Dr O. Ravanko, Dr G. Russell, Professor S. C. Seagrief, Professor G. R. South, Dr L. Terry, Mr I. Tittley and Dr R. T. Wilce.

I am indebted to the Directors and Curators of the following institutions for working facilities, for permission to examine specimens or borrow material:

Botanical Museum and Herbarium, Copenhagen (Denmark) C
British Museum (Natural History), London BM

Department of Botany, The University, St Andrews
Department of Botany, The University, Glasgow GL
Department of Botany, University College, Galway
Department of Biology, Memorial University of Newfoundland, St Johns
Dove Marine Laboratory, University of Newcastle upon Tyne, Cullercoats
Ilfracombe Museum, Ilfracombe ILF
Laboratoire de Cryptogamie, Museum National d'Histoire Naturelle, Paris (France) PC
Marine Biological Association, Plymouth MBA
Marine Science Laboratories, University College of North Wales, Menai Bridge MB
Marine Biological Station, Port Erin
Millport Laboratory, Cumbrae
Rijksherbarium, Leiden (Netherlands) L
Rochester Museum, Rochester
Royal Botanic Gardens, Kew. The algae previously at Kew are now on permanent loan to the British Museum (Natural History) BM-K
School of Plant Biology, University College of North Wales, Bangor UCNW
Scottish Marine Biological Association, Dunstaffnage, Oban
Smithsonian Institution, National Museum of Natural History, Washington (USA)
Ulster Museum, Belfast BEL
Wellcome Marine Laboratory, University of Leeds, Robin Hood's Bay
N.B. The Standard Herbarium Abbreviations are given

All the drawings of microscopic details and some of the habit illustrations are by the present author. Most of the habit illustrations have been prepared by Mrs S. M. Fletcher and Mr N. Thomas to whom grateful acknowledgement is given.

Finally, but most importantly, I would like to thank my wife, Sylvia, for her never-failing patience, encouragement, interest and support during my work on this book.

Introduction

According to Davis & Heywood (1963) a flora should encompass three distinctive elements: it is firstly a classification into families, genera and infra-specific taxa; secondly, it is a means to identification by the provision of keys, synopses and short descriptions; thirdly, it is a compendium of data and information which may be abstracted for further purposes. While accepting these tenets in Davis & Heywood's analysis, it is our opinion that the second of these functions is paramount and that floras are consulted chiefly by those who wish to identify a plant which they have collected. Consequently, with this major objective in mind we have tried to make this volume as utilitarian as possible.

Davis and Heywood (1963) also state that although floras contain the summation of a large number of decisions concerning the identity, circumscriptions, variability and distributions of taxa, seldom do they give much importance to including information on how such decisions were arrived at. We have tried to avoid falling into this error but, if found wanting, we offer our apologies.

In the present work we accept the circumscription of the class as defined by Round (1973), Bold & Wynne (1978), Hoek & Jahns (1978) and Christensen (1980). The last-named author has proposed a nomenclatural change from Phaeophyceae to Fucophyceae. This has not, as yet, been generally adopted but in order to avoid possible eventual confusion both names are cited here.

The present introduction provides a brief summary of the Fucophyceae and includes aspects such as structure, morphology, anatomy, cytology, reproduction and ecology. It is intended to contribute towards a general understanding of the biology of the group and assist in the use of the present brown algal flora of the British Isles. A number of the more critical references are cited; however, for more general information, the reader is referred to the treatments of Smith (1955), Fritsch (1945), Papenfuss (1951), Scagel (1966), Chapman & Chapman (1973), Russell (1973a), Bold & Wynne (1978), Lobban & Wynne (1981) and Clayton & King (1981).

CELL STRUCTURE

The structure of the principal organelles of the brown algal cell had been elucidated by transmission electron microscopy (TEM) by the early 1970s, and since that time relatively little new information has been added (Russell, 1973a; Brawley & Wetherbee, 1981). The occurrence of a heterokont flagellar organisation and the stacking of plastid thylakoids in triplets may be indicative of a common ancestry with other classes in the division Chromophyta *sensu* Christensen (1980). Sperm flagellar fine structure (Manton, 1964) and egg number (Jensen, 1974) have also been used to identify primitive and derived characteristics in the order Fucales, see also Clayton (1984).

The more recent addition of scanning electron microscopy (SEM) to phycology has given less information on cell structure than for other algal groups, although observations on spore/zygote germination (Hardy & Moss, 1978, 1979a, b, c; Fletcher, 1976, 1977, 1981b; Ramon, 1973) and surface structure (Fletcher, 1978) indicate a potential application of this technique.

These, and other investigations of the brown algal cell, are valuable in pointing to possible evolutionary relationships between taxa but have not brought about any major changes to the overall taxonomy of the group.

For a generalised account of the structure of a brown algal cell reference can be suitably made to Bouck (1965), Russell (1973a) and Brawley & Wetherbee (1981) (see Fig. I). Within the multilayered enclosing wall, the protoplast contains a single, usually central, granular nucleus, variable in shape, on average 5–6 μm in diameter (Russell, 1973a) and is usually reported with one, sometimes two, nucleoli. The nucleus is surrounded by a double-membraned, often pored, nuclear envelope in which continuity has been reported with both cytoplasmic endoplasmic reticulum and the plastid endoplasmic reticulum. There are one to several plastids containing the photosynthetic pigments which give the characteristic colour to the brown algal cell. Sometimes associated with the plastids are small hyaline bodies, the pyrenoids, which probably function in carbohydrate synthesis and/or polymerization. The main food storage material is a soluble

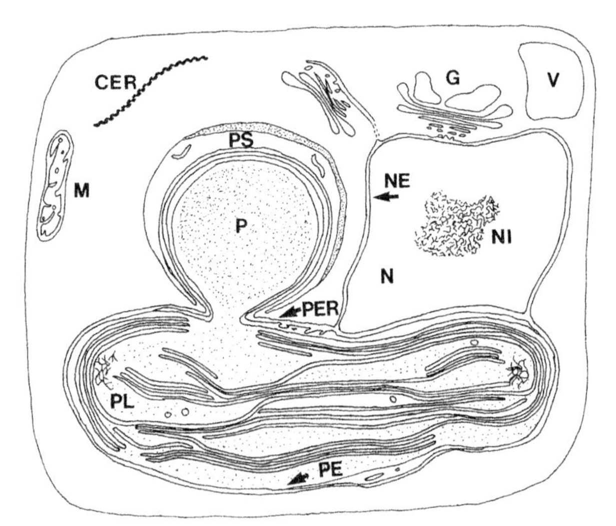

Fig. I Diagram of a hypothetical brown algal cell (reproduced from G. B. Bouck, the *Journal of Cell Biology* 1965, V. 26, 523–537 by copyright permission of the Rockefeller University Press); see text for details of organelle associations. Organelles illustrated include cytoplasmic endoplasmic reticulum (CER), mitochondrion (M), plastid envelope (PE), plastid endoplasmic reticulum (PER), nucleus (N), nuclear envelope (NE), nucleolus (NI), plastid (PL), pyrenoid (P), pyrenoid sac (PS), Golgi bodies (G) and vacuoles (V).

polysaccharide, laminaran, which can constitute up to 34 per cent of the algal dry weight (Craigie, 1974); also present is mannitol. These are not stored in recognisable grains or sheaths, as in the red and green algal groups respectively, but are probably distributed within the cytoplasm.

One to several variously shaped Golgi bodies are often present, either perinuclear in position or scattered throughout the cytoplasm. These are the packaging units within the cell, producing vesicles which are widely variable in shape, chemical composition and function. Not surprisingly, they are, therefore, found to be particularly active in spore differentiation and settlement (Baker & Evans, 1973a, b), sulphated polysaccharide production and secretion (Evans *et al.*, 1973; Evans, 1974) and cell wall formation (Fletcher, 1981b; Brawley & Wetherbee, 1981; Callow *et al.*, 1978).

Other cytoplasmic inclusions include varying numbers of mitochondria, endoplasmic reticula (ER) and a diverse range of vesicles, vacuoles and bodies.

Several of the above mentioned organelles have proved quite valuable features in the assessment of taxonomic interrelationships as well as in both generic and specific identification, and will be considered in more detail. These include cell walls, plastids, pyrenoids, physodes and iridescent bodies.

The cell wall

With the exception of reproductive spore bodies all algal cells are surrounded by a cell wall, commonly of two layers although multiple wall layers have also been described. For example a triple cell wall structure has been reported by Hardy & Moss (1979c) in the female gametangium of *Pelvetia,* whilst Parker & Philpott (1960) reported 25 cell wall layers in their study of a winter population of *Fucus*. In general, the cell wall layers are fibrillar in composition although they may well be arranged in different structural patterns. In the majority of cells studied the outer wall has a reticulate structure and the inner layer a parallel arrangement. Histochemical and autoradiographic studies have revealed the principal components of the cell walls as alginic acid, sulphated polysaccharides, including fucoidan, cellulose, carboxylated polysaccharides and laminaran. In species of the genus *Padina* only, the cell walls are also impregnated with calcium carbonate, giving the thallus a more rigid texture.

Very often the surface cells are covered by a 'cuticle'. The thickness of this cuticle varies for different algae and is also probably determined by inherent characteristics as well as by the surrounding environmental conditions. It is often clearly layered and the outermost layer in some algae, such as *Ascophyllum, Fucus* and *Himanthalia,* has been observed to slough off continually, a process considered to reduce the detrimental build up of surface epiphytes.

An additional surface and intercellular component of cell walls in many brown algae is mucilage, the principal constituent of which is sulphated polysaccharide. In some genera, such as *Laminaria,* mucilage production is so copious as to render the thallus surface very slimy. In this case specialised secretory cells are involved in production of mucilage which is then transported via canals to the thallus surface (Evans *et al.*, 1973). In members of the Fucales, mucilage is also synthesised and secreted by active epidermal and outer cortical cells (Evans & Callow, 1976). The hypothesis of Evans *et al.* (1973) suggests that the amount of sulphated polysaccharide retained by the cell of an alga is related to its degree of exposure to the air. For example, *Pelvetia canaliculata* and *Fucus spiralis*, which are found high on the shore contain large amounts of fucoidan (18–24 per cent on a dry weight basis) compared to the Laminariales and *Fucus serratus*, members of the lower shore community, which only contain 5 per cent fucoidan on a dry weight basis. It was suggested that the hygroscopic nature of sulphated polysaccharide served to reduce desiccation.

The intercellular mucilage might well play an important role in determining the cohesion of cells and filaments of thalli. The cohesive characteristics of cells and filaments are of considerable importance in specific and generic determination. For example, the ease of separation of filaments during squash preparations is a very useful taxonomic character for identification of the crustose members of the Lithodermataceae and Scytosiphonaceae as well as members of the Chordariaceae. Extracellular and intercellular mucilage might also be responsible for the adhesive nature of the attaching rhizoids (see page 21).

An additional feature of cell walls observed throughout the brown algae is the presence of pores which permit some degree of cytoplasmic continuity between cells. In adjacent cells the pores are either variously distributed or grouped in distinct pit-fields. Very often cytoplasmic threads (plasmodesmata) pass through the pores, connecting the cytoplasm of adjacent cells. Not surprisingly, therefore, these pores are considered to be secretory in function. They are particularly numerous in cells of members of the Laminariales.

Plastids

All cell types, including both vegetative and reproductive cells, possess at least one plastid, usually located in the peripheral cytoplasm, although in spermatozoa and in hair cells, plastids may be very reduced in size. The number and shape of the plastids within a cell are important taxonomic characters and are commonly used to assist in the identification of both genera and species (Fig. II, 1–6). Four morphological forms are usually recognised: stellate, discoid, plate-like and ribbon-like. Examples include: one or two axile, stellate plastids in *Bachelotia*, a single, parietal, plate-like plastid in cells of all members of the Scytosiphonaceae, one to several lobed plate-like plastids in various genera such as *Compsonema* and *Myrionema*, and several parietal, ribbon-shaped plastids in *Ectocarpus*. In the great majority of brown algae, however, all cells contain several regularly, or sometimes irregularly, shaped discoid plastids.

The photosynthetic pigments present are chlorophylls a and c; these are associated with carotenoid accessory pigments, the most important being carotene, violoxanthin and fucoxanthin, along with traces of diatoxanthin and diadinoxanthin (Meeks, 1974; Goodwin, 1974). It is the abundance of fucoxanthin which gives the Fucophyceae their characteristic brown coloration although the depth of pigmentation will be variously influenced by the abundance of the other coloured pigments. For example, brown algae range in colour from olive-brown (*Colpomenia peregrina, Punctaria latifolia*), to yellow-brown (*Petroderma maculiforme, Corynophlaea crispa, Leathesia difformis, Bifurcaria bifurcata*) to dark brown/black (*Ralfsia verrucosa, Battersia mirabilis, Halidrys siliquosa*). The colour of an alga will also be determined by its physiological condition; algae in various stages of senescence will often appear green through loss of carotenoids, this effect being particularly well demonstrated in species of *Laminaria* and *Desmarestia*. Also, very often fertile regions of some members of the Fucales are distinctly yellow in colour, e.g. the male receptacles of *Ascophyllum nodosum* and *Fucus serratus*, possibly reflecting a lower chlorophyll content of these tissues.

Pyrenoids

An organelle commonly associated with the plastid in a number of brown algae and usually clearly discernible under the light microscope as a small hyaline disc is the pyrenoid (Fig. II, 1, 2 & 6). Under the electron microscope it comprises a granular, dense body, either embedded in the plastid, or projecting from its side on a stalk. It lacks penetrating thylakoids and is bounded by the plastid endoplasmic reticulum, the plastid envelope and an enclosing sheath or cap. It appears to be involved in polysaccharide synthesis (Kerby & Evans, 1978).

The distribution of pyrenoids in the families of the Fucophyceae is thought to be of phylogenetic importance (Evans, 1966; Hori, 1971, 1972) in that they are present in the lower orders such as the Ectocarpales but generally absent in the higher orders such as the Dictyotales, Laminariales and Fucales. However, their reported presence in these latter orders, either rarely or confined to certain reproductive cell types, casts serious doubt on the phylogenetic relevance of this character (Bold & Wynne, 1978; Evans, 1968; Chi, 1971; Prud'homme van Reine & Star, 1981; Hori, 1972).

Eyespots

Another notable organelle associated with the plastid is the eyespot. Eyespots appear to be confined to the motile reproductive spore and have been described in zoospores of, for example, *Ectocarpus, Pilayella* and *Scytosiphon* (Lofthouse & Capon, 1975; Markey & Wilce, 1975; Manton, 1957), in sperm of *Cutleria, Fucus* and *Ascophyllum* (La Claire & West, 1979; Manton & Clarke, 1951; Bouck, 1970) and in the female gametangium of *Cutleria* (La Claire & West, 1978). They comprise groups of spherical, osmiophilic, carotenoid granules usually projecting slightly from the surface of the plastid and reported to be coloured red or orange under the light microscope (Manton, 1957; Bouck, 1970; Dodge, 1973). Very often these red spots can be seen associated with spores contained within sporangia on the parental thallus. From electron microscope studies there have been reports of a possible association between the eyespot and flagella (Manton & Clarke, 1951; La Claire & West, 1978; Berkaloff & Rousseau, 1979; Bouck, 1970; Dodge, 1969). They are considered probably to function as primitive photoreceptors (Bisalputra, 1974) and are therefore involved in the phototactic responses.

Flagella

The great majority of brown algae produce motile reproductive spore bodies. Examples of non-motile spores are the tetraspores produced by members of the Dictyotales and the female gametes produced by the oogamously reproducing members of the Laminariales, Desmarestiales and Fucales. Most motile cells are pear-shaped and possess two laterally inserted flagella with the anterior directed flagellum longer than the posterior (Fig. II, 7). However, variations have been described; for example, a single flagellum is reported for the sperm cells of *Dictyota* (Manton *et al.,* 1953), whilst in the sperm of *Fucus* and *Ascophyllum* and *Laminaria* it is the posterior flagellum which is the longer of the two (Manton *et al.,* 1953; Maier & Muller, 1982).

A characteristic feature of the anterior flagellum is the presence of variously arranged and orientated, hair-like appendages or mastigonemes. These are also present on the single flagellum of *Dictyota* (Manton *et al.,* 1953). The posterior flagellum is devoid of these hairs although they were reported by Loiseaux & West (1970) on *Giffordia* zoospores. Additional ornamentations of the 'hairy' flagella include terminal filaments and spines. In *Fucus* the sperm have a proboscis-like appendage, extending from the anterior region of the cell, which is considered to play a role in the fertilisation process (Manton & Clarke, 1956). All flagella have the typical 9+2 arrangement of microtubles with a basal body in the cytoplasm (Evans, 1974).

Fig. II 1–5 Examples of different shaped plastids
Ribbon shaped plastids in *Ectocarpus fasciculatus* (1); plate-like plastids in *Colpomenia peregrina* (2); discoid plastids in *Desmarestia ligulata* (3), *'Aglaozonia'* phase of *Cutleria multifida* (4) and *Elachista fucicola* (5). Note hyaline, disc-like pyrenoids drawn in 1 and 2. 6. Portion of paraphysis in *Myriactula chordae* showing cells with discoid plastids and small associated pyrenoids. 7. Typical motile brown algal spore, with discoid plastids and laterally inserted flagella of unequal length. 8. Germinating spores of *Giffordia granulosa.* 9. Unipolar germination sequence in *Petalonia zosterifolia.* 10. Unipolar germination in *Ralfsia verrucosa,* but showing evacuation of contents of original spore cell and part of germ tube. 11. Stellate germination pattern in *Myrionema magnusii.* 12. Young germlings of *Giffordia granulosa* showing growth of erect filaments from original spore cells. 13. Young germling of *Petalonia filiformis* showing growth of erect blade from the proximal region of the branched filamentous base. 14. Young germling of *Giffordia granulosa* showing secondary shoot production (arrowed) from the basal filaments. 15. Young filamentous germling of *Petalonia zosterifolia.* 16. Young germling of *Petalonia fascia* with terminal disc. 17. Discoid germling produced in cultures of *Stragularia clavata.* 18. Young germling of *Sargassum muticum* with small, multicellular, erect shoot giving rise at the base to 7 rhizoidal filaments (one hidden). Bar = 20 μm (1, 2, 3, 4, 6, 7, 8, 9, 10, 11), 50 μm (5, 12, 15, 16, 17, 18), 100 μm (13, 14).

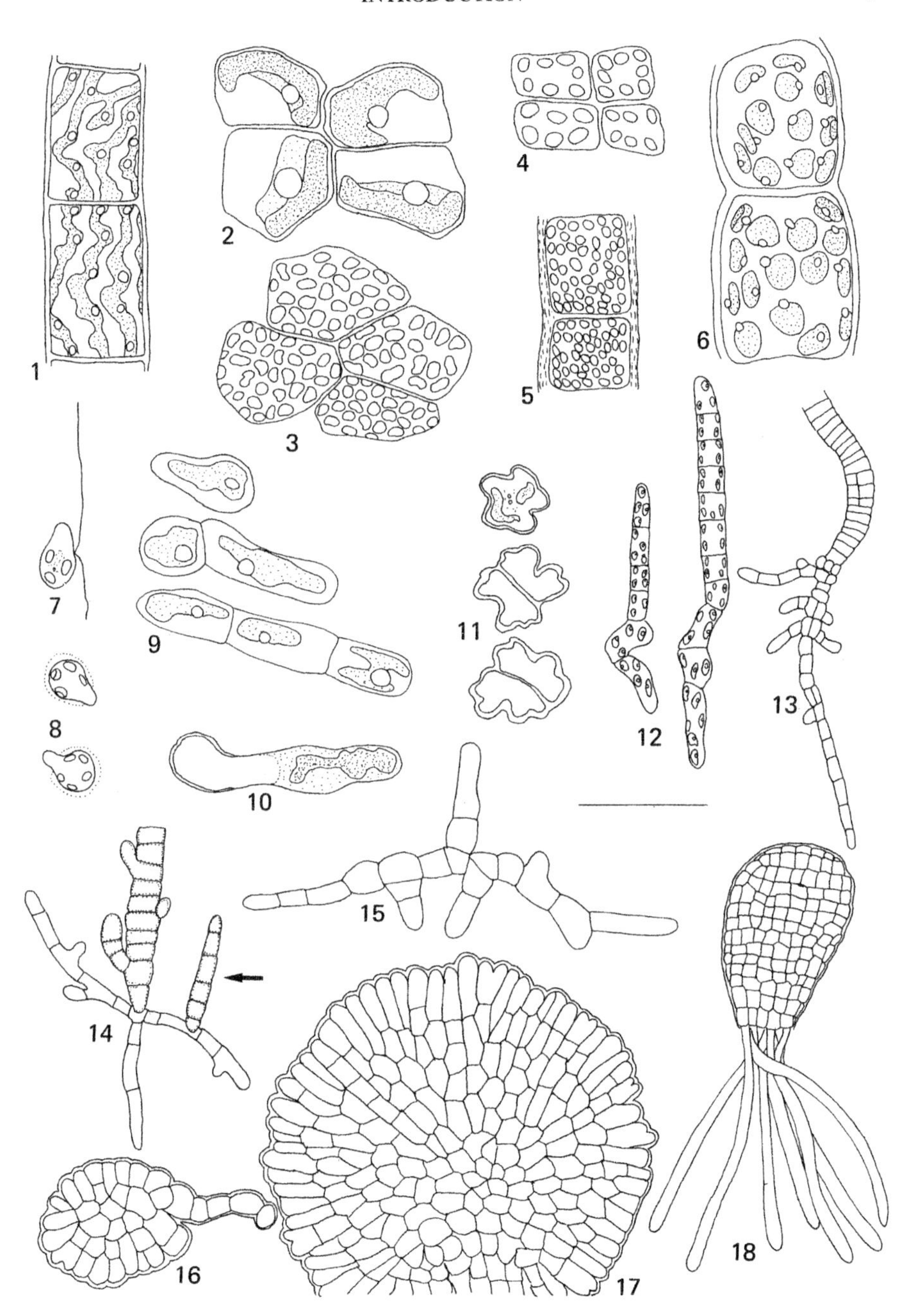
1
2
3
4
5
6
7
8
9
10
11
12
13
14
15
16
17
18

Vacuoles and vesicles

A wide variety of vacuoles and vesicles have been described from brown algal cells. Vacuoles are light-coloured, membrane-bound regions, usually centrally positioned within a cell and causing displacement of other cytoplasmic organelles, such as the nucleus and plastids, to the periphery. They are particularly common in the central, more inactive, cells of an algal thallus, such as medullary, corticating, and hyphal filament cells. Usually Golgi bodies are associated with their formation and they are attributed with various functions, particularly storage of water, inorganic ions and organic metabolites. Vacuoles are likely to be important in the responses of the cell to changes in external water potential.

Of the many vesicles described, particularly by recent E.M. studies, the physodes or fucosan vesicles are the most prominent. These can be colourless and refractile or darkly staining and were originally distinguished by a red coloration with vanillin HCl stains (Kylin, 1938). They appear to contain tannin-like phenolic substances, such as flavanols and catechin, which have been equated with antibiotic properties (Hellebust, 1974; Ragan, 1976); Clayton (1984) suggests they are a mechanism for protection against ultraviolet radiation. Their origin is usually attributed to the plastids, although the Golgi bodies have also been implicated. They are widely distributed in most families of the brown algae but reports of their presence are especially numerous for the higher orders, e.g. Dictyotales, Fucales.

Further noteworthy cellular inclusions are iridescent bodies which give the thalli of certain brown algae a blue or green iridescence when viewed under water. They have been reported in *Dictyota* and some *Cystoseira* spp. Electron microscope studies revealed them to be physode-like, membrane bound vesicles containing electron dense globules (Pellegrini, 1974; Feldmann & Guglielmi, 1972). They are proteinaceous with some polysaccharide in composition and probably originate from the Golgi bodies.

THE BROWN ALGAL THALLUS

Early development

Although the entrapment of vegetative plant portions may play an important role in the dissemination and colonisation processes of marine algae, including the Fucophyceae (Clokie & Boney, 1980) and may be the only reproductive method of algae at the geographical limits of their distribution, the great majority of algae reach new substrata by the production, release and settlement of specialised reproductive spore bodies.

In the great majority of brown algae, motile reproductive spores are produced and settlement of these commonly involves a certain amount of site selection. The non-motile cells formed in oogamously-reproducing algae and members of the Dictyotales (tetraspores) are more at the mercy of the water currents and rely on the more passive process of sinking under gravity. The motile spores appear to be well adapted to their role as substratum colonisers. For example they have been reported to remain motile for periods of up to 24 hours (Wynne, 1969; Baker & Evans, 1973a) and to detect and respond to external stimuli such as light (Baker & Evans, 1973a; Toth, 1976; Fletcher, 1981b) and surface topography (Müller, 1964; Russell & Morris, 1971).

Following settlement most brown algal spores initially become attached to a substratum by the release of an adhesive material (Toth, 1976; Baker & Evans, 1973a; Fletcher, 1981b). The spore adhesive has been shown by electron microscope studies to be derived from small vesicles produced by the Golgi bodies. Upon release through the plasmalemma the glue material, which is probably composed of a polysaccharide-protein complex, spreads out and forms a tenacious attachment to the substratum. Alternative attachment mechanisms of the settling stages recorded in brown algae include direct adhesion of the outer walls, as in *Pelvetia canaliculata* (Hardy & Moss, 1979c) and *Halidrys siliquosa* (Hardy & Moss, 1978) and the immediate production of adhesive rhizoids, as in *Sargassum muticum* (Fletcher, 1980).

Following attachment, the spores undergo a process of germination to form the adult plant (Fig. II, 8). Usually this occurs immediately, although there can be long periods of delay under adverse environmental conditions (Kain, 1979). Essentially germination is a dual process; after cell wall formation the spore initiates the development of two main morphogenetic pathways. These are (1) the formation of a prostrate anchorage system and (2) the formation of the characteristic, erect thallus. During the early stages of this germination process two patterns of development have been described (Pedersen, 1981); these are *unipolar* and *stellate* (Fig. II, 9–11). In unipolar germination a germ tube extends out from the spore and a cross wall is laid down; both cells contain plastids and a nucleus. The younger distal cell then continues growth and division to produce a filamentous germling. Occasionally two or three germ tubes can be produced from the spore cell, all developing in a similar manner. This germination pattern is very common and distributed throughout most families of brown algae. A slight modification of unipolar germination involves the cell contents of the original spore cell migrating into the germ tube; a cross wall is then laid down leaving the original spore cell empty in the proximal area. This germination pattern has been described in various algae, including *Laminaria* spp., (Kain, 1979), *Chorda tomentosa* (Toth, 1976) and *Ralfsia verrucosa* (Nakamura, 1972).

In the stellate germination pattern the spore forms 3–5 lobes which are later closed by oblique walls. Radial growth of these cells then produces a disc-shaped system. This is particularly characteristic of some epiphytic species and has been reported in various myrionematoid algae (Sauvageau, 1897; Loiseaux, 1967b) and the chordariacean algae *Cladosiphon zosterae* (Parke, 1933; Sauvageau, 1924) and *C. okamuranus* (Shinmura, 1974, 1977). These two germination patterns are not, however, mutually exclusive; for example, both types have been reported for some species, such as *Ralfsia verrucosa* (Fletcher, 1978) and even for spores derived from a single sporangium, a phenomenon referred to as *heteroblasty* (see page 30).

Coupled with this basal developmental process is the formation of the characteristic, erect shoots. Usually the latter are produced directly from the upper, polarised region of the original spore cell, although in many algae cells on the basal system can also carry out this function (Fig. II, 12–14). A fundamental aspect of these early stages of development is the extent of growth of the horizontally spreading, primary attachment system. In the least specialised members of the brown algae, development of this basal system is pronounced, either as branched, free filaments or branched, compacted filaments (pseudodisc/disc-like), from various cells of which the erect plant body is initiated (Fig. II, 15–17). This development of the plant body in two perpendicular planes is termed *heterotrichy* (Fritsch, 1945) and has been widely reported in such families as the Ectocarpaceae, Myrionemataceae, Elachistaceae, Chordariaceae and Punctariaceae. An extreme example of the heterotrichous condition can be found in some of the very reduced members of the Ectocarpaceae; in algae such as *Mikrosyphar, Phaeostroma* and *Streblonema* erect development is absent or reduced to a few short branches or hairs.

Non-heterotrichous plant bodies are representative of members of the more advanced brown algal groups. In the Dictyotales and Fucales as well as the sporophytes of the Desmarestiales and Laminariales there is no development of a prostrate vegetative system. There is the immediate development of the erect thallus which attaches only by the basal production of secondary rhizoids (Fig. II, 18). However, traces of the heterotrichous habit do persist in gametophytes of the Desmarestiales and Laminariales and are most evident when they are grown in suboptimal culture conditions.

Mode of construction

All brown algae are branched-filamentous, pseudoparenchymatous or parenchymatous in their mode of construction. There are no unicellular, colonial, unbranched-filamentous or siphonaceous forms such as are variously distributed among the green and red algae. The filamentous algae are generally considered to be less advanced and are common in the lower order Ectocarpales, and, in particular, the Ectocarpaceae (Fig. III). They may be little branched, as in *Compsonema, Acinetospora, Protectocarpus* and *Feldmannia,* or

Fig. III Examples of filamentous thalli
1. Irregularly branched filaments of *Streblonema* sp. 2. Single, branched, erect filament of *Protectocarpus speciosus*. 3. Erect, terminally branched filaments of *Compsonema microspongium*. 4. Portion of erect axis of *Giffordia granulosa* with opposite laterals. 5. Portion of erect axis of *Giffordia sandriana* with secundly arranged laterals. 6. Portion of main axis of *Acinetospora crinita* with infrequently occurring short, opposite laterals. 7. Portion of main axis of *Pilayella littoralis* with opposite laterals. Bar = 50 μm (1, 5), 63 μm (3), 77 μm (2), 100 μm (4, 6, 7).

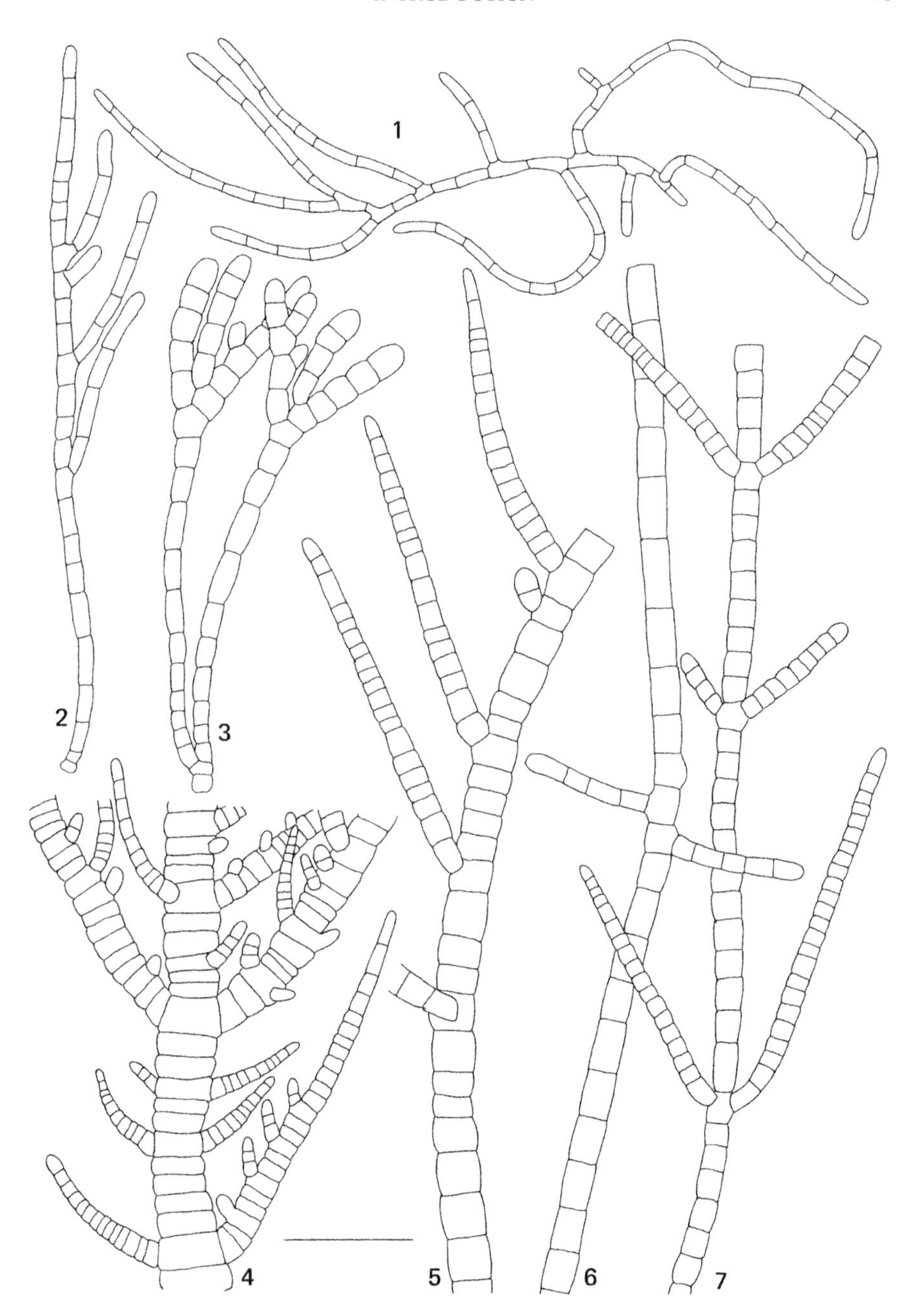
1
2
3
4
5
6
7

copiously branched, as in various species of *Giffordia, Ectocarpus, Pilayella, Streblonema* and *Kuckuckia*. Branching may be opposite, as in *Giffordia granulosa, Pilayella littoralis* and *Feldmannia globifera,* irregular and/or alternate, as in *Ectocarpus siliculosus, Giffordia mitchelliae* and *G. sandriana,* or secund, as in *Protectocarpus speciosus* and *Giffordia hincksiae*. Branches can also be closely set, as in *Compsonema,* or widely divergent, as in *Feldmannia globifera* and *Acinetospora crinita*. The basal systems of these erect filamentous groups are usually also branched and filamentous, as in *Ectocarpus*; less commonly they have a pseudoparenchymatous base (see below).

In many algae the mode of construction of the erect thalli is pseudoparenchymatous (sometimes termed haplostichous) in which there is an aggregation of filaments to give rise to more complex tissue (Fig. IV). The derivation of these filaments can be from either a single initial erect filament (uniaxial) or several erect filaments (multiaxial). Examples of pseudoparenchymatous multiaxial thalli include various members of the families Lithodermataceae, Myrionemataceae, Elachistaceae, Corynophlaeaceae, Chordariaceae and Scytosiphonaceae (crustose members only). In members of the Lithodermataceae and Scytosiphonaceae the erect filaments are short, quite well compacted and little branched to produce the characteristic crustose type of thalli. These erect filaments can be either firmly united and difficult to separate, as in *Ralfsia, Sorapion* and most species of *Pseudolithoderma,* or loosely united and easily separable, as in *Petroderma* and *Microspongium*. In the Myrionemataceae the erect filaments are very loosely united in a gelatinous matrix and easily separable. In the Elachistaceae, Corynophlaeaceae and Chordariaceae the erect filaments are much more developed and branched, resulting in cushion forms *(Leathesia, Petrospongium)* and erect, more macroscopic, cylindrical thalli (e.g. *Cladosiphon, Chordaria, Eudesme*).

Fig. IV Examples of pseudoparenchymatous thalli
1. Crustose thallus of *Ralfsia verrucosa*. 2. Vertical section of *Stragularia clavata* crust showing slightly arched, tightly packed, erect filaments. 3. Squash preparation of the *Microspongium gelatinosum* crustose phase of *Scytosiphon lomentaria* showing easily separated erect filaments. 4. Surface view of discoid thallus of young *Myrionema magnusii*. Note erect ascocysts. 5. Squash preparation of *Myrionema strangulans* showing short, loosely associated, erect filaments arising from a monostromatic base. 6. Cushion-like thallus of *Leathesia difformis*. 7. Basal hemispherical cushion and exerted filaments of *Elachista fucicola*. 8. Vertical section of *Elachista scutulata* showing closely packed filaments of large cells (which constitute the basal cushion) giving rise terminally to paraphyses and exerted filaments (arrowed). 9. Portion of transverse section of *Chordaria flagelliformis* showing compact, multiaxial medulla giving rise peripherally to paraphyses. 10. Branched, cylindrical habit of *Cladosiphon zosterae*. 11. Squash preparation of *Microcoryne ocellata* showing multiaxial medulla of large, elongated cells, giving rise peripherally to paraphyses. 12. Transverse section of *Spermatochnus paradoxus* showing a single, central, axial cell. Bar = 50 µm (2, 3, 4, 5, 12), 10 µm (8, 9), 0·5 mm (11), 1·5 mm (7), 7·5 mm (10), 15 mm (1), 30 mm (6).

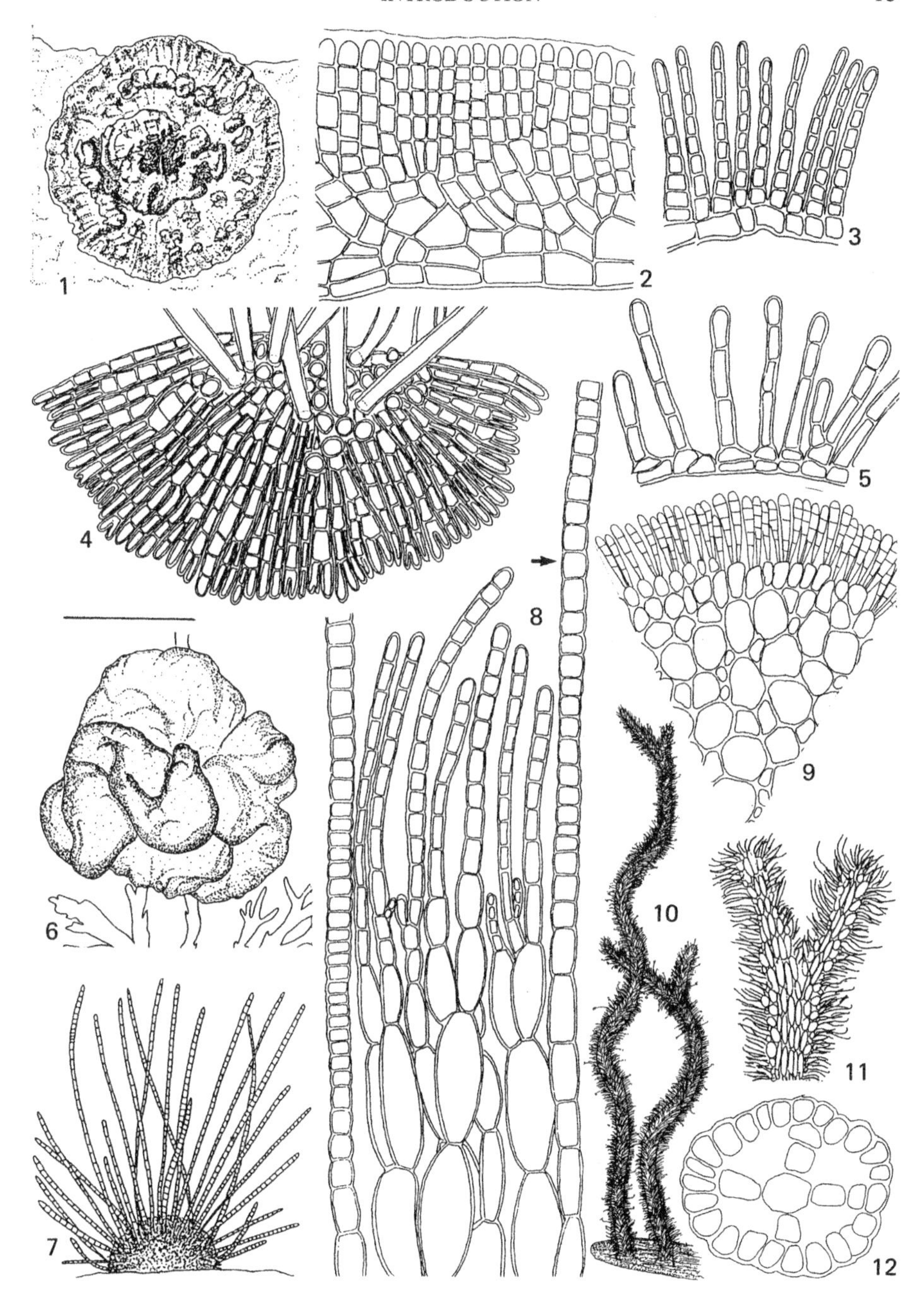
1
2
3
4
5
6
7
8
9
10
11
12

In many algae, however, the pseudoparenchymatous form is based on a uniaxial mode of construction. In the chordariacean genera *Mesogloia* and *Spermatochnus* there is a single erect axis the branch derivatives of which are compacted together to form the solid, gelatinous thallus. The degree of compactness of these Chordariacean genera, including both multiaxial and uniaxial types, does vary; for example *Eudesme* is weakly compacted and gelatinous, whilst *Chordaria* is firmly compacted and solid.

A further elaboration of the pseudoparenchymatous mode of construction can be found in the Desmarestiales (Fig. V, 1–2). In this group both the single erect axis and its lateral branches give rise to rhizoid-like outgrowths which divide and extend outwards and downwards to envelop the axes completely in a solid compacted tissue. The further production of hyphal filaments from the innermost cell layers contributes to the bulk of the thallus.

The third mode of construction of the brown algae is referred to as parenchymatous (sometimes termed polystichous) and involves the cells of a filament undergoing, to a limited or greater extent, longitudinal/periclinal divisions (Fig. V, 3–10). Unlike pseudoparenchyma, which is the result of filament aggregation, parenchyma is the result of cell division in several planes within a filament. The most simple parenchymatous thalli are represented by species such as *Hecatonema maculans, Pilayella littoralis* and *Stictyosiphon griffithsianus* in which occasional cells of the erect filaments have divided longitudinally. In algae such as *Compsonema saxicolum* and some *Chilionema* spp., the occasional longitudinal divisions are confined to cells of the horizontally spreading filaments. Much more elaborate and complex longitudinal divisions occur in the cells of members of the Sphacelariales. Indeed the pattern of internal division is an important and useful taxonomic character. As in the simple parenchymatous types no significant increase in girth of the thallus is associated with this mode of division.

In many brown algae, however, including representatives of the families Myriotrichiaceae, Punctariaceae, Scytosiphonaceae, Cutleriaceae, Chordaceae, Laminariaceae, Alariaceae, Dictyotaceae and Fucaceae, parenchymatous divisions in the early formed erect filament/shoot are much more extensive,

---

Fig. V Examples of both pseudoparenchymatous (1, 2) and parenchymatous thalli (others)
1. Apex of *Desmarestia viridis* showing intercalary meristem with upper, branched hair and lower cortical envelope. 2. Transverse section of *Desmarestia ligulata* showing a large, central, axial cell, a broad cortical zone of large colourless cells and a peripheral layer of smaller, pigmented cells. 3. Portion of erect axis of *Pilayella littoralis* showing a single, longitudinally divided cell. 4. Portion of partly biseriate thallus of *Leblondiella densa*. 5. Portion of *Chilionema ocellatum* showing biseriate basal layer. 6. Apex of *Sphacelaria plumigera* showing segmentation of cells. Note that only a small increase in growth occurs. 7–8. Portions of terete, parenchymatous thalli of *Litosiphon laminariae*. 9. Apical region of *Punctaria tenuissima* showing transformation from uniseriate to biseriate thallus. 10. Portion of young parenchymatous blade of *Punctaria tenuissima*. Bar = 50 μm (3, 4, 6, 7, 8, 10), 63 μm (1, 2, 5, 9).

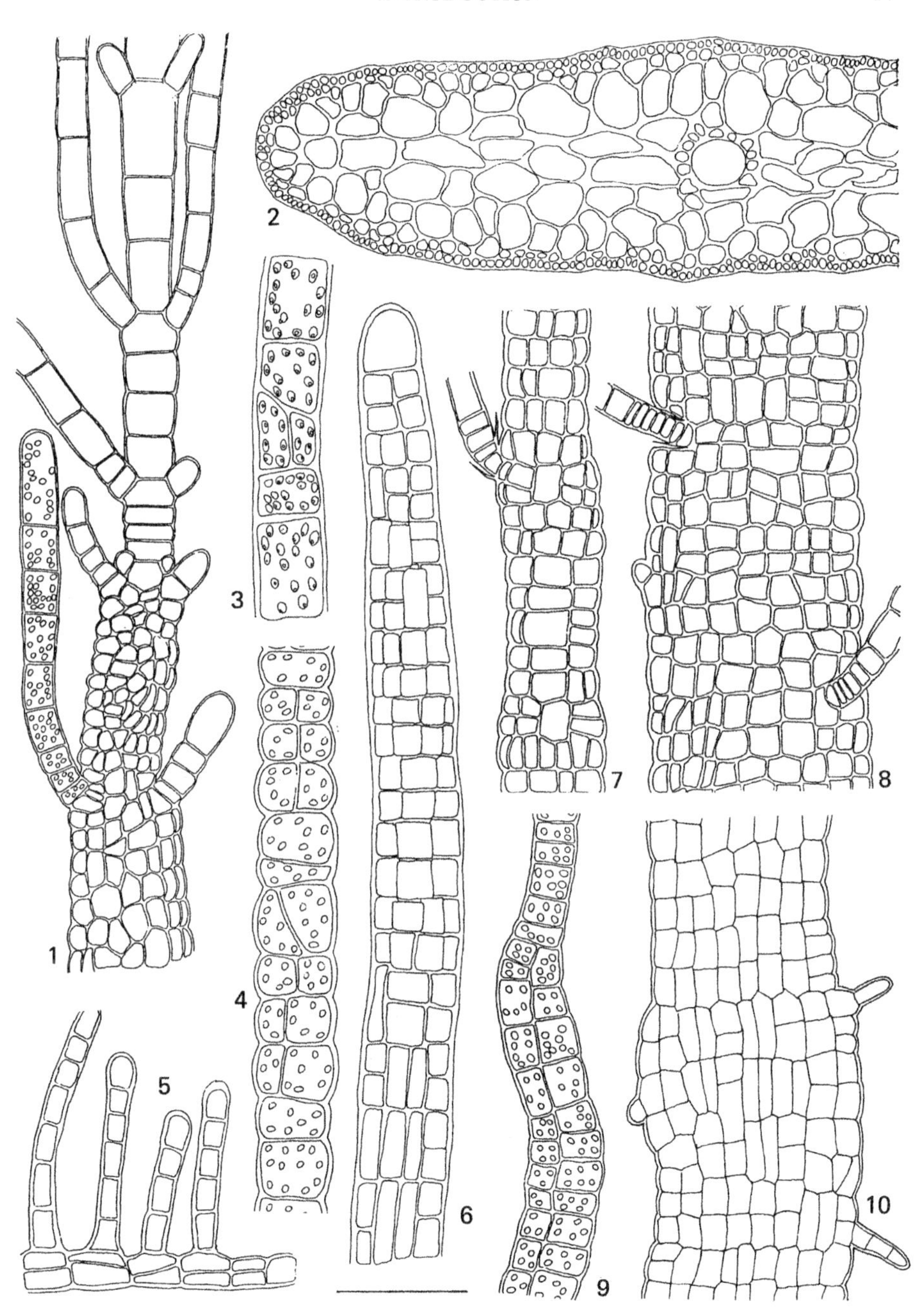
1
2
3
4
5
6
7
8
9
10

producing larger and more elaborate thalli with a wide range of forms (Figs VI, VII). These include thalli which are encrusting/procumbent (*Battersia, Zanardinia, Aglaozonia* phase of *Cutleria*), saccate (*Colpomenia*), unbranched, solid or hollow, and terete (*Asperococcus, Chorda, Litosiphon, Scytosiphon*), branched and terete (*Bifurcaria, Cystoseira*), strap-shaped (*Himanthalia*), unbranched and foliose (*Petalonia, Laminaria, Punctaria*) or branched and foliose (*Dictyota, Cutleria*). Some of these parenchymatous forms, e.g. the kelps, represent the largest of marine plants, with *Laminaria* growing several metres in length and the Pacific giant kelp, *Macrocystis,* reaching over 50 m long.

For both the parenchymatous and pseudoparenchymatous forms, increase in size has usually resulted in considerable structural and anatomical differentiation. For example, in many algae the thallus can be clearly divided into an erect frond, a stipe region and a basal attachment system. The frond consists of the visibly prominent portion of the thallus and contains most of the characteristics/features by which the alga is identified. Apart from the gross morphological components described above, a number of supplementary features are present which characterise the algae. These can include lateral proliferations (*Desmarestia dresnayi, Punctaria latifolia*), surface configurations and bullations (*Laminaria saccharina*), the presence of vesicles/bladders, used for buoyancy and respiratory functions (*Fucus vesiculosus, Sargassum muticum, Cystoseira* spp., *Ascophyllum nodosum*), inflation of thalli (*Scytosiphon lomentaria, Colpomenia peregrina, Asperococcus turneri*), the presence of mid-ribs (*Fucus* spp., *Carpomitra, Desmarestia ligulata*) and the presence of leaf-like structures (*Sargassum muticum*). The frond is the main photosynthetic portion of the plant body and therefore comprises cells with plastids. It is also responsible for the development of the reproductive organs and therefore plays a vital role in the production and dissemination of reproductive bodies. Lastly, the frond is usually the most active portion of the thallus in nutrient uptake and gaseous exchange.

In a number of algae the proximal regions of the frond take the form of a distinct stalk called a stipe which connects below to the attachment system. The extent of development of the stipe varies considerably between algae, e.g. it is only faintly recognisable in *Punctaria* and *Petalonia,* quite well developed in *Fucus,* and especially prominent in *Laminaria* spp. It is usually cylindrical in shape and strong but flexible and probably functions in both a supporting and translocatory role.

Fig. VI Examples of parenchymatous thalli
1. *Zanardinia prototypus;* 2. *Colpomenia peregrina;* 3. *Scytosiphon lomentaria;* 4. *Asperococcus fistulosus;* 5. *Cutleria multifida;* 6. *Leblondiella densa;* 7. *Petalonia fascia.* Bar = 4 mm (6), 13 mm (1), 20 mm (5), 30 mm (2, 4, 7), 33 mm (3).

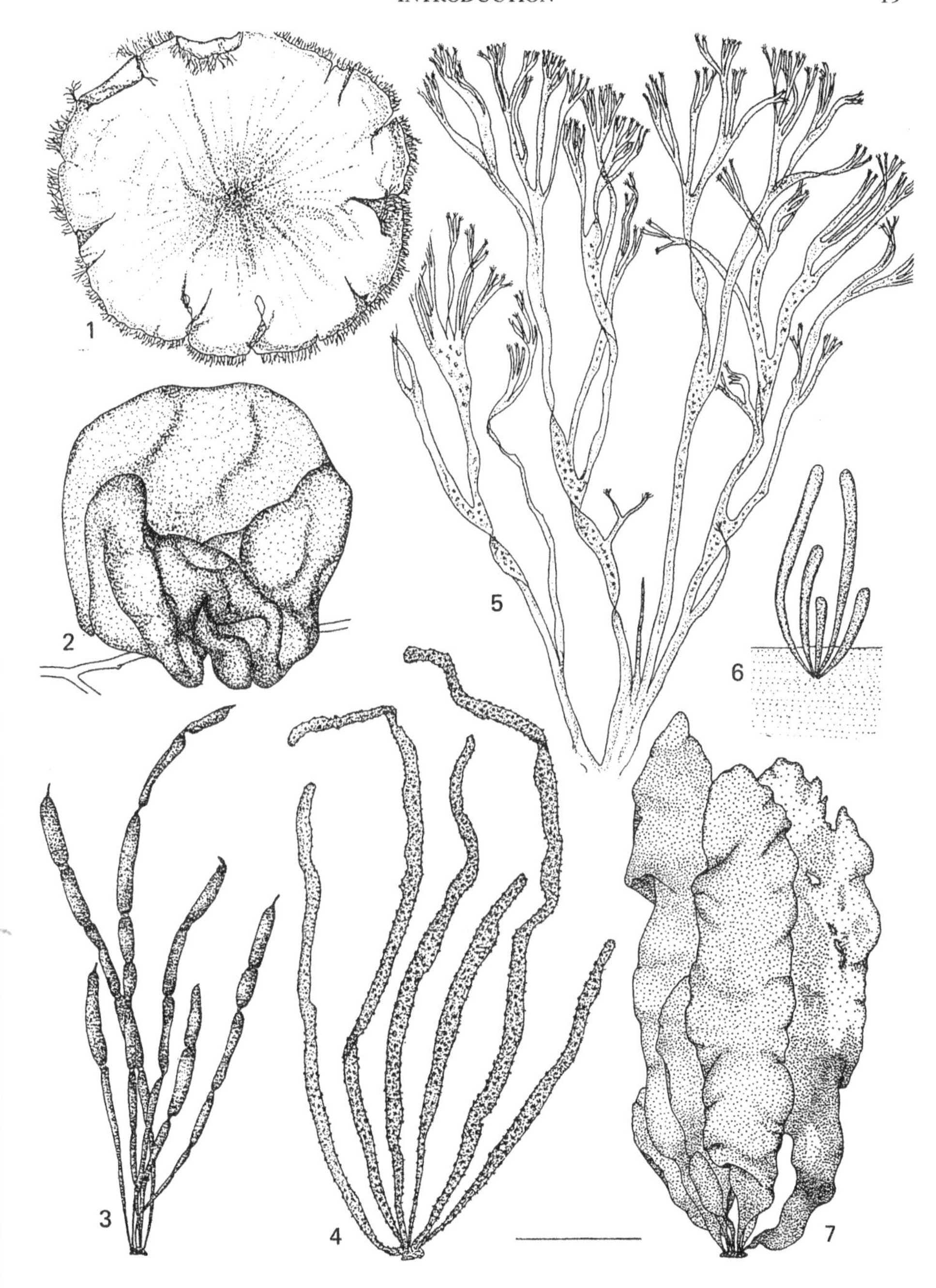
1
2
3
4
5
6
7

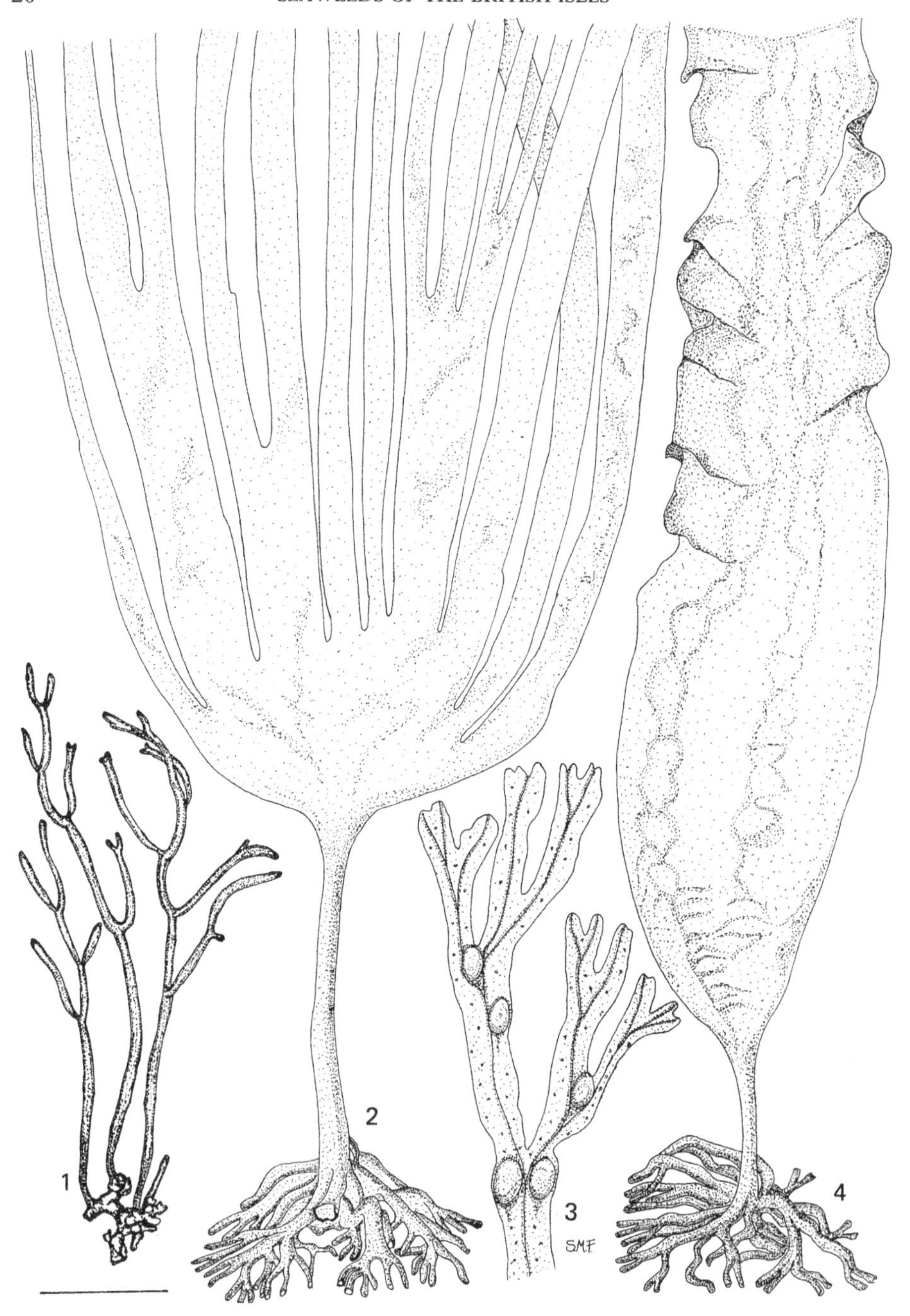
1
2
3
S.M.F.
4

A wide range of attachment systems have been described in the brown algae, the great majority of which are based on the production of secondary rhizoids from vegetative cells on the erect thallus (Fletcher, 1977). These rhizoids grow down, usually in large numbers, to the basal substratum. In algae such as *Giffordia* and *Sphacelaria* they also form an enclosing sheath around more distal parts of the thallus and are then termed corticating filaments. In many thalloid algae, such as *Scytosiphon, Petalonia* and *Fucus*, they are produced both externally, growing down the outside of the thallus, and internally. At the plant base the emergent rhizoidal filaments either remain discrete to produce a fibrous mass (*Petalonia filiformis, Halothrix*) or compact together to form the pseudoparenchymatous holdfast basal systems which are particularly common in the larger brown algae (e.g. *Fucus, Sargassum, Punctaria*). The wide range of holdfast morphologies can be of taxonomic importance, e.g. in *Petalonia* spp. (Fletcher, 1981a) and *Desmarestia* spp. (Chapman, 1972). Two other noteworthy basal systems are the highly specialised finger-like, parenchymatous haptera produced at the base of kelp algae, such as *Laminaria*, and the horizontally spreading, stoloniferous attachment organs produced in the Dictyotales (Fritsch, 1945; Richardson, 1979). Despite this wide range of attachment systems, all fundamentally depend on individual, microscopic, rhizoidal filaments which have adhesive properties (Fletcher, Jones & Jones, 1984).

Concomitant with thallus development is the internal differentiation of the cells into fairly distinct tissue types Fig. VIII). For the majority of algae with thalli in excess of three cells in thickness, this usually comprises an inner 'medulla' and an outer 'cortex'. Medullary cells are usually large, longitudinally elongate, variable in thickness, colourless or with few plastids, and have storage, translocatory and/or mechanical functions. Cortical cells are usually grouped to form a peripheral sub-epidermal layer; they are smaller, fairly thick-walled, pigmented and are responsible for much of the photosynthetic activity of the plant. The relative proportions of medulla and cortex does vary for different species. For example, *Dictyota* has a single layer of medullary cells enclosed by a single layer of cortical cells. In *Desmarestia* the medulla is a single, central, axial cell which is surrounded by a broad zone of cortical cells. In *Petalonia, Punctaria, Leathesia, Asperococcus* and many Chordariacean algae the medulla is a broad zone several cells in thickness whilst the cortex is only 1–3 cells thick. Very often further differentiation of cells occurs in these zones; for example the broad cortical zone in *Desmarestia* has inner, large, colourless cells and outer smaller, pigmented cells, whilst in *Fucus* and *Laminaria* an inner and outer cortex is generally recognised. In many algae, especially members of the Elachistaceae

---

Fig. VII Examples of parenchymatous thalli
1. *Bifurcaria bifurcata;* 2. *Laminaria digitata;* 3. *Fucus vesiculosus* (portion of thallus only); 4. *Laminaria saccharina.* Bar = 30 mm.

and Corynophlaeaceae, some internal differentiation of the medullary tissue is described, with large inner cells and small peripheral cells. However, as for the red algae (Dixon & Irvine, 1977), there is fundamentally very little difference between 'medulla' and 'cortex'; the assignation of tissues to one or other of these is frequently arbitrary.

In the structurally more elaborate thalli of the fucoids and the kelps, the internal organisation of the tissues is more complex and can, in the case of *Laminaria,* be highly specialised, involving a number of cell types and structural modifications. The epidermal and sub-epidermal regions may be meristematic in function, the products of cell division either contributing to the bulk of the thallus or to its surface area, depending upon the plane of division. The superficial meristem is usually termed a meristoderm. In many of the kelps, distinction is made between primary meristoderm, which is located at the junction of frond and stipe and which is responsible for frond growth and stipe elongation, and secondary meristoderm on the stipe which is responsible for increase in stipe growth and haptera formation. Cells arising from tangential cell divisions give rise to the cortex from which, in turn, hypal cells may be borne. These may fuse and interconnect adjacent cells, or spread radially and longitudinally, probably enhancing the conducting and mechanical properties of the thallus, as well as contributing, at the base, to the attachment rhizoidal filaments. Other noteworthy features in *Fucus* and/or *Laminaria* include gelatinisation of the medullary cell walls, the formation of trumpet-hyphae and sieve tubes in the medullary cells, the occurrence of mucilage canals, and the formation of growth rings by seasonal secondary meristoderm activity. More details of these tissue types can be obtained from Fritsch (1945).

Despite the increased anatomical specialisation of the pseudoparenchymatous and parenchymatous thalli, remnants of a more simple, basic construction can still often be associated with both. For example, many erect, parenchymatous thalli (e.g. *Petalonia, Scytosiphon, Punctaria*) and pseudoparenchymatous thalli (e.g. *Desmarestia*) clearly originate from the development of a single, erect

---

Fig. VIII Examples of thallus structure
1. *Punctaria latifolia.* T.S. of blade showing no obvious internal differentiation into medulla and cortex. 2. *Dictyota dichotoma.* T.S. of blade showing a single-celled central medulla enclosed by a single, smaller layer of pigmented cortical cells. 3. *Desmarestia viridis.* T.S. of thallus showing central axial cell and surrounding broad zone of cortical cells. 4–5. *Zanardinia prototypus.* V.S. of thalli showing central region of large, colourless, longitudinally elongate medullary cells enclosed by 1–2 layers of smaller cortical cells. 6. *Petalonia fascia.* T.S. of thallus showing a central medulla of large, colourless cells enclosed by 1–3 layers of smaller, cortical cells. 7. *Leathesia difformis.* V.S. of thallus showing large, irregularly contorted medullary cells which become smaller peripherally and terminate in short, filamentous paraphyses. 8. *Laminaria digitata.* T.S. of blade showing central medulla with longitudinal and transverse filaments, surrounded by a cortex of large, more rounded cells and enclosed by 1–2 layers of epidermal cells. Bar = 50 μm (2, 6), 63 μm (1), 100 μm (4, 5, 7, 8), 125 μm (3).

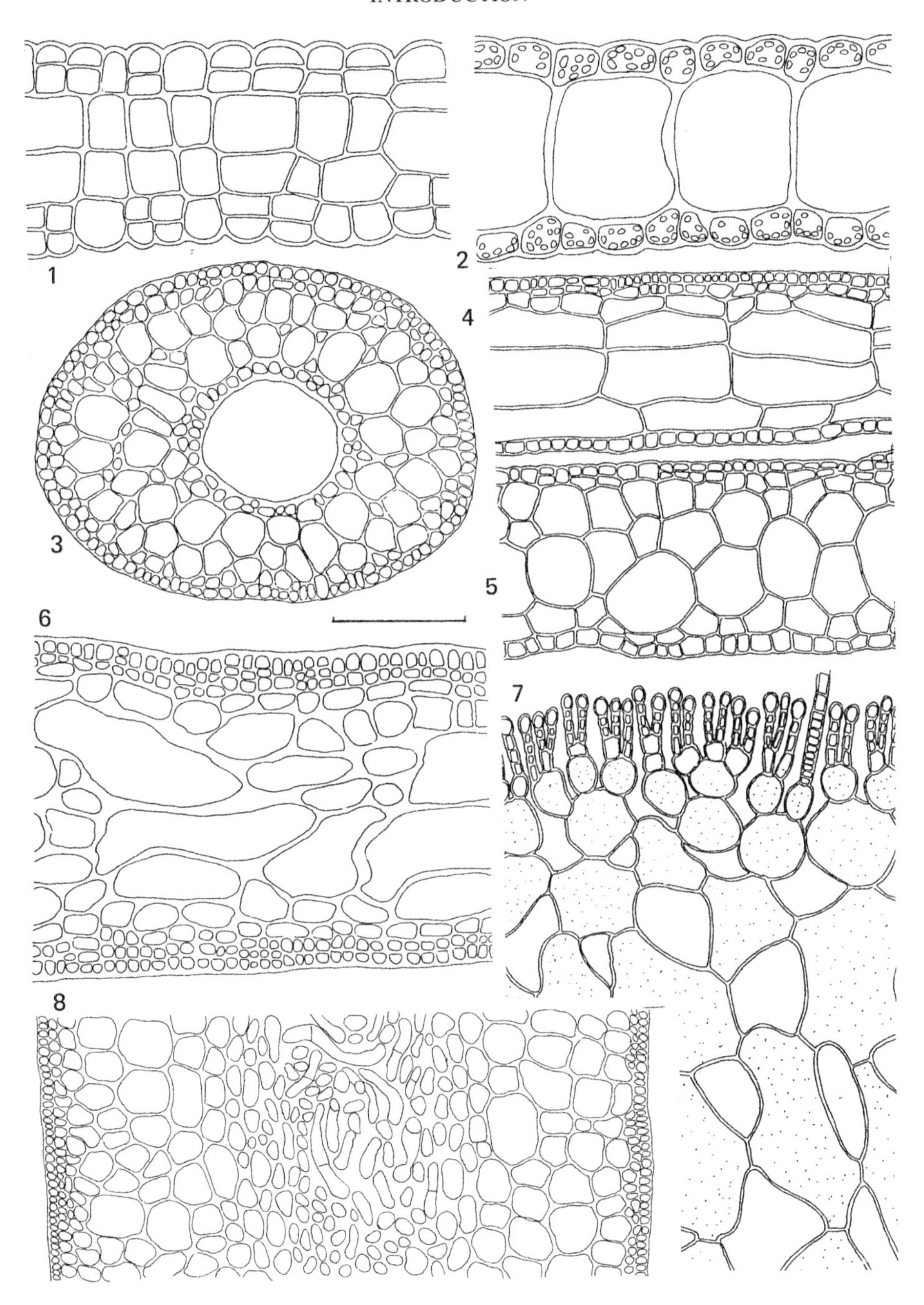
1
2
3
4
5
6
7
8

filament. Moreover, these different modes of construction are not mutually exclusive, and this is well exemplified in those members with a heteromorphic biphasic life history or with pronounced basal development. For example, the large, structurally elaborate parenchymatous and pseudoparenchymatous sporophytes of the Laminariales and Desmarestiales respectively alternate with simple, branched, filamentous, microscopic gametophytes. The life histories of *Petalonia* and *Scytosiphon* include both erect parenchymatous thalli and pseudoparenchymatous *Ralfsia*-like and/or *Microspongium*-like crustose phases. The basal micro-thalli associated with the parenchymatous thalli of *Punctaria* and *Asperococcus* are discoid, pseudoparenchymatous thalli, like those of *Chilionema* and *Hecatonema,* whilst branched, fertile *Streblonema*-like, filamentous thalli have also been associated with the less developed, parenchymatous genera *Myriotrichia* and *Pogotrichum.*

Nevertheless, the mode of construction of the brown algae is a fundamentally important taxonomic characteristic. It is the main criterion used in the delimitation of families and, together with aspects of the reproduction and life history, forms the basis of classification at the ordinal level. For example, members of the Ectocarpaceae are characterised by filamentous thalli, the Lithodermataceae, Myrionemataceae, Elachistaceae, Corynophlaeaceae, Chordariaceae, Desmarestiaceae and Arthrocladiaceae have pseudoparenchymatous thalli, whilst the Scytosiphonaceae, Laminariaceae, Sphacelariaceae, Dictyotaceae and Fucaceae have parenchymatous thalli.

Thallus growth and longevity

Growth is the result of cell expansion which normally follows cell division. In the brown algae cell division may be either diffuse or localised (meristematic), with the meristems apical or intercalary in position (Fig. IX). Examples of algae with diffuse or random growth include filamentous genera in the Ectocarpaceae, such as *Ectocarpus* and *Giffordia,* as well as pseudoparenchymatous genera, such as *Microcoryne* and *Chordaria,* and erect, thalloid genera, such as *Petalonia,*

Fig. IX Examples of growth mechanisms
1. *Herponema velutinum.* Terminal portion of erect filament showing an intercalary meristem. 2. *Arthrocladia villosa.* Plant apex showing trichothallic growth (intercalary meristem with terminal hair). 3. *Zanardinia prototypus.* Surface view of thallus edge showing trichothallic growth with terminal tuft of filaments. 4. *Zanardinia prototypus.* V.S. through plant edge showing trichothallic growth. (Note two superimposed extending hair-like filaments). 5. *Sphacelaria plumigera.* Branch apex showing large apical cell. 6. *Cutleria multifida* (*'Aglaozonia'* phase). V.S. through thallus edge showing large apical cell. 7. *Dictyota dichotoma.* Branch apex showing dome-shaped apical cell. 8. *Chilionema ocellatum.* Surface view of thallus edge showing marginal row of apical cells. 9. *Stragularia clavata.* V.S. through crust edge showing terminal apical cell. Bar = 50 μm (1, 4, 5, 6, 7, 8, 9), 100 μm (3), 200 μm (2).

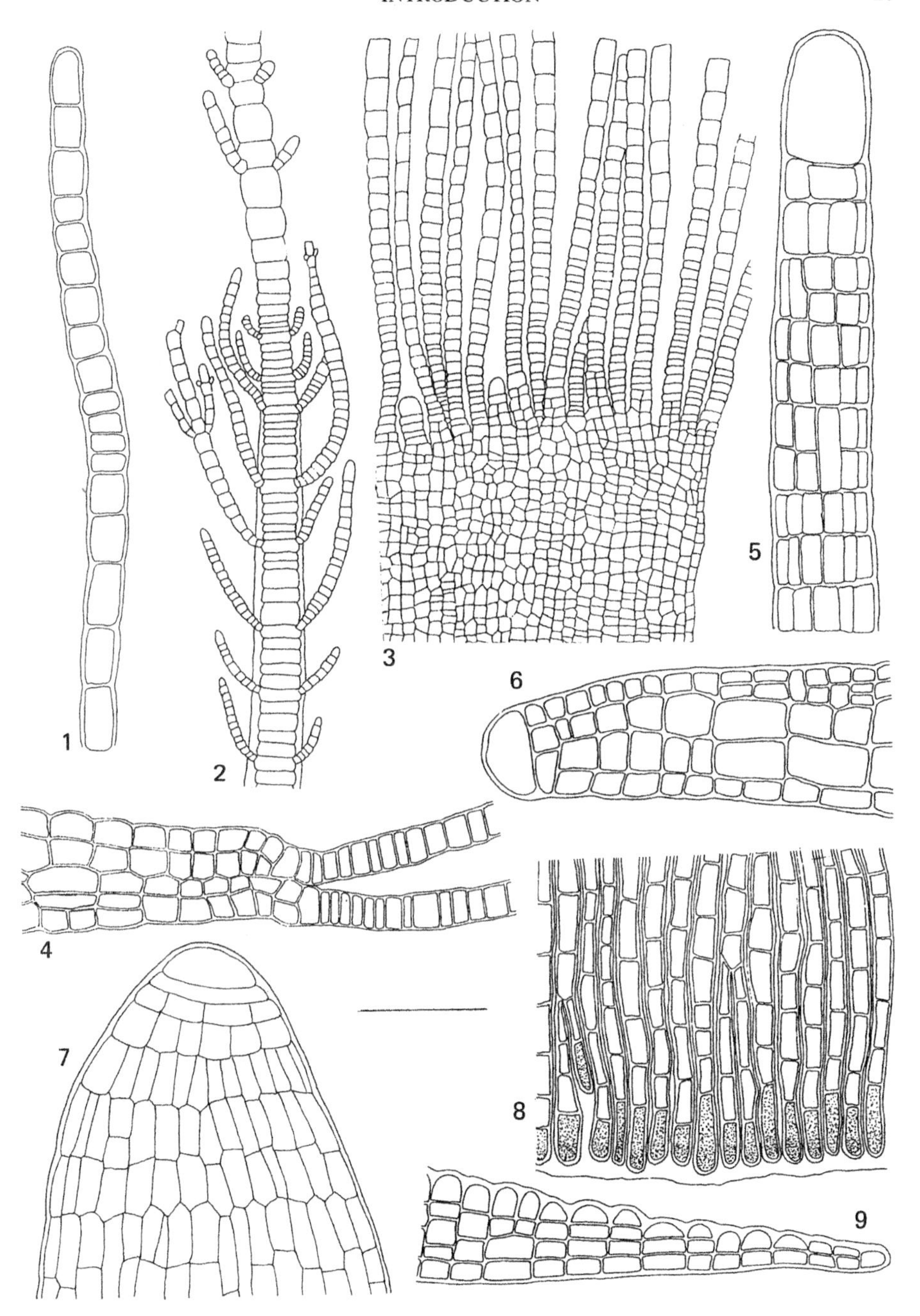
1
2
3
4
5
6
7
8
9

Scytosiphon (Scytosiphonaceae), *Punctaria* and *Asperococcus* (Punctariaceae). Intercalary growth may be trichothallic i.e. resulting, from cell division at the base of a long, terminal, hair-like filament which contributes cells both above and below; sometimes a single filament is involved, as in *Arthrocladia* and *Desmarestia,* or sometimes a tuft of filaments is involved, as in *Sporochnus, Carpomitra, Cutleria* and *Zanardinia.* Algae with a tufted, trichothallic growth mechanism are usually pseudoparenchymatous in construction (e.g. *Desmarestia, Elachista*); however, in *Cutleria* and *Zanardinia*, fusion of the submeristematic cells, followed by longitudinal divisions, results in a parenchymatous type of construction. The intercalary meristems in other species occupy diverse positions in the thallus: in kelps, such as *Laminaria,* the primary meristem (often referred to as the transition zone) is situated between (and contributes cells to) the blade and the stipe. In both the kelp and fucoid algae, an additional peripheral meristem (meristoderm) has been demonstrated in the outer cortical tissues which contributes to both blade and stipe growth (usually referred to as primary and secondary meristoderms).

In the great majority of brown algae, including filamentous, pseudoparenchymatous and parenchymatous forms, growth of the plant body is at the apex. Either a single apical cell is involved, as in *Sphacelaria, Dictyota* and *Fucus,* or several apical cells are involved. These are laterally conjoined, as in pseudoparenchymatous genera such as *Leathesia* and *Ralfsia* and parenchymatous genera such as the *Aglaozonia* phase of *Cutleria,* or they are aggregated into a meristem, as in the Australasian genera *Hormosira* and *Notheia.*

In general, diffuse growth is more characteristic of the structurally simple and reproductively less elaborate algae, whilst apical growth, particularly involving a single apical cell, is characteristic of the more advanced algae. Algae with trichothallic growth are generally intermediate in structural and reproductive complexity. These growth mechanisms are not mutually exclusive. For example, diffuse growth is characteristic of the erect plant portions in heterotrichous ectocarpoid algae, such as *Ectocarpus, Giffordia* and *Acinetospora,* but the basal system expands by apical growth; whilst in algae such as *Desmarestia* and *Arthrocladia,* the erect macroscopic sporophyte exhibits trichothallic growth, but the microscopic gametophyte has apical growth. Also, Scytosiphonaceae members (e.g. *Petalonia, Scytosiphon*) and Punctariaceae members (e.g. *Punctaria, Asperococcus*) have erect blades with diffuse growth but basal systems and/or microthalli with apical growth (*Ralfsia*/*Microspongium* and *Hecatonema*-like respectively). A further example of divergent growth mechanisms is provided by *Cutleria* which has erect thalli with trichothallic growth and prostrate *Aglaozonia*-like thalli with apical growth.

Considerable differences in rates of growth have been reported in brown algae. Growth is particulary slow in crustose algae and very rapid in some of the larger algae, particularly members of the Laminariales and Fucales. For example, Dethier (1981) recorded a slow growth rate for *Ralfsia californica* with the crusts only enlarging to 1 cm during early autumn and winter on the Washington coast of North West America, whilst Fletcher (unpublished) observed the diameter of *Ralfsia verrucosa* crusts in Kent to increase by 4–6 mm over a five-month growth period between March and August. These results are in contrast with the maximum growth rates recorded for Atlantic species of *Laminaria* (1·5 cm/day – Kain, 1979), *Laminaria saccharina* (2·3 cm/day – Parke, 1948), *Sargassum muticum* (4 cm/day – Jephson & Gray, 1977), *Chorda filum* (4·9 cm/day – South & Burrows, 1967) and *Saccorhiza polyschides* (4·7 cm/week – Norton, 1969).

Great variation is also shown in the longevity of brown algae, ranging from short-lived ephemerals of only a few weeks' life span to long-lived perennials, many of which survive for several years and, in some cases, decades. For example, *Laminariocolax tomentosoides* may become fertile at 7 days (Russell, 1964), *Streblonema oligosporum* was reported to reach reproductive maturity in culture after only 9 days at 20°C (Fletcher, 1983), whilst reported life spans include 7–12 months for *Ralfsia californica* (Dethier, 1981), up to 12/13 weeks for *Scytosiphon lomentaria* (Clayton, 1981), a maximum of one year for *Pilayella littoralis* (Knight, 1923), approximately one year for *Saccorhiza polyschides* (Norton & Burrows, 1969), not exceeding 3 years for *Laminaria saccharina*, and 15 years for *L. hyperborea* (Kain, 1979), whilst *Ascophyllum nodosum* plants can possibly survive '25 years' (Dring, 1982). However, for many brown algae determination of longevity of the plants is complicated by the modifying influence of environmental conditions. These can include, for example, substratum availability; the life span of many epiphytic/endophytic algae will be very largely determined by that of the host plant. This particularly applies to species with only a limited range of possible host organisms; examples include *Ulonema rhizophorum* epiphytic on the erect thalli of *Dumontia contorta*, and *Elachista scutulata*, epiphytic on the thongs of *Himanthalia elongata*. Also, for many algae, and in particular short-lived (ephemeral) species and those with distinct seasonal distribution patterns, the longevity of the plant is very largely determined by environmental conditions, such as temperature, light intensity, photoperiod, competition and grazer activity. An increasing number of experimental physiological studies have shown algae able to detect and respond to very subtle changes in a wide range of environmental parameters which very largely determine their life span. Further, many of those algae which apparently disappear during unfavourable periods are now known to die back to perennating basal regions/fragments. These are generally more hardy than the erect portions and retain the ability to initiate the latter with the approach of more favourable conditions. Examples include perennating *Ralfsia*-like thalli in the life history of *Petalonia fascia* (Roeleveld *et al.*, 1974), *Compsonema*-like thalli in the life history of *Petalonia filiformis* (Fletcher, 1981a), *Hecatonema*-like thalli in the life history of *Punctaria* and *Asperococcus* spp. (Fletcher, 1984; Pedersen, 1984), stoloniferous thalli in the life histories of members of the Dictyotales (Richardson, 1979) and basal shoot portions in the life history of *Sargassum muticum* (Jephson & Gray, 1977).

Environmental influences on thallus development

A number of experimental field and laboratory culture studies have examined the response of algae, including the Fucophyceae, to environmental conditions such as temperature, light and salinity (for reviews of earlier papers on these topics see particularly Gessner (1970), Hellebust (1970) and Gessner & Schramm (1971) respectively). Most of these studies were concerned with the environmental control of growth rates. With respect to light/temperature effects, brown algal genera investigated include: *Laminaria* (Kain, 1979; Lüning, 1971; Boden, 1979; Fortes & Lüning, 1980); *Ectocarpus* (Boalch, 1961; Edwards, 1969; Edwards & van Baalen, 1970); *Giffordia* (Edwards, 1969; Edwards & van Baalen, 1970; Fletcher, 1981b); *Desmarestia* (Chapman & Burrows, 1970; Fortes & Lüning, 1980; Nakahara, 1984); *Saccorhiza* (Norton & Burrows, 1969); *Petalonia* (Edwards & Baalen, 1970); *Acinetospora* (Amsler, 1984); *Fucus* (Munda, 1977; Strömgren, 1978; Fortes & Lüning, 1980); *Halidrys* (Moss & Sheader, 1973) and *Ascophyllum* (Sheader & Moss, 1975). Investigations of the effects of salinity on growth have included genera such as *Ectocarpus* (Russell & Bolton, 1975), *Fucus* (Munda, 1977), *Pilayella* (Bolton, 1979), *Scytosiphon* (Ohno, 1969), *Asperococcus* (Pedersen, 1984) and *Litosiphon* (Nygren, 1975).

Of particular interest, however, have been the increasing number of reports concerning the effect of these and other environmental conditions on brown algal development and morphogenesis. Some of the main influential environmental conditions investigated include:

Photoperiod/daylength Photoperiod/daylength was shown to exert a marked influence on erect thallus formation in algae with heteromorphic life histories, for example *Petalonia fascia, P. zosterifolia* and *Scytosiphon lomentaria* (Lüning, 1980a; Dring & Lüning, 1975), *Punctaria tenuissima* (as *Desmotrichum undulatum*) (Lockhart, 1982) and *Analipus japonicus* (Nakahara, 1984). In general the results of these experiments conformed with the phenology of the species in the field, although sometimes this was linked with temperature effects (Lüning, 1980a). In addition, studies on *S. lomentaria* (Lüning, 1981b) and *Punctaria tenuissima* (as *Desmotrichum undulatum*) (Lockhart, 1982) revealed that genetically different (photoperiodic) ecotypes exist with different developmental responses to temperature and photoperiod which appear to be adapted to their geographical positions.

A number of other studies (see particularly Edwards, 1969; Wynne, 1969; Roeleveld *et al.*, 1974; Fletcher, 1978; Nakamura & Tatewaki, 1975) also show that the combination of photoperiod/daylength and temperature not only influences plant development in members of the Scytosiphonaceae but also determines the morphogenetic responses with respect to the production of either the crustose or erect-bladed thalli. A similar temperature/daylength control of morphological expression (microscopic and prostrate or macroscopic and erect) has also been shown in members of the Punctariaceae (Loiseaux, 1969; Rhodes,

1970; Rietema & Hoek, 1981; Lockhart, 1982). However Pedersen (1984) has shown that macrothalli of *Punctaria tenuissima* (as *Desmotrichum undulatum*) may develop from basal systems, irrespective of temperature or photoperiod, suggesting that genetically different strains varying in their morphogenetic response are present in the North Atlantic.

Temperature A few investigations have revealed temperature to act independently of photoperiod in brown algal development and morphogenesis. For example, Pedersen (1984) has demonstrated that macrothallus production in *Asperococcus fistulosus* is apparently more frequent at low temperatures, irrespective of the photoperiod, whilst different forms of *Streblonema immersum* (Levring) Pedersen were produced in low and high temperatures. Rietema & Hoek (1981) also reported temperature to be the main modifier of morphology in *Punctaria tenuissima* (as *Desmotrichum undulatum*).

Light quality Lüning & Dring (1973) observed that the development of *Scytosiphon lomentaria, Petalonia fascia* and *P. zosterifolia* in laboratory culture differed in red and blue light. Sporelings developed into *Ralfsia*-like crustose thalli (profusely branched thalli in *P. zosterifolia*) in blue light but sparsely branched, filamentous thalli in red light. In addition the production of erect thalli from the basal systems was restricted to or stimulated by red light only. However, no requirement for either red or blue light for erect blade formation in *Punctaria tenuissima* (as *Desmotrichum undulatum*) was noted by Lockhart (1982). Light quality was also reported by Lüning & Dring (1972, 1975) to influence the life history of *Laminaria* species. They showed that fertility of the microscopic gametophyte generation required a minimum dose of blue irradiance; Drew (1983) suggested that this might well be limiting at times at certain depths and in murky coastal waters which would seriously impair sporophyte production. With respect to sporophyte growth it is interesting to note that Lüning & Markham (1979) report red and blue light to influence stipe length and blade dimensions in *Laminaria saccharina*. Finally, blue light was shown to be required for normal growth and differentiation in *Dictyota dichotoma* (Müller & Claus, 1976).

Salinity In certain species reduced salinity may be associated with reduction in thallus size and with other morphological and anatomical changes (Munda, 1978; Gessner & Schramm, 1971; Wilkinson, 1980; Norton *et al.*, 1981; Pedersen, 1984). This has been particularly well documented in fucoids, where reduced salinity has been linked with modified branching patterns (Jordan & Vadas, 1972) and vesiculation (Jordan & Vadas, 1972; Alexander *et al.*, 1935; Hartog, 1967). However, it is evident that the morphological impact of reduced salinity is not a simple one as differences in response have been noted between species and between cell types of the same species (Russell, 1985a). The influence of reduced salinity on Baltic algal morphogenesis has been discussed by Waern

(1952) and Russell (1985a). The ways in which algae respond to changes in external salinity have been reviewed by Russell (1985b).

Surface properties of substrata A wide range of surface properties of substrata have been shown to exert an influence on brown algal development and community structure. These include:

i. *Surface free energy* Fletcher, Baier & Fornalik (1984) demonstrated that surfaces of different energy clearly influenced rhizoid development in *Giffordia granulosa*. Fletcher, Baier & Fornalik (1985) also reported surface energy to influence morphogenesis of the crustose phase in the life histories of *Petalonia* and *Scytosiphon* spp.; high energy surfaces produced crustose thalli, low energy surfaces produced filamentous thalli. It was concluded that surface energy properties of substrata probably play an important role in determining the attachment strength of the algae and are likely to affect particularly host/epiphyte relationships (see Linskens, 1966 and Pedersen's (1984) report of variable morphology in *Mikrosyphar polysiphoniae*). It is also likely that heteroblasty is caused by surface energy effects (see Christensen, 1980, p. 135 for reference to heteroblasty).

ii. *Surface texture* A number of field studies have shown material of different surface texture to exert an influence on the algal community structure (Harlin, 1974; Harlin & Lindbergh, 1977). Some species showed a preference for attachment and growth on roughened surfaces, whilst others preferred smooth surfaces. Very few laboratory experimental investigations have been carried out and these generally relate to the morphology of the attachment rhizoids (e.g. Hardy & Moss's (1979b) study on *Fucus serratus*).

iii. *Surface toxicity* Investigations of toxic surfaces, usually panels and ships coated with copper and organo-tin antifouling paints, generally reveal a reduced marine algal flora comprising a small number of brown algae (Harris, 1943, 1946; Igic, 1968; Christie, 1973; Rautenberg, 1960; Fletcher, 1983). The most frequently reported brown alga is *Ectocarpus,* which is often present in a reduced 'brown mat' form; its presence is in accord with reports of its high copper resistance (Russell & Morris, 1970; Hall *et al.*, 1979). One interesting aspect of algal colonisation of these toxic surfaces is the reporting of several brown algae which have become known almost exclusively as epiphytes or epi-endophytes; these include species of *Microspongium, Myrionema, Porterinema, Streblonema* and *Hecatonema* (Harris, 1946; Rautenberg, 1960; Fletcher, 1983).

Degree of exposure to wave action In addition to the well documented effect of wave action on the distribution and composition of benthic algal communities (Lewis, 1964; Jones & Demetropoulos, 1968), wave action also influences the morphology of algae, including species of Fucophyceae. A good example of this influence is the occurrence of the rather distinctive exposed form of North Atlantic *Fucus vesiculosus (F. vesiculosus* f. *linearis*). Other morphological changes in brown algae as a result of wave action have been reported in *Laminaria* spp. (Sundene, 1964; Svendsen & Kain, 1971; Larkum, 1972; Norton *et al.*, 1977), *Chordaria flagelliformis* (Munda, 1979), *Saccorhiza polyschides* (Norton, 1969), and *Sargassum cymosum* (Paula & Oliveira, 1982). In *Laminaria* and *Saccorhiza*, the differences obtained were attributed to phenotypic plasticity; however in *Sargassum* and *Chordaria* genotypic differentiation was suggested to occur, with the evolution of distinct ecotypes. In practice, both sources of variation are likely to occur and, in any given thallus, some morphological attributes will prove to be more plastic than others.

Water movement In addition to reports of water movement affecting colonisation of substrata, either directly by preventing spore attachment (Norton & Fetter, 1981), or indirectly, by influencing silt deposition (e.g. in *Saccorhiza polyschides*, Norton, 1978), water movement can influence the morphological (and sometimes anatomical) development of algae. For example, algal growth can be luxuriant in conditions of strong water movement (Conover, 1968) which could well be attributed to increased nutrient supply and/or dissolved gases (Hiscock, 1983). Also areas which have reduced water movement often support free-living populations of algae which are sometimes morphologically different from the attached forms. For example, in the sublittoral loose-lying populations of brown algae, such as *Halopteris scoparia* (Waern, 1952), *Saccorhiza polyschides* (Norton & Burrows, 1969), *Pilayella littoralis* (Russell, 1967; Wilce *et al.*, 1982) and various brown algae such as *Laminaria saccharina*, *Asperococcus* spp. and *Desmarestia* spp. (Burrows, 1958; Irvine, 1974; Tittley *et al.*, 1977) have been described. Elsewhere, a number of very distinct, dwarf forms of members of the Fucales have also been described. For example, salt marsh areas in the North Atlantic may contain a form (ecad) of *Ascophyllum nodosum*, formerly given taxonomic status as *A. mackaii*, which is free-living, profusely dichotomously branched and either sterile or with reduced fertility (Gibb, 1957; South & Hill, 1970; Newton, 1931; Fritsch, 1945). Another small form, previously referred to as var. *scorpioides*, contributes to loose-lying communities in sheltered sublittoral regions (Chuck & Mathieson, 1976). Widely variant salt marsh ecads of *Fucus* spp. have also been described (e.g. *F. vesiculosus* ecad *volubilis*, ecad *caespitosus* and ecad *muscoides* (Fritsch, 1945), as well as of *Pelvetia canaliculata* (Chapman, 1939; Oliveira & Fletcher, 1980). All the above salt marsh forms of *Ascophyllum*, *Fucus* and *Pelvetia* would appear to be plastic responses to the combined influences of extreme shelter and low salinity, but inherent differences between them and their marine rocky-shore counterparts may also exist (Moss, 1971).

Others Other influential environmental parameters include: *Light intensity*, which can control thallus size (Hellebust, 1970; Norton *et al.*, 1981) and morphological expression, e.g. blade width in *Fucus vesiculosus* (Oltmanns, 1922) and *Laminaria saccharina* (Burrows, 1964); *Nutrients*, which influenced blade/crust development in *Petalonia* (Hsiao, 1969); *Sand burial/scouring*, which will not only select those algae with adaptable/tolerant basal systems (Littler *et al.*, 1983), but will favour survival of the prostrate rather than the erect forms of those algae with a heteromorphic life history, such as *Petalonia* and *Scytosiphon* (Wynne, 1969; Fletcher, 1974); *Ice scouring*, which will determine species composition, as well as the morphology, of individual species in arctic regions (Wilce, 1959); *Grazer activity*, which can selectively pressure algae of different palatability (Moore, 1983; Vadas, 1977; Norton, 1978), different growth types, or with different attachment systems (Littler & Kauker, 1984; Clokie & Norton, 1974), as well as individual algae with dissimilar heteromorphic expressions, e.g. the greater resistance of the crustose phases in the life histories of *Petalonia* and *Scytosiphon* (Littler & Littler, 1983; Lubchenco & Cubit, 1980; Slocum, 1980; Dethier, 1981; Littler & Kauker, 1984). Grazer activity can also act indirectly by removing potential competitors (Dethier, 1981).

Specialised cells

Additional noteworthy morphologial/anatomical features of brown algae include hairs, ascocysts and paraphyses (Fig. X).

Hairs Hairs are specialised filaments comprising hyaline cells. Two types are generally recognised in the brown algae: *false hairs* which are usually slightly pigmented, terminate tips of filaments (in the Ectocarpaceae) and lack a distinct basal meristem and sheath, and *true hairs* (often referred to as 'phaeophycean' and/or 'endogenous' hairs) which are colourless, although sometimes containing vesigial plastids, and have a distinct basal growth zone with or without an enclosing sheath. True hairs, which can be single or grouped, are widely distributed in the Fucophyceae arising endogenously or exogenously. They have been reported to have a functional role in the uptake of nutrients (Sinclair & Whitton, 1977). Several reports indicate that hair production appears to be influenced by environmental conditions, including light intensity (Berthold 1882; Oltmanns 1892), light quality, particularly blue light (Dring & Lüning, 1975; Müller & Claus, 1976; Lockhart, 1982) and nutrient levels (Lockhart 1979, 1982).

Ascocysts Ascocysts are abnormal, usually enlarged, thick-walled, cells, sometimes intercalary, but more often terminal in position, which are initially darkly pigmented and packed with fucosan vesicles, later often empty and hyaline. They are widely distributed in the Fucophyceae, including genera such as *Chilionema, Pseudolithoderma, Petroderma, Symphyocarpus, Laminariocolax, Scytosiphon* and *Colpomenia*. Discoid brown algae with ascocysts were

---

Fig. X Examples of hairs, ascocysts and paraphyses
1. *Ectocarpus siliculosus*. 'False' hair terminating branch apex. 2–4. Examples of true hairs: 2, *Leathesia difformis;* 3, *Chilionema ocellatum;* 4, *Punctaria plantaginea*. 5–8, Ascocysts in *Myrionema magnusii* (5, 6); *Symphyocarpus strangulans* (7) and *Colpomenia peregrina* (8). 9–13. Paraphyses variously associated with both plurilocular (9) and unilocular sporangia (others) in *Microcoryne ocellata* (9), *Stragularia clavata* (10), *Leathesia difformis* (11), *Carpomitra costata* (12) and *Myriactula areschougia* (13). Bar = 50 μm (1, 2, 5, 6, 7, 8, 9, 10, 11, 13), 63 μm (3, 4, 12).

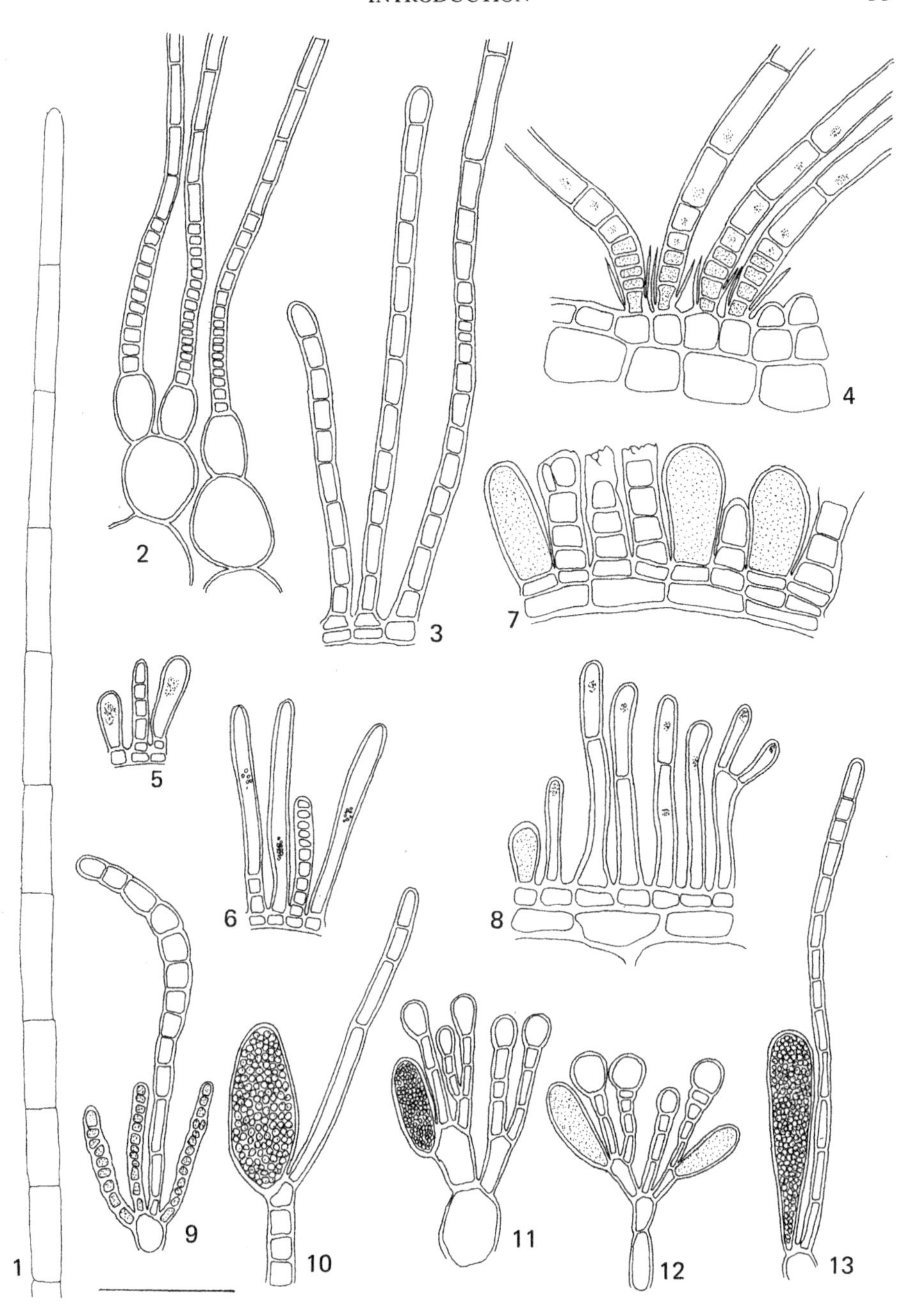
1
2
3
4
5
6
7
8
9
10
11
12
13

previously grouped in the genus *Ascocyclus* Magnus until this was rejected by Loiseaux 1967a (see also Russell, 1964) on the grounds that they represent pathological responses.

Paraphyses Paraphyses is the name given to a variety of sterile structures which are commonly associated with the reproductive organs, usually in sori. They may be unicellular (e.g. in the Laminariales), multicellular and simple (e.g. in *Ralfsia, Stragularia, Leathesia*) or multicellular and branched (e.g. in *Fucus, Petrospongium* and *Corynophlaea*). In a number of families, such as the Lithodermataceae, Elachistaceae, Corynophlaeaceae, Chordariaceae and Scytosiphonaceae, their presence, shape, dimensions, branching pattern, number and shape of constituent cells can be very useful diagnostic characters. They are commonly mucilaginous and perhaps function as protective agents, both from physical damage and desiccation.

REPRODUCTION

Brown algae reproduce vegetatively, asexually and sexually.

Vegetative reproduction

Vegetative reproduction does not involve the release of cell contents (e.g. spores, gametes, etc.) but relies upon the release, distribution and re-establishment of vegetative portions of the plant body. In the majority of brown algae this undoubtedly represents an accidental means of dispersal with the dissemination of plant portions 'released' by environmental agents, such as grazer activity, wave action, water currents, sand scouring, etc. Using bottle-brush collectors placed in the sea, Clokie & Boney (1980) managed to entrap a surprisingly large percentage of the local flora, suggesting that it may represent an important means of dissemination. This would particularly apply to the filamentous genera of brown algae (e.g. in *Acinetospora* (Fritsch, 1945, p. 153) and *Feldmannia* (Etherington, 1964)) which could easily become entrapped and establish firm attachment by secondary rhizoid production. Fragmentation may well be of more importance in estuarine conditions where spore/gamete production may be limited (e.g. in *Pilayella* – Russell, 1967).

Fig. XI Examples of reproductive structures
1. *Sphacelaria cirrosa*. Propagules. 2. *Tilopteris mertensi*. Branches with 2-celled chains of monosporangia. 3. *Dictyota dichotoma*. Surface view of tetrasporangia. 4–9. Examples of unilocular sporangia in *Desmarestia ligulata* (4), *Myriotrichia clavaeformis* (5), *Petroderma maculiforme* (6), *Petrospongium berkeleyi* (7), *Myrionema strangulans* (8) and *Elachista scutulata* (9). 10–16. Examples of plurilocular sporangia in *Petalonia fascia* (10), *Ectocarpus fasciculatus* (11), *Asperococcus turneri* (12), *Petroderma maculiforme* (13), *Giffordia sandriana* (14) and *Hecatonema maculans* (15) and antheridia and oogonia in *Cutleria multifida* (sexual phase) (16). Bar = 20 μm (10), 25 μm (4), 50 μm (5, 6, 7, 8, 9, 11, 13, 14, 16), 63 μm (12, 15), 100 μm (1, 2), 200 μm (3).

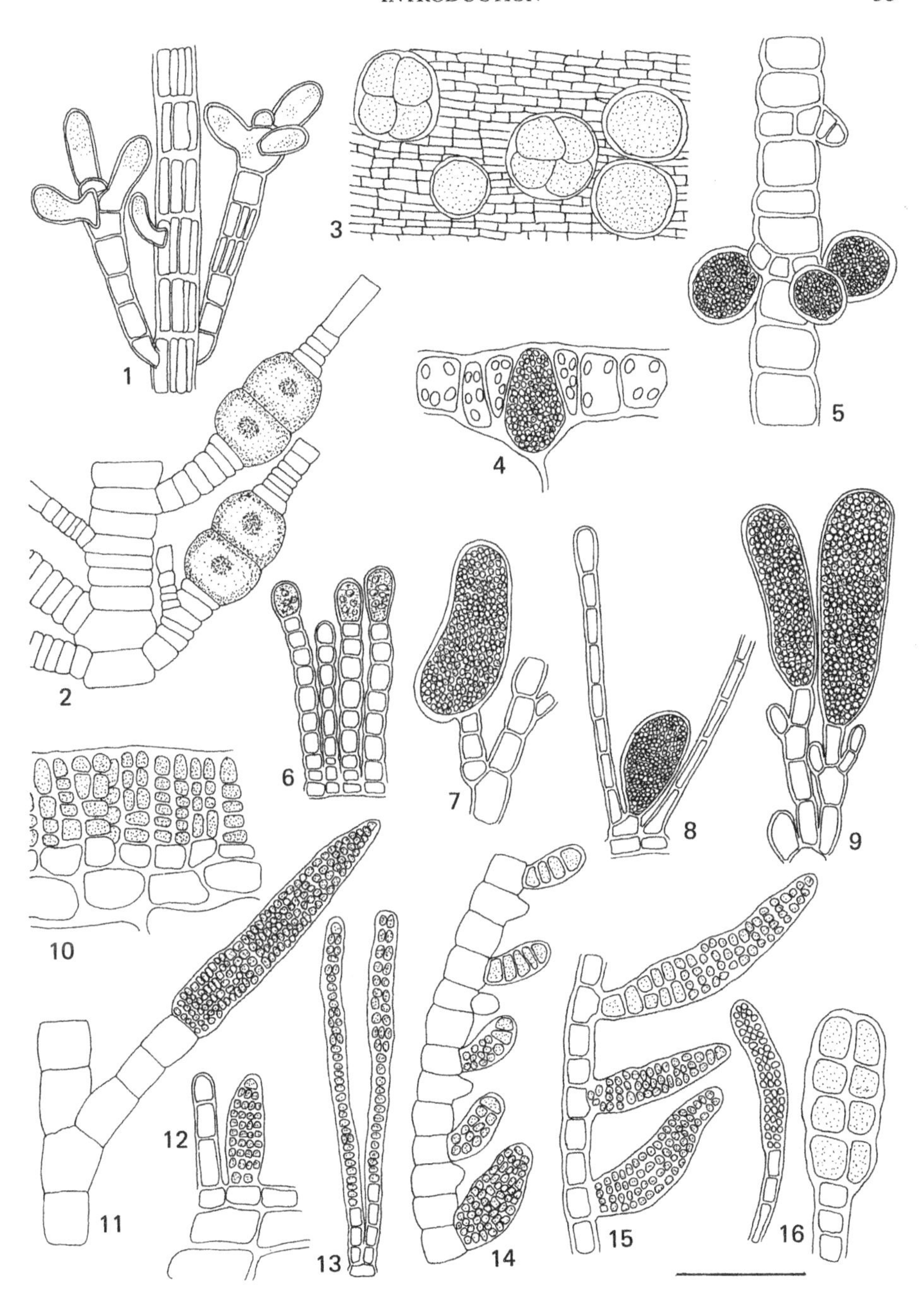
1
2
3
4
5
6
7
8
9
10
11
12
13
14
15
16

There are also a few examples of more specialised vegetative reproductive bodies. These include akinetes, described by Sauvageau (1928) on the thalli of *Tilopteris*, and the variously shaped propagules, described for species of the genus *Sphacelaria* (Prud'homme van Reine, 1982) (Fig. XI, 1). The latter, which represent modified branches, are very effective dispersal agents and, for many species, make a major contribution to the reproduction (Fritsch, 1945).

One additional method of vegetative propagation relies upon the production and growth of new erect shoots from horizontally spreading branches or runners which later often become detached from the parental plant. For example, a number of epiphytic/endophytic algae, such as *Myriactula clandestina, M. stellulata* and *Cylindrocarpus microscopicus,* often proliferate on their respective host plants by the production of creeping, endophytic threads, whilst stolon/adventitious branch formation has been widely recorded (see particularly Fritsch, 1945) in such genera as *Dictyota, Padina, Cladostephus, Bifurcaria* and even *Laminaria.*

Asexual reproduction

In this method of reproduction spores are produced which are non-sexual in their activity; they settle without sexual fusion and germinate directly to form a thallus. For the great majority of algae the spores, often referred to as zoospores or swarmers, are motile. Exceptions include the non-motile monospores (produced in monosporangia) in algae such as *Acinetospora* and *Tilopteris* (Fig. XI, 2), and the tetraspores (produced in modified unilocular sporangia called tetrasporangia) in members of the Dictyotales (e.g. *Dictyota, Padina*) (Fig. XI, 3).

Spores are produced in two principal types of sporangium: *unilocular* and *plurilocular* (Fig. XI, 4–16). In unilocular sporangia, numerous zoospores (often referred to as unispores) are produced within the sporangial mother cell by nuclear divisions which are not accompanied by the formation of partitioning walls. In the sporangial mother cell of plurilocular sporangia, on the other hand, the zoospores (often referred to as plurispores) are formed individually in small, walled cells (loculi). The nuclear divisions which occur in the formation of plurilocular sporangia are probably always mitotic and the plurispores are consequently of the same ploidy level as the parent body. Unilocular sporangia differ in that the initial nuclear division is usually meiotic and results in the formation of haploid unispores. Although unilocular sporangia can function as the seat of meiosis, in many brown algae they are not always so and an increasing number of reports of apomeiotic unilocular sporangia have been published in recent decades. Unilocular sporangia have also more recently been reported on haploid thalli.

Plurilocular sporangia can function either as asexual organs, releasing zoospores which germinate directly into a thallus with the parental genotype, i.e. an accessory reproductive system, or behave as gametangia releasing gametes (see below). Unilocular sporangia are, therefore, usually associated with the 'sporophyte' plant (but see Nakamura & Tatewaki, 1975) whilst plurilocular sporangia can be associated with either sporophyte or gametophyte.

Sexual reproduction

This mode of reproduction involves the production and the release of reproductive bodies (gametes) which behave sexually. These may be motile and identical in appearance (isogamous reproduction), as in *Ectocarpus,* motile and differing in size but not in morphology (anisogamous reproduction), as in *Giffordia* and *Cutleria,* or with the female large and immobile and the male small and motile (oogamous reproduction), as in the Laminariales, Desmarestiales and Fucales. For the sexual process the iso- and anisogametes are usually released into the water and fusion occurs after the female gamete has settled (e.g. in *Ectocarpus*, Müller, 1967); reports of copulating motile gametes have been dismissed as unlikely by Müller (1975). In oogamous reproduction the female gametes are usually fertilised in suspension or when partially emerged from the oogonium (Lüning, 1981a; Müller & Lûthe, 1981; Nakahara, 1984). Fertilisation of eggs retained on the parental thallus has also been reported in *Sargassum muticum* (Fletcher, 1980).

In isogamous and anisogamous reproduction the gametes are produced in plurilocular gametangia. In oogamous reproduction, the female gametes (termed oospores, ova or eggs) are usually produced singly (except in fucoids) in specialised oogonia, whilst the male gametes (termed antherozoids or sperm) are produced singly or in large numbers in antheridia. Male and female gametangia can either be produced on the same thallus (monoecious), as in *Giffordia,* or on different thalli (dioecious), as in *Laminaria.* In *Desmarestia* there appears to be a link between gametophyte sexuality and sporophyte morphology, alternately branched sporophytes having dioecious gametophytes and oppositely branched sporophytes, monoecious gametophytes (Chapman & Burrows (1971), but see Anderson, 1982). Monoecism, in the strict sense, with male and female tissues located in different parts of the same plant, seems to be uncommon, but in *Sargassum* and certain other fucoid genera, conceptacles may be either male or female though within the same receptacle (Fritsch, 1945). There have also been a number of reports of unispores behaving as gametes (Loiseaux, 1967b; Knight *et al.*, 1935; Caram, 1972). However, the evidence for this is not altogether satisfactory and was considered to be extremely dubious by Müller (1975).

An increasing number of reports have revealed the release and sexual union of the gametes to be under pheromone control (Müller, 1981). This process involves the production of a male-attracting substance by the female gametes and has been described in isogamously reproducing genera, such as *Ectocarpus* (Müller *et al.*, 1971), anisogamously reproducing genera, such as *Cutleria* (Boland *et al.,* 1983), as well as oogamously reproducing genera, such as *Fucus* (Müller, 1972b; Müller & Jaenicke, 1973), *Desmarestia* (Müller & Lûthe, 1981) and *Laminaria* (Lüning & Müller, 1978; Müller *et al.*, 1979; Maier & Müller, 1982).

Environmental influences on reproduction

A number of environmental parameters have been shown to influence various aspects of the reproductive process in brown algae. These include:

Light quality An increasing number of reports have revealed light quality, particularly the red and blue wavebands, to exert a control on spore/gamete production and release. In general, red light has been shown to be inhibitory and blue light to promote plurilocular sporangia formation in *Scytosiphon* and *Petalonia* (Lûning & Dring, 1973), and gametogenesis in various members of the Laminariales (Lüning & Dring, 1972, 1975; Lüning 1980b) and *Dictyota dichotoma* in the Dictyotales (Müller & Claus, 1976). There have also been reports of a blue light requirement for spore/gamete release in the brown algae, e.g. plurispore release in *Punctaria tenuissima* (as *Desmotrichum undulatum*) (Lockhart, 1982) and egg release in *Dictyota dichotoma* (Kumke, 1973).

Light intensity A few reports have also shown light intensity to influence sporangia development in some brown algae, including, for example, the ratio of unilocular to plurilocular organs (see studies by Boalch, 1961; Edwards, 1969; Müller, 1962 on *Ectocarpus siliculosus*).

An additional environmental influence on reproduction, probably related to the physiological response to light intensity, is the lunar period. A number of reports have shown that gamete/spore release in various brown algae is periodic and related to the tidal/lunar periods. For example, lunar rhythms have been observed in *Dictyota* (Williams, 1905; Hoyt, 1927; Müller, 1962), *Pelvetia* (Subrahmanyan, 1957), *Sargassum* (Tahara, 1909; Fletcher, 1980), *Nemoderma* (Kuckuck, 1912) and *Himanthalia* (Gibb, 1937). Such a simultaneous release of gametes would certainly increase the probability of fertilisation and be clearly advantageous in the completion of the life history.

Photoperiod/daylength Dring's (1984) recent summary of photoperiodic responses in brown algae includes four relating to reproduction. The processes under photoperiodic control included gametogenesis in *Sphacelaria rigidula* (Hooper *et al.*, 1983), *Desmarestia tabacoides* (Nakahara & Nakamura, 1971), *Ascophyllum nodosum* (Terry & Moss, 1980) and *Fucus distichus* (Bird & McLachlan, 1976).

Daylength responses, as distinct from genuine photoperiodic responses (Dring, 1984), are frequently reported in the literature and sometimes involve aspects of the reproduction of brown algae. For example, Müller (1962) changed the ratio of unilocular to plurilocular sporangia in *Ectocarpus siliculosus* by varying the daylength, and there have been reports of the light/dark ratio controlling gamete release, e.g. in *Dictyota dichotoma* (Bûnning & Müller, 1961) and *Pelvetia canaliculata* (Jaffe, 1954).

However, investigations of the influence of seasonal changes have shown daylength is often linked with temperature. With respect to reproduction, Nakamura & Tatewaki (1975) observed that the cultured crusts or tufts of several Scytosiphonaceae members required short days/cool conditions for unilocular sporangia development, while more recently Nakahara (1984) noted that similar conditions were required for gametogenesis in three *Desmarestia* species. Temperature and daylength were also noted by Colijn & Hoek (1971) to control propagule production in *Sphacelaria furcigera*.

Temperature Under controlled conditions, temperature can by itself influence reproduction of brown algae. For example Müller (1963) noted in *Ectocarpus siliculosus* that plurilocular sporangia formed at high temperatures whilst unilocular sporangia formed at low temperatures; however regional variations did occur. More recently, the extensive publication of Nakahara (1984) investigating *Desmarestia* species and various Laminariales, revealed that temperature played an important influential role in gametogenesis, supporting earlier work by Lüning & Dring (1975) and Lüning (1981a).

Salinity It is likely that reduced salinity will modify or suppress reproduction in brown algae. For example, sexual reproduction in *Pilayella* was reported to be absent in a brackish water population by Russell (1971). Other effects reported include reduced viability of plurilocular sporangia in *Asperococcus fistulosus* and shorter plurilocular sporangia in *Punctaria tenuissima* (as *Desmotrichum undulatum*) (both reported by Pedersen, 1984).

Nutrients A small number of reports have demonstrated that nutrient availability is an important factor in controlling the reproductive processes of algae, including brown algae (see DeBoer, 1981 for short summary).

LIFE HISTORIES

The term life history can be defined as the sum of the reproductive and morphological phases possessed by a species. In the brown algae, the most common life history comprises a cyclic alternation of two chromosomal phases (diploid and haploid) which are linked by meiosis in the diploid phase and fusion of sexual gametes produced by the haploid phase. With the exception of the Fucales, these two phases represent different, free-living generations. Superimposed, however, on this cycle are a wide variety of different pathways which have greatly complicated the interpretations of life histories in this group. In general, the simple and cyclical kind of life history is associated with species that possess morphologically complex thalli such as *Fucus* and *Laminaria*. Flexible, more complicated life histories, on the other hand, are usually associated with the more simple thallus forms. However, exceptions to both generalisations do exist. Nakahara (1984), for example, recently demonstrated apospory, apogamy and polyploidy in various Laminariales growing on the Japanese coast, whilst

Laminaricolax, which has a simple thallus form, also possesses a very simple, reduced life history (Russell, 1964).

Numerous attempts have been made to bring order to chaos and classify and code the wide variety of brown algal life histories which have been recorded. One of the earliest terminologies, originally proposed for the Rhodophyta, was that of Svedelius (1915) who recognised two basic types: *Haplobiontic,* in which one thallus form occurs, and *Diplobiontic,* in which two thallus forms occur. It is interesting to note that this basic subdivision is still widely used today although more consideration is now given to the somatic state of the thallus (e.g. see Edwards, 1976; Bold & Wynne, 1978). Three basic types of life history are usually quoted:

Haplontic Plants are haploid, producing gametes which fuse to form a diploid zygote. During zygotic germination, meiosis occurs and resulting spores reform haploid plants.

Diplontic Plants are diploid, producing haploid gametes by meiosis. Fusion of gametes re-establishes diploid plants.

Diplohaplontic Plants comprise two chromosomal phases – a haploid or gametophytic phase, producing gametes which fuse to form a zygote which germinates into the diploid sporophytic phase. The latter produces haploid spores by meiosis which germinate to reform the haploid phase.

With the exception of the Fucales, to which the second life history applies, all the brown algae have a life history which is basically diplohaplontic. A further subdivision of the latter, proposed by Kylin (1933), recognises that the morphologies of the two phases can be either similar (isomorphic life history, as, for example, in *Padina, Zanardinia* and *Dictyota*) or dissimilar (heteromorphic life history, as, for example, in *Laminaria* and *Desmarestia*). Indeed Kylin classified the brown algae according to their life history, recognising three classes: the Isogeneratae, which possess an alternation of isomorphic generations, the Heterogeneratae, which possess an alternation of heteromorphic generations, and the Cyclosporeae, which lack an alternation of generations (applicable to the Fucales only). Although widely adopted at first, this scheme is no longer supported because of the extreme artificiality of the resulting classification.

Utilising a combination of Svedelius's (1915) 'morphological' classification and his later (Svedelius, 1931) 'somatic state' classification (see Dixon, 1973 for details), an attempt was made by Drew (1955) to define more accurately the life history sequences of algae. For example, mono- and dimorphic life histories were terms applied to algae with a single or two distinct thallus forms respectively. Monomorphic life histories could be haplontic, diplontic or diplohaplontic whilst in dimorphic life histories the two thallus forms could either be at the same ploidy level (i.e. dimorphic haplontic or dimorphic diplontic), or different ploidy levels (i.e. dimorphic, diplohaplontic). Chapman & Chapman's (1961) proposal to extend this terminology has generally been regarded as too cumbersome and unworkable (Russell, 1973a; Dixon, 1973). Dixon's (1963, 1970) proposed 'type' method of discrimination between different life histories (e.g. *Ectocarpus*-type life history, *Fucus*-type life history, etc.) is certainly more simple and quite flexible but is open to misinterpretation in the brown algae because of the widespread occurrence of different types of life history within the same species.

In the present volume we propose to use Drew's terminology when feasible but will not seek to impose it on taxa for which it is evidently ill suited. The chromosomal terminology, e.g. monophasic (either diplontic or haplontic) and biphasic (diplohaplontic) is applied in conjunction with that of the number of morphological forms (e.g. monomorphic, dimorphic) but only on the basis of cytological data.

It is noteworthy in this discussion on life histories of brown algae that problems have arisen even in the interpretation of hitherto well established facts. For example, members of the Fucales have been attributed with a monomorphic, diplontic life history (Bold & Wynne, 1978) (see below). The free-living plant body is a diploid gametophyte which undergoes meiosis at the onset of gametogenesis. In this process the first divisions of the oogonium are meiotic and four haploid nuclei are produced which subsequently divide mitotically to produce an eight nucleate oogonium. Various numbers of these nuclei, according to the different genera, then contribute to egg formation. However, an alternative life history interpretation for the Fucales, revived by Jensen (1974), would be 'dimorphic and diplohaplontic'. In this case the post-meiotic oogonial precursor represents a unilocular sporangium (or megasporangium) and the four nuclei produced by meiosis represent potential reduced gametophytic microthalli (or megagametophytes); subsequently mitotic divisions of these nuclei then produce the haploid eggs. The life history could, therefore, be described as dimorphic (more accurately heteromorphic) with both an enclosed, haploid, gametophytic phase and a free-living, diploid, sporophytic phase (i.e. diplohaplontic).

The following examples illustrate the range of life histories described in the brown algae:

Diplontic

This type of life history is usually attributed only to members of the Fucales. Taking *Fucus* as an example (Fig. XII, a), the large diploid plant produces terminal receptacles on which flask-like conceptacles develop containing male (antheridia) and female (oogonia) gametangia. Production of the male antherozoids and female eggs is preceded by meiosis and fusion of these haploid gametes followed by germination of the resultant zygote re-establishes the diploid macroscopic plant.

a. The life history of _Fucus_ (monomorphic, diplontic)

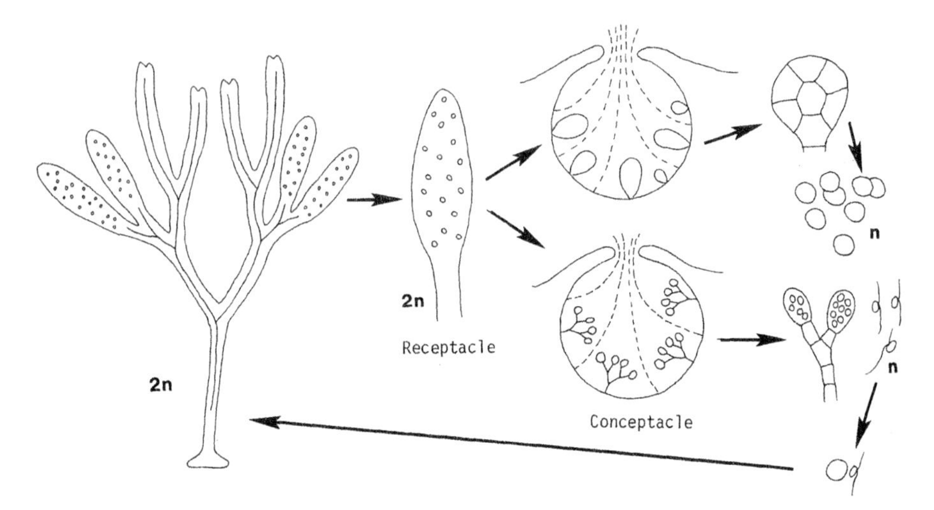

b. The life history of _Ectocarpus_ (monomorphic, diplohaplontic)

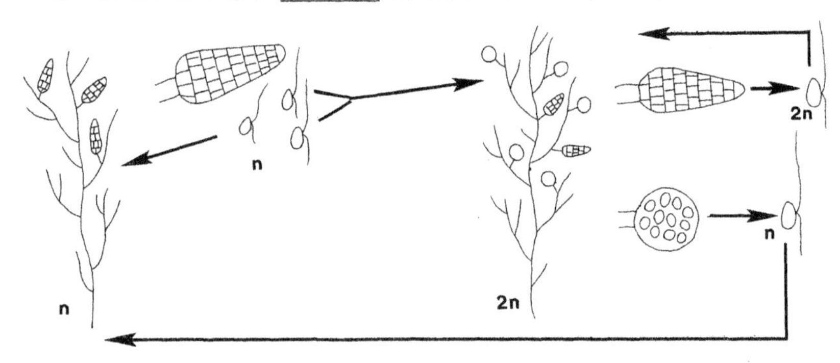

c. The life history of _Laminaria_ (dimorphic, diplohaplontic)

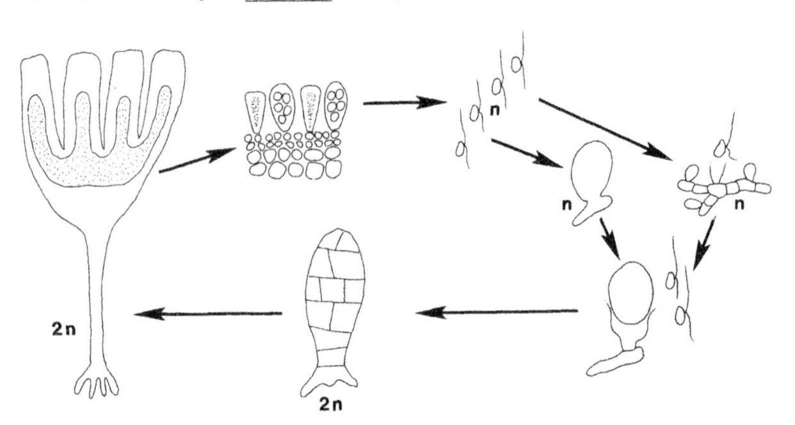

d. The life history of *Scytosiphon* (dimorphic, diplohaplontic)

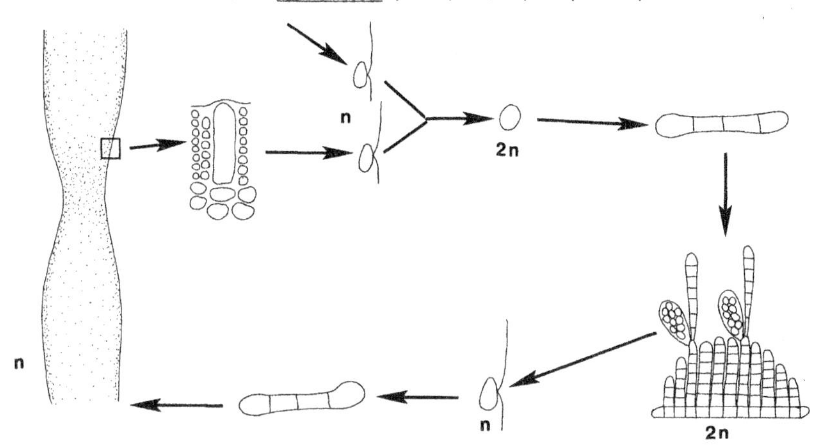

e. The life history of *Streblonema oligosporum* (monomorphic, monophasic)

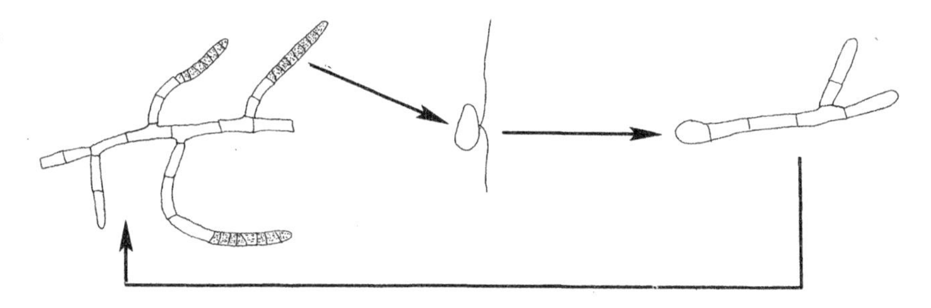

f. The life history of *Stragularia spongiocarpa* (monomorphic, monophasic)

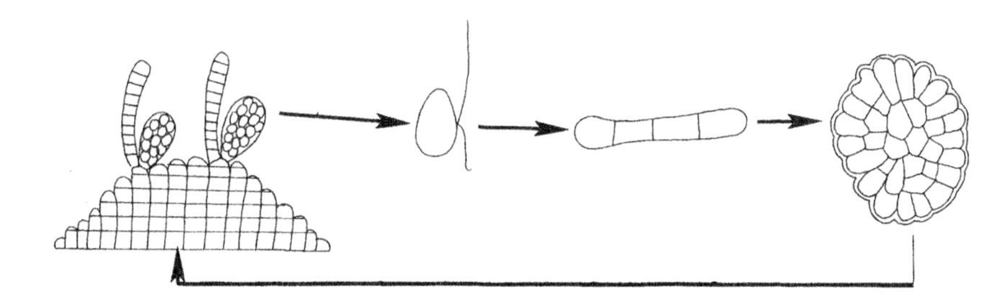

Fig. XII Examples of life histories.

Diplohaplontic

1. *Monomorphic*
In this life history there is an alternation between morphologically identical haploid and diploid plants (often termed isomorphic life history) (Fig. XII, b). It has been reported, for example, in many of the simple, erect members of the Ectocarpales (see particularly the now classical studies on *Ectocarpus* by Knight (1929), Papenfuss (1935) and Müller (1972a) and on *Pilayella* by Knight (1923)), the filiform parenchymatous Sphacelariales (see particularly Hoek & Flinterman's 1968 study of *Sphacelaria furcigera*), and the foliaceous, parenchymatous Dictyotales. The haploid plants produce gametes by mitotic cell division (from plurilocular sporangia) which fuse. The zygotes germinate into morphologically identical plants; asexual spore production by meiosis (in the unilocular sporangia, or tetrasporangia in the Dictyotales) then re-establishes the haploid generation.

A problem in classifying many of these life histories arises from the fact that they may be slightly dimorphic (often termed heteromorphic), e.g. as in *Ectocarpus* (Müller, 1972a), *Giffordia* (Kornmann, 1954), *Feldmannia* (Kornmann, 1953), *Pilayella* (Russell, 1961) and *Sphacelaria* (Hoek & Flinterman, 1968).

2. *Dimorphic (Trimorphic, Polymorphic, etc.)*
In this life history there is an alternation between at least two morphologically dissimilar haploid and diploid plants (Fig. XII, c–d). In the great majority of algae with this type of life history, there is a regular alternation between a large, diploid plant (macrothallus) and a microscopic, haploid plant (microthallus) (Fig. XII, c). Unispores, produced by meiosis in the unilocular sporangia on the diploid plant, germinate into haploid microthalli; the latter then reproduce sexually and the resultant zygote reforms the macrothallus. This life history is particularly characteristic of members of the Laminariales and Desmarestiales, and has also been reported in various members of the two orders (*sensu* Wynne, 1981 and others), Chordariales (e.g. *Sphaerotrichia* (Ajisaka & Umezaki, 1978) and *Cladosiphon* (Shinmura, 1974)) and Dictyosiphonales (e.g. *Striaria* (Caram, 1965) and *Litosiphon* (Dangeard, 1969)).

In a number of algae, however, e.g. *Cutleria* (Cutleriales) and members of the Scytosiphonaceae, the macrothallus is haploid and sexual, while the smaller, prostrate thalli are diploid and asexual (Fig. XII, d). In the life history of *Cutleria*, released gametes fuse and germinate into parenchymatous, procumbent, unilocular sporangia-bearing thalli (the '*Aglaozonia*' phase) which is the diploid, asexual generation; unispores are formed by meiosis which germinate to re-establish the *Cutleria* phase. In certain members of the Scytosiphonaceae, the macrothallus, characteristic of the genera *Colpomenia, Petalonia* and *Scytosiphon,* may be haploid, whilst the encrusting *Ralfsia*-like microthalli are diploid.

Often associated with these diplohaplontic life histories are a number of variations and abbreviations. These include:

a. *Dimorphic, monophasic life histories.* An increasing number of species, particularly in the Chordariales and Dictyosiphonales (*sensu* Wynne, 1981 and others), are being revealed to have a heteromorphic, but asexual life history. There is little evidence that the two distinct, independently existing, morphological growth forms or expressions represent distinct haploid and diploid phases; the life history can, therefore, be described as heteromorphic but 'direct' and monophasic. Such a life history appears to apply to *Asperococcus* and *Punctaria* (Punctariaceae) in which the characteristic, macroscopic, erect, parenchymatous thalli and microscopic, branched, *Hecatonema*-like thalli have been described. Without cytological evidence it is better to adopt the simple terminology of macro- and microthalli for these two growth forms and avoid commitment to such terms as *plethysmothallus* (diploid, sporophytic, usually plurilocular sporangia-bearing, self-perpetuating mictothallus) and *protonema* (plethysmothallus which gives rise as a lateral outgrowth to an erect macroscopic sporophyte) as defined by Sauvageau (1932).

Dimorphic, monophasic, asexual life histories have also commonly been described for genera in the Scytosiphonaceae (Wynne, 1969; Fletcher, 1974; Roeleveld *et al.*, 1974); the biphasic life histories appear to be largely confined to Pacific strains of these algae (Nakamura & Tatewaki, 1975; Clayton, 1980). It is considered that such a life history represents parthenogenetic development of the gametes. A significant aspect of many dimorphic, monophasic life histories is the important role played by environmental conditions in the relationship between the two morphological expressions. Parameters, such as temperature, photoperiod, light intensity, light quality, salinity, etc., can, for example, control the final morphological expression produced (e.g. crustose or erect blade development in the Scytosiphonaceae) and very often the physical connection between the expressions (e.g. environmental control of erect thallus formation from microthalli in the Punctariaceae).

b. *'Vegetative diploidisation'* Such a non-sexual, biphasic, life history has been reported in two algae: *Ectocarpus* (Müller, 1967) and *Elachista stellata* (Wanders *et al.*, 1972), by the direct sprouting of diploid macrothalli from haploid microthalli.

c. *Monomorphic, monophasic life histories* A number of brown algae have been shown to produce spores which germinate directly, without sexual fusion, to repeat the parental thallus form (i.e. a direct, monophasic, monomorphic life history) (Fig. XII, e–f). Such a simple, direct developmental pattern has been observed for plurispores in *Streblonema* (Fletcher, 1983), *Giffordia granulosa* (Fletcher, 1981b), *Laminariocolax* (Russell, 1964) and *Pseudolithoderma roscoffense* (Loiseaux, 1968), as well as for unispores in *Ralfsia verrucosa* (Loiseaux, 1968; Fletcher, 1978) and *Stragularia spongiocarpa* (as *R. spongiocarpa* – Fletcher, 1981c) suggesting that either the parental thallus is haploid, or if diploid, that apomeiosis has occurred in the unilocular sporangia.

STRATEGIES

The ecological importance of the algal life history and, particularly, of the structural characteristics of its morphological phases, has long been recognised. Early attempts to classify life histories in a way which made some ecological sense made use of the concept of 'life-forms' (see Russell & Fielding, 1981). However, in recent years, less emphasis has been given to the events in the life history and more to ecological adaptive considerations and the term 'life-form' is now giving way to that of the 'strategy'.

The classification of higher plant strategies proposed by Grime (1974, 1977, 1979) is based upon a hypothesis that there are three primary selection pressures operating upon plants in nature. These are (1) competition for light, space, nutrients, etc., (2) stress, and (3) disturbance. Stress includes those factors which cause the formation of an unproductive environment which, in the marine environment, may be interpreted in terms of e.g. excessive desiccation, severely fluctuating temperatures, etc. Disturbance, on the other hand, is considered by Grime to include factors which reduce production in an otherwise productive environment. In the marine environment this may be considered to include herbivory and possibly trampling (Boalch *et al.*, 1974) or even wave action.

Grime's classification has received some approval from Littler & Littler (1980) and Dring (1982) and there is some evidence from the literature to support its use with regard to marine algae (see Russell & Fielding, 1981 for discussion). For example, from the work of Dayton (1975) and Lubchenco (1978), etc., it would appear that canopy-forming brown algae, such as *Ascophyllum* and *Laminaria,* are strongly competitive algae. It is equally evident that many of the encrusting, calcified and tough cartilaginous algae are disregarded by grazing animals (Lubchenco & Cubit, 1980; Littler & Kauker, 1984). Finally, it seems likely that the lichens, Cyanobacteria and other permanent occupants of the higher shore levels are very tolerant to stress conditions. However, this system is unlikely to prove a completely successful means of identifying algal strategies for the reason that it fails to accommodate those species that are adapted to cope with more than one selection pressure, e.g. crustose coralline algae are grazing resistant but under conditions of dim light also stress resistant. Similarly, Grime's classification cannot easily accommodate those species which are characterised by populations with many different adaptive features (e.g. heavy metal tolerant and sensitive strains, stenohaline and euryhaline strains).

We shall not adopt the Grime or any other strategies' classification but will indicate, where appropriate, the existence of special adaptive or functional characteristics of the algae.

COMMUNITY ECOLOGY

During the past decade, intertidal community ecology has developed from a mainly descriptive science to one which is experimental with a variety of field and laboratory techniques. The descriptive phase was marked by much preoccupation with zonation and its causes. Numerous methods for the description of intertidal zonation patterns have been proposed (see Sundene, 1953; Lewis, 1964; Russell, 1972, 1973a, b, c, 1977, 1980). In the first volume of the present work, Dixon & Irvine (1977) introduced a zonal system and terminology as shown below. This scheme is plainly comparable with that of Lewis (1964) in the way the zones are defined, and, by dividing Lewis's Eulittoral into upper and lower zones, it is arguably more precise. However, the Dixon and Irvine terminology has not gained widespread acceptance in the way that Lewis's has. It also incorporates the term Midlittoral which already existed in the literature as a synonym for the Eulittoral zone, Stephenson & Stephenson (1949, 1972). The possibility of confusion over what is meant by Midlittoral should be avoided. For these reasons we shall adopt the Lewis (1964) system.

Zonal classification

Lewis, 1964		*Dixon & Irvine, 1977*
Littoral	Littoral fringe	Upper littoral
	Eulittoral	Mid littoral + Lower littoral
Sublittoral		Sublittoral

In general, the three-zone classification of Lewis has been validated by more recent and objective analytical methods (see Russell, 1972, 1973a, b, 1977, 1980; Jones *et al.*, 1980) but other patterns are, of course, possible. For example, estuarine localities frequently lack a sublittoral algal zone (Wilkinson, 1980; Russell, 1973b; Tittley & Price, 1977).

Zonation has been attributed mainly to tidal (emersion-immersion) factors and differences in adaptation to the associated stresses. Colman (1933) and Doty (1946) have taken this interpretation to the point of identifying critical tidal levels where zonal boundaries occur. Other workers, e.g. Lewis (1964), have been unable to locate such levels and point out that on exposed rocky shores, tidal amplitude is of secondary importance to wave action in determining emersion-immersion times. Nevertheless, the algae of higher shore levels have been shown by Dring (1982) to have superior powers of recovery from desiccation than lower shore species.

The problem of identifying causes of zonation have arisen chiefly from the fact that the physico-chemical factors operating on a rocky shore tend to do so as gradients whereas zonal boundaries are sharp discontinuities. Therefore, a number of recent workers have preferred to invoke biological interactive effects as the major factor determining zonal frontiers (Chapman, 1974). Others, e.g. Schonbeck & Norton (1978), have preferred to combine physico-chemical effects with biological interactions in identifying the causes of zonation.

The effects of competition between marine algae have been measured in field experiments by Dayton (1975), Lubchenco (1978), Hawkins & Hartnoll (1980) and others by removal of hypothetical dominants. In most cases the large, brown canopy-forming algae have proved to be dominant in the sense of determining overall algal species composition. Occasionally, however, ephemeral algae have been found to be capable of competitively excluding perennial forms, *Enteromorpha* having been found capable of outcompeting *Chondrus* (Lubchenco & Menge, 1978). Annual algae, such as *Ectocarpus,* have also been found to compete successfully under certain conditions against other annual algae (Russell & Fielding, 1974) in laboratory culture whilst Fletcher (1975) has observed heteroantagonism between crustose species in culture.

Other influential factors in algal ecology include '*herbivory*' and '*succession*'. The importance of herbivory has been established by various field experiments/observations extending principally over the past thirty years, since the earlier studies of Burrows & Lodge (1950) to more recent work by Hawkins (1981). Field experimentation, particularly involving observations of the recolonisation of cleared areas, has also played a major role in our understanding of successional changes in plant community structure (see particularly Burrows & Lodge, 1950; Lodge, 1948; Southward, 1956; Dayton, 1975; Hawkins, 1981). Recent attempts at elucidating the mechanisms of succession have generally failed although particular interest has been given to the three 'models' proposed by Connell & Slatyer (1977). These are (1) the '*facilitation*' model which states that only certain pioneering species can colonise and then modify the environment, rendering it more suitable for later succession species, (2) the '*tolerance*' model which suggests that any species capable of living in the habitat as an adult can colonise but that this will not exclude later succession species and so the more environmentally stress tolerant species will eventually become established, and (3) the '*inhibition*' model which proposes that any species capable of living in the habitat as an adult can colonise and these will exclude all other species until their death or disturbance creates space. Evidence giving clear support for one or other of these interpretations is still lacking. For example, Hawkins's (1981) observation that the early diatom and green algal colonisers are not an essential prerequisite for recruitment and growth of *Fucus* is counter to the facilitation model, whilst Knight & Parke's (1950) observation of *Fucus* germlings growing up through *Enteromorpha* is counter to the inhibition model. It, therefore, seems unlikely that any single model will be completely true for all successional stages.

FLORISTICS

Table 1 shows the comparative numbers of brown algae recorded for several North Atlantic regions. It can be seen that floristically the British Isles comprises the largest number (197) of brown algae; also relatively rich are Norway (175 species), N.W. France (160) and Denmark (128), despite the latter

TABLE 1
Distribution of brown algae in the North Atlantic

Region	No. Fucophyceae	Reference
Southernmost Greenland	49	Pedersen, 1976
Faeroes	74	Irvine, 1982
Iceland	72	Caram & Jónsson, 1972
Norway	175	Rueness, 1977
Denmark	128	Christensen & Thomsen, 1974
British Isles	197	Parke & Dixon, 1976
Roscoff	160	Feldmann & Magne, 1964
Portugal	98	Ardré, 1970

having a limited range of shore types. The similarity in the brown algal floras of these regions is not surprising; as Guiry (1978) pointed out, these regions, which would all be classified by Hoek (1975) as 'cold-temperate Atlantic-Boreal', are likely to be similar phytogeographically and comprise a fairly rich transition flora. Certainly, it can be seen that passage northwards and southwards to southernmost Greenland and Portugal respectively brings about a corresponding marked diminution in the brown algal flora.

SYSTEMATICS

Table 2 compares the numbers of supra-generic taxa, genera and species listed in Harvey (1849) and Parke & Dixon (1976). There has been a remarkable increase in numbers of higher taxa which has not been matched by a comparable increase in species. The resulting complication of brown algal classification is not always necessary or justifiable, nor does it facilitate species determination to any obvious extent. The delimitation of certain orders in the brown algae remains unresolved. We have preferred to retain the broad Fritschian concept of the Ectocarpales, whereas others (Christensen, 1980; Bold & Wynne, 1978; Abbott & Hollenberg, 1976; Wynne, 1981) have recognised as Orders groups such as the Scytosiphonales, Chordariales, Dictyosiphonales, Ralfsiales, Tilopteridales and Punctariales. The criteria on which these are based are single or

TABLE 2
Comparison of Harvey and Parke & Dixon classifications

<table>
<tr><th></th><th colspan="2">Supra-generic Taxa</th><th></th><th></th></tr>
<tr><th></th><th>Orders</th><th>Families</th><th>Genera</th><th>Species</th></tr>
<tr><td>Harvey (1949)</td><td></td><td>8</td><td>35</td><td>95</td></tr>
<tr><td>Parke & Dixon (1976)</td><td>7</td><td>32</td><td>96</td><td>197</td></tr>
<tr><td>% increase</td><td colspan="2">488</td><td>274</td><td>207</td></tr>
</table>

few in number and lead us to the suspicion that the resulting groups are somewhat artificial. However, the orders which we have adopted may not prove to stand the tests of time either and so we shall concentrate upon the families as the higher taxonomic ranks in this text. These are on the whole less controversial than orders and seems to us to be more natural units. However, recent discoveries on brown algal life histories, etc., have made it necessary to remove some genera and species from families in which they have been placed by other authors.

Figure XIII illustrates some of the principal characteristics and affinities of the families recorded for the British Isles. It is a modification of the diagram used by Hoek & Jahns (1978) to show the ordinal interrelationships in the brown algae, and adopts similar criteria. However, it is essentially a phenetic diagram and, therefore, is not strictly comparable with the phylogenetic proposals of Kylin (1933), Papenfuss (1951), Wynne & Loiseaux (1976) and others. It also necessarily involves generalised statements about families which may not apply to every species within them.

With respect to *species*, these remain, as for most algae, largely morphological entities. The validity of some of these may be seriously doubted but they are retained. Doubts about these, when felt, have been expressed. During the 1950s and 1960s most *infra specific taxa* reported for British coasts were removed from checklists. This was done mostly for the reason that the variants were insufficiently distinct to merit recognition, a standpoint which has to some extent been supported by culture studies. However, variation, even if self-evidently continuous in character, seems to be worth recording and where sufficient evidence is at hand to reveal the incidence and extent of such variation we have recorded it. Some infraspecific variants have recently received formal

---

Fig. XIII Diagrammatic representation of the characteristics and interrelationships among the brown algal families recorded for the British Isles (modified from Hoek & Jahns, 1978). The thirty-three families listed (see below) and displayed are those currently included in Parke & Dixon (1976), with the exception that the Lithodermataceae replace the Ralfsiaceae and the new family Pogotrichaceae is included.

Key to Families

1. Ectocarpaceae
2. Lithodermataceae
3. Myrionemataceae
4. Elachistaceae
5. Corynophlaeaceae
6. Chordariaceae
7. Acrotrichaceae
8. Spermatochnaceae
9. Pogotrichaceae
10. Striariaceae
11. Myriotrichiaceae
12. Giraudiceae
13. Punctariaceae
14. Buffhamiaceae
15. Dictyosiphonaceae
16. Tilopteridaceae
17. Scytosiphonaceae
18. Cutleriaceae
19. Arthrocladiaceae
20. Desmarestiaceae
21. Sporochnaceae
22. Chordaceae
23. Laminariaceae
24. Alariaceae
25. Choristocarpaceae
26. Sphacelariaceae
27. Stypocaulaceae
28. Cladostephaceae
29. Dictyotaceae
30. Fucaceae
31. Himanthaliaceae
32. Cystoseiraceae
33. Sargassaceae

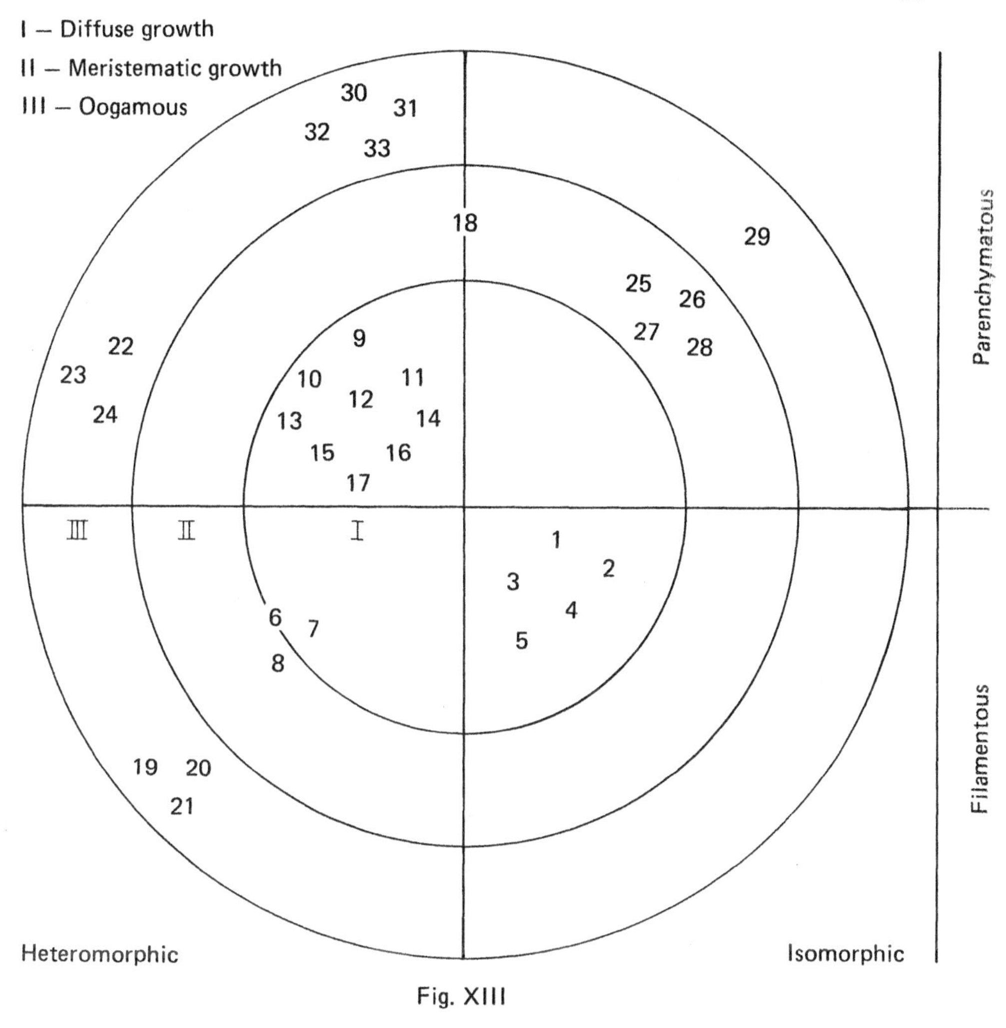

Fig. XIII

recognition within the framework of orthodox taxonomy, e.g. *Cladostephus spongiosus* f. *verticillatus* (Prud'homme van Reine, 1972), *Laminaria hyperborea* f. *cucullata* (Svendsen & Kain 1971); these epithets we have retained. Other variants which may have received no formal description we have referred to using the informal terms from plant biosystematics (Stace, 1980) as seems appropriate to the particular taxon.

Table 3 lists the taxa treated in the present Part 1, Volume 3 of the Fucophyceae. It can be seen that taxonomic treatment is presented for 12 families, 36

TABLE 3

Check-list of taxa included in Volume 3, Part 1 of the Fucophyceae

Lithodermataceae
Petroderma Kuckuck, 1897
maculiforme (Wollny) Kuckuck
Pseudolithoderma Svedelius in Kjellman & Svedelius, 1910
extensum (Crouan frat.) Lund
roscoffense Loiseaux
Sorapion Kuckuck, 1894
simulans Kuckuck
Symphyocarpus Rosenvinge, 1893
strangulans Rosenvinge

Myrionemataceae
Compsonema Kuckuck, 1899
microspongium (Batters) Kuckuck
minutum (C. Agardh) Kuckuck
Microspongium Reinke, 1988
globosum Reinke
Myrionema Greville, 1827
corunnae Sauvageau
liechtensternii Hauck
magnusii (Sauvageau) Loiseaux
papillosum Sauvageau
strangulans Greville
Protectocarpus Kuckuck, 1955
speciosus (Børgesen) Kuckuck
Ulonema Foslie, 1894
rhizophorum Foslie

Elachistaceae
Elachista Duby, 1830
flaccida (Dillwyn) Areschoug
fucicola (Velley) Areschoug
scutulata (Smith) Duby
stellaris Areschoug
Halothrix Reinke, 1888
lumbricalis (Kützing) Reinke
Leptonematella Silva, 1959
fasciculata (Reinke) Silva

Corynophlaeaceae
Corynophlaea Kützing, 1843
crispa (Harvey) Kuckuck
Leathesia Gray, 1821
difformis (Linnaeus) Areschoug
Microcoryne Strömfelt, 1888
ocellata Strömfelt
Myriactula Küntze, 1898
areschougii (Crouan frat.) Hamel
chordae (Areschoug) Levring
clandestina (Crouan frat.) J. Feldmann
haydenii (Gatty) Levring
rivulariae (Sühr in Areschoug) J. Feldmann
stellulata (Harvey) Levring
Petrospongium Nägeli in Kützing, 1858
berkeleyi (Greville in Berkeley) Nägeli in Kützing

Pogotrichaceae
Pogotrichum Reinke, 1892
filiforme (Reinke) Batters

Myriotrichiaceae
Leblondiella Hamel, 1939
densa (Batters) Hamel
Litosiphon Harvey, 1849
laminariae (Lyngbye) Harvey
Myriotrichia Harvey, 1834
clavaeformis Harvey

Punctariaceae
Asperococcus Lamouroux, 1813
compressus Griffiths ex Hooker
fistulosus (Hudson) Hooker
scaber Kuckuck
turneri (Smith) Hooker
Chilionema Sauvageau, 1897
foecundum (Strömfelt) Fletcher
hispanicum (Sauvageau) Fletcher
ocellatum (Kützing) Kuckuck
reptans (Crouan frat.) Sauvageau

Hecatonema Sauvageau, 1897
maculans (Collins) Sauvageau
Punctaria Greville, 1830
crispata (Kützing) Batters
latifolia Greville
plantaginea (Roth) Greville
tenuissima (C. Agardh) Greville

Scytosiphonaceae
Colpomenia (Endlicher) Derbès et Solier, 1851
peregrina (Sauvageau) Hamel
Compsonema saxicolum (Kuckuck) Kuckuck phase of *Petalonia/Scytosiphon*
Microspongium gelatinosum Reinke phase of *Scytosiphon lomentaria*
Petalonia Derbès et Solier, 1850
fascia (O. F. Müller) O. Kuntze
filiformis (Batters) O. Kuntze
zosterifolia (Reinke) O. Kuntze
Ralfsia Berkeley in Smith & Sowerby, Suppl. III pl. 2866 (1st Nov. 1841)
verrucosa (Areschoug) J. Agardh
Scytosiphon C. Agardh, 1820
dotyi Wynne
lomentaria (Lyngbye) Link

Stragularia Strömfelt, 1886
clavata (Harvey in Hooker) Hamel
spongiocarpa (Batters) Hamel

Cutleriaceae
Cutleria Greville, 1830
multifida (Smith) Greville
Zanardinia Nardo ex Crouan frat., 1857
prototypus (Nardo) Nardo

Arthrocladiaceae
Arthrocladia Duby, 1830
villosa (Hudson) Duby

Desmarestiaceae
Desmarestia Lamouroux, 1813
aculeata (Linnaeus) Lamouroux
dresnayi Lamouroux ex Leman
ligulata (Lightfoot) Lamouroux
viridis (Müller) Lamouroux

Sporochnaceae
Carpomitra Kützing, 1843
costata (Stackhouse) Batters
Sporochnus C. Agardh, 1817
pedunculatus (Hudson) C. Agardh

genera, 66 species and two 'phases' (*Compsonema saxicolum* phase of *Petalonia/Scytosiphon* and *Microspongium gelatinosum* phase of *Scytosiphon lomentaria*). The order of presentation is similar to that of Parke & Dixon (1976). Notable changes from the latter authors' checklist include:

At the family level
1. With the removal of the genus *Ralfsia* to the Scytosiphonaceae, the family Lithodermataceae is revived (replacing the Ralfsiaceae) to include the majority of crustose brown algae.
2. The family Pogotrichaceae is recognised following the procedure of Pedersen (1978).

At the genus level
1. *Ralfsia* is transferred from the Ralfsiaceae to the Scytosiphonaceae.
2. *Chilionema* and *Hecatonema* are removed from the Myrionemataceae to the Punctariaceae.
3. *Leptonematella* is removed from the Myrionemataceae to the Elachistaceae.
4. *Pleurocladia* is provisionally removed from the Myrionemataceae pending further studies revealing its true taxonomic position.

5. *Cylindrocarpus* (along with the type species *C. microscopicus*) is provisionally removed from the Myrionemataceae pending further studies to reveal its true taxonomic position. The genus *Petrospongium* is revived to include *P. berkeleyi* (formerly *Cylindrocarpus berkeleyi*).
6. The genus *Pogotrichum* is recognised, placed in the family Pogotrichaceae.
7. The genus *Litosiphon* is transferred to the Myriotrichiaceae.
8. In the Punctariaceae the genus *Desmotrichum* is included within the synonomy of *Punctaria*.
9. In the Scytosiphonaceae the genus *Stragularia* is revived to include two species formerly placed in *Ralfsia*.

At the species level

1. In the Myrionemataceae

a) Two species of *Compsonema* are recognised; *C. microspongium* and a new addition to the flora, *C. minutum*. *C. saxicolum* is removed to the Scytosiphonaceae and described as a microthallus in the life histories of *Petalonia/Scytosiphon*.

b) *Hecatonema liechtensternii* has been transferred to the genus *Myrionema* (pending further studies) whilst *H. foecundum* and *H. hispanicum* are transferred to the genus *Chilionema*.

c) *Microspongium gelatinosum* is transferred to the Scytosiphonaceae and described as a microthallus in the life history of *Scytosiphon lomentaria*.

d) *Myrionema aecidioides* is provisionally assigned to the genus *Gononema* and will be treated in Part 2.

e) *Myrionema polycladum* is included within the synonomy of *Microspongium globosum*.

2. In the Myriotrichiaceae, *Myriotrichia filiformis* and *M. repens* are included within the synonomy of *M. clavaeformis* whilst *Litosiphon pusillus* is included within the synonomy of *L. laminariae*.
3. In the Punctariaceae, *Desmotrichum undulatum* is included within the synonomy of *Punctaria tenuissima*.
4. In the Scytosiphonaceae

a) *Ralfsia clavata* and *R. spongiocarpa* are transferred to the genus *Stragularia*. *Ralfsia disciformis* and *R. pusilla* are doubtful records and have been provisionally removed from the flora.

b) *Scytosiphon* includes an additional species for the British Isles, *S. dotyi*.

ARRANGEMENT OF THE WORK

The format adopted here for the presentation of Volume 3, Part 1 of the Fucophyceae follows quite closely that of Volume 1 (Rhodophyta) of this series. Taxa are arranged largely according to the Parke & Dixon (1976) check-list although a few changes have been incorporated (see above). The families are not arranged according to any higher taxon; rather they reflect the primary interests of the present author.

Full formalised diagnoses are given for each family, genus and species, which is usually followed by additional notes/discussions. The latter can be of an explanatory nature, a summary of the salient features of the taxon, a comparison of closely related taxa, or details of life history features, etc. Particular emphasis is placed on the latter information in view of the important role played (or potentially played) by many of the presently included microscopic taxa in the life histories of various macroalgae.

The references and proposals for conservation of the family names follow Silva (1980). For each genus, the type species is stated along with the major synonomy and, where appropriate, keys to the species are included. For each species the major synonomy is also listed, followed by a detailed description and notes on habitat characteristics, distributional patterns around the British Isles and seasonal aspects of growth and reproduction. Where possible, the species description follows the common sequence of morphology, anatomy, cytology and reproduction. Distributional data, which are given on the basis of the pre-1974 county system (see inside cover map), include only verified records and, in view of self-imposed time pressures in the production of this volume, must be considered as conservative. World-wide distribution patterns of each species, based on published reports, are also presented in tabular format at the end of the text (pp. 297). Particular emphasis has been given to the provision of illustrations and at least one compound plate of line drawings and/or black and white photographs is provided for each species.

The rapid advance in our knowledge of the brown algae generally and in particular, those present around the British Isles, over the past few decades, has supported the need for a new revised and up to date Flora of the British Isles. Newton's (1931) volume, now over 50 years old, has long been out of date with respect to the nomenclature, taxonomy and distributional data. There is also a particular need for a more detailed text on the smaller, brown algae which received very little consideration by Newton with respect to both descriptions and illustrations. Emphasis was, therefore, given to the production, as early as possible, of a basic working text on the brown algae of the British Isles which could subsequently be updated regularly. By implication this will result in the inadequate coverage of certain aspects such as distributional data, typification etc.; for this we offer our apologies.

ROBERT L. FLETCHER AND GEORGE RUSSELL

REFERENCES FOR INTRODUCTION

ABBOTT, I. A. & HOLLENBERG, G. J. 1976. *Marine algae of California*. Stanford University Press, Stanford.

AJISAKA, T. & UMEZAKI, I. 1978. The life history of *Sphaerotrichia divaricata* (Ag.) Kylin (Phaeophyta, Chordariales) in culture. *Jap. J. Phycol.* **26:** 53–59.

ALEXANDER, W. B., SOUTHGATE, B. A. & BASSINDALE, B. 1935. Survey of the River Tees. Part II. The estuary – chemical and biological. D.S.I.R., London. *Tech. Pap. Wat. Pollut. Res.* No. 5.

AMSLER, C. D. 1984. Culture and field studies of *Acinetospora crinita* (Carmichael) Sauvageau (Ectocarpaceae, Phaeophyceae) in North Carolina, U.S.A. *Phycologia* **23:** 377–382.

ANDERSON, R. J. 1982. The life history of *Desmarestia firma* (C.Ag.) Skottsb. (Phaeophyceae, Desmarestiales). *Phycologia* **21:** 316–322.

ARDRÉ, F. 1970. Contribution a l'étude des algues marines du Portugal. *Port. Acta biol.* sér.B **10:** 137–555. (reprint 1–423).

BAKER, J. R. J. & EVANS, L. V. 1973a. The ship fouling alga *Ectocarpus* I. Ultrastructure and cytochemistry of plurilocular reproductive stages. *Protoplasma* **77:** 1–13.

BAKER, J. R. J. & EVANS, L. V. 1973b. The ship fouling alga *Ectocarpus* II. Ultrastructure of the unilocular reproductive stages. *Protoplasma* **77:** 181–189.

BERKALOFF, C. & ROUSSEAU, B. 1979. Ultrastructure of male gametogenesis in *Fucus serratus* (Phaeophyceae). *J. Phycol.* **15:** 163–173.

BERTHOLD, G. 1882. Über die Verteilung der Algen im Golf von Neapel nebst einem Verziechnis der bisher daselbst beobachteten Arten. *Mitt. zool. Stn Neapel* **3,** 393–536.

BIRD, N. L. & MCLACHLAN, J. 1976. Control of formation of receptacles in *Fucus distichus* L. subsp. *distichus* (Phaeophyceae, Fucales). *Phycologia* **15:** 79–84.

BISALPUTRA, T. 1974. Plastids. *In* Stewart, W. D. P. (Ed.) *Algal physiology and biochemistry* pp. 124–160. Botanical Monographs **10** Blackwell Scientific Publications, Oxford.

BOALCH, G. T. 1961. Studies on *Ectocarpus* in culture. II. Growth and nutrition of a bacteria-free culture. *J. mar. biol. Ass. U.K.* **41:** 287–304.

BOALCH, G. T., HOLME, N. A., JEPHSON, N. A. & SIDWELL, J. M. C. 1974. A resurvey of Colman's intertidal traverses at Wembury, South Devon. *J. mar. biol. Ass.* U.K. **54:** 551–553.

BODEN, G. T. 1979. The effect of depth on summer growth of *Laminaria saccharina* (Phaeophyta, Laminariales). *Phycologia* **18:** 405–408.

BOLAND, W., MARNER, F.-J. & JAENICKE, L. 1983. Comparative receptor study in gamete chemotaxis of the seaweeds *Ectocarpus siliculosus* and *Cutleria multifida*. *Eur. J. Biochem.* **134:** 97–103.

BOLD, H. C. & WYNNE, M. J. 1978. *Introduction to the algae*. Prentice-Hall, New Jersey.

BOLTON, J. 1979. Estuarine adaptations in populations of *Pilayella littoralis* (L.) Kjellm. (Phaeophyta, Ectocarpales). *Estuar. & Coast. Mar. Sci.* **9:** 273–280.

BOUCK, G. B. 1965. Fine structure and organelle associations in brown algae. *J. Cell Biol.* **26:** 523–537.

BOUCK, G. B. 1970. The development and postfertilization fate of the eyespot and the apparent photoreceptor in *Fucus* sperm. *Ann. N.Y. Acad. Sci.* **175:** 673–685.

BRAWLEY, S. H. & WETHERBEE, R. 1981. Cytology and ultrastructure. *In* Lobban, C. & Wynne, M. J. (Eds) *The biology of seaweeds* pp. 248–299. Botanical Monographs **17.** Blackwell Scientific Publications, Oxford.

BÜNNING, V. E. & MÜLLER, D. 1961. Wie messen organismen lunare Zyklen. *Z. Naturf.* **16B:** 391–395.

BURROWS, E. M. 1958. Sublittoral algal populations in Port Erin Bay, Isle of Man. *J. mar. biol. Ass. U.K.* **37:** 687–703.

BURROWS, E. M. 1964. An experimental assessment of some of the characters used for specific delimitation in the genus *Laminaria*. *J. mar. biol. Ass. U.K.* **44:** 137–143.

BURROWS, E. M. & LODGE, S. M. 1950. A note on the inter-relationships of *Patella, Balanus* and *Fucus* on a semi-exposed coast. *Rep. mar. biol. Stn Port Erin* **62:** 30–34.

CALLOW, M. E., COUGHLAN, S. J. & EVANS, L. V. 1978. The role of the golgi bodies in polysaccharide sulphation in *Fucus* zygotes. *J. Cell Sci.* **32:** 337–356.

CARAM, C. 1965. Recherches sur la reproduction et le cycle sexue de quelques Phéophycées. *Vie Milieu* **16:** 21–221.

CARAM, B. 1972. Le cycle de reproduction des Phéophycées-Phéosporées et ses modifications. *Mém. Soc. bot. Fr.* **1972:** 151–160.

CARAM, B. & JONSSON, S. 1972. Nouvel inventaire des algues marines de l'Islande. *Act. Bot. Isl.* **1:** 5–31.

CHAPMAN, A. R. O. 1972. Morphological variation and its taxonomic implications in the ligulate members of the genus *Desmarestia* occurring on the west coast of North America. *Syesis* **5:** 1–20.

CHAPMAN, A. R. O. 1974. The ecology of macroscopic marine algae. *Ann. Rev. Evol. Syst.* **5:** 65–80.

CHAPMAN, A. R. O. & BURROWS, E. M. 1970. Experimental investigations into the controlling effects of light conditions on the development and growth of *Desmarestia aculeata* (L.) Lamour. *Phycologia* **9:** 103–108.

CHAPMAN, A. R. O. & BURROWS, E. M. 1971. Field and culture studies of *Desmarestia aculeata* (L.) Lamour. *Phycologia* **10:** 63–76.

CHAPMAN, D. J. & CHAPMAN V. J. 1961. Life histories in the algae. *Ann. Bot., N.S.* **25:** 547–561.

CHAPMAN, V. J. 1939. Studies in salt-marsh ecology. Sections IV and V. *J. Ecol.* **27:** 160–201.

CHAPMAN, V. J. & CHAPMAN, D. J. 1973. *The algae.* 2nd edn. MacMillan, London.

CHI, E. Y. 1971. Brown algal pyrenoids. *Protoplasma* **72:** 101–104.

CHUCK, J. S. & MATHIESON, A. C. 1976. Ecological studies of the salt marsh ecad scorpioides (Hornemann) Hauck of *Ascophyllum nodosum* (L.) Le Jolis. *J. exp. mar. Biol. Ecol.* **23:** 171–190.

CHRISTENSEN, T. 1980. *Algae – a taxonomic survey.* AiO Tryk, Odense.

CHRISTENSEN, T. & THOMSEN, H. A. 1974. *Algefortegnelse.* Universitetsbogladen/Naturfagsbogladen, København.

CHRISTIE, A. O. 1973. Spore settlement in relation to fouling by *Enteromorpha.* *In* Acker, R. F., Floyd Brown, B., De Palma, J. R. & Iverson, W. P. (Eds) *Proc. 3rd Int. Congr. Mar. Corrosion and Fouling* pp. 674–681. Northwestern University Press, Evanston.

CLAYTON, M. N. 1980. Sexual reproduction – a rare occurrence in the life history of the complanate form of *Scytosiphon* (Scytosiphonaceae, Phaeophyta) from Southern Australia. *Br. phycol. J.* **15:** 105–118.

CLAYTON, M. N. 1981. Correlated studies on seasonal changes in the sexuality, growth rate and longevity of complanate *Scytosiphon* (Scytosiphonaceae: Phaeophyta) from Southern Australia growing in situ. *J. exp. mar. Biol. Ecol.* **51:** 87–98.

CLAYTON, M. N. 1984. Evolution of the Phaeophyta with particular reference to the Fucales. *In* Round, F. E. & Chapman, D. J. (Eds) *Progress in phycological research* **3** pp. 11–46. Biopress Ltd, Bristol.

CLAYTON, M. N. & KING, R. J. (Eds) 1981. *Marine botany: an Australasian perspective.* Longman Cheshire, Melbourne.

CLOKIE, J. J. P. & BONEY, A. D. 1980. The assessment of changes in intertidal ecosystems following major reclamation work: framework for interpretation of algal-dominated biota and the use and misuse of data. *In* Price, J. H., Irvine, D. E. G. & Farnham. W. F. (Eds) *The shore environment* **2:** *ecosystems* pp. 609–675. Systematics Association Special Volume **17(b).** Academic Press, London and New York.

CLOKIE, J. J. P. & NORTON, T. A. 1974. The effects of grazing on the algal vegetation of pebbles from the Firth of Clyde. *Br. phycol. J.* **9:** 216.

COLIJN, F. & HOEK C. VAN DEN, 1971. The life history of *Sphacelaria furcigera* Kütz. (Phaeophyceae) II. The influence of daylength and temperature on sexual and vegetative reproduction. *Nova Hedwigia* **11:** 899–922.

COLMAN, J. 1933. The nature of the intertidal zonation of plants and animals. *J. mar. biol. Ass. U.K.* **18:** 435–476.

CONNELL, J. H. & SLATYER, R. O. 1977. Mechanisms of succession in natural communities and their role in community stability and organization. *Am. Nat.* **111:** 1119–1144.

CONOVER, J. T. 1968. The importance of natural diffusion gradients and transport of substances related to benthic marine plant metabolism. *Botanica mar.* **11:** 1–9.

CRAIGIE, J. S. 1974. Storage products. *In* Stewart, W. D. P. (Ed.) *Algal physiology and biochemistry* pp. 206–235. Botanical Monographs **10.** Blackwell Scientific Publications, Oxford.

DANGEARD, P. 1969. A propos des travaux récents sur le cycle évolutif de quelques Phéophycées, Phéosporées. *Botaniste* **52:** 59–88.

DAVIS, P. H. & HEYWOOD, V. H. 1963. *Principles of angiosperm taxonomy.* Oliver & Boyd, Edinburgh.
DAYTON, P. K. 1975. Experimental evaluation of ecological dominance in a rocky intertidal algal community. *Ecol. Monogr.* **45:** 137–159.
DEBOER, J. A. 1981. Nutrients. *In* Lobban, C. & Wynne, M. J. (Eds) *The biology of seaweeds* pp. 356–392. Botanical Monographs **17.** Blackwell Scientific Publications, Oxford.
DETHIER, M. N. 1981. Heteromorphic algal life histories: the seasonal pattern and response to herbivory of the brown crust, *Ralfsia californica.* *Oecologia (Berl.)* **49:** 333–339.
DIXON, P. S. 1963. The Rhodophyta: some aspects of their biology. *Oceanogr. mar. Biol. ann. Rev.* **1:** 177–196.
DIXON, P. S. 1970. The Rhodophyta: some aspects of their biology. II. *Oceanogr. mar. Biol. ann. Rev.* **8:** 307–352.
DIXON, P. S. 1973. *Biology of the Rhodophyta.* Hafner Press, New York.
DIXON, P. S. & IRVINE, L. M. 1977. *Seaweeds of the British Isles.* Vol 1. *Rhodophyta.* Part 1, *Introduction, Nemaliales, Gigartinales.* British Museum (Natural History), London.
DODGE, J. D. 1969. A review of the fine structure of algal eyespots. *Br. phycol. J.* **4:** 199–210.
DODGE, J. D. 1973. *The fine structure of algal cells.* Academic Press, London and New York.
DOTY, M. S. 1946. Critical tide factors that are correlated with the vertical distribution of marine algae and other organisms along the Pacific coast. *Ecology* **27:** 315–328.
DREW, E. A. 1983. Light. *In* Earll, R. & Erwin, D. G. (Eds) *Sublittoral ecology The ecology of the shallow sublittoral benthos* pp. 10–57. Clarendon Press, Oxford.
DREW, K. M. 1955. Life histories in the algae with special reference to the Chlorophyta, Phaeophyta and Rhodophyta. *Biol. Rev.* **30:** 343–390.
DRING, M. J. 1982. *The biology of marine plants.* Edward Arnold, London.
DRING, M. J. 1984. Photoperiodism and phycology. *In* Round, F. E. & Chapman, D. J. (Eds) *Progress in phycological research* **3** pp. 159–192. Biopress Ltd, Bristol.
DRING, M. J. & LÜNING, K. 1975. Induction of two-dimensional growth and hair formation by blue light in the brown alga *Scytosiphon lomentaria.* *Z. pflanzenphysiol.* **75:** 107–117.
EDWARDS, P. 1969. Field and cultural studies on the seasonal periodicity of growth and reproduction of selected Texas benthic marine algae. *Contr. mar. Sci. Univ. Tex.* **14:** 59–114.
EDWARDS, P. 1976. *Illustrated guide to the seaweeds and sea grasses in the vicinity of Port Aransas, Texas.* University of Texas Press, Austin and London.
EDWARDS, P. & BAALEN, C. VAN 1970. An apparatus for the culture of benthic marine algae under varying regimes of temperature and light intensity. *Botanica mar.* **13:** 42–43.
ETHERINGTON, J. 1964. Rhizoid formation and fragmentation in *Feldmannia globifera* (Kütz.) Hamel. *Br. phycol. Bull.* **2:** 373–375.
EVANS, L. V. 1966. Distribution of pyrenoids among some brown algae. *J. Cell Sci.* **I:** 449–454.
EVANS, L. V. 1968. Chloroplast morphology and fine structure in British fucoids. *New Phytol.* **67:** 173–178.
EVANS, L. V. 1974. Cytoplasmic organelles. *In* Stewart, W. D. P. (Ed.) *Algal physiology and biochemistry* pp. 86–123. Botanical Monographs **10.** Blackwell Scientific Publications, Oxford.
EVANS, L. V. & CALLOW, M. E. 1976. Secretory processes in seaweeds. *In* Sunderland, N. (Ed.) *Perspectives in experimental biology* **2** pp. 487–499. Pergamon, Oxford and New York.
EVANS, L. V., SIMPSON, M. & CALLOW, M. E. 1973. Sulphated polysaccharide synthesis in brown algae. *Planta* **110:** 237–252.
FELDMANN, G. & GUGLIELMI, M. G. 1972. Les physodes et les corps irisants du *Dictyota dichotoma* (Hudson) Lamouroux. *C. r. hebd. Séanc. Acad. Sci., Paris* **275:** 751–754.
FELDMANN, J. & MAGNE, F. 1964. Additions à l'inventaire de la flore marine de Roscoff. Algues, Champignons. Lichenes. *Trav. Stn biol. Roscoff* N.S. **15:** 1–28.
FLETCHER, R. L. 1974. Studies on the life history and taxonomy of some members of the Phaeophycean families Ralfsiaceae and Scytosiphonaceae. *Ph.D. Thesis, University of London.*
FLETCHER, R. L. 1975. Heteroantagonism observed in mixed algal cultures. *Nature, Lond.* **253:** 534–535.

FLETCHER, R. L. 1976. Post-germination attachment mechanisms in marine fouling algae. *In* Sharpley, J. M. & Kaplan, A. M. (Eds) *Proc. 3rd Int. Symp. Biodegradation* pp. 443–464. Applied Science Publishers, London.

FLETCHER, R. L. 1977. Observations on secondary attachment mechanisms in marine fouling algae. *In* C.R.E.O. (Ed.) *Proc. 4th Int. Congr. Mar. Corrosion and Fouling* pp. 169–177. C.R.E.O., Paris.

FLETCHER, R. L. 1978. Studies on the family Ralfsiaceae (Phaeophyta) around the British Isles. *In* Irvine, D. E. G. & Price, J. H. (Eds) *Modern approaches to the taxonomy of red and brown algae* pp. 371–398. Systematics Association Special Volume **10.** Academic Press, London and New York.

FLETCHER, R. L. 1980. Studies on the recently introduced brown alga *Sargassum muticum* (Yendo) Fensholt. III. Periodicity in gamete release and 'incubation' of early germling stages. *Botanica mar.* **23:** 425–532.

FLETCHER, R. L. 1981a. Studies on the ecology, structure and life history of the brown alga *Petalonia filiformis* (Batt.) Kuntze (Scytosiphonaceae) around the British Isles. *Phycologia* **20:** 103–104.

FLETCHER, R. L. 1981b. Studies on the marine fouling brown alga *Giffordia granulosa* (Sm.) Hamel in the Solent (south coast of England). *Botanica mar.* **24:** 211–221.

FLETCHER, R. L. 1981c. Observations on the ecology and life history of *Ralfsia spongiocarpa* Batt. *Proc. int. Seaweed Symp* **8:** 323–330.

FLETCHER, R. L. 1983. The occurrence of the brown alga *Streblonema oligosporum* Strömfelt in Britain. *Br. phycol. J.* **18:** 415–423.

FLETCHER, R. L. 1984. Observations on the life history of the brown alga *Hecatonema maculans* (Coll.) Sauv. (Ectocarpales, Myrionemataceae) in laboratory culture. *Br. phycol. J.* **19:** 193.

FLETCHER, R. L., BAIER, R. E. & FORNALIK, M. S. 1984. The influence of surface energy on spore development in some common marine fouling algae. *In Proc. 6th Int. Congr. Mar. Corrosion and Fouling* pp. 129–144. Athens.

FLETCHER, R. L., BAIER, R. E. & FORNALIK, M. S. 1985. The effects of surface energy on the development of some marine macroalgae. *Br. phycol. J.* **20:** 184–185.

FLETCHER, R. L., JONES, A. M. & JONES, E. B. G. 1984. The attachment of fouling macroalgae. *In* Costlow, J. D. & Tipper, R. C. (Eds) *Marine biodeterioration: an interdisciplinary study* pp. 172–182. Naval Institute Press, Annapolis.

FORTES, M. D. & LÜNING, K. 1980. Growth rates of North Sea macroalgae in relation to temperature, irradiance and photoperiod. *Helgoländer wiss. Meeresunters.* **34:** 15–29.

FRITSCH, F. E. 1945. *The structure and reproduction of the algae,* Vol. 2. Cambridge University Press, Cambridge.

GESSNER, F. 1970. Temperature *In* Kinne, O. (Ed.) *Marine ecology* Vol. 1, Part 1. pp. 363–406. Wiley & Sons, Chichester.

GESSNER, F. & SCHRAMM, W. 1971. Salinity: plants *In* Kinne, O. (Ed.) *Marine ecology,* Vol. 1, Part 2. pp. 705–820. Wiley & Sons, Chichester.

GIBB, D. C. 1937. Observations on *Himanthalia lorea* (L.) Lyngb. *J. Linn. Soc., Bot.* **51:** 11–21.

GIBB, D. C. 1957. The free-living forms of *Ascophyllum nodosum* (L.) Le Jol. *J. Ecol.* **45:** 49–83.

GOODWIN, T. W. 1974. Carotenoids and biliproteins. *In* Stewart, W. D. P. (Ed.) *Algal physiology and biochemistry* pp. 176–205. Botanical Monographs **10.** Blackwell Scientific Publications, Oxford.

GRIME, J. P. 1974. Vegetation classification by reference to strategies. *Nature, Lond.* **25:** 26–31.

GRIME, J. P. 1977. Evidence for the existence of three primary strategies in plants and its relevance to ecological and evolutionary theory. *Am. Nat.* **111:** 1169–1194.

GRIME, J. P. 1979. *Plant strategies and vegetation processes.* Wiley & Sons, Chichester.

GUIRY, M. D. 1978. A concensus and bibliography of Irish seaweeds. *Bibl. phycol.* **44:** 1–287.

HALL, A., FIELDING, A. H. & BUTLER, M. 1979. Mechanisms of copper tolerance in the marine fouling alga *Ectocarpus siliculosus* – evidence for an exclusion mechanism. *Mar. Biol.* **54:** 195–199.

HARDY, F. G. & MOSS, B. L. 1978. The attachment of zygotes and germlings of *Halidrys siliquosa*. *Phycologia* **17:** 69–78.

HARDY, F. G. & MOSS, B. L. 1979a. Notes on the attachment of zygotes and germlings of *Bifurcaria bifurcata* Ross (Phaeophyceae, Fucales). *Phycologia* **18:** 164–170.

HARDY, F. G. & MOSS, B. L. 1979b. The effects of the substratum on the morphology of the rhizoids of *Fucus* germlings. *Estuar. & Coast. Mar. Sci.* **9:** 577–584.

HARDY, F. G. & MOSS, B. L. 1979c. Attachment and development of the zygotes of *Pelvetia canaliculata* (L.) Dcne et Thur. (Phaeophyceae, Fucales). *Phycologia* **18:** 203–212.

HARLIN, M. M. 1974. The surfaces seaweeds grow on may be a clue to their control. *Maritimes* (Univ. R. I. Grad. Sch. Ocean) **18:** 7–8.

HARLIN, M. M. & LINDBERGH, J. M. 1977. Selection of substrata by seaweeds: optimal surface relief. *Mar. Biol.* **40:** 33–40.

HARRIS, J. E. 1943. First report of the marine corrosion sub-committee. Section C. Antifouling investigations. *J. Iron Steel Inst.* **147:** 405–420.

HARRIS, J. E. Report on antifouling research, 1942–44. *J. Iron Steel Inst.* **154:** 297–333.

HARTOG, C. DEN 1967. Brackish water as an environment for algae. *Blumea* **15:** 31–43.

HARVEY, W. H. 1849. *A manual of the British marine algae.* John van Hoorst, London.

HAWKINS, S. J. 1981. The influence of season and barnacles on the algal colonization of *Patella vulgata* exclusion areas. *J. mar. biol. Ass. U.K.* **61:** 1–15.

HAWKINS, S. J. & HARTNOLL, R. G. 1980. Small-scale relationship between species number and area on a rocky shore. *Estuar. & Coast. Mar. Sci.* **10:** 201–214.

HELLEBUST, J. A. 1970. Light. *In* Kinne, O. (Ed.) *Marine ecology* Vol. **1,** Part 1 pp. 125–158. Wiley & Sons, Chichester.

HELLEBUST, J. A. 1974. Extracellular products. *In* Stewart, W. D. P. (Ed.) *Algal physiology and biochemistry* pp. 838–863. Botanical Monographs **10.** Blackwell Scientific Publications, Oxford.

HISCOCK, K. 1983. Water movement. *In* Earll, R. & Erwin, D. G. (Eds) *Sublittoral ecology – the ecology of the shallow sublittoral benthos* pp. 58–96. Clarendon Press, Oxford.

HOEK, C. VAN DEN 1975. Phycological Reviews 3. Phytogeographic provinces along the coasts of the Northern Atlantic Ocean. *Phycologia* **14:** 317–330.

HOEK, C. VAN DEN & FLINTERMAN, A. 1968. The life history of *Sphacelaria furcigera* Kütz. (Phaeophyceae). *Blumea* **16:** 193–243.

HOEK, C. VAN DEN & JAHNS, H. M. 1978. *Algen, Einführung in die Phykologie.* Thieme, Stuttgart.

HOOPER, A. TEN, BOS, S. & BIEEMAN, A. M. 1983. Photoperiodic response in the formation of gametangia of the long-day plant *Sphacelaria rigidula* (Phaeophyceae) *Mar. Ecol. Prog. Ser.* **13:** 285–289.

HORI, T. 1971. Survey of pyrenoid distribution in brown algae. *Bot. Mag., Tokyo* **84:** 231–242.

HORI, T. 1972. Further survey of the pyrenoid distribution in Japanese brown algae. *Bot. Mag., Tokyo* **85:** 125–134.

HOYT, W. D. 1927. The periodic fruiting of *Dictyota* and its relations to the environment. *Am. J. Bot.* **14:** 529–619.

HSAIO, S. I. C. 1969. Life history and iodine nutrition of the marine brown alga, *Petalonia fascia* (O.F. Müll.) Kuntze. *Can. J. Bot.* **47:** 1611–1616.

IGIC, L. 1968. The fouling on ships as the consequence of their navigation in the Adriatic and other world seas. In *Proc. 2nd Int. Congr. Mar. Corrosion and Fouling* pp. 571–577. Athens.

IRVINE, D. E. G. 1974. The marine vegetation of the Shetland Isles. *In* Goodier, R. (Ed.) *The natural environment of Shetland* pp. 107–113. The Nature Conservancy Council, Edinburgh.

IRVINE, D. E. G. 1982. Seaweeds of the Faroes. I. The flora. *Bull. Br. Mus. nat. Hist.* (Bot.) **10:** 109–131.

JAFFE, L. 1954. Stimulation of the discharge of gametangia from a brown alga by a change from light to darkness. *Nature, Lond.* **174:** 743.

JENSEN, J. B. 1974. Morphological studies in Cystoseiraceae and Sargassaceae (Phaeophyceae) with special reference to apical organisation. *Univ. Calif. Publs Bot.* **68:** vi + 61 pp.

JEPHSON, N. A. & GRAY, P. W. G. 1977. Aspects of the ecology of *Sargassum muticum* (Yendo) Fensholt in the Solent region of the British Isles. I. The growth cycle and epiphytes. *In* Keegan, B. F., Ceidigh, P. O. & Boaden, P. J. S. (Eds) *Biology of benthic organisms* (Proc. XIth Europ. Symp. Mar. Biol. Galway, 1976) pp. 367–375. Pergamon Press, Oxford.

JONES, W. E. & DEMETROPOULOS, A. 1968. Exposure to wave action: measurements of an important ecological parameter on rocky shores of Anglesey. *J. exp. mar. Biol. Ecol.* **2:** 46–63

JONES, W. E., BENNELL, S., BEVERIDGE, C., MCCONNELL, B., MACK-SMITH, S. & MITCHELL, J. 1980. Methods of data collection and processing in rocky intertidal monitoring. *In* Price, J. H., Irvine, D. E. G. & Farnham, W. F. (Eds) *The shore environment* **1:** *methods* pp. 137–170. Systematics Association Special Volume **17**(a). Academic Press, London and New York.

JORDAN, A. J. & VADAS, R. L. 1972. Influence of environmental parameters on intraspecific variation in *Fucus vesiculosus*. *Mar. Biol* **14:** 248–252.

KAIN, J. M. 1979. A view of the genus *Laminaria*. *Oceanogr. mar. Biol. ann. Rev.* **17:** 101–161.

KERBY, N. W. & EVANS, L. V. 1978. Isolation and partial characterization of pyrenoids from the brown alga *Pilayella littoralis* (L.) Kjellm. *Planta* **142:** 91–95.

KNIGHT, M. 1923. Studies in the Ectocarpaceae. I. The life history and cytology of *Pylaiella littoralis*, Kjellm. *Trans. R. Soc. Edinb.* **53:** 343–360.

KNIGHT, M. 1929. Studies in the Ectocarpaceae. II. The life history and cytology of *Ectocarpus siliculosus*, Dillw. *Trans. R. Soc. Edinb.* **56:** 307–332.

KNIGHT, M., BLACKLER, M. C. H. & PARKE, M. W. 1935. Notes on the life cycle of species of *Asperococcus*. *Proc. Trans. L'pool biol. Soc.* **48:** 79–97.

KNIGHT, M. & PARKE, M. W. 1950. A biological study of *Fucus vesiculosus* L. and *F. serratus*. *J. mar. biol. Ass. U.K.* **29:** 439–514.

KORNMANN, P. 1953. Der Formekreis von *Acinetospora crinita* (Carm.) nov. comb. *Helgoländer wiss. Meeresunters.* **4:** 205–224.

KORNMANN, P. 1954. *Giffordia fuscata* (Zan.) Kuck. nov. comb. eine Ectocarpaceae mit heteromorphen, homophasischen Generationen. *Helgoländer wiss. Meeresunters.* **5:** 41–52.

KUCKUCK, P. 1912. Beiträge zur Kenntnis der Meeresalgen. 10. Neue Untersuchunge über *Nemoderma* Schousboe. *Helgoländer wiss. Meeresunters.* **5:** 119–152.

KUMKE, J. 1973. Beiträe zur Periodizität der Oogon-Entleerung bei *Dictyota dichotoma* (Phaeophyta). *Z. Pflanzenphysiol.* **70:** 191–210.

KYLIN, H. 1933. Über die Entwicklungsgeschichte der Phaeophyceen. *Acta Univ. lund* **29:** 1–102.

KYLIN, H. 1938. Bemerkungen über die Fucosanblasen der Phaeophyceen. *K. fysiogr. Sällsk. Lund Förh.* **8:** 1–10.

LA CLAIRE, J. W. II & WEST, J. A. 1978. Light and E.M. studies of growth and reproduction in *Cutleria*. I. Gametogenesis in the female plant of *C. hancockii*. *Protoplasma* **97:** 93–110.

LA CLAIRE, J. W. II & WEST, J. A. 1979. Light and E.M. studies of growth and reproduction in *Cutleria*. II. Gametogenesis in the male plant of *C. hancockii*. *Protoplasma* **101:** 247–267.

LARKUM, A. W. D. 1972. Frond structure and growth in *Laminaria hyperborea*. *J. mar. biol. Ass. U.K.* **52:** 405–418.

LEWIS, J. R. 1964. *The ecology of rocky shores*. English Universities Press, London.

LINSKENS, H. F. 1966. Adhäsion von Fort pflanzungszellen Benthontischer. *Planta* **68:** 99–110.

LITTLER, M. M. & KAUKER, B. J. 1984. Heterotrichy and survival strategies in the red alga *Corallina officinalis* L. *Botanica mar.* **27:** 37–44.

LITTLER, M. M. & LITTLER, D. S. 1980. The evolution of thallus form and survival strategies in benthic marine macroalgae: field and laboratory tests of a functional form model. *Am. Nat.* **116:** 25–44.

LITTLER, M. M. & LITTLER, D. S. 1983. Heteromorphic life history strategies in the brown alga *Scytosiphon lomentaria* (Lyngb.) Link. *J. Phycol.* **19:** 425–431.

LITTLER, M. M., MARTZ, D. R. & LITTLER, D. S. 1983. Effects of recurrent sand deposition on rocky intertidal organisms: importance of substrate heterogeneity in a fluctuating environment. *Mar. Ecol. Prog. Ser.* **11:** 129–139.

LOBBAN, C. A. & WYNNE, M. J. (Eds) 1981. *The biology of seaweeds.* Botanical Monographs **17.** Blackwell Scientific Publications, Oxford.

LOCKHART, J. C. 1979. Factors determining various forms in *Cladosiphon zosterae* (Phaeophyceae). *Am. J. Bot.* **66:** 836–844.

LOCKHART, J. C. 1982. Influence of light, temperature and nitrogen on morphogenesis of *Desmotrichum undulatum* (J. Agardh) Reinke. *Phycologia* **21:** 264–272.

LODGE, S. M. 1948. Algal growth in the absence of *Patella* on an experimental strip of foreshore, Port St. Mary, Isle of Man. *Proc. Trans. L'pool biol. Soc.* **56:** 78–85.

LOFTHOUSE, P. F. & CAPON, B. 1975. Ultrastructural changes accompanying mitosporogenesis in *Ectocarpus parvus* *Protoplasma* **84:** 83–99.

LOISEAUX, S. 1967a. Morphologie et cytologie des Myrionémacées. Critères taxonomiques. *Rev. gén. Bot.* **74:** 329–347.

LOISEAUX, S. 1967b. Recherches sur les cycles de développement des Myrionématacées (Phéophycées) I–II. Hecatonématées et Myrionématées. *Rev. gén. bot.* **74,** 529–576.

LOISEAUX, S. 1968. Recherches sur les cycles de développement des Myrionématacées (Phéophycées) III. Tribu des Ralfsiées. IV. Conclusions générales. *Rev. gén. Bot.* **75:** 295–318.

LOISEAUX, S. 1969. Sur une espèce de *Myriotrichia* obtenue en culture à partir de zoïdes d'*Hecatonema maculans* Sauv. *Phycologia* **8:** 11–15.

LOISEAUX, S. & WEST, J. A. 1970. Brown algal mastigonemes: comparative ultrastructure. *Trans. Am. microsc. Soc.* **89:** 524–532.

LUBCHENCO, J. 1978. Plant species diversity in a marine intertidal community: importance of herbivore food preference and algal competitive abilities. *Am. Nat.* **112:** 23–39.

LUBCHENCO, J. & CUBIT, J. 1980. Heteromorphic life histories of certain marine algae as adaptations to variations in herbivory. *Ecology* **61:** 676–687.

LUBCHENCO, J. & MENGE, B. A. 1978. Community development and persistence in a low rocky intertidal zone. *Ecol. Monogr.* **59:** 67–94.

LÜNING, K. 1971. Seasonal growth of *Laminaria hyperborea* under recorded underwater light conditions near Helgoland. *In* Crisp, D. J. (Ed.) *Proc. 4th. Eur. Mar. Biol. Symp.* pp. 347–361. Cambridge University Press, Cambridge.

LÜNING, K. 1980a. Control of algal life history by daylength and temperature. *In* Price. J. H., Irvine, D. E. G. & Farnham, W. F. (Eds) *The shore environment* **2:** ecosystems pp. 915–945. Systematics Association Special Volume **17(b).** Academic Press, London and New York.

LÜNING, K. 1980b. Critical levels of light and temperature regulating the gametogenesis of three *Laminaria* species (Phaeophyceae). *J. Phycol.* **16:** 1–15.

LÜNING, K. 1981a. Egg release in gametophytes of *Laminaria saccharina:* induction by darkness and inhibition by blue light and u.v. *Br. phycol. J.* **16:** 379–393.

LÜNING, K. 1981b. Photomorphogenesis of reproduction in marine macroalgae. *Ber. dt. bot. Ges.* **94:** 401–417.

LÜNING, K. & DRING, M. J. 1972. Reproduction induced by blue light in female gametophytes of *Laminaria saccharina.* *Planta (Berl.)* **104:** 252–256.

LÜNING, K. & DRING, M. J. 1973. The influence of light quality on the development of the brown algae *Petalonia* and *Scytosiphon.* *Br. phycol. J.* **8:** 333–338.

LÜNING, K. & DRING, M. J. 1975. Reproduction, growth and photosynthesis of gametophytes of *Laminaria saccharina* grown in blue and red light. *Mar. Biol.* **29:** 195–200.

LÜNING, K. & MARKHAM, J. W. 1979. Morphogenetic responses of *Laminaria saccharina* sporophytes to red and blue light. *Br. phycol. J.* **14:** 125–126.

LÜNING, K. & MÜLLER, D. G. 1978. Chemical interaction in sexual reproduction of several Laminariales (Phaeophyceae): release and attraction of spermatozoids. *Z. Pflanzenphysiol.* **89:** 333–341.

MAIER, I. & MÜLLER, D. G. 1982. Antheridium fine structure and spermatozoid release in *Laminaria digitata* (Phaeophyceae). *Phycologia* **21:** 1–8.

MANTON, I. 1957. Observations with the electron microscope on the internal structure of the zoospore of a brown alga. *J. exp. Bot.* **8:** 294–303.
MANTON, I. 1964. The possible significance of some details of flagellar bases in plants. *Jl R. microsc. Soc.* **82:** 279–285.
MANTON, I. & CLARKE, B. 1951. An electron microscope study of the spermatozoid of *Fucus serratus*. *Ann. Bot. N.S.* **15:** 461–471.
MANTON, I. & CLARKE, B. 1956. Observations with the electron microscope on the internal structure of the spermatozoid of *Fucus*. *J. exp. Bot.* **7:** 416–432.
MANTON, I., CLARKE, B. & GREENWOOD, A. D. 1953. Further observations with the electron microscope on spermatozoids in the Brown algae. *J. exp. Bot.* **4:** 319–329.
MARKEY, D. R. & WILCE, R. T. 1975. The ultrastructure of reproduction in the brown alga *Pylaiella littoralis* I Mitosis – Cytokinesis in the plurilocular gametangia. *Protoplasma* **85:** 219–241.
MEEKS, J. C. 1974. Chlorophylls. *In* Stewart, W. D. P. (Ed.) *Algal physiology and biochemistry* pp. 161–175. Botanical Monographs **10.** Blackwell Scientific Publications, Oxford.
MOORE, P. G. 1983. Biological interactions. *In* Earll, R. & Erwin, D. G. (Eds) *Sublittoral ecology – the ecology of the shallow sublittoral benthos* pp. 125–143. Clarendon Press, Oxford.
MOSS, B. 1971. Meristems and morphogenesis in *Ascophyllum nodosum* ecad *mackaii* (Cotton). *Br. phycol. J.* **6:** 187–193.
MOSS, B. & SHEADER, A. 1973. The effect of light and temperature upon the germination and growth of *Halidrys siliquosa* (L.) Lyngb. (Phaeophyceae, Fucales) *Phycologia* **12:** 63–68.
MÜLLER, D. G. 1962. Uber jahres – und lunarperiodische Erscheinungen bei einigen Braunalgen. *Botanica mar.* **4:** 140–155.
MÜLLER, D. G. 1963. Die Temperaturabhängigkeit der Sporangienbildung bei *Ectocarpus siliculosus* von verschiedenen standorten. *Pubbl. Staz. zool. Napoli* **33:** 310–314.
MÜLLER, D. G. 1964. Die Beteiligung eines Beruhrungsreizes heim Festsetzen von Algenschwarmen auf dem Substrat. *Z. Bot.* **52:** 193–198.
MÜLLER, D. G. 1967. Generationswechsel, Kernphasenwechsel und Sexualität der Braunalge *Ectocarpus siliculosus* im Kulturversuch. *Planta* **75:** 39–54.
MÜLLER, D. G. 1972a. Studies on reproduction in *Ectocarpus siliculosus*. *Mém. Soc. bot. Fr.* **1972:** 87–98.
MÜLLER, D. G. 1972b. Chemotaxis in Brown algae. Detection and isolation of the attractant released by eggs of *Fucus serratus* L. *Naturwissenschaften* **4**; 166–168.
MÜLLER, D. G. 1975. Experimental evidence against sexual fusions of spores from unilocular sporangia of *Ectocarpus siliculosus* (Phaeophyta). *Br. phycol. J.* **10:** 315–321.
MÜLLER, D. G. 1981. Sexuality and sex attraction. *In* Lobban, C. & Wynne, M. J. (Eds) *The biology of seaweeds* pp. 661–674. Botanical Monographs **17.** Blackwell Scientific Publications, Oxford.
MÜLLER, S. & CLAUS, H. 1976. Aspects of photomorphogenesis in the brown alga *Dictyota dichotoma*. *Z. Pflanzenphysiol.* **78:** 461–465.
MÜLLER, D. G., GASSMANN, G. & LÜNING, K. 1979. Isolation of a spermatozoid-releasing and attracting substance from female gametophytes of *Laminaria digitata*. *Nature, Lond.* **279:** 430–431.
MÜLLER, D. G. & JAENICKE, L. 1973. Fucoserraten, the female sex attractant of *Fucus serratus* L. (Phaeophyta). *F.E.B.S. Lett.* **30:** 127–139.
MÜLLER, D. G., JAENICKE, L., DONIKE, M. & AKINTOBI, T. 1971. Sex attractant in a brown alga: chemical structure. *Science* **171:** 815–817.
MÜLLER, D. G. & LÜTHE, N. M. 1981. Hormonal interaction in sexual reproduction of *Desmarestia aculeata* (Phaeophyceae). *Br. phycol. J.* **16:** 351–356.
MUNDA, I. M. 1977. Combined effects of temperature and salinity on growth rates of germlings of three *Fucus* species from Iceland, Helgoland and the North Adriatic Sea. *Helgoländer wiss. Meeresunters.* **29:** 302–310.
MUNDA, I. M. 1978. Salinity dependent distribution of benthic algae in estuarine areas of Icelandic fjords. *Botanica mar.* **21:** 451–468.

MUNDA, I. M. 1979. A note on the ecology and growth forms of *Chordaria flagelliformis* (O.F. Müll.) C. Ag. in Icelandic waters. *Nova Hedwigia* **31:** 567–591.

NAKAHARA, H. 1984. Alternation of generations of some brown algae in unialgal and axenic cultures. *Scient. Pap. Inst. algol. Res. Hokkaido Univ.* **7:** 77–194.

NAKAHARA, H. & NAKAMURA Y. 1971. The life history of *Desmarestia tabacoides* Okamura. *Bot. Mag., Tokyo* **84:** 69–75.

NAKAMURA, Y. 1972. A proposal on the classification of the Phaeophyta. *In* Abbott, I. A. & Kurogi, M. (Eds) *Contributions to the systematics of marine algae of the North Pacific* pp. 147–156. Japanese Soc. Phycology, Kobe.

NAKAMURA, Y. & TATEWAKI, M. 1975. The life history some species of Scytosiphonales. *Scient. Pap. Inst. algol. Res. Hokkaido Univ.* **6:** 57–93.

NEWTON, L. 1931. *A handbook of the British seaweeds.* British Museum (Natural History), London.

NORTON, T. A. 1969. Growth form and environment in *Saccorhiza polyschides. J. mar. biol. Ass. U.K.* **49:** 1025–1045.

NORTON, T. A. 1978. The factors influencing the distribution of *Saccorhiza polyschides* in the region of Lough Ine. *J. mar. biol. Ass. U.K.* **58:** 528–536.

NORTON, T. A. & BURROWS, E. M. 1969. Studies on marine algae of the British Isles 7. *Saccorhiza polyschides* (Light.) Batt. *Br. phycol. J.* **4:** 19–53.

NORTON, T. A. & FETTER, R. 1981. The settlement of *Sargassum muticum* propagules in stationary and flowing water. *J. mar. biol. Ass. U.K.* **61:** 929–940.

NORTON, T. A., HISCOCK, K. & KITCHING, J. A. 1977. The ecology of Lough Ine. XX. The *Laminaria* forest at Carrigathorna. *J. Ecol.* **65:** 919–941.

NORTON, T. A., MATHIESON, A. C. & NEUSHUL, M. 1981. Morphology and environment. *In* Lobban, C. & Wynne, M. J. (Eds) *The biology of seaweeds* pp. 421–451. Botanical Monographs **17.** Blackwell Scientific Publications, Oxford.

NYGREN, S. 1975. Influence of salinity on the growth and distribution of some Phaeophyceae on the Swedish west coast. *Botanica mar.* **18:** 143–147.

OHNO, M. 1969. A physiological ecology of the early stage of some marine algae. *Rep. Usa mar. biol. stn Kochi Univ.* **16:** 1–46.

OLIVEIRA, E. C. DE & FLETCHER, A. 1980. Taxonomic and ecological relationships between rocky-shore and saltmarsh populations of *Pelvetia canaliculata* (Phaeophyta) at Four Mile Bridge, Anglesey, U.K. *Botanica mar.* **23:** 409–417.

OLTMANNS, F. 1892. Über die kultur – und Lebensbedingungen der Meeresalgen. *Jb. wiss. Bot* **23:** 349–440.

OLTMANNS, F. 1922. *Morphologie und Biologie der Algen,* ed. 2, II. Jena.

PAPENFUSS, G. F. 1935. Alternation of generations in *Ectocarpus siliculosus. Bot. Gaz.* **96:** 421–446.

PAPENFUSS, G. F. 1951. Phaeophyta. *In* Smith, G. M. (Ed.) *Manual of phycology* pp. 119–158. Chronica Botanica, Waltham.

PARKE, M. W. 1933. A contribution to the knowledge of the Mesogloiaceae and associated families. *Publs Hartley bot. Labs L'pool Univ.* **9:** 1–43.

PARKE, M. 1948. Studies on British Laminariaceae. I. Growth in *Laminaria saccharina* (L.) Lamour. *J. mar. biol. Ass. U.K.* **27:** 651–709.

PARKE, M. & DIXON, P. S. 1976. Check list of British marine algae – third revision. *J. mar. biol. Ass. U.K.* **56:** 527–594.

PARKER, J. & PHILPOTT, D. E. 1960. E.M. studies of *F. vesiculosus* cytoplasm in summer and winter. *Biol. Bull. mar. biol. Lab., Woods Hole* **119:** 330–331.

PAULA, R. J. DE & OLIVEIRA, E. C. DE 1982. Wave exposure and ecotypical differentiation in *Sargassum cymosum* (Phaeophyta-Fucales). *Phycologia* **21:** 145–152.

PEDERSEN, P. M. 1976. Marine, benthic algae from southernmost Greenland. *Meddr Grønland* **199:** 1–80.

PEDERSEN, P. M. 1978. Culture studies on marine algae from West Greenland III The life histories and systematic positions of *Pogotrichum filiforme* and *Leptonematella fasciculata* (Phaeophyceae). *Phycologia* **17:** 61–68.

PEDERSEN, P. M. 1981. Life histories of brown algae. *In* Lobban, C. & Wynne, M. J. (Eds) *The biology of seaweeds* pp. 194–217. Botanical Monographs **17.** Blackwell Scientific Publications, Oxford.

PEDERSEN, P. M. 1984. Studies on primitive brown algae (Fucophyceae). *Opera Bot.* **74:** 1–76.

PELLEGRINI, L. 1974. Origine et modifications ultrastructurales du matériel osmiophile contenu dans les physodes et dans certains corps iridescents des cellules végétatives apicales chez *Cystoseira stricta* Sauvageau (Phéophycée, Fucale). *C.r. hebd. Séanc. Acad. Sci., Paris.* sér. D. **279:** 903–906.

PRUD'HOMME VAN REINE, W. F. 1972. Notes on Sphacelariales (Phaeophyceae) II. On the identity of *Cladostephus setaceus* Suhr and remarks on European *Cladostephus. Blumea* **20:** 138–144.

PRUD'HOMME VAN REINE, W. F. 1982. A taxonomic revision of the European Sphacelariaceae (Sphacelariales, Phaeophyceae). *Leiden Botanical Series* **6:** 10 + 1–293.

PRUD'HOMME VAN REINE, W. F. & STAR, W. 1981. Transmission electron microscopy of apical cells of *Sphacelaria* spp. (Sphacelariales, Phaeophyceae). *Blumea* **27:** 523–546.

RAGAN, M. A. 1976. Physodes and the phenolic compounds of brown algae. Composition and significance of physodes *in vivo. Botanica mar.* **19:** 145–154.

RAMON, E. 1973. Germination and attachment of zygotes of *Himanthalia elongata* (L.) S. F. Gray. *J. Phycol.* **9:** 445–449.

RAUTENBERG, E. 1960. Zur Morphologie und Ökologie einiger epiphytischer und epi-endophytischer Algen. *Botanica mar.* **2:** 133–145.

RHODES, R. G. 1970. Relation of temperature to development of the macrothallus of *Desmotrichum undulatum. J. Phycol.* **6:** 312–314.

RICHARDSON, J. P. 1979. Overwintering of *Dictyota dichotoma* (Phaeophyceae) near its northern distribution limit on the east coast of North America. *J. Phycol.* **15:** 22–26.

RIETEMA, H. & HOEK, C. VAN DEN 1981. The life history of *Desmotrichum undulatum* (Phaeophyceae) and its regulation by temperature and light conditions. *Mar. Ecol. Prog. Ser.* **4:** 321–335.

ROELEVELD, J. G., DUISTERHOF, M. & VROMAN, M. 1974. On the year cycle of *Petalonia fascia* in the Netherlands. *Neth. J. Sea. Res.* **8:** 410–426.

ROUND, F. E. 1973. *The biology of the algae.* Edward Arnold, New York.

RUENESS, J. 1977. *Norsk Algeflora.* Scandinavian University Books, Oslo, Bergen and Trondheim.

RUSSELL, G. 1961. The autecology and life history of *Pilayella littoralis* (L.) Kjellm. *Rep. Challenger Soc.* **3:** 30–31.

RUSSELL, G. 1964. *Laminariocolax tomentosoides* on the Isle of Man. *J. mar. biol. Ass. U.K.* **44:** 601–612.

RUSSELL, G. 1967. The ecology of some free-living Ectocarpaceae. *Helgoländer wiss. Meeresunters.* **15:** 155–162.

RUSSELL, G. 1971. Marine algal reproduction in two British estuaries. *Vie Milieu,* Suppl. **22:** 219–230.

RUSSELL, G. 1972. Phytosociological studies on a two-zone shore I Basic pattern. *J. Ecol.* **60:** 539–545.

RUSSELL, G. 1973a. The Phaeophyta: a synopsis of some recent developments. *Oceanogr. mar. Biol. ann. Rev.* **11:** 45–88.

RUSSELL, G. 1973b. The 'litus' line: a re-assessment. *Oikos* **24:** 158–161.

RUSSELL, G. 1973c. Phytosociological studies on a two-zone shore. II. Community structure. *J. Ecol.* **61:** 525–536.

RUSSELL, G. 1977. Vegetation on rocky shores at some North Irish Sea sites. *J. Ecol.* **65:** 485–495.

RUSSELL, G. 1980. Applications of simple numerical methods to the analysis of intertidal vegetation. *In* Price, J. H., Irvine, D. E. G. & Farnham, W. F. (Eds) *The shore environment* **1:** *methods* pp. 171–192. Systematics Association Special Volume **17(a)** Academic Press, London and New York.

RUSSELL, G. 1985a. Some anatomical and physiological differences in *Chorda filum* from coastal waters of Finland and Great Britain. *J. mar. biol. Ass. U.K.* **65:** 343–349.

RUSSELL, G. 1985b. Recent evolutionary changes in the algae of the Baltic Sea. *Br. phycol. J.* **20:** 87–104.

RUSSELL, G. & BOLTON, J. J. 1975. Euryhaline ecotypes of *Ectocarpus siliculosus* (Dillw.) Lyngb. *Estuar. & Coast. Mar. Sci.* **3:** 91–94.

RUSSELL, G. & FIELDING, A. H. 1974. The competitive properties of marine algae in culture. *J. Ecol.* **62:** 689–698.

RUSSELL, G. & FIELDING, A. H. 1981. Individuals, populations and communities. *In* Lobban, C. S. & Wynne, M. J. (Eds) *The biology of seaweeds* pp. 393–420. Botanical Monographs **17.** Blackwell Scientific Publications, Oxford.

RUSSELL, G. & MORRIS, O. P. 1970. Copper tolerance in the marine fouling alga *Ectocarpus siliculosus. Nature, Lond.* **228:** 228–289.

RUSSELL, G. & MORRIS, O. P. 1971. A ship model in antifouling research. *Sea Breezes July:* 512–513.

SAUVAGEAU, C. 1897. Sur quelques Myrionématacées. *Annls. Sci. nat. Bot.*, sér. 8, **5:** 161–288.

SAUVAGEAU, C. 1924. Sur quelques exemples d'heteroblastie dans le development des algues phéosporées. *C.r. hebd. Séanc. Acad. Sci., Paris*, sér. D. **179:** 1576–1579.

SAUVAGEAU, C. 1928. Sur la végétation et la sexualité des Tilopteridales. *Bull. Stn biol. Arcachon* **25:** 51–94.

SAUVAGEAU, C. 1932. Le plethysmothalle. *Bull. Stn biol. Arcachon* **29:** 1–16.

SCAGEL, R. F. 1966. The Phaeophyceae in perspective. *Oceanogr. mar. Biol. ann. Rev.* **4:** 123–194.

SCHONBECK, M. & NORTON, T. A. 1978. Factors controlling the upper limits of fucoid algae on the shore. *J. exp. mar. Biol. Ecol.* **31:** 303–313.

SHEADER, A. & MOSS, B. L. 1975. The effects of light and temperature on germination and growth of *Ascophyllum nodosum* (L.) Le Jol. *Estuar. & Coast. Mar. Sci.* **3:** 125–132.

SHINMURA, I. 1974. Studies on the cultivation of an edible brown alga *Cladosiphon okamuranus* III Development of zoospores from plurilocular sporangium. *Bull. Jap. Soc. scient. Fish.* **40:** 1213–1222.

SHINMURA, I. 1977. Life history of *Cladosiphon okamuranus* Tokida from southern Japan. *Bull. Jap. Soc. Phycol.* **25:** 333–340.

SILVA, P. C. 1980. *Names of classes and families of living algae* Regnum Vegetabile vol. 103. Utrecht & The Hague.

SINCLAIR, C. & WHITTON, B. A. 1977. Influence of nutrient deficiency on hair formation in the Rivulariaceae. *Br. phycol. J.* **12:** 297–313.

SLOCUM, C. J. 1980. Differential susceptibility to grazers in two phases of an intertidal alga: advantages of heteromorphic generations. *J. exp. mar. Biol. Ecol.* **46:** 99–110.

SMITH, G. M. 1955. *Cryptogamic botany 1. Algae and fungi.* McGraw-Hill, New York.

SOUTH, G. R. & BURROWS, E. M. 1967. Studies on marine algae of the British Isles 5. *Chorda filum* (L.) Stackh. *Br. phycol. Bull.* **3:** 379–402.

SOUTH, G. R. & HILL, R. D. 1970. Studies on marine algae of Newfoundland. I. Occurrence and distribution of free-living *Ascophyllum nodosum* in Newfoundland. *Can. J. Bot.* **48:** 1697–1701.

SOUTHWARD, A. J. 1956. The population balance between limpets and seaweeds on wave-beaten rocky shores. *Rep. mar. Biol. Stn Port Erin* **68:** 20–29.

STACE, C. A. 1980. *Plant taxonomy and biosystematics* Edward Arnold, London.

STEPHENSON, T. A. & STEPHENSON, A. 1949. The universal features of zonation between tide-marks on rocky coasts. *J. Ecol.* **37:** 289–305.

STEPHENSON, T. A. & STEPHENSON, A. 1972. *Life between tidemarks on rocky shores.* Freeman & Company, San Francisco.

STRÖMGREN, T. 1978. The effect of photoperiod on the length growth of five species of intertidal Fucales. *Sarsia* **63:** 155–158.

SUBRAHMANYAN, R. 1957. Observations on the anatomy, cytology, development of the reproductive structures, fertilization and embryology of *Pelvetia canaliculata* Dcne. et Thur. *J. Indian bot. Soc.* **36:** 373–395.

SUNDENE, O. 1953. The algal vegetation of Oslofjord. *Skr. norske Vidensk-Akad. Mat.-naturv. Kl.* **2:** 1–244.

SUNDENE, O. 1964. The ecology of *Laminaria digitata* in Norway in view of transplant experiments. *Nytt Mag. Bot.* **11:** 83–107.

SVEDELIUS, N. 1915. Zytologisch-entwicklungsgeschichtliche Studien über *Scinaia furcellata. Nova Acta R. Soc. Scient. upsal.,* ser. 4, **4:** 1–55.

SVEDELIUS, N. 1931. Nuclear phases and alternation in the Rhodophyceae. *Beih. bot. Zbl.* **48:** 38–59.

SVENDSEN, P. & KAIN, J. M. 1971. The taxonomic status, distribution and morphology of *Laminaria cucullata* sensu Jorde and Klavestad. *Sarsia* **46:** 1–22.

TAHARA, M. 1909. On the periodical liberation of oospheres in *Sargassum* (prelim). *Bot. Mag., Tokyo* **23:** 151–153.

TERRY, L. A. & MOSS, B. L. 1980. The effect of photoperiod on receptacle initiation in *Ascophyllum nodosum* (L.) Le Jol. *Br. phycol. J.* **15:** 291–301.

TITTLEY, I., IRVINE, D. E. G. & JEPHSON, N. A. 1977. The infralittoral marine algae of Sullom Voe, Shetland. *Trans. Proc. bot. Soc. Edinb.* **42:** 397–419.

TITTLEY, I. & PRICE, J. H. 1977. The marine algae of the tidal Thames. *Lond. Nat.* **56:** 10–17.

TOTH, R. 1976. The release, settlement and germination of zoospores in *Chorda tomentosa* Phaeophyceae, Laminariales. *J. Phycol.* **12:** 222–233.

VADAS, R. L. 1977. Preferential feeding: an optimization strategy in sea urchins. *Ecol. Monogr.* **47:** 337–371.

WAERN, M. 1952. Rocky shore algae in the Öregrund Archipelago. *Acta phytogeogr. suec.* **30:** 1–298.

WANDERS, J. B. W., HOEK, C. VAN DEN & SCHILLERN, VAN NES, E. N. 1972. Observations on the life history of *Elachista stellaris* (Phaeophyceae) in culture. *Neth. J. Sea Res.* **5:** 458–491.

WILCE, R. T. 1959. The marine algae of the Labrador Peninsula and northwest Newfoundland (ecology and distribution). *Bull. natn. Mus. Can.* **158:** 1–103.

WILCE, R. T., SCHNEIDER, C. W., QUINLAN, A. V. & VAN DEN BOSCH, K. 1982. The life history and morphology of free-living *Pilayella littoralis* (L.) Kjellm. (Ectocarpaceae, Ectocarpales) in Natiant Bay, Massachusetts. *Phycologia* **21:** 336–354.

WILKINSON, M. 1980. Estuarine benthic algae and their environment: a review. *In* Price, J. H., Irvine, D. E. G. & Farnham, W. F. (Eds) *The shore environment* **2:** *Ecosystems* pp. 425–486. Systematics Association Special Volume. **17(b).** Academic Press, London and New York.

WILLIAMS, J. L. L. 1905. Studies in the Dictyotaceae III The periodicity of the sexual cells in *Dictyota dichotoma. Ann. Bot.* **19:** 531–560.

WYNNE, M. J. 1969. Life history and systematic studies of some Pacific North American Phaeophyceae (brown algae). *Univ. Calif. Publs Bot.* **50:** 1–88.

WYNNE, M. J. 1981. Phaeophyta: morphology and classification. *In* Lobban, C. & Wynne, M. J. (Eds). *The biology of seaweeds* pp. 52–85. Botanical Monographs **17.** Blackwell Scientific Publications, Oxford.

WYNNE, M. J. & LOISEAUX, S. 1976. Recent advances in the life history studies of the Phaeophyta. *Phycologia* **15:** 435–452.

KEY TO GENERA

The following key refers only to those genera which are included in the present Volume 3, Part 1 of the Fucophyceae of the British Isles. It must, therefore, be considered as provisional and subject to replacement by a full generic key in Volume 3, Part 2. Despite the inherent limitations of such an incomplete key, it was felt that it would still usefully help towards the identification of at least, a number of the brown algae, recorded around the British Isles. To this end the algae are initially grouped into one of eleven morphologically different growth forms which can be reached by progression through the dichotomous key. This is then followed by a key, utilising both macroscopic and microscopic features, for the identification of those genera which are included in the present Part 1. For some genera such as *Asperococcus* and *Desmarestia,* which can include more than one growth form, identification is sometimes given at the species level. Finally, an explanatory note is inserted at the end of most of the groups, drawing attention to those algae of similar morphology but which will be treated in Part 2. For some groups, particularly those with the smaller discoid, crustose and pulvinate growth forms, most, if not all, of the genera recorded for the British Isles are included. However, the more macroscopic, branched, filamentous, filiform and bladed genera will largely be included in Part 2 and it is recommended that the reader refers to other published keys. The following selection of keys is suggested for this purpose.

FLETCHER, R. L. 1980. *Catalogue of main marine fouling organisms.* Vol. 6 Algae. Office d'Etudes Marines et Atmosphériques, for Comité International Permanent pour la Recherche sur la Préservation des Materiaux en Milieu Marin, Bruxelles. (Key to Phaeophyta. pp. 19–21).

GAYRAL, P. 1966. *Les algues des côtes françaises (Manche et Atlantique).* Paris (See pp. 66–83 for characteristics of orders, families and genera.)

HISCOCK, S. 1979. A field key to the British brown seaweeds (Phaeophyta). *Fld. Stud.* **5:** 1–44.

JONES, W. E. 1962. A key to the genera of the British seaweeds. *Fld. Stud.* **1:** 1–32. (N.B. Revised reprint published in 1964.)

KORNMANN, P. & SAHLING, P. H. 1977. Merresalgen von Helgoland. Benthische Grun-Braun-und Rotalgen. *Helgoländer wiss. Meeresunters.* **29:** 1–289. (See pp. 9–11 for key to the genera of the brown algae.)

NEWTON, L. 1931. *A handbook of the British seaweeds.* British Museum (Natural History), London (Phaeophyceae – Key to Genera, pp. 106–112.)

PRUD'HOMME VAN REINE, W. F. A taxonomic revision of the European Sphacelariaceae (Sphacelariales, Phaeophyceae). *Leiden Botanical Series* **6:** x + 1–293. (See pp. 57–58 for key to Sphacelariales and some ecads.)

ROBERTS, M. 1967. Studies on marine algae of the British Isles. 3. The genus *Cystoseira. Br. phycol. Bull.* **3:** 345–366. (see p. 363 for key.)

SOUTH, G. R. & HOOPER, R. G. 1980. A catalogue and atlas of the benthic marine algae of the Island of Newfoundland. *Mem. Univ. Nfld. Occas. Pap. Biol.* **3:** 1–136. (See pp. 8–13 for key to the genera of brown algae.)

KEY

1 Thallus small, usually orbicular discs, spots or stains; epiphytic rarely epilithic 2
Thallus of another form; epiphytic, epizoic or epilithic 8

2(1) Cells of erect filaments with a single, large, plate-like plastid; ascocysts usually present 3
Cells of erect filament with 1–3 small, plate-like/discoid plastids; ascocysts rare 4

3(2) Ascocysts conical; plurilocular sporangia subcylindrical, biseriate or triseriate, to 17 μm in diameter . . . . . .*Chilionema (C. hispanicum)*
Ascocysts elongate-cylindrical or slightly clavate; plurilocular sporangia in 2–4 vertical columns, rarely irregularly clustered, to 24 μm in diameter . . . . *Symphyocarpus*

4(2) Basal cells of erect filaments frequently longitudinally divided (biseriate); plurilocular sporangia multiseriate . . . . *Chilionema*
Basal cells of erect filaments not longitudinally divided (uniseriate); plurilocular sporangia uniseriate, rarely biseriate 5

5 (4) Erect filaments pseudodichotomously branched, bearing terminal and/or lateral plurilocular sporangia 6
Erect filaments unbranched, secundly branched or with short protuberances; plurilocular sporangia sessile or shortly stalked on basal layer, more rarely terminal on erect filaments 7

6 (5) Thallus epiphytic . . . . *Microspongium*
Thallus epilithic . . . .*Myrionema (liechtensternii)*

7 (5) Thallus epiphytic on the red alga *Dumontia contorta* only; cells of basal layer producing downwardly extending, branched, rhizoidal filaments . . . . *Ulonema*
Thallus reported on various hosts but not *Dumontia*; cells of basal layer without extending rhizoidal filaments. . . . . *Myrionema*

All the genera in the above section comprise epiphytes which form rounded, sometimes irregularly shaped, prostrate, closely encrusting patches, usually easily visible when the hosts are held up in front of a light source. With the exception of *Ulonema*, which has deeply penetrating rhizoids, all the thalli are easily scraped intact from the surface of the host to reveal, under the microscope, short, erect, usually simple, compacted but easily separated, gelatinous, erect filaments arising from a discoid or pseudodiscoid basal system. The sporangia, if present, also arise vertically from the basal cells and are sessile, stalked or lateral/terminal on the erect filaments. In habit the algae are similar to young thalli of larger genera such as *Hecatonema* and *Protectocarpus*. They are also similar to endophytic genera such as *Cylindrocarpus, Endodictyon, Gononema, Microsyphar* and *Streblonema* and the epiphytic genus *Phaeostroma*, all of which are treated in Volume 3, Part 2. Unlike the epiphytic genera included here the greater part of the thalli of all the endophytic genera is usually retained within the host tissue with only limited 'emergence' at the surface, usually as small pustules of sporangia and/or hairs. The epiphytic genus *Phaeostroma* also differs from those considered here in having a filamentous/pseudodiscoid base which does not give rise to erect filaments or erect plurilocular sporangia, the latter being formed as irregularly shaped clusters by internal division of the basal cells.

8 (1) Thallus forming closely adherent, crust-like patches 9
Thallus of another form 15

9 (8) Erect filaments of crust tightly adjoined and not easily separated in squash preparations 10
Erect filaments of crust moderately or weakly adjoined and easily separated in squash preparations 12

10 (9) Crusts relatively thick and well developed, loose to moderately well attached, olive to dark brown/black in colour; margin prominent, usually quite easily lifted by a scalpel, central regions verrucose, bullate and often brittle; vertical sections through margin reveal arched erect filaments with no obviously terminal, apical cell; crusts common throughout the littoral, exposed or immersed in pools, frequently epizoic on limpets and barnacles. . . . . *Ralfsia*
Crusts comparatively thin and strongly attached, light to dark brown/black with discrete, gradually, delimiting, closely adherent, margin not lifted by a scalpel; vertical sections through margin reveal a single, large apical cell giving rise to a basal layer from which vertical or only slightly arched erect filaments develop; crusts usually confined to the lower eulittoral or sublittoral, not reported on limpets and barnacles. 11

11 (10) Cells with a single, large plate-like plastid, erect filaments not exceeding 8 cells in length; rare, only known from the sublittoral in Plymouth, Devon . . . . *Sorapion*
Cells with several, discoid plastids; erect filaments commonly exceeding 8 cells in length; common and widespread in the lower eulittoral and sublittoral all around the British Isles. . . . . *Pseudolithoderma*

12 (9) Crusts smooth, firm and slightly subcoriaceous; erect filaments not obviously gelatinous, moderately adjoined and not readily separated by light pressure under a coverslip; unilocular sporangia in prominent, raised, gelatinous sori which are obvious in surface view . . . *Stragularia*
Crusts soft, sponge-like; erect filaments markedly gelatinous, very loosely adjoined and easily separated by light pressure under a cover-slip; unilocular sporangia (if present) in sori not obvious externally. 13

13 (12) Crust slightly pulvinate; unilocular sporangia up to 100×27 µm, each arising laterally at the base of a multicellular paraphysis; plurilocular sporangia unknown for the British Isles; ascocyst-like cells not present. . . . *Microspongium gelatinosum* phase of *Scytosiphon lomentaria*
Crust comparatively thin; unilocular sporangia, if present, up to 23×14 µm, arising terminally on erect filaments, unaccompanied by paraphyses; plurilocular sporangia and ascocyst-like cells common . 14

14 (13) Crust epilithic, common, reported all around the British Isles, littoral fringe to lower eulittoral; unilocular sporangia common; plurilocular sporangia uniseriate, rarely bi-triseriate . . . . *Petroderma*
Crust epiphytic, rare, reported from a few scattered localities around the British Isles, lower eulittoral and sublittoral; unilocular sporangia unknown; plurilocular sporangia multiseriate, sometimes clustered . . . . *Symphyocarpus*

The above section comprises all the crustose, predominantly lithophytic brown algae recorded for the British Isles except the genus *Battersia* Reinke ex Batters which will be included in Volume 3, Part 2. A combination of the following characteristics should enable *Battersia* to be distinguished from the above genera: it forms black, very thin, closely encrusting and frequently superimposed thalli in shaded, tidal pools in the lower eulittoral (also recorded in shaded pools at the base of overhanging cliffs); the erect, adjoined filaments of the thalli are initially derived from parenchymatous divisions of the basal cells; the cells contain several discoid plastids (a character shared only with *Pseudolithoderma*); unilocular sporangia only have been described – they are terminal on erect, clustered stalks which arise from the thallus surface as small, scattered, discrete, whitish sori just visible to the unaided eye; thalli have only been recorded from the Shetland Isles, Fife and Northumberland.

15 (8) Thallus procumbent, membranous, loosely attached underneath and easily removed intact 16
Thallus essentially erect, of another form 17
16 (15) Thallus more or less orbicular, radially ridged, lobate, usually with an outer fringe of hair-like filaments; reproductive sori on thallus surface comprising either gametangia (oogonia, antheridia) or unilocular sporangia . . . . *Zanardinia*
Thallus orbicular at first, becoming irregularly spreading later, frequently with overlapping lobes, with entire and rounded margin without an outer fringe of hair-like filaments; reproductive sori on thallus surface comprising unilocular sporangia only
. . . . *Aglaozonia* phase of *Cutleria multifida*

The above brown algae are the only representatives in the British Isles with procumbent thalli and are easily distinguished from the crust forming genera in being clearly membranous and bladed in appearance, loosely attached underneath by rhizoids and easily raised intact from the substratum by forceps, etc., and, being parenchymatous in basic construction rather than pseudoparenchymatous, (a character shared only with *Battersia*), they are not recognisably composed throughout of erect columns of cells in vertical section.

17 (15) Thallus forming microscopic tufts of erect, compacted filaments more or less of equal height, either discrete or confluent to form a fine felt-like layer 18
Thallus of another form, or if tufted with filaments not obviously compacted but free. 22
18 (17) Erect filaments simple, or rarely branched towards the base 19
Erect filaments frequently branched above 20
19 (18) Erect filaments arising from a basal layer of cells which are frequently longitudinally divided (biseriate); thallus epilithic, forming microscopic confluent tufts in the littoral fringe
. . . . . . *Compsonema saxicolum* phase of *Petalonia*/*Scytosiphon*
Erect filaments arising from a basal layer of uniseriate cells, or more commonly from a basal cushion of compacted, erect filaments of large, colourless cells; thallus epiphytic forming microscopic confluent tufts on hosts in the eulittoral and shallow sublittoral *Myriactula*

20 (18) Cells of basal layer with longitudinal divisions (biseriate) in part; branching of erect filaments irregularly lateral, with both branches and lateral plurilocular sporangia widely divergent at first, later recurving towards parental filament . . . . *Hecatonema*
Cells of basal layer uniseriate; branching of erect filaments pseudodichotomous or secund, with branches and lateral plurilocular sporangia erect and adpressed closely towards parental filament 21
21 (20) Plurilocular sporangia commonly terminal on erect filaments and frequently branched, usually in a characteristic cock's comb (secund) fashion. *Protectocarpus*
Plurilocular sporangia rarely terminal on erect filaments and never branched in a cocks comb fashion . . . . *Compsonema*

Genera which might also key out in the above section but are to be treated in Volume 3, Part 2 include *Herponema, Laminariocolax, Pleurocladia, Sphacelaria* and *Waerniella.*

22 (17) Thallus forming small, erect, brush-like tufts of filaments, either microscopic or more commonly just visible to the unaided eye; cells of erect filaments uniseriate, or rarely with longitudinal divisions 23
Thallus of another form or if tufted, not filamentous throughout but more than one cell wide 27
23 (22) Erect filaments frequently branched . . . . *Hecatonema*
Erect filaments unbranched 24
24 (23) Erect filaments arising from a wart-like, usually hemispherical cushion of compacted, erect filaments of large, colourless cells 25
Erect filaments arising from a discoid/pseudodiscoid base of outwardly radiating filaments or from a fibrous network of rhizoidal filaments 26
25 (24) Erect filaments up to 2 mm in length . . . . *Myriactula*
Erect filaments 3–40 mm in length . . . . *Elachista*
26 (24) Cells of erect filaments 25–65 µm in diameter; plurilocular sporangia clustered, forming densely packed, vertical tiers enclosing vegetative cells; unilocular sporangia unknown . . . . *Halothrix*
Cells of erect filaments 7–17 µm in diameter; plurilocular sporangia formed by subdivision of vegetative cells; unilocular sporangia rare, borne laterally on the base of the erect filaments . . . . *Leptonematella*

This section comprises all those genera recorded for the British Isles which form brush-like tufts of erect, *unbranched,* uniseriate filaments. Also included is *Hecatonema* which has branched, erect filaments and some longitudinally segmented cells. However there does remain a large number of genera which form generally small, erect tufts of branched, uniseriate filaments (e.g. *Ectocarpus, Feldmannia, Giffordia, Spongonema*), branched filaments with, in part, longitudinally segmented cells (e.g. *Haplospora, Isthmoplea, Pilayella, Tilopteris*) and branched filaments with extensive longitudinal segmentation of cells (e.g. *Cladostephus, Halopteris, Sphacelaria*); all these genera will be treated in Volume 3, Part 2.

27 (22) Plants forming small tufts, less than 5 cm in length, comprising short, erect, solid, filiform thalli. 28
Plants of another form 32

28 (27) Erect thalli unbranched, or rarely branched towards the base 29
Erect thalli laterally branched; branches radially produced, either irregular and discrete or confluent and closely-packed 31

29 (28) Erect thalli not exceeding 115 μm in width and without laterally produced hairs . . . . *Pogotrichum*
Erect thalli exceeding 115 μm in width with hairs commonly arising from surface cells 30

30 (29) Hairs common, with basal meristem and sheath; plurilocular sporangia cylindrical or conical, projecting from surface cells, sometimes with associated short, multicellular, paraphyses . . . *Asperococcus* (*A. scaber*)
Hairs abundant, with basal meristem but lacking sheath; plurilocular sporangia formed directly in surface cells following vegetative divisions, without paraphyses . . . . *Litosiphon*

31 (28) Erect thalli linear, sharply delimited below and arising from a discoid base; hairs uncommon; epiphytic on *Zostera* . . . . . . *Leblondiella*
Erect thalli linear becoming nodose, and then elongate-clavate, attached at the base by spreading rhizoidal filaments; hairs abundant; epiphytic on various hosts in particular *Scytosiphon* . . . . *Myriotrichia*

The above section comprises only those genera in the British Isles with short, erect, solid, filiform thalli which are either unbranched (*Asperococcus* (*scaber*), *Litosiphon, Pogotrichum*) or have short, lateral branches of limited growth clothing the axis (*Leblondiella, Myriotrichia*). The considerable number of larger genera with branched, filiform thalli variously distributed in such families as the Cystoseiraceae, Dictyosiphonaceae, Fucaceae, Sargassaceae, Spermatochnaceae and Sphacelariaceae will be treated in Volume 3, Part 2.

32 (27) Thallus forming a globose, hemispherical or small finger-like (< 5 mm in length) gelatinous cushion, pseudoparenchymatous in structure and easily squashed into constituent filaments under pressure 33
Thallus of another form or if (1) globose/hemispherical, then parenchymatous in structure and not squashed into constituent filaments, if (2) erect, finger-like and gelatinous then exceeding 5 mm in length . . 36

33 (32) Thallus epiphytic, forming small (< 5 mm in length) discrete, erect, cylindrical, clavate, more rarely forked, finger-like extensions on hosts hosts . . . . *Microcoryne*
Thallus epiphytic or epilithic, forming discrete, less commonly confluent, hemispherical cushions on substrata 34

34 (33) Thallus epilithic or epiphytic on crusts of the brown alga *Ralfsia verrucosa* . . . . *Petrospongium*
Thallus epiphytic, reported on various hosts in particular *Chondrus* and *Corallina* 35

35 (34) Thallus hemispherical and smooth, light/dark brown in colour, solid throughout, not exceeding 3 mm in diameter; terminal paraphyses 5–17 cells in length; epiphytic on *Chondrus crispus* . . . . . *Corynophlaea*
Thallus hemispherical becoming globose, irregular and convoluted later, yellow brown in colour, solid becoming hollow, to 8 cm in diameter; terminal paraphyses 2–5 cells in length; epiphytic on various hosts, especially *Corallina officinalis* . . . . *Leathesia*

Common features of the above genera include the cushion-like growth form of the thalli, the gelatinous nature of the constituent filaments and the ease with which they can be separated during squash preparations. All the genera with a similar gelatinous nature, but more obviously erect, macroscopic, filiform and branched are almost exclusively placed in the family Chordariaceae (e.g. *Eudesme, Mesogloia*) and will be treated in Volume 3, Part 2.

36 (32) Thallus globose, sac-like and membranous, olive brown, drying to green, solid when young, becoming hollow later and frequently collapsed and irregularly split; membrane smooth and slightly lubricous outside, slightly roughened inside; structure parenchymatous, not separating into filaments under pressure; surface cells polygonal, each with a single, large plate-like plastid with pyrenoid; in transverse section membrane with inner large, colourless medullary cells and outer smaller pigmented cortical cells; plurilocular sporangia in closely packed vertical columns arising from surface cells; unilocular sporangia unknown on thallus . . . . *Colpomenia*
Thallus of another form 37

37 (36) Thallus erect, tubular and unbranched, up to 40 cm in length 38
Thallus of another form 39

38 (37) Thallus epilithic, linear, inflated or collapsed, usually with regular constrictions at intervals and smooth in texture; surface cells small, to 9 μm in diameter, usually irregularly arranged, each with a single plate-like plastid with pyrenoid; plurilocular sporangia in closely packed vertical columns arising from surface cells; unilocular sporangia unknown on blade; littoral fringe to shallow sublittoral . . . . *Scytosiphon*
Thallus epiphytic, less commonly epilithic, linear or clavate, inflated or collapsed, without constrictions, and slightly roughened in texture; surface cells large, to 26 μm in diameter, usually arranged in rows, each with several discoid plastids and pyrenoids; unilocular and plurilocular sporangia in punctiferous sori, produced in surface cells, usually with associated short, multicellular, paraphysis-like filaments; plurilocular sporangia rare . . . . *Asperococcus*

The only other erect, tubular and unbranched genus in the British Isles is *Chorda,* which will be treated in Volume 3, Part 2.

39 (37) Thallus an erect, flattened, unbranched blade 40
Thallus of another form 43

40 (39) Blades with a prominent conical holdfast and long stipe extending above into a distinct midrib, which gives rise to a network of primary and secondary veins, terminating at the margin in tufts of branched hair-like filaments; rare, only recorded from some scattered localities in the sublittoral. . . . . *Desmarestia (D. dresnayi)*
Blades either acuminate below or arising from a short stipe, and attached by a fibrous mat of rhizoidal filaments, crust or small discoid holdfast; blades without a midrib, venation or marginal tufts of filaments; common and widely recorded around the British Isles 41

41 (40) In surface view cells small, not exceeding 18 μm in diameter and irregularly arranged, each with a single plate-like plastid with pyrenoid; plurilocular sporangia in closely packed vertical columns, extensive in terminal blade region; unilocular sporangia unknown on blade surface . . . . *Petalonia*

In surface view cells large, commonly exceeding 18 μm in diameter, often arranged in longitudinal rows each with several discoid plastids with pyrenoids; sporangia widely scattered on blade surface, solitary or grouped, formed directly within surface cells, slightly immersed or projecting; unilocular sporangia common; plurilocular sporangia rare, multiseriate 42

42 (41) In transverse section cells of equal size or showing slight differentiation into an inner medulla of large colourless cells and an outer cortex of smaller pigmented cells; sporangia scattered or grouped, without associated paraphyses . . . . *Punctaria*

In transverse section, thalli showing marked differentiation into an inner medulla of large, colourless, sometimes ruptured cells and an outer cortex of smaller, pigmented cells; sporangia discrete, more commonly crowded in punctate sori, associated with short multicelluar paraphyses . . . . *Asperococcus (A. compressus)*

The above section comprises all the genera treated in Volume 3, Part 1 which form erect, unbranched blades. Other blade-like algae treated in Volume 3, Part 2 which could also be interpreted as unbranched include the fan-shaped and lobed *Padina pavonia*, the laminate *Laminaria saccharina* and *Alaria esculenta* and the digitate *Saccorhiza* and species of *Laminaria*

43 (39) Thallus erect, flattened and branched 44

Thallus erect, filiform or terete and branched 46

44 (43) Thallus of main axis without a distinct midrib; gametangia (oogonia, antheridia) borne on short branches and grouped in punctate sori, scattered on blade surface; unilocular sporangia unknown on blade surface . . . . *Cutleria*

Thallus of main axis with a distinct midrib; only unilocular sporangia known on blade surface, either scattered and discrete, formed in surface cells or grouped in specialised terminal receptacles 45

45 (44) Main axis prominent, up to 2 m long, oppositely branched to several orders with ultimate branches short and spine-like and ending in tufts of branched filaments; unilocular sporangia formed in slightly enlarged surface cells scattered on blade surface; transverse sections reveal a single axial cell surrounded by a broad cortical zone . . . . *Desmarestia (D. ligulata)*

Main axis not prominent, up to 0·2 m long, branching pseudodichotomous, occasionally triradiate or alternate, without terminal spine-like branches but with a basal collar and/or emergent tuft of hair-like, simple filaments; transverse sections reveal a central region of thick-walled axial cells surrounded by a broad cortical zone; unilocular sporangia borne in sori on the surface of specialised conical or cylindrical receptacles terminal on branches or in axes . . . . . *Carpomitra*

The above section contains only a relatively small number of the genera recorded for the British Isles with erect, branched dorsi-ventrally flattened thalli. Genera of similar habit which are treated in Volume 3, Part 2 include *Ascophyllum, Dictyopteris, Dictyota, Fucus, Halidrys, Himanthalia, Pelvetia,* and *Taonia.*

46 (43) Main axis and branches bearing regular whorls of 3–4 tufts of short, laterally projecting, hair-like filaments; unilocular sporangia borne in chains on the whorled filaments . . . . *Arthrocladia*

Main axis and branches without whorls of hair-like filaments; unilocular sporangia formed in little modified surface cells or borne in sori on short lateral branches of limited growth 47

47 (46) Branching of main axis to one order only, all branches with numerous short, divaricate branches of limited growth terminated by a tuft of pigmented, hair-like filaments; in transverse section thalli with a broad central region of large, fairly thick-walled cells enclosed by 1–2 layers of small pigmented cells; unilocular sporangia borne laterally on short, multicellular paraphyses in continuous mucilaginous sori enclosing the short lateral branches . . . . *Sporochnus*

Thalli much branched, to several orders, without short, divaricate branches of limited growth; in transverse section thalli with a distinct large, central axial cell, a broad cortical zone of large colourless cells enclosed by 1–2 layers of small pigmented cells; unilocular sporangia scattered, in little modified surface cells . . . . *Desmarestia (D. aculeata, D. viridis)*

Taxonomic treatment

LITHODERMATACEAE Hauck

LITHODERMATACEAE Hauck (1883–85), pp. 318, 402 [as Lithodermaceae].

Thallus encrusting, more rarely pulvinate, discrete or expansive, firm and subcoriaceous, rarely soft and sponge-like, closely adherent to substratum, with or without rhizoids; epilithic, epizoic or epiphytic; thallus structure pseudoparenchymatous, composed of prostrate, radiating, branched filaments, laterally adjoined and disc-like with marginal row of apical cells, giving rise behind to erect or slightly curved, compacted filaments; erect filaments, strongly or weakly united, firm or gelatinous, little branched or unbranched; vegetative cells with one plate-like or several discoid plastids, with or without pyrenoids; hairs present or absent, with basal meristem and sheath; ascocyst-like cells present or absent, intercalary or terminal, on erect filaments. Plurilocular and unilocular sporangia in sori, either discrete or expansive, slightly raised on thallus surface, terminating erect filaments, formed by either direct transformation or extension of surface cells, with or without ascocyst-like cells.

In many earlier treatments of the crustose brown algae (e.g. Feldmann, 1937; Taylor, 1937*a*, 1957; Parke, 1953; Pankow, 1971) up to three families were generally recognised; these were the Lithodermataceae Hauck, the Ralfsiaceae Hauck and the Nemodermataceae Feldmann. The Lithodermataceae was usually characterised by terminally borne unilocular and plurilocular sporangia and several plastids in each cell; the Ralfsiaceae had lateral unilocular sporangia, intercalary plurilocular sporangia and a single, plate-like plastid in each cell; the monotypic Nemodermataceae had intercalary unilocular sporangia, lateral plurilocular sporangia and several plastids in each cell. However, these differences were not always considered clear-cut and, for example, in many treatments, the Lithodermataceae was included in the circumscription of the Ralfsiaceae (Hollenberg, 1969; Abbott & Hollenberg, 1976; Parke & Dixon, 1976; Rueness, 1977). Newton (1931), Fritsch (1945) and Loiseaux (1967*a*) went further and included all the crusts together with the discoid myrionematoid and hecatonematoid algae within the single, large family Myrionemataceae Foslie. Within this family Loiseaux grouped all the crustose genera in a tribe 'Ralfsiées'.

Recognition of the three families, Lithodermataceae, Ralfsiaceae and Nemodermataceae, distinguished on the basis of the position of the reproductive organs, was again proposed by Nakamura (1972). He placed these families in a new order, Ralfsiales, members of which are characterised by an *Ectocarpus*-type (isomorphic) life history, a discal type of spore germination pattern and a single, parietal, plate-like plastid which lacks a pyrenoid. This new order has not been recognised by Russell & Fletcher (1975), Parke & Dixon (1976), South (1976) and Nelson (1982) but has been accepted by Bold & Wynne (1978) and Tanaka & Chihara (1980*a*, 1982) who erected and included in the order a new family Mesosporaceae.

In the present treatment the crustose genus *Ralfsia* is removed to the Scytosiphonaceae, in agreement with Pedersen (1976), principally on the basis of life history studies (see p. 217). The family Lithodermataceae is, therefore, revived and characterised principally by the possession of terminally situated unilocular and plurilocular sporangia. In the British Isles, the family is represented by the four genera *Petroderma*,

Pseudolithoderma, Sorapion and *Symphyocarpus*, which can be distinguished by combinations of the following three characters: numbers of plastids in each cell, strength of cohesion of the erect filaments and morphology of the reproductive sporangia.

PETRODERMA Kuckuck

PETRODERMA Kuckuck (1897), p. 382.

Type species: *P. maculiforme* (Wollny) Kuckuck (1897), p. 382.

Thallus encrusting, thin or comparatively thick, discrete or extensive, sponge-like and slightly gelatinous, firmly adherent by undersurface to substratum, usually without rhizoids; consisting of a monostromatic discoid base, giving rise to vertical, little branched, mucilaginous filaments loosely bound together except at base and easily separable under pressure; cells with single, lobed, plate-like plastid without obvious pyrenoid; hairs single, terminal on erect filaments, with basal meristem and sheath.

Plurilocular and unilocular sporangia terminal on erect filaments without accompanying paraphyses, in extensive sori, not obvious externally; unilocular sporangia club-shaped or cylindrical, often present within old sporangial husks; plurilocular sporangia uni- to multiseriate, occasionally branched.

One species in the British Isles:

Petroderma maculiforme (Wollny) Kuckuck (1897), p. 382. Fig. 1

Lithoderma maculiforme Wollny (1881), p. 31.
Lithoderma lignicola Kjellman (1883), p. 256.

Thallus encrusting, ranging from small, thin specks to comparatively thick, confluent extensions up to several centimetres across, light to dark brown, quite firmly attached by undersurface to substratum, rhizoids usually absent; surface smooth, spongy but firm, gelatinous, surface cells rounded, quite closely packed, 7–10 μm in diameter, with prominent single, parietal, ring-shaped, plastid without obvious pyrenoid; thallus structure pseudoparenchymatous, consisting of a monostromatic base of laterally united, coherently spreading filaments giving rise to simple or little branched, erect, gelatinous filaments, weakly joined and easily separable under pressure; filaments to 25 (–37) cells, 180 (–400) μm long, comprising cells 1–3 diameters long, 5–18 × 6–10 μm, quadrate to rectangular, with single, parietal, lobed, plate-like plastid in upper cell region; long intercalary and terminal ascocyst-like cells frequent; hairs single, with basal meristem and sheath, arising from surface cells.

Plurilocular and unilocular sporangia terminal on erect filaments, without paraphyses, crowded in extensive sori on crust surface, not obvious externally; unilocular sporangia

Fig. 1 *Petroderma maculiforme*
A. Surface view of vegetative crust. B. R.V.S. of crust margin. C. V.S. of vegetative crusts, showing loosely adjoined erect filaments and intercalary ascocyst-like cells. Note lobed, plate-like plastid drawn in cells of one section. D. V.S. of crusts with terminal unilocular sporangia. E. V.S. of crusts with terminal plurilocular sporangia. Bar = 50 μm.

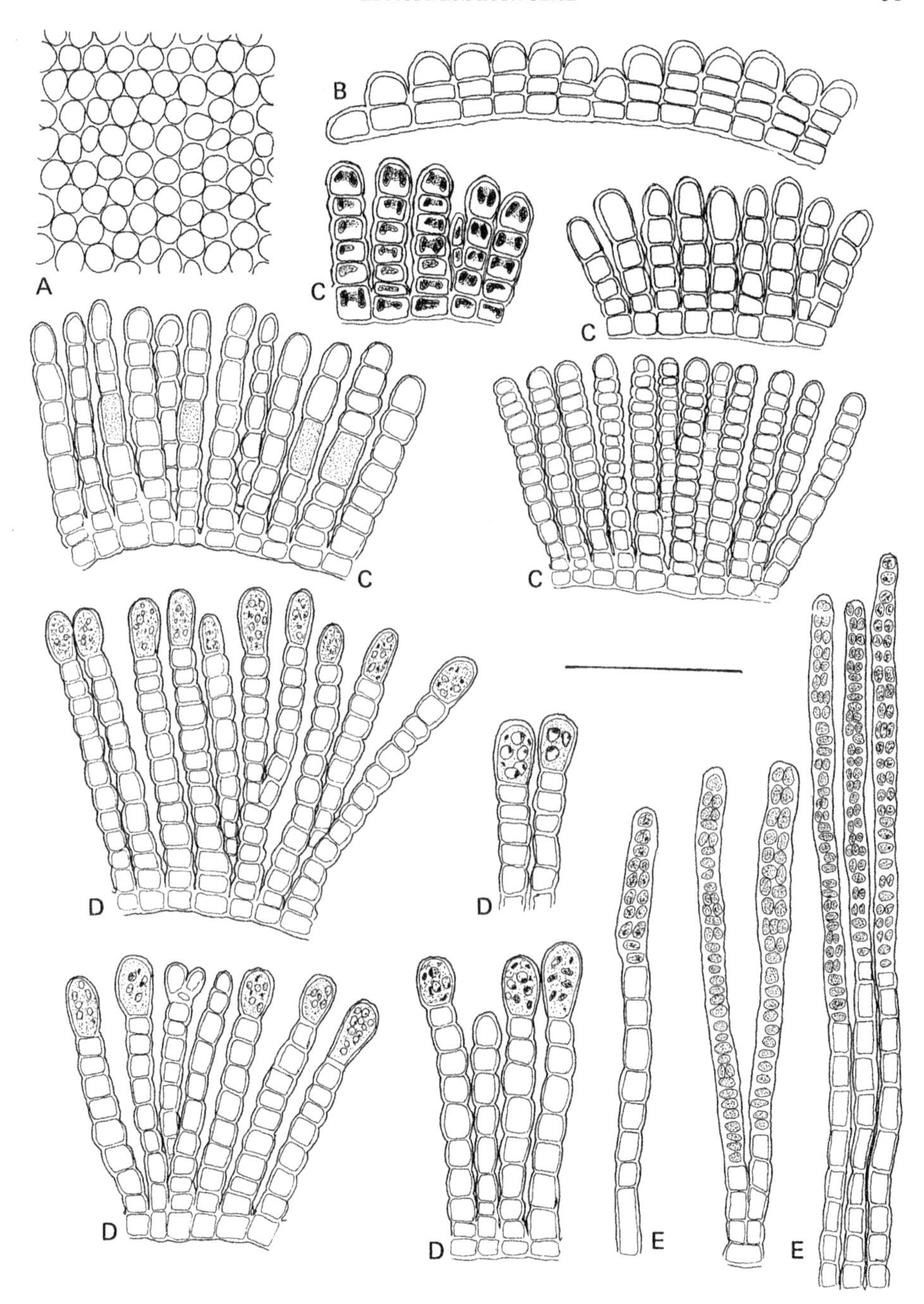
A
B
C
C
C
C
D
D
D
D
E
E

club-shaped or cylindrical, 13–23 × 12–14 µm, often formed within empty sporangial husks; plurilocular sporangia uni- to multiseriate, occasionally branched to 135 µm (–32 cells) long and 10 µm in diameter.

Epilithic, from littoral fringe to lower eulittoral, usually in shallow pools and channels, although emergent and extensive in littoral fringe under exposed conditions; appears to show a preference for hard substrata, such as flint stones.

Recorded for scattered localities all around the British Isles (Shetland Isles, Berwick, Yorkshire, Kent, Hampshire, Dorset, Devon, Cornwall, Pembroke; in Ireland, Mayo and Galway).

Probably perennial although thalli tend to be most conspicuous during the winter months. Plurilocular and unilocular sporangia recorded October to May.

This is a common and widespread epilithic alga. It was first recorded for the British Isles by Cotton (1912) on Clare Island, Mayo, Ireland, but surprisingly was not included in Newton (1931). It is a very closely encrusting alga, occurring as either solitary, fine specks or, more usually as expansive confluent crusts several centimetres in diameter. It is found on a wide variety of substrata, extending from the littoral fringe to the lower eulittoral region, both exposed and in pools. It is particularly found on hard siliceous flint nodules, often in company with crusts of the red alga *Hildenbrandia*. On the soft chalk substrata of the Kent and Sussex coasts it is very abundant and expansive on the vertical cliff faces in the littoral fringe, often in association with barnacles and *Enteromorpha* spp.

Culture studies on Pacific and North Atlantic populations of *P. maculiforme* by Wynne (1969) and Fletcher (1978) respectively found no evidence of a sexual cycle in the life history, which appeared to be of the 'direct' type. However, in agreement with Wilce *et al.* (1970) more information based on culture studies is required before a further understanding of the life history of this species is possible.

Wollny (1881), p. 31, Table II, figs 1–4; Kjellman (1883), p. 256, pl. 26, figs 8–11; Kuckuck (1897), p. 382, figs 9 & 10; Waern (1949), pp. 663–666, pl. 1; (1952), pp. 141–143, fig. 75; Edelstein & McLachlan (1969*b*), pp. 561–563, figs 1–19; Wynne (1969), pp. 9–10, fig. 3, pl. 3; Wilce *et al.* (1970), pp. 119–135, figs 1–5; Abbott & Hollenberg (1976), p. 174, fig. 140; Fletcher (1978), pp. 371–398, figs 8, 9, 20 & 21.

PSEUDOLITHODERMA Svedelius

PSEUDOLITHODERMA Svedelius in Kjellman & Svedelius (1910), p. 175.

Type species: *P. fatiscens* Svedelius in Kjellman & Svedelius (1910), p. 176.

Thallus encrusting, thin to moderately thick, discrete or extensive, light brown to black, firm, smooth, subcoriaceous to coriaceous, firmly attached to the substratum by the whole undersurface, normally without rhizoids; thallus structure pseudoparenchymatous, base monostromatic and discoid, giving rise to erect, little-branched, very tightly bound filaments, covered by a surface cuticle, with or without terminal ascocyst-like cells; cells with several small discoid plastids without obvious pyrenoids; hairs unknown in the British Isles.

Plurilocular sporangia terminal on erect filaments, with or without unicellular paraphysis-like cells, in slightly raised mucilaginous sori, extensive on crust surface, uni- to multiseriate, with straight or oblique cross walls; unilocular sporangia present or absent, terminal on erect filaments.

Two species of the genus *Pseudolithoderma* are presently recognised for the British Isles: *P. extensum* and the more recently discovered *P. roscoffense*. It is possible that other species of this genus might also be present but await identification. To date three other species have been described in the North Atlantic; these are *Pseudolithoderma rosenvingii* (Waern) Lund known from the Swedish and Finnish Baltic coasts, East and West Greenland, the White Sea and North East Canada (Waern, 1949, 1952; Lund, 1959; Sears & Wilce, 1973), *Pseudolithoderma subextensum* (Waern) Lund known only from the Swedish Baltic coast (Waern, 1949, 1952) and *Pseudolithoderma paradoxum* Sears & Wilce known from North East America (Sears & Wilce, 1973). For further information on these species and comparison with crusts of *P. extensum* reference should be made to the literature cited.

KEY TO SPECIES

Plurilocular sporangia uniseriate, rarely biseriate; loculi with oblique cross walls . . . . *P. extensum*
Plurilocular sporangia biseriate to multiseriate; loculi with straight cross walls . . . . *P. roscoffense*

Pseudolithoderma extensum (Crouan frat.) Lund (1959), p. 84. Fig. 2

Ralfsia extensa Crouan frat. (1867), p. 166.
Lithoderma fatiscens sensu Kuckuck (1894), p. 238, non Areschoug (1875), p. 22.
Pseudolithoderma fatiscens Svedelius in Kjellman & Svedelius (1910), p. 176.
Lithoderma extensum (Crouan frat.) Hamel (1931–39), p. 110.

Thallus encrusting, light brown to black, thin to thick, usually confluent to 10 cm or more in diameter, firm, coriaceous, very firmly attached by whole undersurface to substratum, rhizoids usually absent; surface cells polygonal, quite closely packed together, 6–10 × 4–6 μm containing up to 6 discoid, peripherally placed plastids; thallus structure pseudoparenchymatous, consisting of a monostromatic discoid base giving rise to erect rows of little branched, very firmly adjoined filaments, to 22 cells, 165 μm long, covered by a thick surface cuticle; cells usually markedly wider than high below, quadrate or subquadrate above, 4–11 × 6–20 μm, with discoid, terminally placed plastids without obvious pyrenoids; hairs not observed.

Plurilocular and unilocular sporangia terminal on erect filaments, in very slightly raised, mucilaginous sori extensive on crust surface; plurilocular sporangia common, uniseriate rarely biseriate, to 8 loculi, 40 μm long × *c.* 9 μm in diameter, attenuate at apex with characteristic oblique cross walls, frequently with accompanying club-shaped or cylindrical ascocyst-like cells, to 33 μm long × *c.* 7 μm in diameter; unilocular sporangia rare slightly globose, to 33 μm long × *c.* 20 μm in diameter.

Epilithic on stones and bedrock, in lower eulittoral pools, more common and extensive in the sublittoral, to a depth of at least 20 m.

Abundant and widely distributed around the British Isles.
Perennial, sporangia recorded during winter (November to March).

This is the only crustose brown alga which is commonly and perennially found in the sublittoral around the British Isles; other crustose species are only rarely reported and are usually confined to the littoral and shallow sublittoral. It occurs on bedrock, boulders, shells, various artificial substrata mainly contributing to the underflora of *Laminaria* spp. and often in association with crustose coralline and non-coralline red algae.

Culture studies by Kuckuck (1912) indicate that probably an isomorphic alternation of generations type of life history occurs. The plurispores were reported to behave like isomorphic gametes with sexual fusion occurring to produce a zygote. However, mature fertile crusts with unilocular sporangia were not produced from these zygotes, and no cytological data were presented. Further studies are, therefore, required on this species to confirm the above proposed life history pattern.

Kuckuck (1894), pp. 238–240, fig. 11 A–B; Batters (1895a), p. 275; Kuckuck (1912), pp. 167–176, figs 2–4, pl. 7 (1–18); Newton (1931), p. 127, fig. 75 A–C; Hamel (1931–39), pp. 110–111, fig. 26 D–E; Lund (1938), pp. 1–18, figs 4–6; Waern (1949), pp. 658–659, fig. 2 i–j; Taylor (1957), p. 137; Lund (1959), p. 84; Pankow (1971), pp. 172–173, fig. 206; Rueness (1977), p. 126, fig. 51 A–B; Fletcher (1978), pp. 371–398; Stegenga & Mol (1983), pp. 81–82.

Pseudolithoderma roscoffense Loiseaux (1968), p. 308 (as *P. roscoffensis*). Fig. 3

? Lithoderma adriaticum Hauck (1883–85), p. 403.

Thallus encrusting, thin and light brown to moderately thick and black, firm, smooth, subcoriaceous, circular and discrete or irregular and confluent to 3 cm or more in extent, very firmly attached to the substratum by the undersurface, usually without rhizoids, surface cells polygonal, closely packed and irregularly arranged, 6–13 × 5–9 μm containing up to 10 discoid, peripherally placed plastids; thallus structure pseudoparenchymatous, consisting of a monostromatic discoid base giving rise to erect rows of little-branched, firmly adjoined filaments, to 30 cells, 190 μm long, enclosed by a thick surface cuticle, rarely with terminal, long (to 200 μm) unicellular, colourless, cylindrical ascocysts present; erect filament cells usually subquadrate rarely rectangular, $\frac{1}{4}$–$2\frac{1}{2}$ diameters long, 4–12 × 4–6 μm with 3–6 (–10) small discoid plastids, without obvious pyrenoids; hairs not observed.

Plurilocular sporangia terminal, without paraphyses, in slightly raised, mucilaginous sori, extensive on crust surface, in surface view rounded or irregular in shape, with 4, rarely 2, loculi, in vertical section clavate to slightly pyriform in shape, bi- to multiseriate, 28–45 × 11–18 μm, to 8 loculi long, with straight dividing walls and covered by a thick surface cuticle; plurilocular sporangial initials look like paraphyses or unilocular sporangia, darkly coloured, elongate to slightly pyriform, 15–24 × 6–10 μm; unilocular sporangia unknown.

Fig. 2 *Pseudolithoderma extensum*
A. Surface view of vegetative crust. B–C. V.S. of vegetative crusts showing closely adjoined erect filaments. D. A single erect filament showing cells with 1–4 discoid plastids. E–F. V.S. of crusts with terminal immature plurilocular sporangia. G–H. V.S. of crusts with terminal plurilocular sporangia. I. V.S. of crust with terminal unilocular sporangia. Bar = 20 μm (D), 50 μm (others).

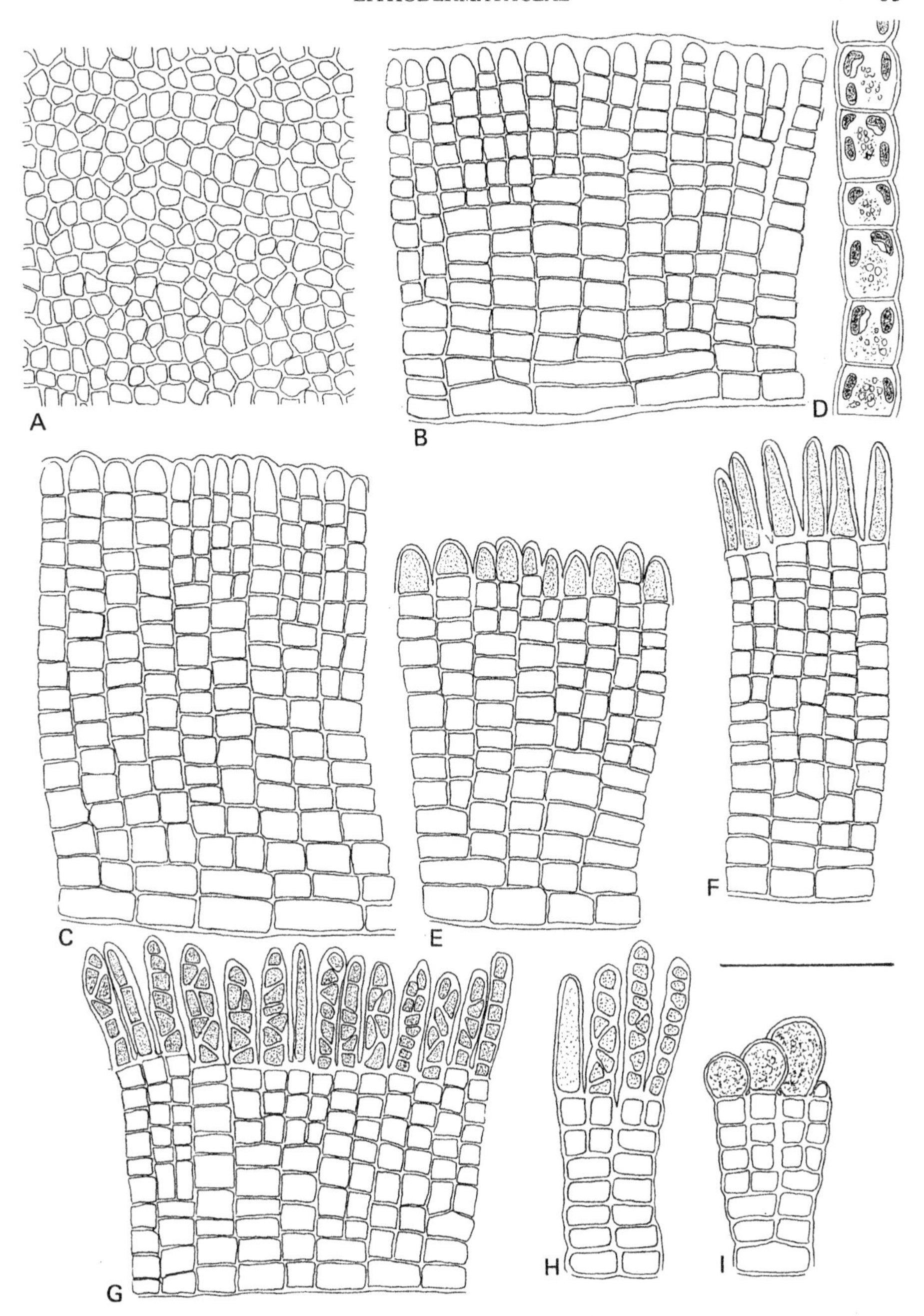
A
B
C
D
E
F
G
H
I

Epilithic in the lower eulittoral, covering upper (and frequently lower) surface of small flint stones in pools and exposed, in open sheltered places.

Apparently restricted to the south coast of England (Sussex, Hampshire, Dorset, Devon and Cornwall) and east coast of Ireland (Dublin).

Plurilocular sporangia have been recorded throughout the year on the south coast of England.

This alga was first described by Loiseaux (1968) based on material collected from Roscoff in Brittany. It was first discovered in the British Isles at Bembridge, Isle of Wight in 1973 (Fletcher unpublished) and subsequently reported in Parke & Dixon (1976). It possibly represents a recent addition to the south coast (as is likely in the case of *Zanardinia prototypus*) although because of its restricted distribution it may merely have been overlooked. It is a littoral species and can be distinguished from the other *Pseudolithoderma* species (*P. extensum*) in having the plurilocular sporangia grouped in two or four vertical columns rather than one and in having straight cross walls rather than oblique. Furthermore, unilocular sporangia have not been described to date. Since its initial discovery it has been found to be very common in the Solent region of the south coast of England, where it dominates the exposed, flint stones in the lower eulittoral in the sheltered muddy harbours of Langstone and Portsmouth. It has subsequently been found at other isolated localities in Sussex, Hampshire, Dorset, Devon and Cornwall and more recently on stones collected from County Dublin in Ireland (Fletcher & Maggs, 1985). Despite regular field searches it has not been found on the Kent coast. It appears, therefore, to be restricted to south west shores, although it is likely to occur more extensively on the west coast.

Laboratory culture studies have been carried out on this species by Loiseaux (1968) and Fletcher (1978). Both authors reported the plurispores to germinate directly, without sexual fusion, to repeat the parental crustose thallus; Loiseaux additionally reported the development of plurilocular sporangia on the crusts. These studies indicate that the life history is of the 'direct' type.

The above description of *P. roscoffense* appears to show a marked similarity to that of *Lithoderma adriaticum* Hauck given by Hamel (1931–39, p. xxxi, fig. 62–19). This species, originally described for the Adriatic Sea and widely distributed in the Mediterranean, was also reported by Hamel (1931–39, p. xxxi & 111) for the North Atlantic (Tatihou and Iles Chausey). It seems possible, therefore, that *Lithoderma adriaticum* Hauck *sensu* Hamel is conspecific with *Pseudolithoderma roscoffense* Loiseaux, although such a connection was not discussed by Loiseaux (1968).

Further reference must also be made to Mediterranean material of *Lithoderma adriaticum*, particularly the type material, to determine the possible relationship with *Pseudolithoderma roscoffense*.

? Hamel (1931–39, p. xxxi, 111, fig. 62(19); Loiseaux (1968), pp. 308–311, fig. 6; Fletcher (1978), pp. 371–398, figs 18, 19 & 28; Fletcher & Maggs (1985), pp. 523–526, fig. 2.

Fig. 3 *Pseudolithoderma roscoffense*
A. Surface view of vegetative crust. B. Surface view of crust cells with several discoid plastids. C. V.S. of crust margin. D–F. V.S. of vegetative crusts. G. Portion of two erect filaments showing enclosed discoid plastids. H. V.S. of crust showing terminal immature plurilocular sporangia. I. Surface view of fertile crust showing plurilocular sporangia. J. V.S. of fertile crust showing terminal plurilocular sporangia. Bar = 20 µm (B, G), 50 µm (others).

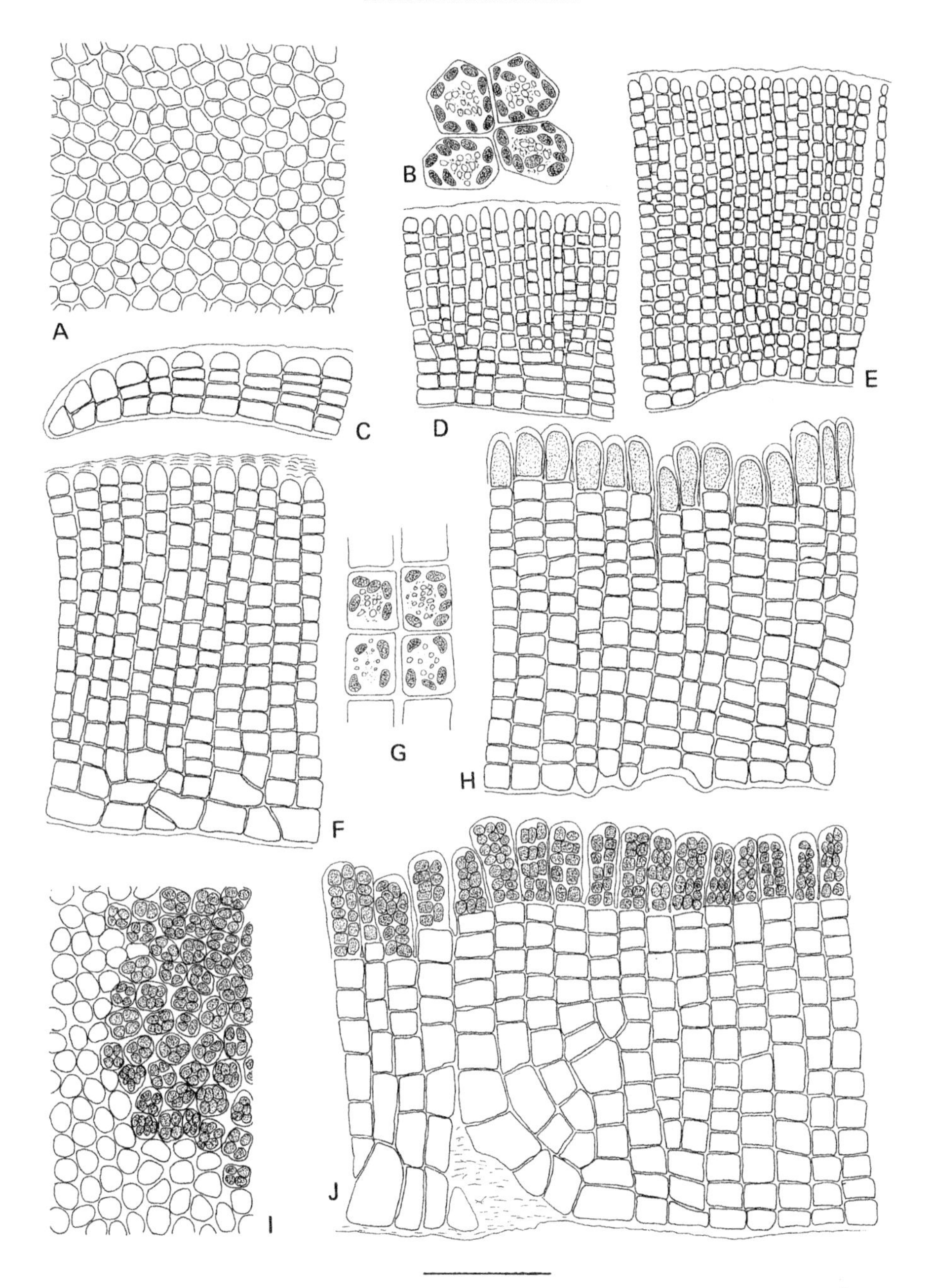
A
B
C
D
E
F
G
H
I
J

SORAPION Kuckuck

SORAPION Kuckuck (1894), p. 236.

Type species: *S. simulans* Kuckuck (1894), p. 236.

Thallus encrusting, thin, orbicular or indefinite in outline, closely adherent to substratum; thallus structure pseudoparenchymatous, consisting of a monostromatic basal layer, the cells of which give rise to simple or little-branched erect filaments, quite firmly united; vegetative cells with a single, plate-like, parietal plastid and pyrenoid; hairs not observed on British Isles material.

Unilocular sporangia terminal on erect filaments, in sori, unaccompanied by paraphyses; plurilocular sporangia unknown

One species in the British Isles:

Sorapion simulans Kuckuck (1894), p. 236. Fig. 4

Lithoderma simulans (Kuckuck) Batters (1896), p. 385.

Thallus crustose, to 1–2 mm in diameter, light brown, thin, firmly attached to substratum by whole undersurface, rhizoids not observed; in central, thicker regions, surface cells polygonal, closely packed, 9–12 × 7–11 µm, each containing a single plate-like parietal plastid with pyrenoid, in peripheral regions more obviously one cell thick, consisting of outwardly spreading branched, firmly united filaments of cells which are mainly rectangular, 7–19 × 6–9 µm; thallus structure pseudoparenchymatous, in vertical section the monostromatic base giving rise to straight, vertical, tightly joined filaments, to 52 µm long (6 cells), comprising cells quadrate or more often, especially at base, broader than high, 5–10 × 8–19 µm, each with a single plate-like plastid and pyrenoid in upper cell region; hairs not observed.

Unilocular sporangia terminal, in small groups, approximately rounded in surface view, 13–20 µm in diameter, slightly raised above vegetative cells, more distinctly pyriform in section, 20–27 × 15–18 µm; plurilocular sporangia unknown.

Epilithic on stones, in the sublittoral.
Only recorded from South Devon.
Insufficient data to comment on seasonal distribution.

British material of this species is represented by a single specimen collected by George Brebner in Plymouth Sound, Devon and reported by Batters (1896). Three slides are present in the British Museum (Natural History) from which the above description and illustrations have been made.

Fig. 4 *Sorapion simulans*
A. Surface view of vegetative crust. B. Surface view of crust margin showing outwardly spreading, laterally united filaments with terminal apical cells. C. Surface view of fertile crust showing unilocular sporangia. D. V.S. of vegetative crust showing cells containing a single, large, plate-like plastid with occasional pyrenoid. E. V.S. of fertile crust showing large, terminal pyriform, unilocular sporangia. Note associated 2-celled filament. Bar = 50 µm.

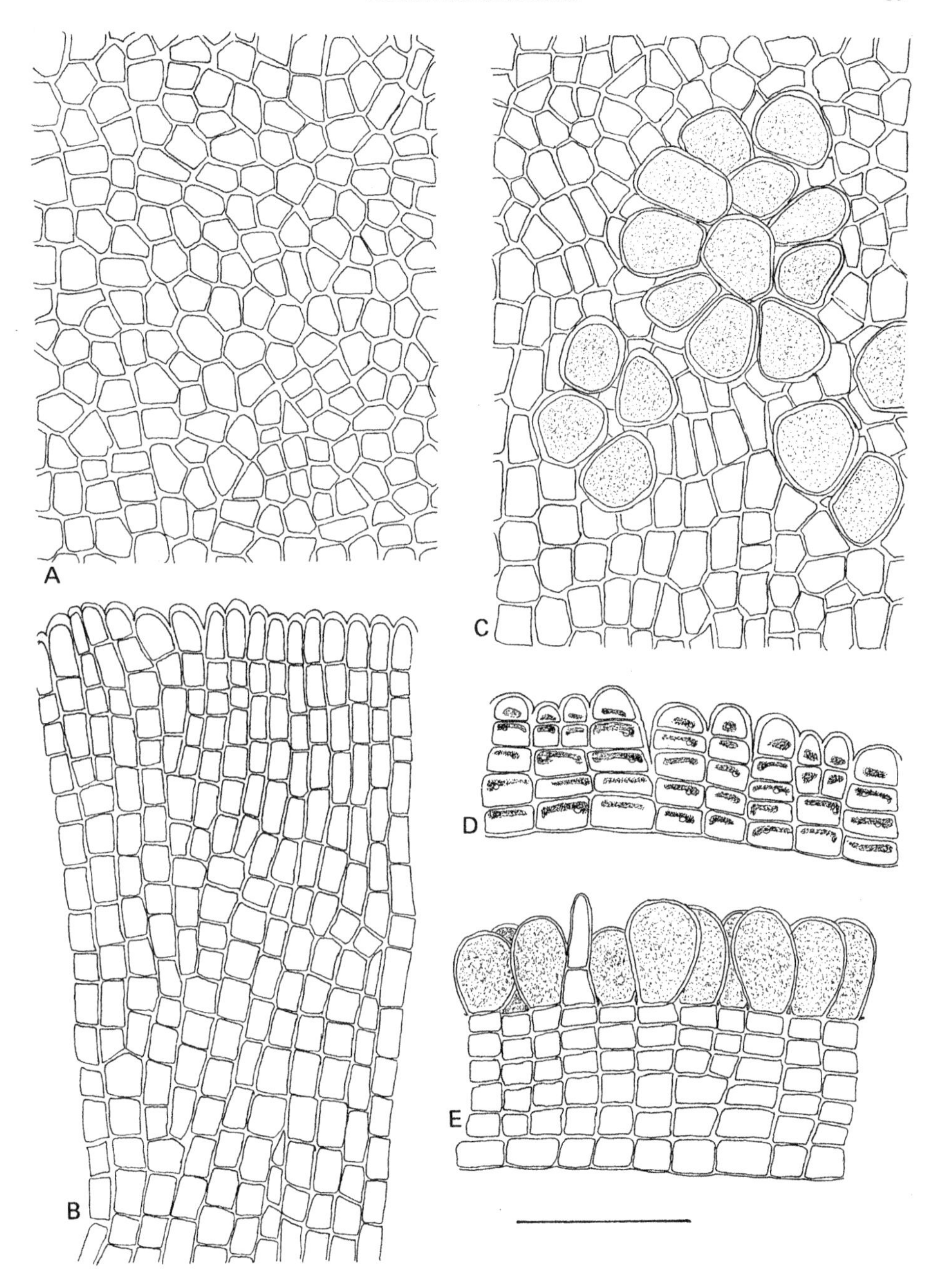
A
B
C
D
E

The characteristic features of this species (erect filaments of cells with a single plate-like plastid and terminal, pyriform unilocular sporangia) are not unlike those of an early stage in the development of another crustose alga, *Stragularia spongiocarpa* (Batters) Hamel and it was speculated that the two might, therefore, be conspecific (Fletcher, 1981*b* reported earlier in Parke & Dixon, 1976, p. 563 note 20 as *Ralfsia spongiocarpa*). However, subsequent detailed examination of material of both species reveals this to be unlikely and they are, therefore, described as separate entities in the present treatment.

Another species closely related to *S. simulans* which might also be present around the British Isles is *Sorapion kjellmanii* (Wille) Rosenvinge (1898, p. 95). It has been widely reported in the North Atlantic from countries such as America (Taylor, 1957; this was, however probably *S. simulans*), Canada (Sears & Wilce, 1973; South & Hooper, 1980), Greenland (Rosenvinge, 1898; Lund, 1959; Pedersen, 1976), Russia (Wille & Rosenvinge, 1885), the Faeroes (Børgesen, 1902), Sweden (Waern, 1949) and Denmark (Kristiansen, 1978). It differs from *S. simulans* in that the plate-like plastid lacks a pyrenoid, the erect filaments are branched towards the apex, hairs are present (placed in hollows) and the unilocular sori are less well defined. However, as Rosenvinge (1898) indicated, these points of difference are not great and the two species might well be conspecific.

Pedersen (1981*b*) linked *Sorapion kjellmanii* as an alternate phase in the life history of the relatively little known brown alga *Porterinema fluviatile* (Porter) Waern. Unispores from the unilocular sporangia in the *Sorapion* plants germinated directly without sexual fusion, to form new crustose *Porterinema*-like plants with plurilocular sporangia. Plurispores from the latter also behaved asexually and germinated directly to produce plants with plurilocular sporangia identical with the parents. The absence of cytological data, however, prevented a more precise understanding of the relationship between the two crustose phases.

Kuckuck (1894), pp. 236–237, fig. 10; Batters (1896), pp. 385–386; Rosenvinge (1898), pp. 95–97; Newton (1931), p. 127; Hamel (1931–39), pp. 111–112, fig. 26F; Waern (1949), pp. 662–663; Waern (1952), pp. 136–141; Lund (1959), pp. 79–81, fig. 14; Pedersen (1981*b*), pp. 203–208, figs 1–9.

SYMPHYOCARPUS Rosenvinge

SYMPHYOCARPUS Rosenvinge (1893), p. 896.

Type species: *S. strangulans* Rosenvinge (1893), p. 896.

Thallus epiphytic, epizoic, encrusting, pseudoparenchymatous consisting of a monostromatic basal layer, the cells of which give rise to loosely united and gelatinous erect filaments; cells with a single plate-like plastid and pyrenoid; ascocyst-like cells common arising from basal layer or terminating erect filaments.

Plurilocular sporangia terminal on erect filaments, multiseriate, clustered.

One species in the British Isles:

Fig. 5 *Symphyocarpus strangulans*
A. Surface view of vegetative crust showing scattered, hyaline ascocysts. B–D. V.S. of vegetative crusts showing erect filaments, large, hyaline ascocysts and cells with a single plate-like plastid. E. Surface view of fertile crust showing ascocysts and plurilocular sporangia with 2–4 loculi. F. V.S. of fertile crusts showing erect filaments, ascocysts and terminal plurilocular sporangia. Bar = 50 μm.

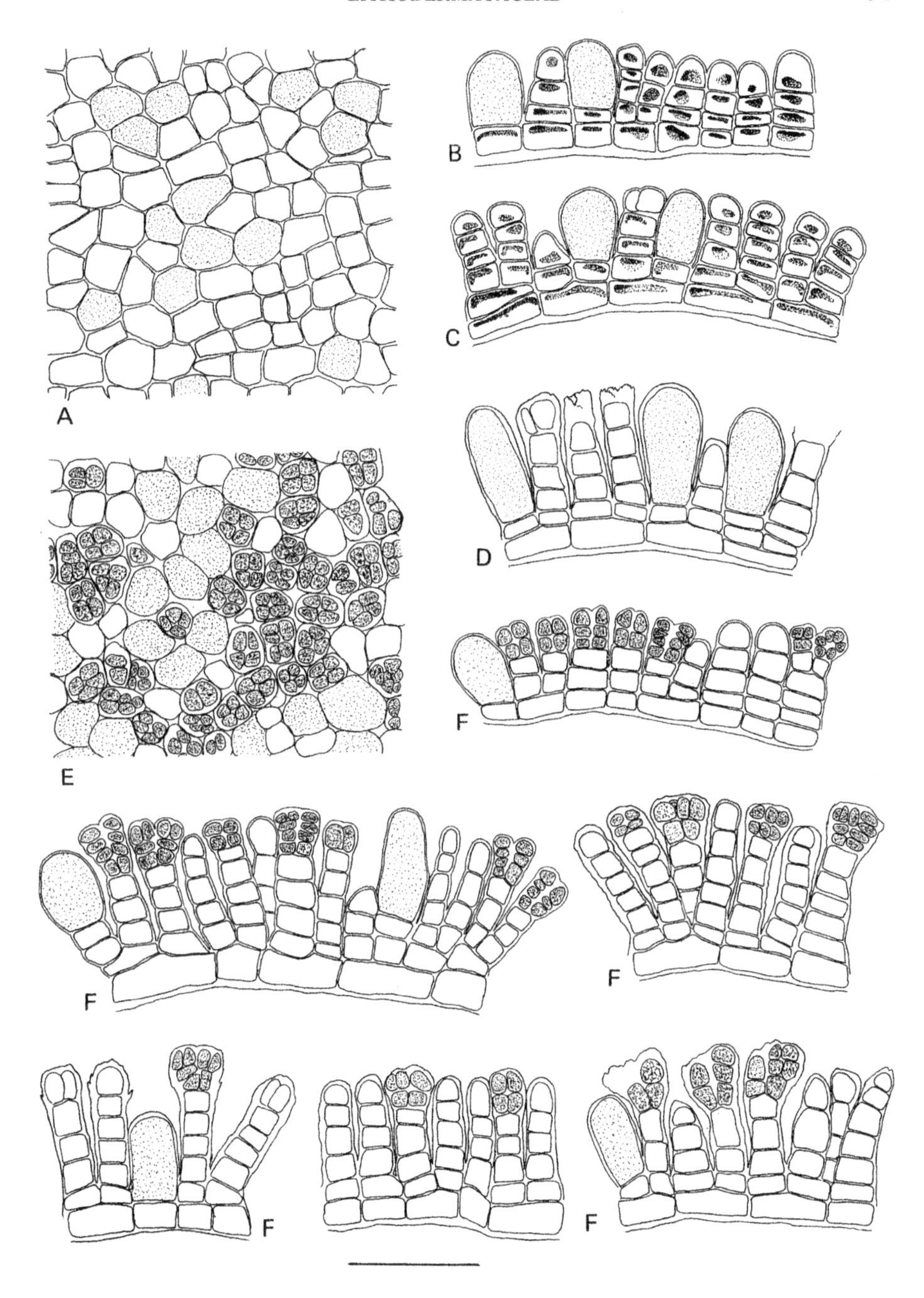
A
B
C
D
E
F
F
F
F
F

Symphyocarpus strangulans Rosenvinge (1893), p. 896. Fig. 5

Thallus forming small, thin, closely adherent crusts, 1–5 mm in diameter, light brown in colour; thallus structure pseudoparenchymatous, comprising a monostromatic base of laterally adjoined, firmly attached, outwardly spreading filaments of cells, with or without downwardly growing rhizoids, giving rise to numerous compacted, erect filaments; in surface view peripheral vegetative cells rectangular, 11–24 × 7–16 μm, central vegetative cells irregular or rounded 11–22 × 7–17 μm, all cells with a single, large plate-like plastid, without obvious pyrenoid; in vertical section erect filaments straight, little branched, loosely united and easily separable under pressure, slightly gelatinous in texture, to 11 cells, 85 μm long, basal cells rectangular in section usually much wider than high 10–16 × 15–37 μm, upper cells usually subquadrate or quadrate, rarely rectangular, 4–21 × 9–20 μm, each with a large, single, plate-like plastid occupying upper cell region; hairs infrequent, usually terminal on erect filaments, *c.* 8 μm in diameter, with basal meristem but lacking obvious sheath; ascocysts common, either colourless or dark yellow pigmented, borne either on the basal cells or terminating erect filaments, in surface view, rounded in shape 12–32 × 18–25 μm, discrete or in sori, in section elongated and cylindrical or slightly clavate, 23–52 × 16–26 μm.

Plurilocular sporangia common, usually in sori, terminal on erect filaments; in surface view rounded or irregular in shape, slightly raised above vegetative cells, often with 4 loculi, sometimes multilocular, 14–27 × 15–24 μm; in section multiseriate, arranged in 2 or 4 vertical columns, rarely irregularly clustered, 4–7 loculi, 14–40 × 17–24 μm, loculi 4–5 × 8–11 μm; unilocular sporangia unknown.

Epiphytic on *Laminaria* and maerl, also epizoic on dogfish cases, lower eulittoral and sublittoral to 20 m depth.

Only known from Berwick, Argyll and the Outer Hebrides; in Ireland, Galway.

Data inadequate to comment on seasonal behaviour.

Rosenvinge (1893), p. 896, figs 28–29; Batters (1895*a*), p. 275; Newton (1931), p. 126, fig. 74; Hamel (1931–39), p. XXX, fig. 62 (12–15); Lund (1959), pp. 70–73, fig. 11; Jaasund (1963), pp. 6–7, fig. 4; Jaasund (1965), p. 61; Pedersen (1976), pp. 39–40, pl. 4 d–f; Rueness (1977), p. 130.

MYRIONEMATACEAE Nägeli

MYRIONEMATACEAE Nägeli (1847), pp. 145, 252 [as Myrionemeae]

Thallus epiphytic or more rarely epilithic, forming small discs, hemispherical cushions or tufts, gelatinous or more rarely fibrous in texture; thallus structure pseudoparenchymatous, comprising a prostrate system of outwardly radiating, pseudodichotomously branched, monostromatic filaments, either free and diffuse or laterally adjoined, irregularly into pseudodiscs or regularly into discs; basal cells giving rise above to 1 (–2) erect filaments, hairs, ascocyst-like cells and/or sporangia; erect filaments short, of equal length, simple or branched; growth apical, cells with 1–4 plate-like, often lobed, plastids with pyrenoids; hairs sessile or stalked, on basal or erect filament cells, with basal meristem and sheath; plurilocular and/or unilocular sporangia present, sessile or stalked, on basal layer or terminal/lateral on erect filaments; plurilocular sporangia common, uniseriate to multiseriate; unilocular sporangia uncommon, usually lateral at the base of multicellular paraphysis-like filaments.

This is a poorly circumscribed family and needs urgent attention. Members basically consist of small epiphytes, comprising a basal layer of outwardly radiating, laterally adjoined filaments forming discs/pseudodiscs, most cells of which give rise to erect, short, primarily simple, loosely associated and often gelatinous filaments, sporangia, true hairs and/or ascocysts. As Setchell and Gardner (1925 p. 453) observed, the family is positioned between the Ectocarpaceae and the Elachistaceae. Members differ from the Ectocarpaceae in having a discoid base, relatively short, erect, unbranched filaments, sporangia mainly associated with the basal layer, and true hairs. Their minute size and the undifferentiated nature of the erect filaments, on the other hand, distinguish them from the more elaborate pustules and cushions of the members of the two families Elachistaceae and Corynophlaeaceae. Further, their epiphytic habit, loosely coalescent discoid base and erect filaments distinguish them from members of the Lithodermataceae and encrusting Scytosiphonaceae. However, clear demarcation of this family is difficult and some workers have recognised its close affinity with these other families. For example Hamel (1931–39) postulated a broader concept of the family Myrionemataceae to include three tribes, the 'Streblonemées' (members of which are usually retained in the Ectocarpaceae *sensu stricto*), the 'Ralfsiées' containing the crustose species usually placed in the Lithodermataceae and Ralfsiaceae, and the 'Myrionemées' containing the smaller epiphytes usually placed in the Myrionemataceae *sensu stricto*. A broad concept of the Myrionemataceae was also adopted by Loiseaux (1967*a*), in which three tribes were recognised, the Myrionemées, Hecatonemées and Ralfsiées; the Streblonemées were not included.

The poor circumscription of the family has resulted over the years in the inclusion of a rather mixed bag of entities, many of which would be more suitably placed elsewhere. Genera included in the family by Parke & Dixon (1976) are *Chilionema* Sauvageau, *Compsonema* Kuckuck, *Hecatonema* Sauvageau, *Leptonematella* Silva, *Pleurocladia* A. Braun, *Protectocarpus* Kuckuck, *Microspongium* Reinke, *Myrionema* Greville and *Ulonema* Foslie. In the present work only the genera *Compsonema, Microspongium, Myrionema, Protectocarpus* and *Ulonema* are included in the family Myrionemataceae. Two species of *Compsonema* are described, *C. microspongium* and the recently discovered *C. minutum; C. saxicolum,* included in Parke & Dixon (1976), has been transferred to the Scytosiphonaceae on the basis of structural and life history features (see p. 221). However, the taxonomic position of *Compsonema* is difficult to establish. The epilithic habit, the absence of a basal disc (in British material), the occurrence of quite well developed, erect, branched filaments and the development of ectocarpoid plurilocular sporangia on the erect filaments rather than the basal layer, are all features more characteristic of members of the Ectocarpaceae, the family in which it was originally placed (Kuckuck, 1953). Its transfer to the Myrionemataceae by Loiseaux (1967a) is, therefore, only provisionally accepted here pending further studies.

Only one species of *Microspongium, M. globosum,* is included in the Myrionemataceae in the present work. The species *M. gelatinosum,* included in Parke & Dixon (1976), has been transferred to the Scytosiphonaceae as a result of culture studies (see p. 224). The genus is quite closely related to *Myrionema* and *Ulonema*, both of which are included here. However, the status of *Myrionema* as a natural phylogenetic unit has, more recently, been questioned by Pedersen (1984) in view of *Myrionema*-like microthalli occurring in the life histories of some members of the families Chordariaceae and Giraudiaceae.

Finally, *Protectocarpus* is included, although with some reservation. Originally described by Kuckuck (1955) and included in the Ectocarpaceae, it was removed to the Myrionemataceae by Loiseaux (1967*a*), a position accepted by a number of authors (e.g. Parke & Dixon, 1968, 1976; Kornmann & Sahling, 1977; Rueness, 1977). The occurrence of a discoid base and short, little-branched, erect filaments with both terminal and lateral plurilocular sporangia are certainly features which closely ally it with other members of the Myrionemataceae.

Genera included in the Myrionemataceae by Parke & Dixon (1976) but excluded here are *Chilionema, Hecatonema, Leptonematella* and *Pleurocladia*. The two genera *Chilionema* and *Hecatonema* have been transferred to the Punctariaceae, on the basis of life history studies (see p. 178) whilst *Leptonematella* has been transferred to the Elachistaceae (see p. 136). The British record of *Pleurocladia* is based on a reported finding of *Pleurocladia lacustris* (as *Pilinia maritima*) by Anand (1937) on the Kent coast. It was also more recently reported for the Kent coast by Tittley & Shaw (1980) (British Museum, Nat. Hist. slide no. 1892). However, the occurrence of a branched, filamentous base, rather than a discoid/pseudodiscoid one and the absence of true sheathed hairs with distinct basal meristem are features which suggest it is more suitably placed in the Ectocarpaceae rather than the Myrionemataceae, in agreement with Wilce (1966). It is, therefore, removed from the latter family in the present work, pending further studies.

COMPSONEMA Kuckuck

COMPSONEMA Kuckuck (1899*a*), p. 58.

Type species: *C. gracile* Kuckuck (1899*a*), p. 58 (= *C. minutum* (Agardh) Kuckuck)).

Thallus epiphytic or epilithic, forming either small hemispherical cushions or dense pile-like tufts; comprising a prostrate system of branched, outwardly radiating, monostromatic filaments, usually remaining free although sometimes reputed to be adjoined laterally into discs, giving rise to short, erect, simple or little-branched, gelatinous, later fibrous filaments usually of equal length; plastids single, plate-like and lobed with pyrenoids; hairs with basal meristem and sheath. Plurilocular and unilocular sporangia lateral or terminal on erect filaments, more rarely borne directly on basal filaments; plurilocular sporangia multiseriate.

The genus *Compsonema*, established by Kuckuck in 1899*a*, is cosmopolitan and characterised by the following combination of features: a monostromatic basal layer of branched, outwardly spreading filaments, which can either be free or laterally united into a discoid structure; fairly short, simple or little branched, free, erect filaments; cells with one or two plate-like plastids; laterally situated sporangia, which in the case of the plurilocular sporangia are multiseriate. With the exception of Kuckuck (1953), Parke & Dixon (1964), Jaasund (1965) and Dangeard (1970) who placed the genus in the Ectocarpaceae, it is usually placed in the Myrionemataceae. It resembles the other Myrionematoid genera *Myrionema* and *Microspongium* in having a monostromatic base and the genera *Chilionema* and *Hecatonema* in having multiseriate plurilocular sporangia.

However, in agreement with Abbott and Hollenberg (1976) it is a 'poorly defined genus' and has, over the years, included quite a disparate group of entities, such as species for which only unilocular sporangia are known, species with sporangia which are borne on the basal layer rather than positioned on the erect filaments, species with variable numbers and shapes of plastids and, more significantly, species which have been reported with a distromatic rather than a monostromatic base.

Particularly pertinent, however, to any discussion about the status of the genus *Compsonema*, have been the results of some life history studies. In three investigations, a direct, monophasic life history has been shown. Fletcher (unpublished) and Dangeard (1970) observed a repetition of the parental phase via plurispores in both *C. microspongium* (collected from the south coast of England) and *C. minutum* (collected from the Bay of Biscay) respectively whilst a similar life history via unispores was reported by Pedersen (1981c) for *C. saxicolum* (collected from Denmark). In two other investigations, however, *Compsonema* species have been implicated as phases, probably diploid, in the life histories of members of the erect-bladed family Scytosiphonaceae. Loiseaux (1970*b*) cultured the Californian species *C. sporangiferum* Setchell & Gardner and produced a small, erect, *Scytosiphon*-like blade, which she tentatively identified as *S. pygmaeus* Reinke, while more recently there have been reports of *C. saxicolum* collected from the south coast of England occurring as a phase in the life histories of *Petalonia* and/or *Scytosiphon* species (Fletcher, 1981a, and unpublished).

The occurrence of the above described heteromorphic life history does question the status of the genus *Compsonema* and the extent of its involvement with Scytosiphonaceae members. The indications are, however, that the revealed heteromorphic life histories may not be typical of members of this genus. Both *C. saxicolum* and the closely related *C. sporangiferum* appear to possess a unique combination of characters i.e. a single plate-like plastid with large pyrenoid, partly distromatic base and basally positioned unilocular sporangia. Indeed out of 18 species of *Compsonema* described for California by Setchell & Gardner (1925) only *C. sporangiferum* possessed a distromatic base. On the basis of these results *C. saxicolum* is transferred to the Scytosiphonaceae, following the proposal of Pedersen (1981*c*) (see p. 221), while *Compsonema* is retained as an autonomous genus within the Myrionemataceae pending further studies.

Two species are recognised for the British Isles: *C. microspongium* and the newly recorded *C. minutum*.

KEY TO SPECIES

Thallus epiphytic, forming hemispherical cushions up to 1 mm high on *Ralfsia verrucosa*, in late summer, autumn . . . . *C. microspongium*

Thallus epilithic, forming confluent, pile-like tufts, up to 0·5 mm high, in winter . . . . *C. minutum*

Compsonema microspongium (Batters) Kuckuck (1953), p. 347. Fig. 6

Ectocarpus microspongium Batters (1897), p. 436.

Thallus epiphytic, forming small, yellow-brown, hemispherical cushions on host, usually solitary and rounded, 1–4 mm wide to 1 mm high, occasionally oblong or confluent

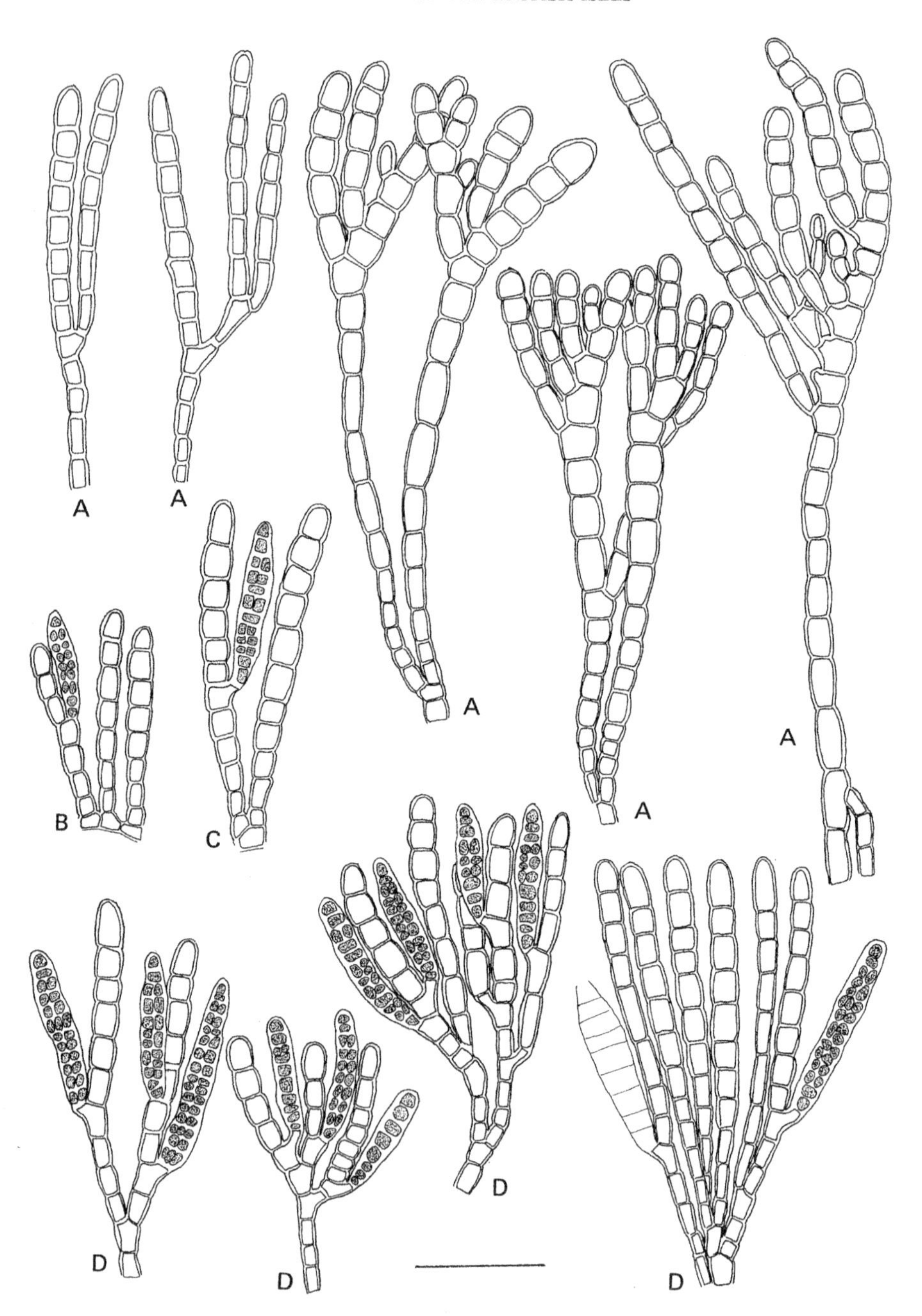
A
A
A
A
A
A
B
C
D
D
D
D

and netlike; consisting of an endophytic base of branched outwardly spreading, monostromatic filaments ramifying through host tissue, giving rise to numerous erect filaments; emergent filaments free, loosely aggregated, slightly gelatinous at first, later becoming more fibrous in texture, frequently pseudodichotomously branched; lower branches of older thalli spreading, matted and rhizoid-like, comprising cells irregularly nodose, 2–4 diameters long, 12–32 × 5–11 μm, upper branches erect, often secund, comprising quadrate to rectangular cells, 1 (–2) diameters long, 8–18 × 9–16 μm; plastids single, plate-like and lobed, with pyrenoids, usually occupying upper cell region; hairs common, terminal on erect filaments with basal meristem and sheath.

Plurilocular sporangia common on upper branches, bi- to multiseriate, solitary, or more often in short secund series, sessile or on 1–3 celled stalks, usually erect and closely adpressed to parental filament, oblong-lanceolate in shape, 65–90 × 10–20 μm; unilocular sporangia unknown.

Epiphytic, on crusts of *Ralfsia verrucosa,* in the upper eulittoral and littoral fringe.

Only recorded for the south west (south Cornwall and north and south Devon); in Ireland, Mayo. Probably more widely distributed.

Appears to be a summer annual, June to September, more rarely found in October, November.

Laboratory culture studies by Fletcher (unpublished) revealed the plurispores to behave asexually and germinate directly to repeat the fertile parental thallus. The life history appears, therefore, to be of the 'direct' type.

Batters (1897), p. 436–437; Newton (1931), p. 117; Kuckuck (1953), p. 347–350, fig. 15A–B.

Compsonema minutum (C. Agardh) Kuckuck (1953), p. 341. Fig. 7

Ectocarpus minutus C. Agardh (1827), p. 639.
Ectocarpus monocarpus C. Agardh (1820–28), p. 48.
Compsonema gracile Kuckuck (1899a), p. 56.

Thallus epilithic, forming small, light brown tufts to 0·5 mm high, densely crowded and pile-like in appearance, spreading to 5 cm or more in extent; consisting of a basal layer of outwardly spreading, irregularly branched, monosiphonous, free filaments, not obviously laterally adjoined and disc-like, giving rise to erect filaments, hairs and/or sporangia; in squash preparation basal layer cells rectangular or more commonly irregularly nodose, usually 1–2 diameters long, 9–13 × 5–9 μm, with erect portion arising from mid-cell wall region; erect filaments short, to 400 μm long (–35 cells), simple or with 1 (–3) branches, linear or slightly attenuate towards base, free or loosely aggregated in a mucilaginous matrix; branching pseudodichotomous, more rarely unilateral, to 1 (–2) orders, usually restricted to upper region, with branches strongly recurved towards main

Fig. 6 *Compsonema microspongium*
A. Terminal portions of erect, vegetative filaments. B–C. Short, erect filaments arising from a monostromatic base with lateral plurilocular sporangia. D. Terminal portions of erect, fertile filaments with lateral plurilocular sporangia. Bar = 50 μm.

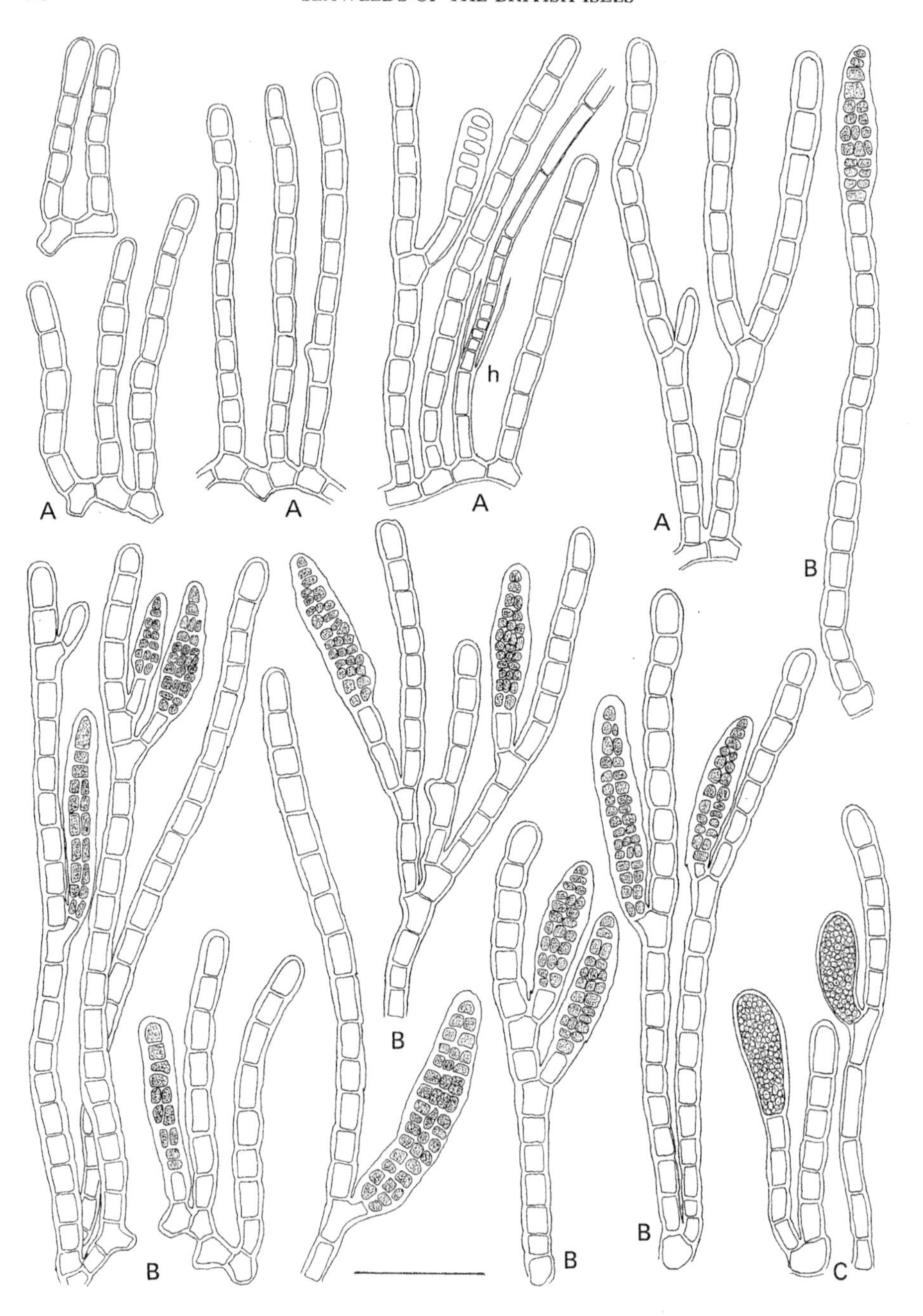
h
A
A
A
A
B
B
B
B
B
B
C

axis; erect filament cells rectangular, or slightly barrel-shaped, 1 (–2) diameters long, 9–25 × 9–16 μm, with thick hyaline walls and enclosing a single, irregularly lobed, plate-like plastid with pyrenoids; outer cell wall regions often with short, closely set, upwardly pointing, spine-like protuberances (remains of dehisced outer cell wall portion?); hairs common, usually terminal on erect filaments and branches, c. 6–7 μm in diameter, with basal meristem and sheath.

Sporangia borne on erect filaments, more rarely arising from basal cell layer; plurilocular sporangia common, usually lateral and midway on erect filaments, sessile or on 1–2 celled stalks and strongly recurved towards parental filament, more rarely terminal on erect filaments or arising from basal cell layers, lanceolate in shape, irregularly margined, bi- to multiseriate, (50–) 65–100 × 15–21 μm; unilocular sporangia uncommon, usually associated with the plurilocular sporangia, lateral or terminal on erect filaments, elongate-pyriform in shape, 40–45 × 15–20 μm.

Epilithic, on bedrock in the littoral fringe.

Only recorded for Hampshire (Isle of Wight); probably more widely distributed on the south coast of England.

Collections made in December and January; insufficient data to comment on seasonal distribution.

Laboratory culture studies by Dangeard (1970) on *C. minutum* collected at Mimizan, in the Bay of Biscay, revealed the occurrence of a 'direct' type life history with the plurispores behaving asexually and germinating directly to repeat the fertile parental thallus.

Kuckuck (1899a), pp. 56–60, figs 6–9; Schiffner (1916), p. 154, figs 70–74; Børgesen (1926), pp. 59–62, fig. 31; Kuckuck (1953), pp. 341–343, fig. 13; Dangeard (1970), pp. 63–65, figs. 1–13, pls. 1–2.

MICROSPONGIUM Reinke

MICROSPONGIUM Reinke (1888), p. 20.

Type species: *M. gelatinosum* Reinke (1888), p. 20.

Thallus epiphytic, crustose, pulvinate or globular, sponge-like and gelatinous in texture, consisting of a monostromatic base of outwardly spreading filaments, laterally united and discoid, giving rise to short, erect, densely crowded, branched filaments, easily separable under pressure; cells with 1–3 plate-like plastids; plurilocular sporangia terminal or lateral on erect filaments, uniseriate, sometimes part biseriate; unilocular sporangia unknown.

Fig. 7 *Compsonema minutum*
A. Erect, vegetative filaments arising from a monostromatic base. Note terminal hair with sheath (h). B. Erect, fertile filaments with terminal and lateral plurilocular sporangia. C. Erect, fertile filaments with terminal and lateral unilocular sporangia. Bar = 50 μm.

The genus *Microspongium* was established by Reinke (1888) based on material collected in the western Baltic; two species were described, *M. globosum* and *M. gelatinosum*. Only plurilocular sporangia were observed; unilocular sporangia were stated as unknown. However, later (Reinke, 1889a, b) a more detailed description and illustration of these two species was given, with unilocular sporangia described for *M. gelatinosum*. It now seems likely that three entities are involved within Reinke's concept of the two species (see Fletcher, 1974a and Kristiansen and Pedersen, 1979 for discussion on this subject). These are:

1. Material of *M. globosum* described by Reinke as epiphytic on *Cladophora gracilis, Polysiphonia nigrescens* and *Zostera*. Characteristics include a monostromatic base, branched erect filaments, terminal and lateral, uniseriate plurilocular sporangia and the absence of unilocular sporangia.
2. Material of *M. gelatinosum* described Reinke as epiphytic on *Fucus vesiculosus* and *F. serratus*. Characteristics include a partly distromatic base, commonly branched erect filaments, 1–4 plastids per cell, lateral uniseriate plurilocular sporangia and the absence of unilocular sporangia.
3. Material of *M. gelatinosum* described by Reinke as epizoic on *Mytilus*. Characteristics include a monostromatic base, little-branched erect filaments with the basal cells closely associated, a single plastid per cell, lateral unilocular sporangia and the absence of plurilocular sporangia.

With respect to the two entities described under *M. gelatinosum*, these undoubtedly represent two different taxa and should, therefore, be separated, thus following the procedure of Kylin (1947), Fletcher (1974a) and Kristiansen and Pedersen (1979). To date only the *M. gelatinosum* material described with unilocular sporangia has been identified for the British Isles (but see note on *M. globosum*) and is thus considered in the present work. It has been included in the Scytosiphonaceae on the basis of its plastid number, close similarity with species of *Stragularia* and reports of its occurrence as a phase in the life history of the erect tubular alga *Scytosiphon lomentaria* (see p. 224).

With respect to *M. globosum*, although this does show some similarity with the plurilocular sporangia bearing material of *M. gelatinosum* (designated the lectotype of the genus) it does differ in one important characteristic—the absence of a partly distromatic base. Its continued placement in the genus *Microspongium* should possibly, therefore, be questioned. The uncertainty of its systematic position has undoubtedly been reflected by the numerous genera into which it has been placed, including *Ascocyclus* (by Reinke, 1889a), *Myrionema* (by Foslie, 1894 and Taylor, 1957), *Phycocelis* (by Rosenvinge, 1898) and *Hecatonema* (by Batters, 1902). However, pending further studies it is herein retained within the genus *Microspongium* in agreement with various authors such as Kylin (1947), Lund (1959), Parke & Dixon (1976) South & Hooper (1980) and Kornmann & Sahling (1983). The genus *Microspongium* is very closely related to *Myrionema*.

One species in the British Isles.

Fig. 8 *Microspongium globosum*
A. Surface view of peripheral thallus region showing branched outwardly spreading, closely associated vegetative filaments. B. Squash preparation (S.P.) of vegetative thallus showing erect filaments arising from a monostromatic base. Note occasional hair (h) and enlarged terminal cells. C. S.P. of fertile thallus showing more extensively branched, erect filaments and both terminal and lateral plurilocular sporangia. Bar = 20 μm. (top right hand figure only), 50 μm (others).

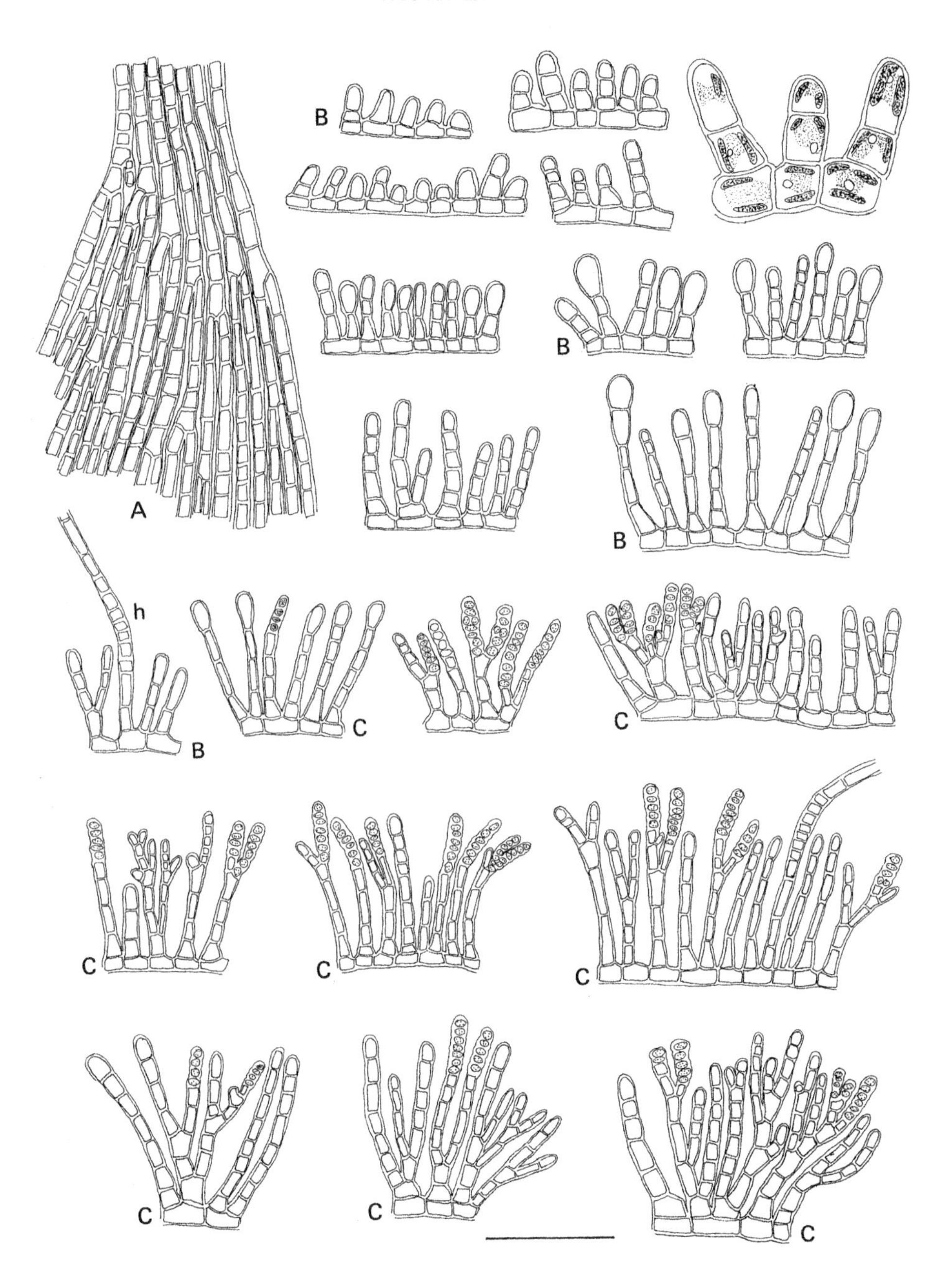
A
B
B
B
h
B
C
C
C
C
C
C
C
C

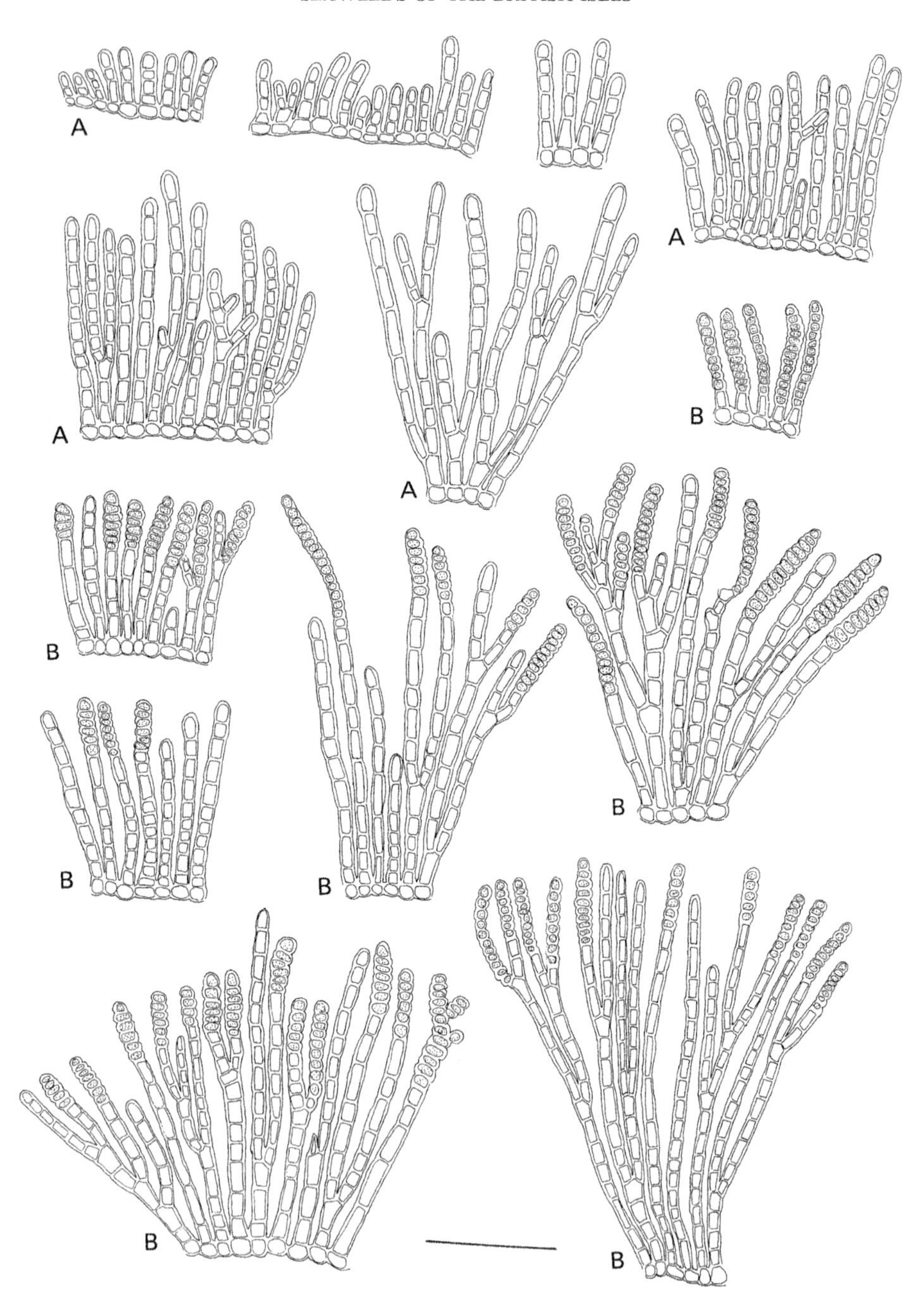
A
A
A
A
B
B
B
B
B
B
B

Microspongium globosum Reinke (1888), p. 20. Figs 8, 9

Ascocyclus globosus Reinke (1889a), p. 20.
Hecatonema globosum (Reinke) Batters (1902), p. 41.
Myrionema globosum (Reinke) Foslie (1894), p. 130.
Phycocelis globosus (Reinke) Rosenvinge (1898), p. 86.
Myrionema polycladum Sauvageau (1897a), p. 233.
Hecatonema globosum var. *nanum* Cotton (1907), p. 370.
Myrionema subglobosum Kylin (1907), p. 37.

Thallus epiphytic, forming light to dark brown spots on hosts, usually solitary, globular or circular, 2–3 (–4) mm in diameter, occasionally confluent, slightly pulvinate, smooth or radially ridged; mid region obviously thickened, surface cells rounded, densely packed, 6–11 μm in diameter, marginal region thin, of loosely adjoined, frequently branched, outwardly radiating filaments, comprising rectangular cells 1–2 diameters long, 10–16 × 8–10 μm, each with 1–3 plate-like multilobed plastids and associated pyrenoids; in squash preparation, consisting of a loosely attached, monostromatic base giving rise to erect filaments, hairs and/or plurilocular sporangia; erect filaments short, branched linear or sometimes distinctly clavate with enlarged, terminal, more darkly staining cell, to 75 μm (4–9 cells) long, mucilaginous comprising mainly rectangular cells, 1–2 diameters long, 5–15 × 5–7 μm (terminal cell to 22 × 12 μm) with 1–3 terminal plate-like, lobed plastids and pyrenoids; hairs frequent, arising from basal cells or terminal on erect filaments with basal meristem and sheath.

Plurilocular sporangia common, terminal or lateral on erect filaments, frequently formed within empty sporangial husks, uniseriate, subcylindrical, to 35 μm (–10 cells) long × 5–7 μm in diameter, with occasional oblique cross wall; unilocular sporangia unknown.

Epiphytic on various hosts including *Palmaria palmata*, *Chaetomorpha* spp., *Cladophora* spp., *Laminaria* spp. and *Fucus* spp., eulittoral in pools and sublittoral to 5 m.

Recorded for scattered localities around the British Isles (East Lothian, Northumberland, Hampshire, Dorset, Devon, Cheshire, Argyll; in Ireland, Mayo).

Summer annual; April to September.

Particularly noteworthy is the material of *M. globosum* observed on *Fucus* spp., (see Fig. 9) which has been collected from two localities:
(1) County Mayo, Ireland by A. D. Cotton in 1909; British Museum (Nat. Hist.) slide no's 2642–2644; on *F. vesiculosus* and
(2) Peveril Pt., Swanage, Dorset, by the present author, 21/4/85; on *F. serratus*. The squash preparations of the above material reveal long, erect, branched filaments arising from a monostromatic base and terminal plurilocular sporangia. In many respects this material is very similar to Reinke's *M. gelatinosum* described with plurilocular sporangia; however the latter was described by Reinke with a distromatic base and pending further studies it would be better to maintain these as separate entities. (But note that Rautenberg

Fig. 9 *Microspongium globosum*
A. S.P. of vegetative thallus showing erect, branched filaments. B. S.P. of fertile thallus showing terminal and lateral plurilocular sporangia. (Form typical of plants epiphytic on *Fucus* spp). Bar = 50 μm.

(1960) and Kylin (1947) described *M. gelatinosum* with a monostromatic base.) It is possible that Reinke was mistaken on this point but there would subsequently be no criterion left to distinguish the plurilocular sporangia bearing material of *M. globosum* and *M. gelatinosum*.

Reinke (1889a), p. 20, pl. 17; Reinke (1889b), p. 46; Batters (1892b), p. 21; Sauvageau (1897a), pp. 223–237, fig. 13; Rosenvinge (1898), pp. 86–89, figs. 19–20; Batters (1900), p. 371; Batters (1902), p. 41; Børgesen (1902), pp. 419–421, figs 76–77; Cotton (1907), pp. 369–370; Kylin (1907), pp. 37–38, fig. 8; Newton (1931), pp. 151, 156; Hamel (1931–39), pp. 92–93, 96; Kylin (1947), p. 39, fig. 31; Jaasund (1951), pp. 134–135, fig. 4a–e; Lund (1959), p. 111; Taylor (1957), pp. 154–155; Jaasund (1965), p. 60; Rueness (1977), p. 135, fig. 60; Kristiansen & Pedersen (1979), p. 55; Kornmann & Sahling (1983), p. 50, figs. 30–31.

MYRIONEMA Greville

MYRIONEMA Greville (1827), pl. 300.

Type species *M. strangulans* Greville (1827), pl. 300.

Thallus epiphytic or rarely epilithic, forming small circular, less frequently irregular spots on hosts; consisting of a basal prostrate layer of outwardly spreading, pseudodichotomously branched, monostromatic, filaments of cells, laterally adjoined and disc-like in appearance, often with a distinct marginal row of synchronously growing apical cells, less frequently irregularly associated and pseudodiscoid; basal cells without rhizoids, each giving rise, except near the periphery to 1 (–2) erect filaments, hairs, ascocysts and/or sporangia; erect filaments commonly short, simple, more rarely branched or with short protuberances, loosely compacted, often in a mucilaginous matrix, comprising cells with 1–3 plate-like plastids and pyrenoids; hairs, ascocysts and sporangia usually sessile or stalked on basal layer, less commonly terminal or lateral on erect filaments; hairs with basal meristem and sheath.

Plurilocular and unilocular sporangia arising from basal layer, sessile or shortly stalked, or terminal/lateral on erect filaments; plurilocular sporangia common, simple or rarely branched, uniseriate or rarely biseriate; unilocular sporangia uncommon, ovate or pyriform.

The genus *Myrionema* is characterised by a monostromatic discoid base, from the central cells of which arise short, unbranched filaments. Culture studies on a small number of species reveal this discoid morphology to originate in a 'stellate' spore germination pattern. Recently Pedersen (1984) has pointed out that such a similar 'stellate' germination pattern followed by disc formation are characteristic of the early developmental processes in some members of the families Chordariaceae and Giraudiaceae (see particularly Sauvageau, 1924; Shinmura, 1977; Lockhart, 1979). This suggests that *Myrionema* plants do not form a natural phylogenetic group and that at least some species may represent, or have represented, microthallial phases of various algae presently included in other diverse families. It will be interesting to await the results of further confirmatory work, particularly relating to life history and developmental patterns before the true taxonomic identities of these algae can be fully assessed.

Five species of *Myrionema* are recognised for the British Isles viz. *M. corunnae, M. liechtensternii, M. magnusii, M. papillosum* and *M. strangulans*. Two other species of

Myrionema recognised by Parke & Dixon (1976) i.e. *M. aecidioides* and *M. polycladum* are transferred to other genera. *M. aecidioides* has been transferred to *Gononema* in agreement with Pedersen (1981a) whilst *M. polycladum* is here included within the circumscription of *Microspongium globosum*. Out of the five species of *Myrionema* recognised here, *M. corunnae* appears anomalous and is only retained in this genus pending further studies. It differs from the other species of *Myrionema* in having a more irregularly structured, pseudodiscoid rather than discoid basal layer, more slender basal and erect filaments, hairs with reduced basal meristem lacking sheath, and the occurrence of plurilocular sporangia only. *M. liechtenstermii* is transferred from the genus *Hecatonema* and only retained here pending further studies; British Isles material of this species does come rather close in description to *Microspongium globosum*.

KEY TO SPECIES

1 Thallus epilithic . . . . *M. liechtensternii*
Thallus epiphytic 2
2 Thalli with ascocysts. . . . . *M. magnusii*
Thalli without ascocysts 3
3 Erect filaments with short, lateral protuberances . . . . . *M. papillosum*
Erect filaments without short, lateral protuberances 4
4 Erect filaments and plurilocular sporangia clavate; erect filaments usually $>$ 5 µm diameter; plurilocular sporangia uniseriate, frequently biseriate, up to 50 µm long; unilocular sporangia common; epiphytic mainly on species of *Ulva* and *Enteromorpha*, rarely on *Laminaria* blades . . . . *M. strangulans*
Erect filaments and plurilocular sporangia linear; erect filaments usually $<$ 5 µm diameter; plurilocular sporangia uniseriate rarely biseriate, up to 125 µm long; unilocular sporangia unknown; epiphytic on *Laminaria* blades only . . . . *M. corunnae*

Myrionema corunnae Sauvageau (1897a), p. 237. Fig. 10

Thallus epiphytic, forming small spots on hosts, usually solitary and circular, 0·5–1·0 mm across, occasionally confluent to 5 mm or more in extent; in surface view, peripheral region composed of a monostromatic, pseudodiscoid layer of branched, outwardly radiating, irregularly associated filaments comprising cells quadrate, rectangular or irregularly shaped, 1–3 diameters long, 5–13 µm × 4–6 µm, each with 1 (–3) plate-like plastids, outer central region composed of closely packed, rounded cells, 5–8 µm in diameter, inner central region turf-like, similar to a carpet pile; in squash preparations central basal cells usually 1 (–2) diameters long, 5–12 × 5–8 µm each giving rise to one, rarely two, erect filaments, hairs and/or plurilocular sporangia; erect filaments short, linear, loosely associated and gelatinous, multicellular, to 13 cells (150 µm), simple or very rarely branched, comprising cells quadrate or more commonly rectangular, 1–3 (–4) diameters long, 4–21 × 4–5 µm with 1 (–2) plate-like plastids and pyrenoids; hairs common, usually arising from basal cells, rarely terminal on erect

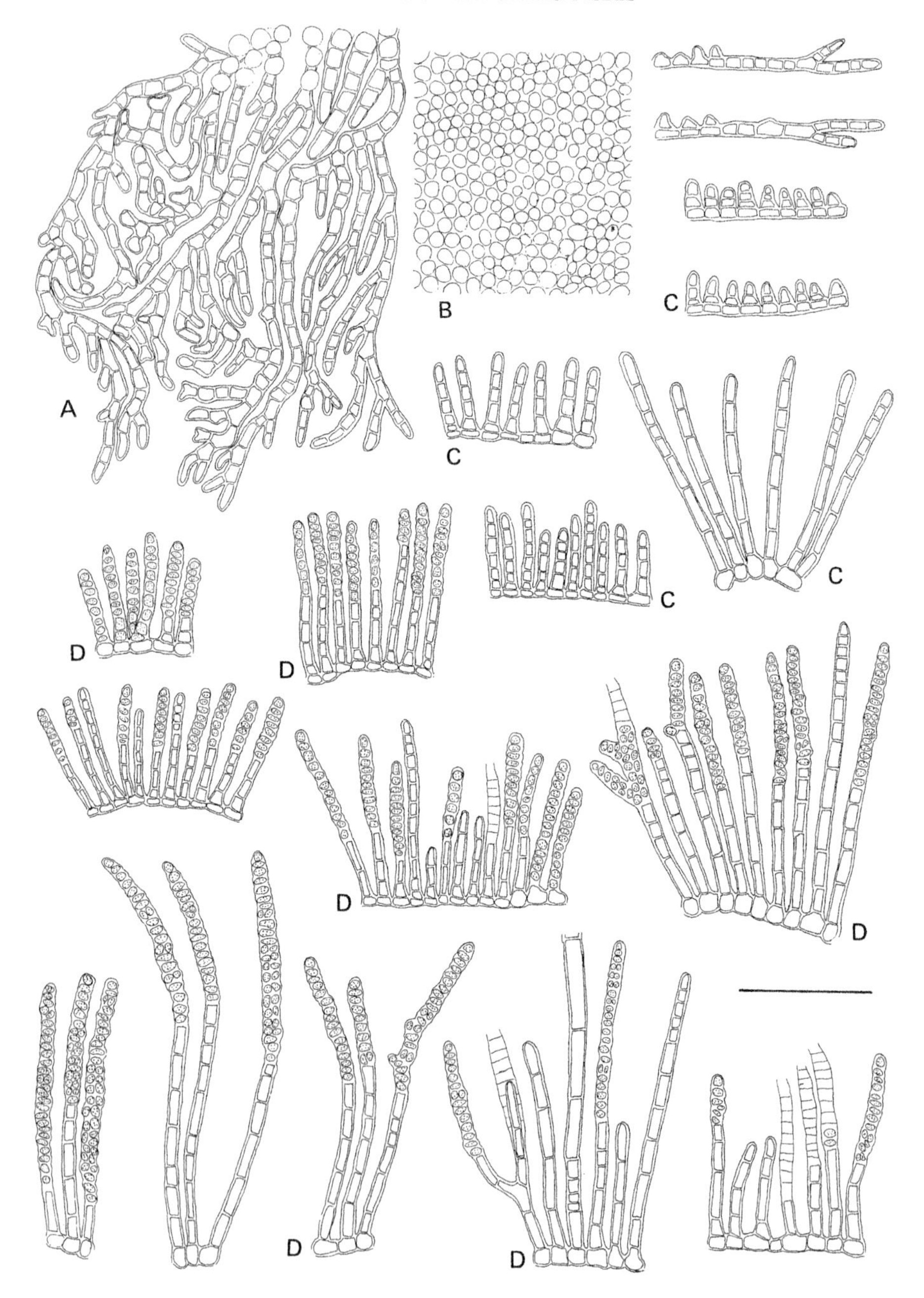
A
B
C
C
C
C
D
D
D
D
D
D

filaments, without sheath and with reduced basal meristem, *c.* 6 μm in diameter; rhizoids not observed.

Plurilocular sporangia abundant, simple or rarely branched, sessile or more usually terminal on erect filaments, uniseriate, rarely biseriate, with straight, sometimes oblique cross walls, slightly torulose, to 25 loculi (125 μm) long × 5–8 μm in diameter; unilocular sporangia unknown.

Epiphytic on *Laminaria* blades, lower eulittoral, in pools and shallow sublittoral.

Recorded for scattered localities (Berwick, Kent, Hampshire, Dorset, Devon, Cornwall, Channel Isles; in Ireland, Wexford); probably more common around the British Isles.

No data on seasonal growth. Fertile thalli have been recorded throughout the year.

A number of authors have linked *M. corunnae* as a stage in the life history of the more macroscopic ectocarpacean alga *Ectocarpus fasciculatus* Harvey. Intimate association of the two species on the host alga *Laminaria* was noted by both Jaasund (1961, 1965) in Norway and Sauvageau (1897b) in Spain. *M. corunnae* was also proposed as a myrionemoid variant of *Ectocarpus fasciculatus* (collected from Canada and the British Isles) based on laboratory culture studies by Baker & Evans (1971). The evidence for such a life history connection is not, however, convincing and more extensive culture studies are required; the myrionemoid variants may well represent contaminant organisms as proposed by Clayton (1972). Culture studies of *M. corunnae* from California (Loiseaux, 1970a) revealed a 'direct' life history to be operating, with plurispores directly repeating the fertile parental thallus.

Sauvageau (1897a), pp. 237–242, fig. 14; Setchell & Gardner (1925), p. 458; Newton (1931), p. 151; Hamel (1931–39) p. 91, fig. 24 (16–18); Levring (1937), p. 49, fig. 7A–B; Kylin (1947), pp. 38–39, fig. 30; Jaasund (1957), pp. 223–228, fig. 8; Taylor (1957), p. 155; Jaasund (1961), pp. 239–241, fig. 1; Jaasund (1965), pp. 56–57; Loiseaux (1970a), p. 251; Abbott & Hollenberg (1976), p. 158, fig. 123; Rueness (1977), p. 137, fig. 62; Stegenga & Mol (1983), p. 85, pl. 26, fig. 2.

Myrionema liechtensternii Hauck (1877), p. 185. Fig. 11

Ascocyclus lichtensteinii (Hauck) Holmes and Batters (1890), p. 82.
Hecatonema liechtensternii (Hauck) Batters (1902), p. 42.

Thallus epilithic, forming thin, closely adherent, light/olive brown patches on substrata, either solitary and circular, to 1 mm in diameter or confluent and irregular to several millimetres in extent; in surface view, peripheral region monostromatic, discoid/pseudodiscoid composed of outwardly radiating, closely packed filaments, central region of rounded cells or squashed pile-like in appearance; in squash preparation consisting of a basal layer of uniseriate cells, 6–14 × 4–8 μm usually quadrate, rectangular or irregular

Fig. 10 *Myrionema corunnae*
A. Surface view of peripheral thallus region showing irregularly spreading, vegetative filaments. B. Surface view of central thallus region (vegetative). C. S.P. of vegetative thallus showing erect filaments arising from a monostromatic base. D. S.P. of fertile thallus showing terminal plurilocular sporangia. Bar = 50 μm.

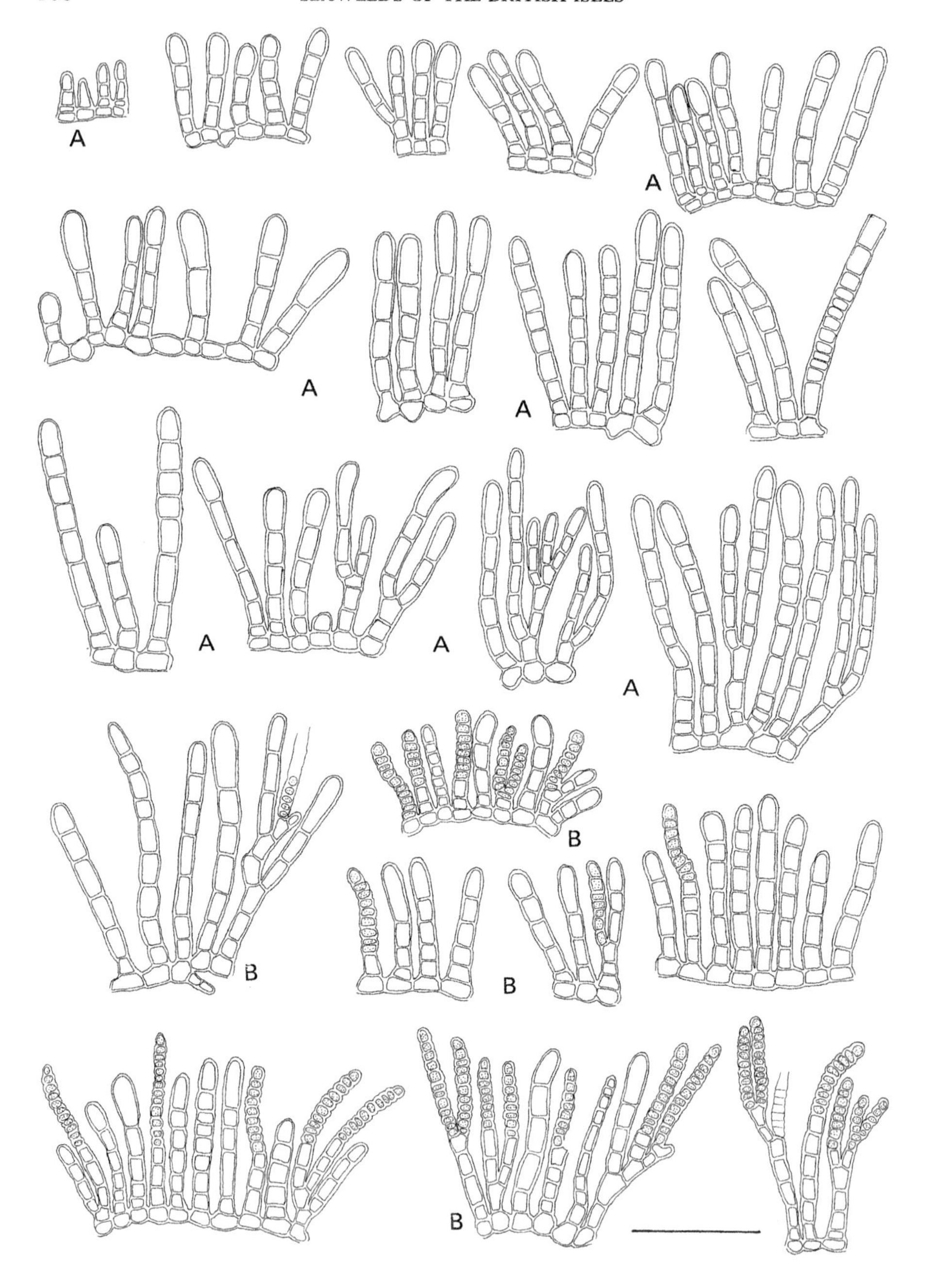
A
A
A
A
A
A
A
B
B
B
B

in shape, sometimes with short, 2–3 celled, rhizoidal filaments giving rise in the central regions to one, rarely two, erect filaments, hairs and/or plurilocular sporangia; erect filaments short, linear or elongate-clavate, simple or dichotomously branched to one (–2) orders, loosely associated and slightly mucilaginous to 7 (–12) cells (50–106 μm) long, comprising cells often subquadrate at the base, but quadrate to rectangular above, 1–3·5 diameters long, 4–13 (–18) × 4–10 μm often with enlarged, sometimes pyriform and ascocyst-like apical cell, 16–26 × 6–16 μm; cells with 1 (–2) multi-lobed, plate-like plastids with pyrenoids; hairs uncommon arising from basal layer, approximately 10 μm in diameter, with basal meristem but sheath not observed.

Plurilocular sporangia common, sessile or stalked on basal layer, more commonly terminal or lateral on erect filaments, frequently formed in old sporangial husks, simple, cylindrical, uniseriate to 7 (–14) loculi, 24–48 × 4–7 μm; loculi subquadrate, less commonly quadrate, 3–6 × 5–7 μm with straight, occasionally oblique cross walls; unilocular sporangia unknown.

Epilithic on bedrock, in upper eulittoral and littoral fringe, associated with algae such as *Petalonia filiformis*, *Rivularia* spp. and *Phymatolithon lenormandii*.

Only reported from Berwick-upon-Tweed, Northumberland.

Material collected in October; information insufficient to comment on seasonal distribution.

The above description and notes are based entirely upon examination of herbarium specimens deposited in the British Museum (Natural History) and originally collected by E. A. L. Batters in 1889. The material comes very close in description to the epiphytic alga *Microspongium globosum* (in particular the material formally described as *Myrionema polycladum* Sauvageau) and is only retained here as an independent taxon pending further studies. It differs from *M. globosum* in being epilithic (British material only), has longer erect filaments and slightly larger vegetative cells.

Hauck (1877), pp. 185–186, figs. 1–2; (1883–85), pp. 321–322; Newton (1931), p. 157; Coppejans and Dhondt (1976), pp. 112–117, figs. 1–9; Verlaque and Boudouresque (1981), p. 147, figs 12–13; Athanasiadis (1985), p. 465, fig. 16.

Myrionema magnusii (Sauvageau) Loiseaux (1967a), p. 338. Fig. 12; Pl. 2a

Ascocyclus magnusii Sauvageau (1927b), p. 13.
Non *Myrionema orbiculare* J. Agardh (1848–76), p. 48.
Ascocyclus orbicularis (J. Agardh) DeToni (1895), p. 583.
Ascocyclus orbicularis (J. Agardh) Magnus (1875), p. 73; (see Dixon & Russell, 1964).
?*Ascocyclus affinis* Svedelius (1901), p. 107.

Thallus epiphytic, forming light to dark brown spots on host, usually solitary and circular, 150–500 μm in diameter, occasionally confluent to 2 mm or more in extent; in surface view, peripheral region light brown, monostromatic, discoid, composed of

Fig. 11 *Myrionema liechtensternii*
A. S.P. of vegetative thallus showing erect filaments arising from a monostromatic base.
B. S.P. of fertile thallus showing terminal plurilocular sporangia. Bar = 50 μm.

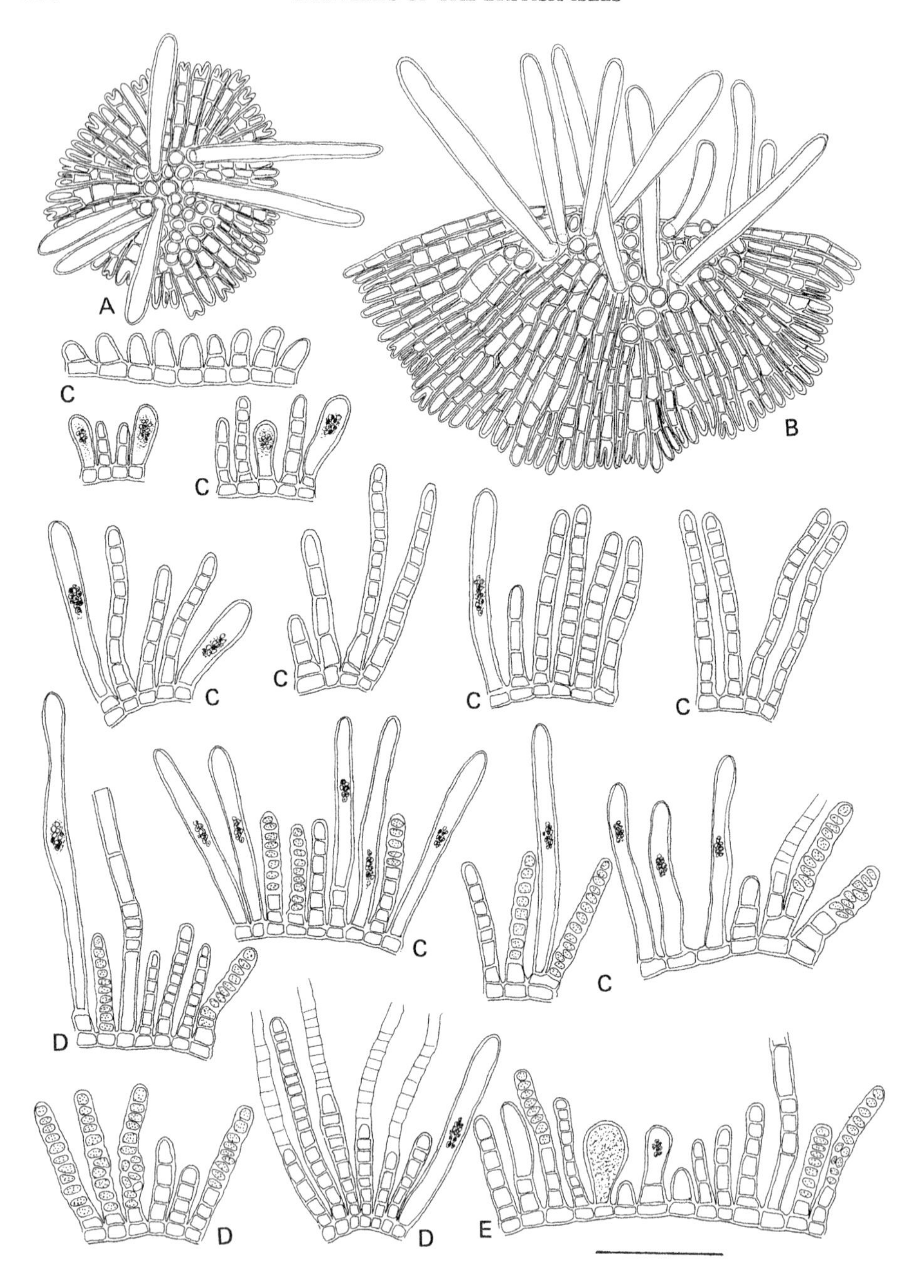
A
B
C
C
C
C
C
C
C
C
C
D
D
D
E

branched, outwardly radiating, quite firmly united filaments enclosed within outer cuticle, comprising rectangular cells 1–3 (–4) diameters long, 6–13 × 3–7 µm, each with a single, multilobed plastid and pyrenoid, central region dark brown, of rounded cells, 4–9 µm in diameter, later with colourless ascocysts emerging; structure consisting of a basal layer of uniseriate cells giving rise in the central regions to erect filaments, hairs, ascocysts and/or sporangia; erect filaments short, simple, loosely associated and slightly mucilaginous, to 4–12 cells (40–90 µm) long, comprising cells quadrate to rectangular, usually 1–2 (–3) diameters long, 5–20 × 7–10 µm each with a single, multilobed plastid with pyrenoid; ascocysts extremely common, either produced from basal cells or on short 1–3 celled stalks, darkly pigmented and slightly globose when young becoming linear, elongated, 45–105 (–120) × 8–11 µm, thick-walled and colourless later; hairs frequent, arising singly from basal layer with basal meristem and sheath, 8–10 µm in diameter.

Unilocular and plurilocular sporangia borne on the same plants; plurilocular sporangia abundant, uniseriate, rarely biseriate, simple, cylindrical, 34–58 × 7–9 µm, to 7–13 loculi, on 1–4 celled stalks, loculi subquadrate, often lenticular, 3–5 × 5–9 µm, with straight, occasionally oblique cross walls and quite distinct red eye spot; unilocular sporangia rare, sessile or shortly stalked on basal layer, ovate to pyriform, 28–33 × 14–16 µm.

Epiphytic, on *Zostera* leaves, in lower eulittoral, and shallow sublittoral.
Generally distributed around the British Isles.
Summer annual, May to September, with sporangia.

Culture studies by Loiseaux (1964) revealed *M. magnusii* (recorded as *Acocyclus magnusii*) to have a 'direct' type of life history. The plurispores behaved asexually and, by the process of heteroblasty, germinated directly to form two morphologically different thalli: ectocarpoid filamentous growths and stellate, later discoid growths. Both growth forms produced plurilocular sporangia, the plurispores from which germinated to repeat the above described developmental processes. A direct life history was also shown for a Swedish isolate by Kylin (1933) (as *Ascocyclus orbicularis*). To date no evidence supports Parkes idea (1933) that *M. magnusii* (as *Ascoryclus orbicularis*) is connected in life history with *Cladosiphon contortus* (as *Castagnea contorta*).

Kylin (1907), p. 39, fig. 9; Sauvageau (1927b), p. 13; Newton (1931), p. 159, fig. 99; Hamel (1931–39), pp. 100–101; Kylin (1933), pp. 22–25, fig. 4; Feldmann (1937), pp. 251–258, fig. 37B, 38; Levring (1940), pp. 40–43, fig. 8; Kylin (1947), p. 40, figs 32–33; Waern (1952), pp. 149–151, fig. 68 A–B & E; Christensen (1958), pp. 129–132, pl. 4; Taylor (1957), p. 158; Dixon & Russell (1964), pp. 281–282; Loiseaux (1964), pp. 2903–2905, figs 1–2; Loiseaux (1967a), pp. 329–347, figs 1, 2a, 4, 5b, pl. 1; Loiseaux (1967b), p. 567; Abbott & Hollenberg (1976), p. 158, fig. 124; Rueness (1977), p. 127, fig. 63; Price *et al.* (1978), pp. 137–138; Stegenga & Mol (1983), p. 85, pl. 25, fig. 7.

Fig. 12 *Myrionema magnusii*
A–B. Surface view of young thalli. Note central elongated ascocysts. C. S.P. of vegetative thalli showing erect filaments and ascocysts arising from a monostromatic base. D. S.P. of fertile thalli showing ascocysts and plurilocular sporangia. E. S.P. of fertile thallus showing ascocysts, plurilocular sporangia and a single unilocular sporangium. Bar = 50 µm.

Myrionema papillosum Sauvageau (1897a), p. 242. Fig. 13

Thallus epiphytic, forming light brown spots on hosts, usually solitary and circular, 0·5–1·0 mm across; in surface view, peripheral region consisting of laterally united, outwardly radiating, branched, filaments of quadrate, more usually rectangular, cells, 1–3 diameters long, 8–20 × 5–10 µm, each with 1–3 plate-like plastids with pyrenoids, outer central region composed of closely packed, rounded cells, 6–12 µm in diameter, inner central region turf-like, similar to a carpet pile; in squash preparations comprising a monostromatic basal layer of quadrate to rectangular cells, 1–5 diameters long, 8–22 × 4–7 µm, each giving rise to 1–2 erect filaments, hairs and/or sporangia; erect filaments short, linear or slightly elongate-clavate, loosely associated and gelatinous, multicellular, to 12 cells (170 µm) long, simple, or rarely branched, later commonly giving rise to numerous 1–2 (–3) celled lateral protuberances, comprising cells 1–3 (–4) diameters long, 5–21 × 5–8 µm, each with 1–3 plate-like plastids and associated pyrenoids; hairs rare, arising from basal layer, with basal meristem and sheath; rhizoids not observed.

Plurilocular and unilocular sporangia often associated on same thallus; plurilocular sporangia common, terminal or lateral on erect filaments, simple, cylindrical or slightly moniliform, uniseriate, to 5–9 loculi (22–42 µm) long × 7–9 µm wide, with straight, occasionally oblique cross walls; unilocular sporangia uncommon, lateral on erect filaments, sessile, ovate to pyriform, 28–65 × 17–34 µm.

Epiphytic, on *Laminaria saccharina* blades, lower eulittoral pools and sublittoral to 5 m.

Only recorded from Dorset, probably of more general occurrence around the British Isles.

Summer annual, July to October.

Sauvageau (1897a), pp. 242–247, figs 15–17; Batters (1900), p. 371; Newton (1931), p. 151; Hamel (1931–39), pp. 91–92, fig. 24 (11–15).

Myrionema strangulans Greville (1827), pl. 300. Fig. 14; Pls 1, 2b

Myrionema punctiforme Harvey in Hooker (1833), p. 391.
Myrionema maculiforme Kützing (1845–71), p. 264.
Myrionema leclancheri (Chauvin) Harvey (1846–51), pl. 41.
Myrionema vulgare Thuret in Le Jolis (1863), p. 82.

Thallus epiphytic, forming light brown spots on hosts, usually solitary and circular, 0·5–1 mm in diameter, occasionally confluent to 2–3 mm; in surface view, peripheral region light brown and disc-like in appearance, composed of monostromatic, frequently

Fig. 13 *Myrionema papillosum*
A. Surface view of peripheral thallus region showing outwardly spreading, closely united filaments. B. S.P. of vegetative thallus showing erect filaments arising from a monostromatic base. Note lateral, usually one-celled, protruberances on erect filaments. C. S.P. of fertile thallus showing erect filaments with lateral unilocular sporangia. D. S.P. of fertile thallus showing erect filaments with lateral and terminal plurilocular sporangia. Bar = 50 µm.

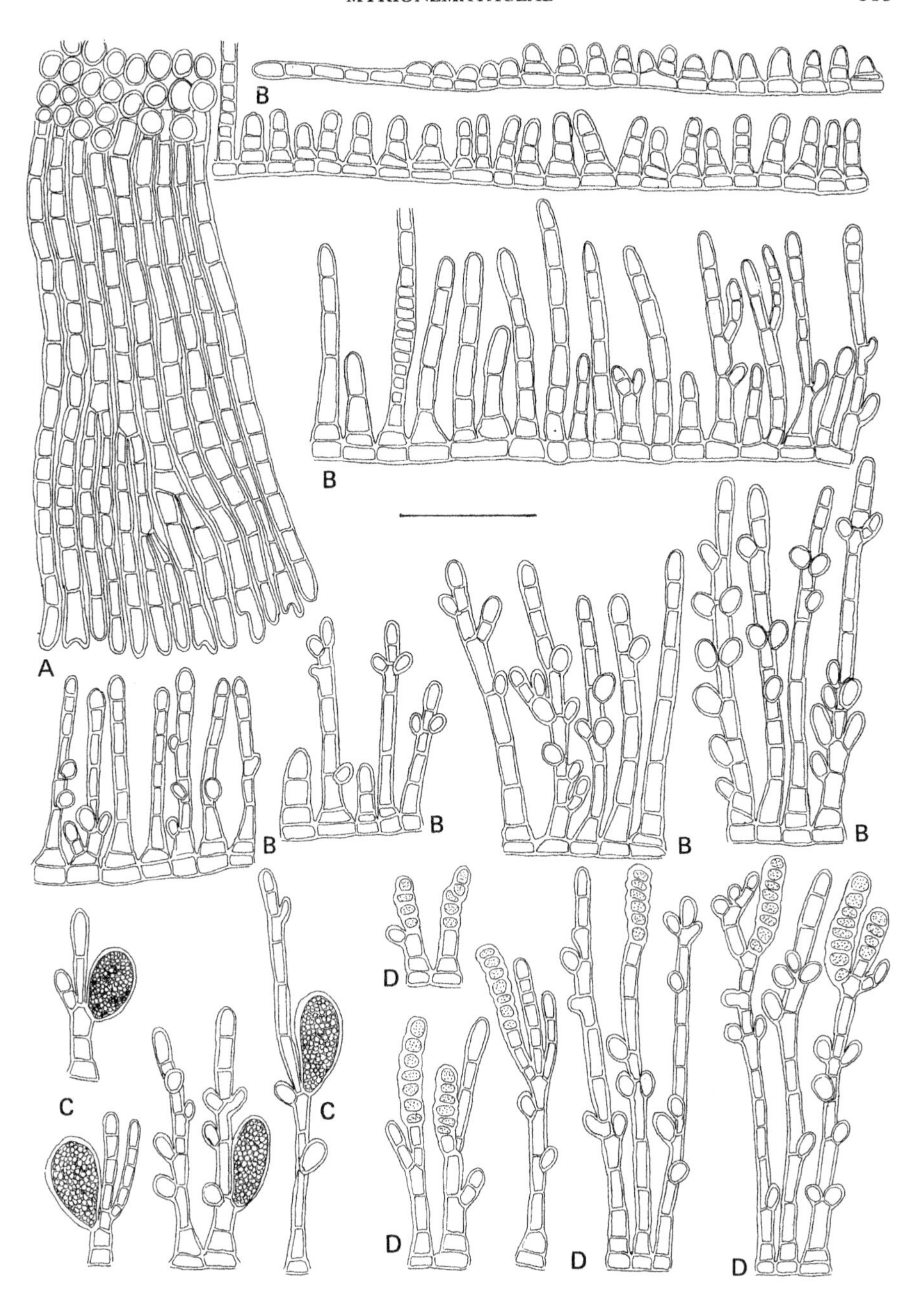

A
B
B
B
B
B
B
C
C
D
D
D
D

branched, outwardly radiating, loosely united filaments, with terminal apical cells often enclosed within outer cuticle, central region dark brown, pile-like, composed of rounded cells, variously packed, 7–11 μm in diameter, later becoming turf-like, similar to a carpet pile; in structure consisting of a basal layer of uniseriate cells, 1–4 diameters long, 11–24 × 5–6 μm, each giving rise over the greater part of the thallus to 1–2, erect filaments, hairs and/or sporangia; erect filaments short, simple, loosely associated and gelatinous, elongate-clavate or elongate-cylindrical, 3–7 (–9) cells, (55–85 μm) long, comprising rectangular cells, 1–5 diameters long, 7–24 × 4–9 μm, each with 1–3 discoid plastids and associated pyrenoids; hairs frequent, arising singly from basal layer, with basal meristem and sheath, *c.* 6–8 μm in diameter.

Unilocular sporangia abundant, spherical to pyriform, 39–60 × 22–30 μm, sessile or stalked on basal layer or more commonly lateral at the base of vegetative filaments; plurilocular sporangia less frequent, cylindrical, uniseriate, occasionally biseriate, sessile on basal layer or terminal on erect filaments, 20–45 × 7–11 μm, to 8 loculi long.

Epiphytic on various hosts, in particular *Ulva* and *Enteromorpha* spp., littoral fringe to lower eulittoral, in pools or (more rarely) emergent.

Common and widely distributed throughout the British Isles.

Summer annual, found occasionally throughout the year.

M. strangulans has been investigated in culture by Kylin (1934), Loiseaux (1967b) and Pedersen (1981c). A heteromorphic life history appears to be operating with a diploid macrothallus alternating with a haploid, filamentous microthallus. Plurispores released from the latter were reported to be sexual and fused to reform the sporophyte (Loiseaux). The life history was further supplemented by asexual reproduction of unfused gametes and sexual reproduction by the fusion of unispores (Loiseaux). Heteroblasty was also reported with stellate and filamentous germlings produced from spores originating in the same sporangium; Pedersen suggested that this phenomenon might be determined by the initial degree of germling/surface contact.

Sauvageau (1897a), pp. 185–229, figs 1–11; Setchell & Gardner (1925), p. 471, pl. 35, fig. 12; pl. 40, fig. 51; Newton (1931), p. 150, fig. 93; Hamel (1931–39), pp. 88–90, fig. 24, 1–10; Kylin (1934), pp. 5–9, figs 1–3; Kylin (1947), p. 36, fig. 28; Smith (1955), pp. 245–246, fig. 141A–C; Taylor (1957), p. 156, pl. 11, figs. 13–14; Jaasund (1965), p. 57, fig. 17; Loiseaux (1967a), pp. 329–347, figs 2B, 5A; Loiseaux (1967b), pp. 529–576; Abbott & Hollenberg (1976), pp. 158–159, fig. 125; Rueness (1977), pp. 137–138, fig. 64 A & B; Pedersen (1981c), pp. 194–217, fig. 5.1 (D–G); Kornmann & Sahling (1983), pp. 46–50, figs 27–29; Stegenga & Mol (1983), p. 85, pl. 26, figs 3–5.

Fig. 14 *Myrionema strangulans*
A. Surface view of peripheral thallus region showing outwardly spreading, closely united filaments. B. S.P. of vegetative thallus showing erect filaments and hair (h) arising from a monostromatic base. C. S.P. of fertile thallus showing unilocular sporangia sessile or shortly stalked on basal layer. D. S.P. of fertile thallus showing plurilocular sporangia. Bar = 50 μm.

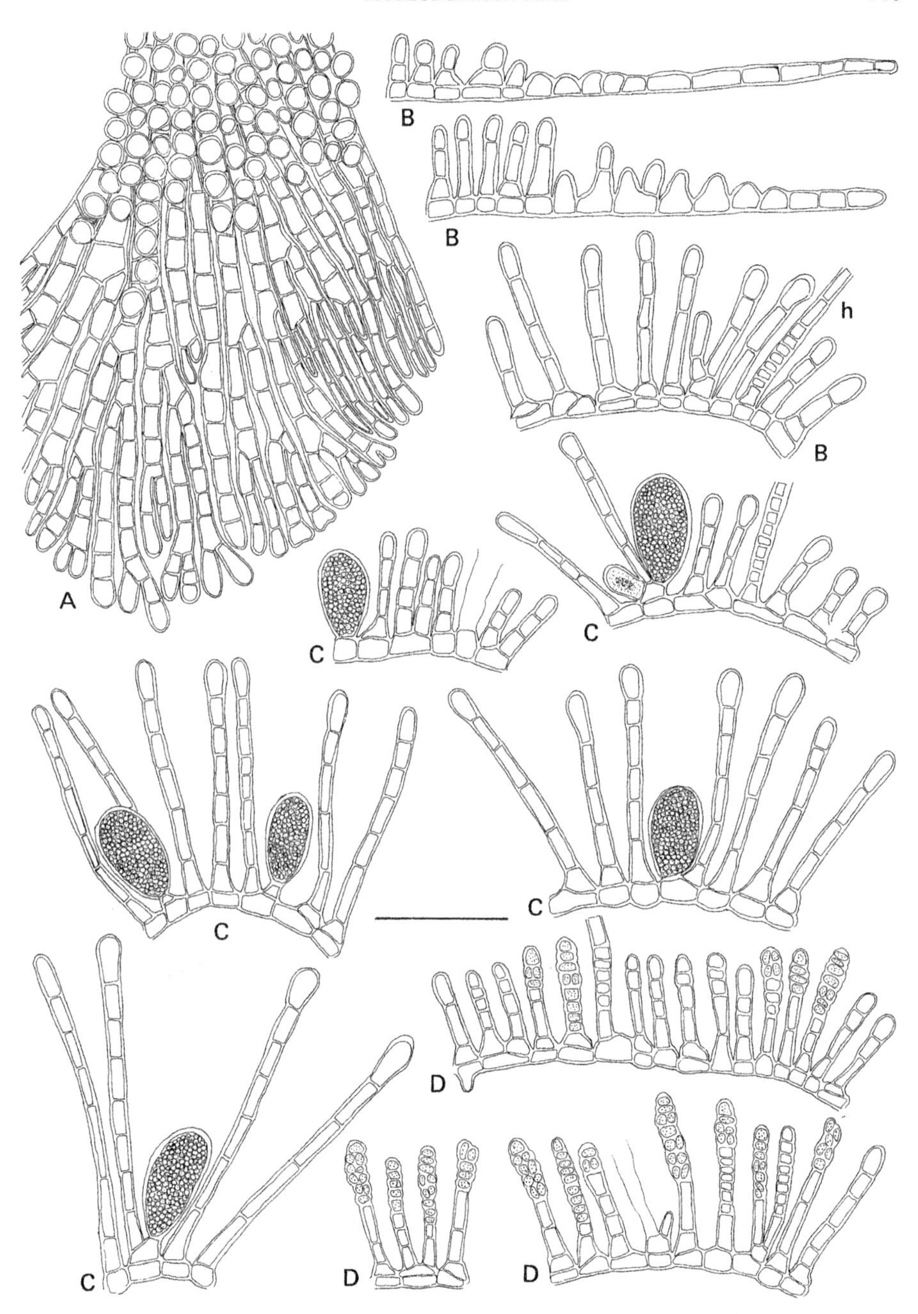
A
B
B
h
B
C
C
C
C
C
D
D
D

PROTECTOCARPUS Kuckuck

PROTECTOCARPUS Kuckuck (1955), p. 119.

Type species: *P. speciosus* (Børgesen) Kuckuck (1955), p. 120.

Thallus epilithic or epizoic, forming small, gregarious tufts up to 1 (–3) mm high; consisting of a monostromatic basal layer of outwardly spreading, pseudodichotomously branched filaments which are laterally adjoined and discoid or irregularly associated and pseudodiscoid in appearance, most cells of which, except at the periphery, give rise to erect filaments, hairs and/or sporangia; erect filaments simple or branched to the first order, with branches often in unilateral rows and recurved towards parental filament; cells with 1–2 plate-like, lobed plastids with pyrenoids; hairs terminal or lateral on erect filaments, with basal meristem and sheath; plurilocular sporangia uniseriate to multiseriate, either sessile or stalked on basal layer more commonly terminal or lateral, often in unilateral rows, on the erect filaments, simple or with unilateral, branch-like extensions; unilocular sporangia sessile or shortly stalked on the basal layer, or lateral on the erect filaments.

One species in the British Isles.

Protectocarpus speciosus (Børgesen) Kuckuck (1955), p. 120. Figs 15, 16

Myrionema speciosum Børgesen (1902), p. 421.
Hecatonema diffusum Kylin (1907), p. 39.
Hecatonema speciosum (Børgesen) Cotton (1912), p. 121.
Ectocarpus speciosus (Børgesen) Kuckuck in Oltmanns (1922), p. 13.
?Hecatonema faeroeense (Børgesen) Levring (1937), p. 47.

Thallus epiphytic, more rarely epilithic and epizoic, forming small solitary or confluent tufts, just visible to the naked eye; comprising a monostromatic base of outwardly spreading, frequently branched, filaments, more or less laterally adjoined and discoid/pseudodiscoid in appearance, the uniseriate cells of which give rise in the central regions to erect filaments, hairs and/or sporangia; in surface view outer basal cells rectangular 1–2 (–3) diameters long, 6–22 × 5–10 μm, in squash preparation inner basal cells, rectangular, quadrate less frequently ovate, slightly longer than wide, 8–13 × 4–7 μm; erect filaments free, loosely associated, slightly gelatinous, simple or more commonly sparingly branched, to 570 μm high (38 cells), comprising uniseriate cells 1–3 (–5) diameters long, 7–39 × 6–10 μm, each with a single, terminal, parietal, plate-like plastid with pyrenoid; branching usually to one order only, with branches short, pinnate and secund, later often being transformed into plurilocular sporangia; hairs not uncommon,

---

Fig. 15 *Protectocarpus speciosus*
A. Surface view of portion of peripheral thallus region showing outwardly spreading basal filaments. B. S.P. of fertile thallus showing erect filaments arising from a monostromatic base, with terminal and lateral plurilocular sporangia. C. S.P. of fertile thallus showing unilocular sporangia. Bar = 50 μm.

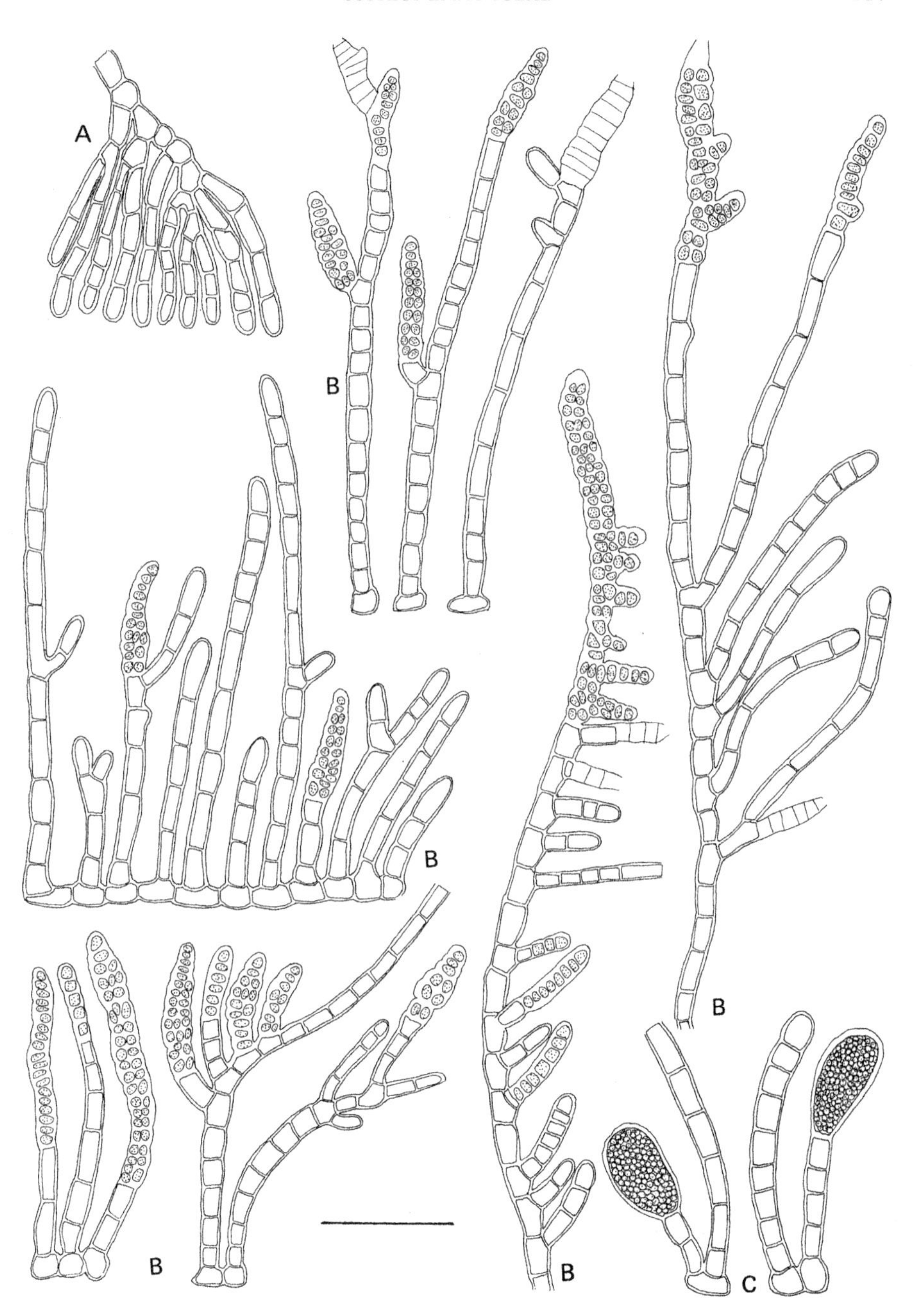
A
B
B
B
B
B
B
C

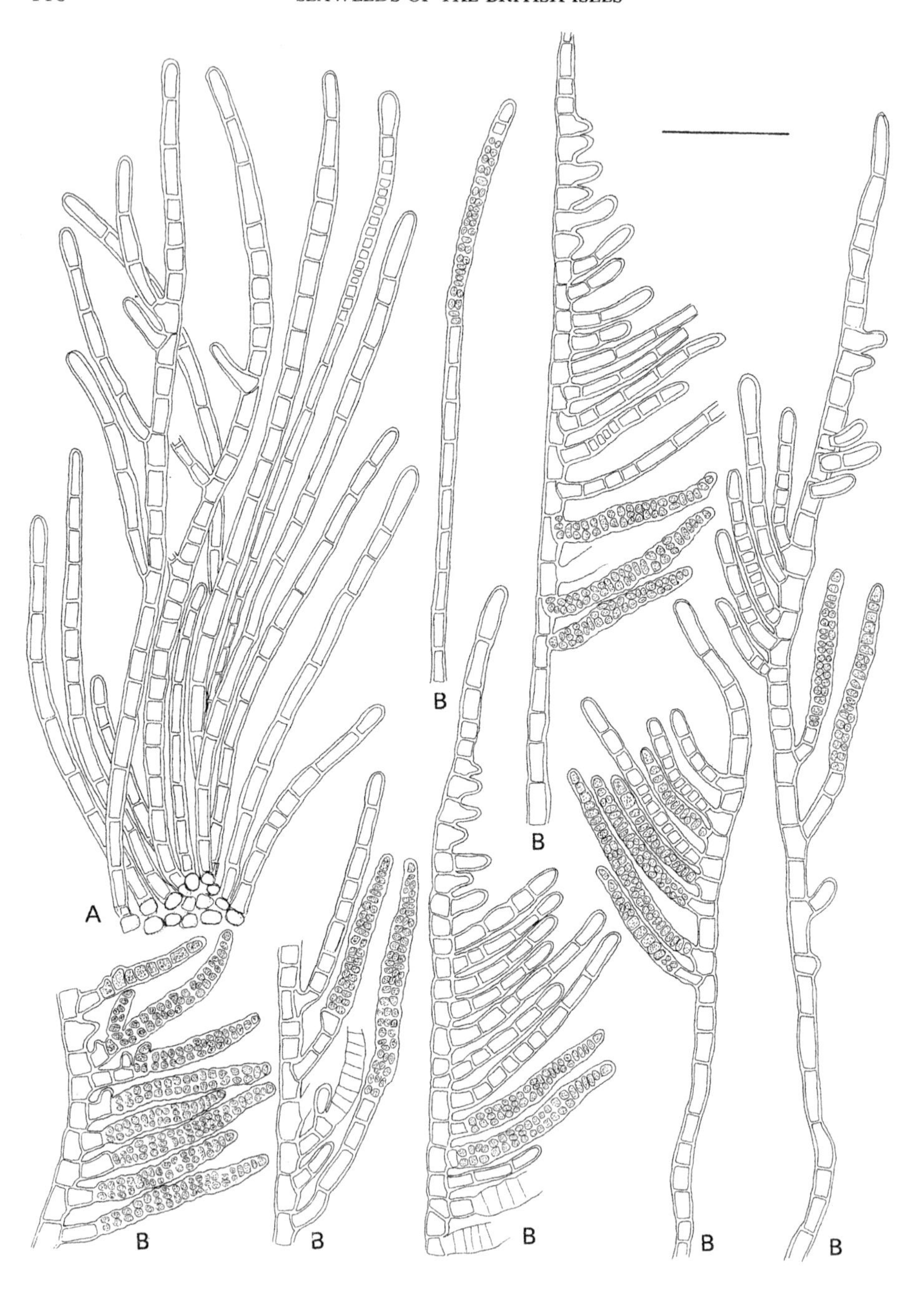
A
B
B
B
B
B
B
B
B

arising from basal cells or more commonly terminal or lateral on erect filaments and branches, with basal meristem and sheath.

Plurilocular sporangia abundant, sessile or stalked on basal cells, more commonly terminal and lateral on erect filaments, lateral sporangia sessile or stalked, usually arranged in terminal regions of filament; sporangia elongate-cylindrical, tapering upwards, simple or branched, with branches usually short, secund, widely divergent and arranged in a cock's comb fashion, uniseriate or more commonly bi-multiseriate, with straight or oblique cross walls, 32–100 (–130) × 9–16 μm; unilocular sporangia common, on basal layer or more commonly lateral at the base of erect filaments, usually stalked, elongate-ovate, 40–60 × 22–25 μm.

Epiphytic on various algae, less commonly epizoic on shells and epilithic on stones, eulittoral and shallow sublittoral.

Recorded for scattered localities around the British Isles (Shetland Isles, Orkney, East Lothian, Essex, Hampshire, Channel Isles, Isles of Scilly, Inner Hebrides, Outer Hebrides; in Ireland, Cork, Wexford and Mayo).

Spring and summer annual.

Børgesen (1902), pp. 421–424, fig. 78; Kylin (1907), pp. 39–41; Børgesen (1926), pp. 55–59, figs 29–30; Newton (1931), p. 157; Kylin (1947), pp. 17–18, fig. 13; Kuckuck (1955), pp. 120–135, figs 1–8; Cardinal (1964), p. 77; Jaasund (1965), pp. 53–55, fig. 16; Kornmann & Sahling (1977), p. 275, fig. 158; Rueness (1977), pp. 138–139, fig. 65; Stegenga & Mol (1983), p. 89, pl. 27, figs 1–3.

ULONEMA Foslie

Ulonema Foslie (1894), p. 131.

Type species: *U. rhizophorum* Foslie (1894), p. 131.

Thallus epiphytic, forming small circular spots on host; consisting of a monostromatic basal layer of outwardly spreading, pseudodichotomously branching filaments, not obviously laterally adjoined and discoid but irregular and diffuse, giving rise in the central regions to erect filaments, hairs and sporangia and downwardly penetrating, branched, multicellular, rhizoidal filaments; erect filaments short, linear or clavate, simple or more frequently secundly branched to one order, with branches remaining short, 1–3 celled; cells with 1–3 plate-like plastids with pyrenoids; hairs common, with basal meristem and sheath.

Plurilocular sporangia either sessile or shortly stalked on basal layer, unilocular sporangia commonly lateral at the base of erect filaments; unilocular sporangia ovate to pyriform; plurilocular sporangia uniseriate, simple.

Fig. 16 *Protectocarpus speciosus*
A. S.P. of vegetative thallus. B. S.P. of fertile thallus showing portions of erect filaments with plurilocular sporangia. Bar = 50 μm.

The genus *Ulonema* is very closely related to *Myrionema* from which it differs in having a diffuse and irregularly spreading filamentous base with deeply penetrating, branched rhizoidal filaments rather than a discoid base without rhizoids. However, these may well be insufficient criteria for separating the two genera. Some *Myrionema* species have been reported with rhizoids whilst the diffuse, filamentous base of *Ulonema* might merely be an adaptation to the surface type of the host alga *Dumontia*. In his original diagnosis for *Ulonema*, Foslie (1894) also stated that plurilocular sporangia were absent. This has been shown not to be the case and plurilocular sporangia have been described in *Ulonema* by Sauvageau (1897a, p. 233, fig. 12D) and Hamel (1931–39, p. 34). These observations lend support to the inclusion of *Ulonema* as a synonym of *Myrionema*; however pending confirmatory experimental studies the autonomy of *Ulonema* is retained here.

One species in the British Isles.

Ulonema rhizophorum Foslie (1894), p. 132. Fig. 17

Thallus epiphytic, forming light brown spots on hosts, usually solitary and circular, 0·25–0·75 mm in diameter, occasionally confluent and irregular; in surface view peripheral region thin and prostrate comprising outwardly radiating, frequently pseudodichotomously branched, diffused filaments, not laterally cohered into discs, with inner region raised, hemispherical and pile-like comprising cells variously packed in surface view, 7–12 µm in diameter; in squash preparations consisting of a monostromatic basal layer of uniseriate cells, 1–4 diameters long, 8–21 × 7–9 µm, all cells, except at periphery, giving rise to erect filaments, hairs and/or sporangia, with many cells giving rise below to single or multicellular, occasionally branched, rhizoidal filaments; erect filaments short, loosely associated in a gelatinous matrix, simple or sparingly branched, linear or clavate, to 4–7 cells (60–85 µm) long, comprising cells 2–6 diameters long, 16–23 × 4–7 µm each enclosing 1–3 plate-like plastids with pyrenoids; hairs frequent, arising singly from basal layer, with basal meristem and sheath, *c.* 7–8 µm in diameter.

Unilocular sporangia abundant, either sessile on basal layer or on one-celled stalks and usually lateral at the base of the erect filaments, spherical to pyriform in shape, 26–46 × 17–25 µm; plurilocular sporangia rare, sessile or shortly stalked on basal layer, to 25 µm long, 8 loculi, uniseriate.

Epiphytic on the red alga *Dumontia contorta*, especially on mature thalli, littoral fringe to lower eulittoral pools.

Common and widely distributed around the British Isles.

Annual; mainly spring and early summer.

Fig. 17 *Ulonema rhizophorum*
A. Habit of plant, forming small spots on the red alga *Dumontia contorta*. B. Surface view of peripheral thallus region showing irregularly spreading basal filaments. C. Surface view of central thallus region showing terminal cells of erect filaments. D. S.P. of vegetative thallus showing erect, later branched filaments arising from a monostromatic base. Note rhizoids emerging from basal layer. E. S.P. of fertile thallus showing unilocular sporangia. F. S.P. of fertile thallus showing plurilocular sporangia. Bar = 15 mm (A), 50 µm (others).

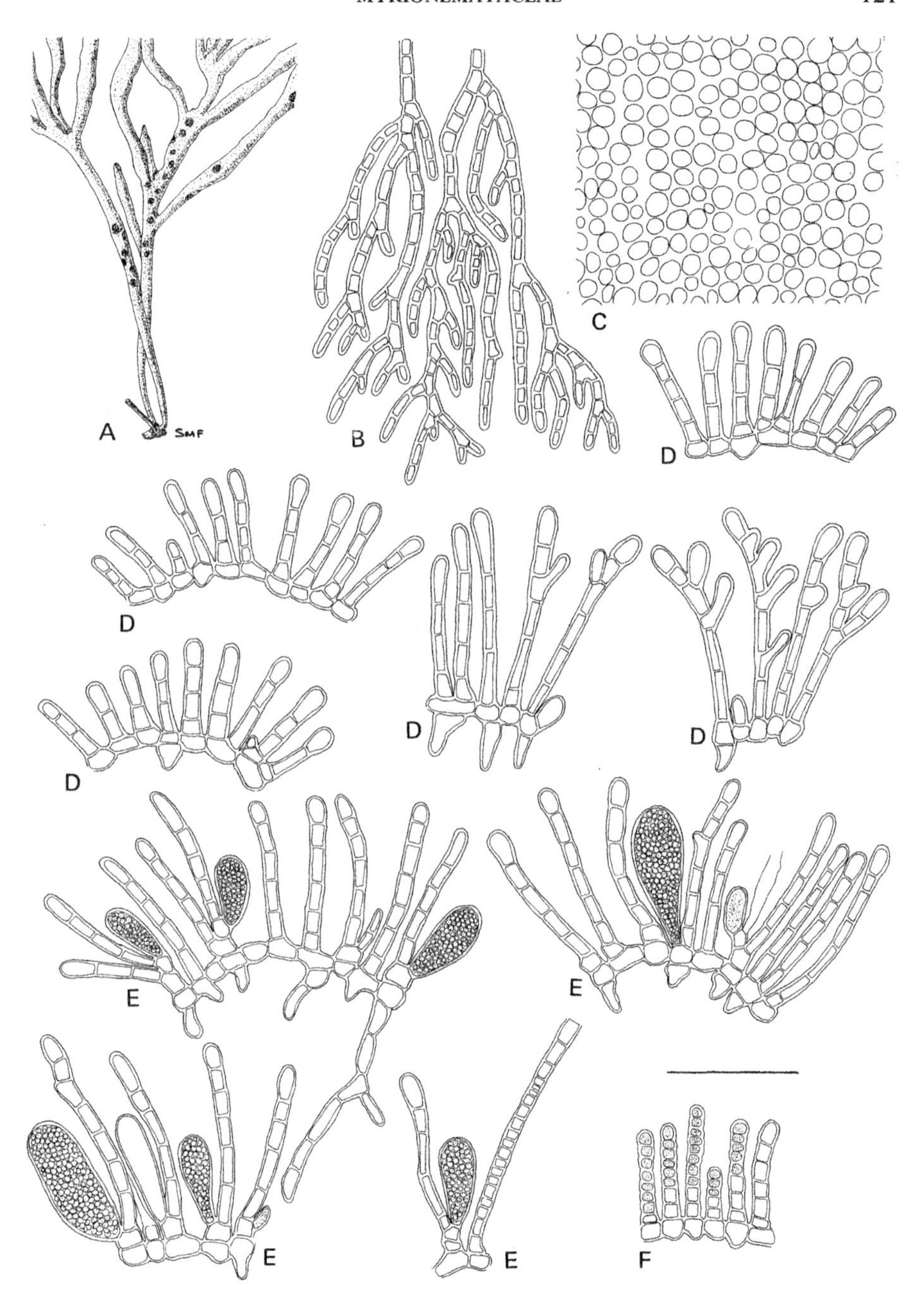
A
SMF
B
C
D
D
D
D
D
E
E
E
E
F

In many floristic studies, *U. rhizophorum* has either been included within the circumscription of *Myrionema strangulans* (Levring, 1937; Jønsson, 1903) or authors have commented on their possible conspecificity (Jaasund, 1951, p. 130; Parke & Dixon, 1976, p. 562; South & Hooper, 1980, p. 35). However, agreement is expressed with Jaasund (1951, p. 131) that *Ulonema rhizophorum* does appear to differ from *Myrionema strangulans* and its autonomy is thus retained in the present work pending further experimental studies.

Foslie (1894), pp. 131–134, pl. III, figs 11–17; Batters (1895a), p. 275, pl. III; Sauvageau (1897a), pp. 229–233, fig. 12; Knight & Parke (1931), p. 68, pl. IX, figs 7, 10, 13; Newton (1931), pp. 154–155, fig. 96; Hamel (1931–39), p. 94, fig. 24 (19–21); Kylin (1947), pp. 40–41, fig. 34; Jaasund (1951), pp. 130–131, fig. 1 (e–f); Jaasund (1965), pp. 58–59, fig. 18A–B; Edelstein & McLachlan (1969a), pp. 555–557, figs 5–12; Stegenga & Mol (1983), p. 89, pl. 27, figs 4–5.

ELACHISTACEAE Kjellman

ELACHISTACEAE Kjellman (1890), p. 41.

Elachistaceae Reinke (1889b), p. 49.

Thallus epiphytic, epilithic or epizoic, forming small brush-like tufts with or without basal hemispherical cushions, arising from either a monostromatic layer of outwardly radiating filaments (free or laterally united and pseudodiscoid/discoid in appearance) or from a fibrous mat of rhizoidal filaments, sometimes penetrating below into host tissue; filaments of brush-like tufts erect, linear, flaccid, free, multicellular, monosiphonous, simple or with limited branching towards base, accuminate sharply below more gradually above, with intercalary, usually basally positioned meristem, comprising cells with numerous discoid plastids without pyrenoids; erect tufts arising either directly from basal system or from bulbous/hemispherical cushions; cushions solid, pseudoparenchymatous, soft and gelatinous or firm and cartilaginous, comprising erect, branched, compacted filaments of large, thin-walled, colourless cells, in the basal regions giving rise to downwardly penetrating rhizoidal filaments, in the terminal regions giving rise directly to the erect filaments and/or paraphyses and sporangia; paraphyses short, simple, multicellular, photosynthetic; hairs unknown; unilocular sporangia pyriform, globular or elongate-cylindrical, lateral at the bases of the paraphyses and/or erect filaments, more rarely on the upper regions of the erect filaments; plurilocular sporangia either lateral or terminal on the paraphyses, simple and uniseriate, more rarely lateral or intercalary, single or in chains, on the erect filaments.

This family is represented in the British Isles by three genera: *Elachista, Halothrix* and *Leptonematella*. Common characteristics include a predominantly epiphytic habit, a thallus comprising brush-like tufts of long, flaccid, generally simple filaments (often termed 'assimilatory hairs'), predominantly pyriform unilocular sporangia and the absence of hairs. Developmental studies on a small number of species indicate the following mode of thallus construction. There is an initial development of a discoid/ pseudodiscoid, monostromatic, basal layer of branched filaments, from the central cells of which directly arise the erect, linear filaments constituting the brush-like tufts. Unlimited growth of these filaments is provided by a basally positioned zone of meristematic cells, whilst enhanced attachment is provided by the downward growth of supplementary rhizoidal filaments from a small number of submeristematic cells. Lateral branching of the usually submeristematic cell results in the production of secondary erect filaments. At a later stage of development in *Elachista* only, a number of these secondary erect filaments remain limited in growth and resemble paraphyses (often termed 'assimilators'). *Elachista* also differs from the two other genera reported for the British Isles in the later development of cushion-like thalli associated with the base of the erect filaments. These originate by a process of enlargement and compacting together of the submeristematic cells of the erect filaments. This results in the formation of a usually pulvinate or hemispherical shaped, solid, cartilaginous tissue (often termed a 'tubercle') which gradually enlarges by an acropetal movement of the meristematic zone and the consequential 'availability' of new submeristematic cells. *Elachista* is, therefore, considered to show more anatomical differentiation than *Halothrix* and *Leptonematella* and comprises a lower/mid region of erect, compacted filaments of enlarged, colourless cells (usually termed the 'medulla') and an upper region of less compacted, photosynthetic, erect filaments, with or without paraphyses (usually termed the 'cortex').

Some differences between the three genera are also reported with respect to the formation of the reproductive organs. Unilocular sporangia have been described in *Elachista* and *Leptonematella* only, and occur laterally at the base of the free, erect filaments (usually associated with the paraphyses in the 'cortex' in *Elachista*), more rarely additionally lateral in the upper regions of the erect filaments in *Elachista stellaris*. The plurilocular sporangia also occur at the base of the free, erect filaments in *Elachista*, usually lateral and/or terminal on the paraphyses, whilst they occur in the upper regions of the erect filaments in *Halothrix* and *Leptonematella*, formed by internal division of vegetative cells.

It is likely that the above three genera do not form a natural phylogenetic grouping and that *Halothrix* and *Leptonematella* would be more suitably placed elsewhere. *Leptonematella* has, for example, long been recognised as a problematical genus and has been placed in a number of different families. Most investigators have included it in the Elachistaceae (Knight & Parke, 1931; Newton, 1931; Taylor, 1957; Lund, 1959; Jaasund, 1965; Wynne, 1969; Pankow, 1971; Pedersen, 1976) although Hamel (1931–39) considered that it belonged to the Ectocarpaceae. More recently on the basis of culture studies, Dangeard (1966b) transferred it to the Myrionemataceae and this has been accepted by a number of workers including Parke & Dixon (1976), Rueness (1977) and Kornmann & Sahling (1977). The basis for Dangeard's decision was his observation of branched, erect filaments (which removes the genus from the Elachistaceae) and the presence of a discoid base and true hairs (which removes the genus from the Ectocarpaceae). However, in agreement with Wynne (1969) and Pedersen (1976) the removal of this genus to the Myrionemataceae is considered unjustified and in the present treatment it is transferred back to the Elachistaceae. More recently Pedersen (1978a) allied the development and life history of *Leptonematella fasciculata* with *Pogotrichum filiforme* which was placed into a new family, Pogotrichaceae/Dictyosiphonales, and suggested a closer taxonomic relationship exists between the two species.

Like *Leptonematella*, the genus *Halothrix* is only provisionally assigned to the Elachistaceae (in agreement with most previous authors) pending further investigation. Particularly interesting and requiring consideration is the recent publication of Pedersen

(1984) in which he allies *Halothrix* with *Giraudia* Derbès & Solier and places it in the family Giraudiaceae with a question mark.

ELACHISTA Duby

ELACHISTA Duby (1830), p. 972 (*nom. cons.*).

Type species: *E. scutulata* (Smith) Duby (1830), p. 972.

Thallus epiphytic, forming small brush-like tufts on hosts, with or without basal hemispherical cushions, arising from a prostrate, monostromatic layer of outwardly radiating, branched filaments, free or laterally united and discoid/pseudodiscoid; filaments of tufts erect, free, linear, flaccid, multicellular, monosiphonous, simple or branched at base, accuminate sharply below, more gradually above, with basal meristem, comprising thick-walled cells with numerous discoid plastids without pyrenoids; hemispherical cushions, if present, solid, firm and cartilaginous or slightly soft and gelatinous, pseudoparenchymatous, comprising vertical, compacted, dichotomously branched filaments of large, colourless, thin-walled cells, giving rise terminally to the free erect filaments, paraphyses and/or sporangia, basally to downwardly growing rhizoidal filaments which often penetrate host tissue; paraphyses multicellular, simple, linear or curved, photosynthetic; hairs unknown.

Unilocular and plurilocular sporangia usually formed at the base of free erect filaments, usually lateral and/or terminal on the paraphyses, more rarely borne on the upper regions of the erect filaments; unilocular sporangia pyriform, plurilocular sporangia filiform, uniseriate, simple.

Species of the genus *Elachista* can be generally characterised by an epiphytic habit and small brush-like tufts of free, erect, simple filaments which arise initially from a discoid/pseudodiscoid basal layer, later from a solid, cushion-like, pseudoparenchymatous thallus comprising vertical rows of compacted, branched, filaments of large colourless cells. The reproductive sporangia are usually borne at the base of the free, erect filaments, associated with paraphysis-like filaments.

In the British Isles this genus is represented by four fairly distinct species. *Elachista stellaris* has been placed into two separate genera *Symphoricoccus* and *Areschougia* by Kuckuck (1929) and Meneghini (1844) respectively, on the bases of sporangial development on the erect filaments and the ability of the erect filaments to act as stolons (see Wanders *et al.*, 1972 and Dangeard, 1968a for discussion on this subject). However, in agreement with Wanders *et al.* these are not considered sufficient criteria for placing *Elachista stellaris* in a separate genus.

KEY TO SPECIES

1 Thallus epiphytic on the thongs of the brown alga *Himanthalia*; unilocular sporangia elongate-cylindrical, pedicellate . . . . *E. scutulata*

Thallus not epiphytic on *Himanthalia;* unilocular sporangia pyriform, not pedicellate 2

2 Erect filaments commonly 50–135 µm in diameter; epiphytic on *Cystoseira* spp, more rarely on the brown alga *Halidrys siliquosa* . . . . *E. flaccida*

Erect filaments usually less than 50 µm in diameter; not epiphytic on *Cystoseira* and *Halidrys* 3

3 Epiphytic on *Fucus* spp., rarely other algae; paraphyses obvious and abundant; unilocular sporangia borne at base of erect filaments only; plurilocular sporangia not known on macrothallus . . . . . *E. fucicola*

Not recorded on *Fucus* spp.; paraphyses not obvious; unilocular sporangia borne laterally on the basal and upper regions of the erect filaments; plurilocular sporangia recorded on macrothallus . . . . . *E. stellaris*

Elachista flaccida (Dillwyn) Areschoug (1843), p. 262. Fig. 18; Pl. 2c

Converva flaccida Dillwyn (1802–09), p. 52.
Elachista breviarticulata Areschoug (1842), p. 234.

Plants epiphytic, forming light to yellow-brown, brush-like tufts on hosts, to 1 (–2) cm in height, single or gregarious, arising at the base from a hemispherical, solid, wart-like cushion, 2–4 mm in diameter; cushions firm and cartilaginous or slightly gelatinous, comprising vertical rows of dichotomously branched, closely packed filaments of large, thin-walled, colourless cells, 6–30 µm in diameter, elongate-cylindrical below becoming more globose above, penetrating below into host tissue, giving rise terminally to the erect brush-like tufts of filaments, paraphyses and/or sporangia; erect filaments 7–10 (–16) mm long, free, linear, attenuate at base and apex, monosiphonous, comprising thick-walled cells, 0·5–1 (–2) diameters long, 70–125 µm × 55–160 µm, each with large numbers of small discoid plastids without pyrenoids; paraphyses abundant, closely packed, multicellular, simple, to 160 (–280) µm long, 10 (–15) cells, slightly curved, markedly clavate, with upper cells becoming moniliform, 13–47 × 15–60 µm; hairs unknown.

Unilocular sporangia lateral at the base of the paraphyses, pyriform, 70–90 × 23–33 µm; plurilocular sporangia unknown.

Epiphytic on *Cystoseira* spp., more rarely on *Halidrys siliquosa*, in lower eulittoral pools and shallow sublittoral.

Generally distributed around the British Isles, but apparently absent from south east England (Sussex and Kent) and much of eastern England and eastern Ireland. Recorded for the Channel Isles, Isles of Scilly, eastwards to the Isle of Wight, northwards to Mull, in Ireland northwards to Donegal, eastwards to Wexford, on *Cystoseira* spp. Recorded for scattered localities on *Halidrys* (Isle of Man; Outer Hebrides, Orkneys, Shetland Isles, Berwick).

Annual; spring and summer, April to September, rarely recorded at other times.

Areschoug (1842), p. 234, figs 4–5; Kuckuck (1929), p. 23; Newton (1931), p. 134; Hamel (1931–39), pp. 121–122, fig. 27c; Gayral (1958), p. 237, fig. 35; Gayral (1966), p. 267, fig. 34.

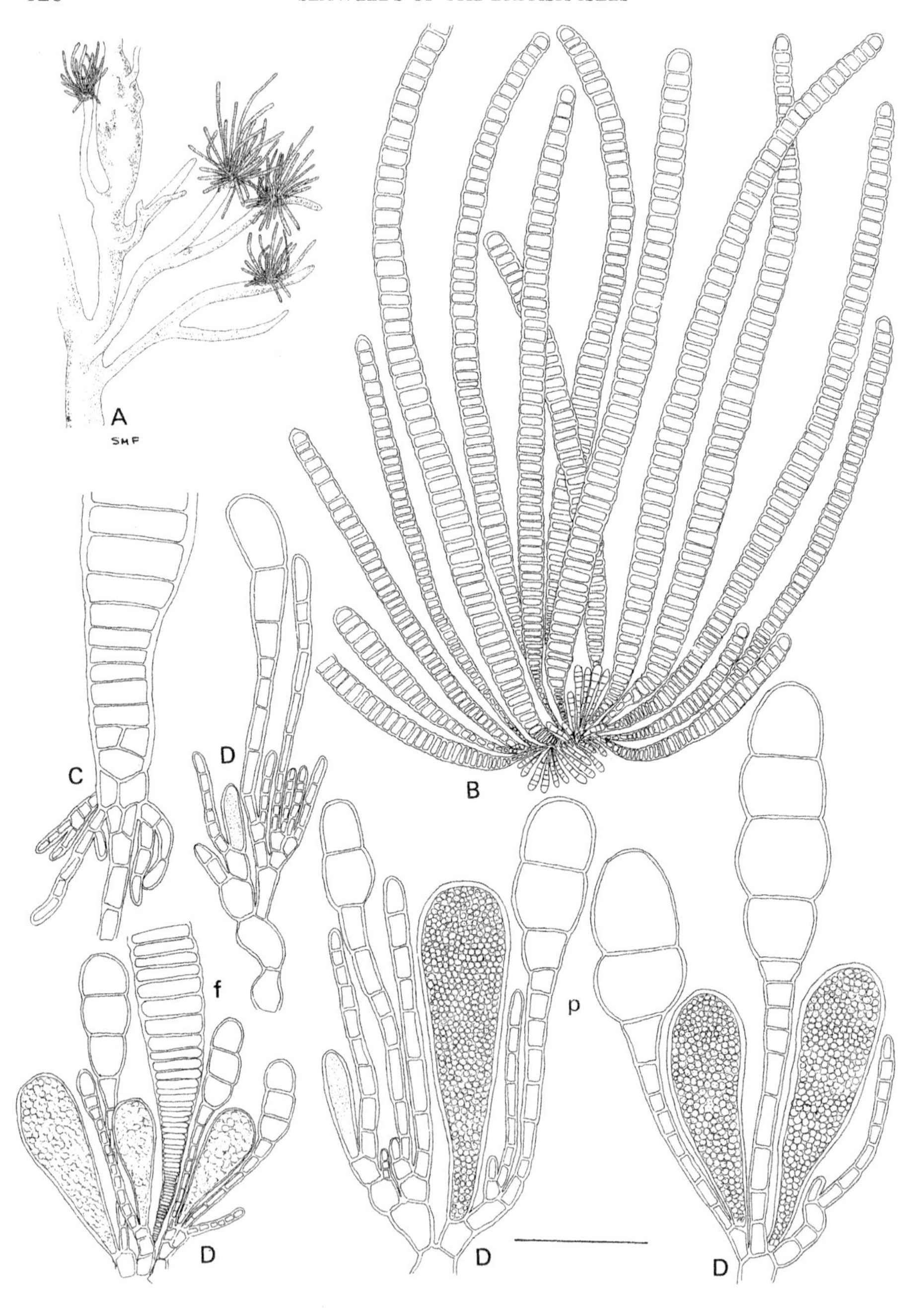
A
SMF
B
C
D
f
D
p
D
D
D

Elachista fucicola (Velley) Areschoug (1842), p. 235. Fig. 19

Conferva fucicola Velley (1795), pl. 4.
Conferva fucorum Roth (1797), p. 190,
Elachista globosa Ørsted (1844), p. 50.
Elachista grevillei Arnott in Harvey (1857), p. 202.
Elachista fucicola f. *grevillei* Hamel (1931–39), p. 119.

Plants epiphytic, forming light to dark brown, brush-like tufts on hosts, to 2 (–4) cm long, single or gregarious, arising either directly from a monostromatic basal layer of outwardly radiating filaments, laterally adjoined and discoid/pseudodiscoid, or from a basal hemispherical, solid, wart-like cushion; basal cushions, if present, cartilaginous or slightly mucilaginous, comprising vertical rows of closely packed, dichotomously branched filaments of large, thin-walled, colourless cells, elongate-cylindrical below and penetrating host tissue, gradually becoming globose above and terminating in the brush-like tufts of free, erect filaments, paraphyses and/or unilocular sporangia; young plants without obvious basal cushion and erect filaments, paraphyses and/or unilocular sporangia arising directly from monostromatic basal layer; erect filaments to 4 cm long, simple, linear, with basal meristem, acute below, acuminate above, uniseriate, comprising thick-walled cells, quadrate to rectangular, 10–43 × 23–40 μm, each with numerous discoid plastids without pyrenoids; paraphyses multicellular, simple, clavate, slightly curved, to 120 μm long, 7–11 cells, basal cells rectangular, terminal cells moniliform to 18 μm in diameter; hairs unknown.

Unilocular sporangia common, lateral at the base of the paraphyses, pyriform, 52–90 × 24–37 μm; plurilocular sporangia not known on macroscopic plant (see below).

Epiphytic on *Fucus* spp., especially *F. vesiculosus*, more rarely on *F. spiralis, F. serratus* and various other algae, eulittoral.

Common and widely distributed all around the British Isles.

Recorded throughout the year, with unilocular sporangia, although usually more common mid-summer to early winter. Young plants develop in spring and early summer, producing unilocular sporangia during autumn and winter as the brush-like tufts of erect filaments are gradually shed from the basal cushion. It is possible that the basal cushions can survive the winter period and produce new erect filaments in the spring.

Culture studies have been carried out on European isolates of *E. fucicola* by Kylin (1937), Kornmann (1962b), Blackler & Katpitia (1963) and Koeman & Cortel-Breeman (1976). In the culture studies of Kylin, Kornmann and Blackler & Katpitia the unispores were observed to grow directly into microscopic branched filaments from which new macroscopic plants developed. Kornmann and Blackler & Katpitia additionally reported the presence of plurilocular sporangia in the cultures, which Kornmann specifically described on the microscopic filaments; plurispores from these sporangia also

Fig. 18 *Elachista flaccida*
A. Habit of plant epiphytic on *Cystoseira*. B. S.P. of thallus showing erect filaments. C. Basal region of erect filament. D. S.P. of fertile thallus showing erect filaments (f), paraphyses (p) and large, pyriform unilocular sporangia. Bar = 18 mm (A), 0·5 mm (B), 50 μm (C–D).

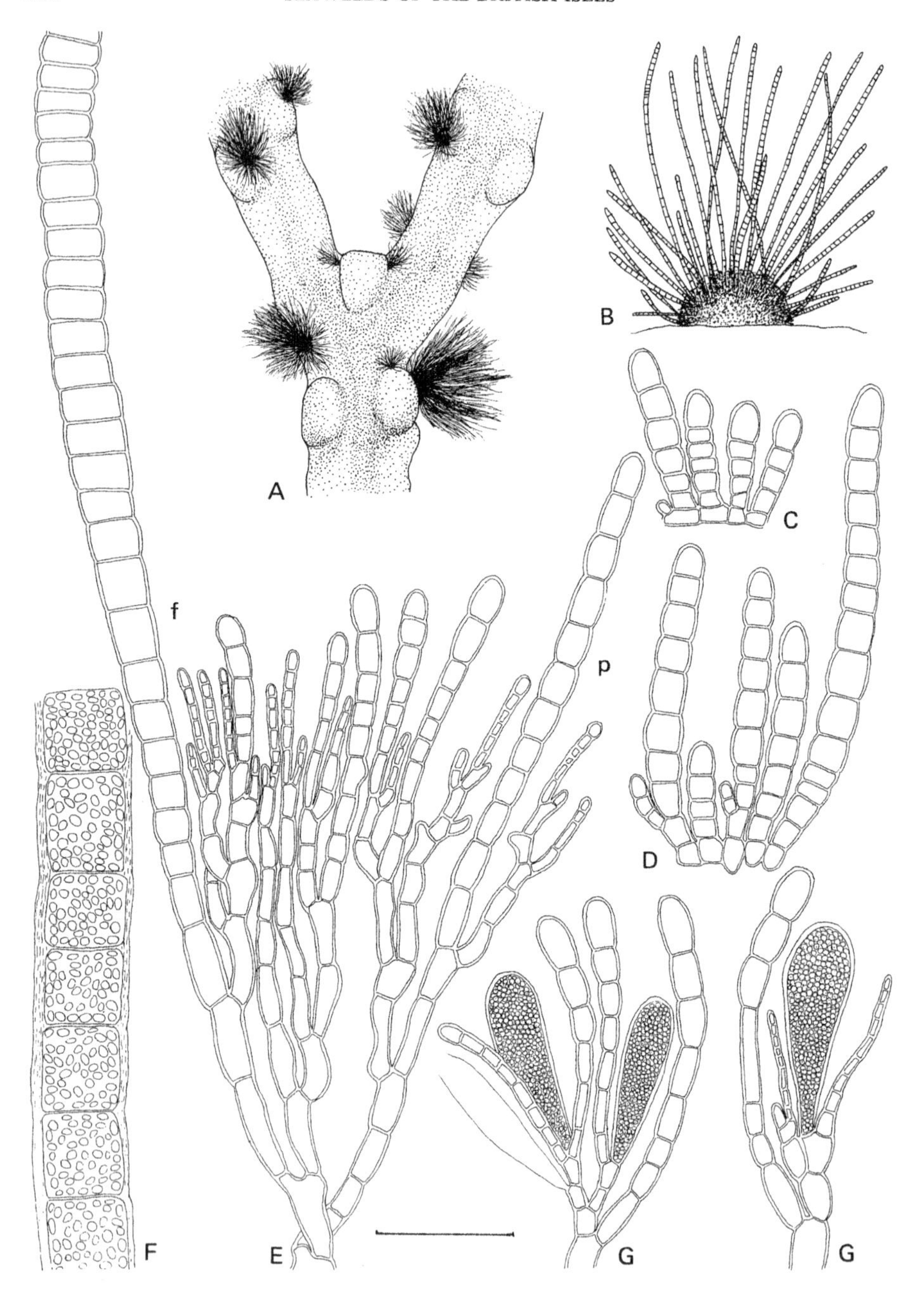
A
B
C
D
f
p
E
F
G
G

germinated directly, without sexual fusion and repeated the parental phase. The species was, therefore, attributed with a 'direct' type of life history with apparently apomeiotic unilocular sporangia. Essentially similar results were obtained by Koeman & Cortel-Breeman, although they did provide additional information about the influence of environmental factors on the life history and about the ploidy level of the two heteromorphic phases (microthallus and macrothallus). For example, higher temperatures (16 and 20°C) promoted the direct budding of the typical erect, macroscopic tufts from the microthallus, whilst under low temperatures (4°C) the microthalli developed plurilocular sporangia. They also observed a haploid number of chromosomes in a few young unispore-derived germlings only, which suggested that meiosis may take place in the unilocular sporangia and that vegetative diploidisation could occur during the development of macrothalli from the microthalli as described in *Elachista stellaris* (Wanders, *et al.*, 1972). However, they conclude that the relationship between the two phases is not an obligate alternation of ploidy levels but a facultative one, and more likely to be controlled by environmental conditions (e.g. temperature and daylength).

Areschoug (1842), p. 235, pl. 8, figs 6–7; Harvey (1849), pl. 240; Harvey (1857), p. 202, pl. 12B, figs 1–3; Batters (1883), p. 110; Setchell & Gardner (1925), pp. 503–504, pl. 38, figs 33–35; Printz (1926), pp. 159–161; Kuckuck (1929), pp. 22–23; Newton (1931), p. 133, fig. 80; Hamel (1931–39), pp. 117–118, fig. 27b; Rosenvinge (1935), pp. 19–24; Kylin (1947), p. 51, fig. 44A; Waern (1952), pp. 153–155; Sundene (1953), pp. 163–164; Taylor (1957), p. 140, pl. 10, figs 1–3; Jaasund (1960), p. 105; Kornmann (1962b), pp. 293–297, figs 1–4; Blackler & Katpitia (1963), pp. 392–395; Jaasund (1965), pp. 64–65; Abbott & Hollenberg (1976), pp. 178–179, fig. 145; Koeman & Cortel-Breeman (1976), pp. 107–117, figs 1–22; Kornmann & Sahling (1977), p. 121, fig. 64; Rueness (1977), pp. 139–140, fig. 67; Pedersen (1979), pp. 151–159, figs 3–7; Stegenga & Mol (1983), p. 92, pl. 29, fig. 1.

Elachista scutulata (Smith) Duby (1830), p. 972. Fig. 20; Pl. 2d

Conferva scutulata Smith (1790–1814), pl. 2311.

Plants epiphytic, forming dark brown/black, low-lying, closely packed tufts on host, to 5 (–10) mm in height, usually discrete and circular, commonly completely enveloping host tissue, sometimes gregarious and irregularly spreading to 2 (–3) cm, arising from a well developed, hemispherical, solid cushion; cushions fairly soft and gelatinous, quite easily squashed, composed of horizontally and vertically spreading, di-trichotomously branched, closely packed filaments of large, thin-walled colourless cells, usually elongate-cylindrical, occasionally moniliform and barrel shaped, 30–70 × 15–45 μm, giving rise below to downwardly or backwardly extending rhizoidal filaments penetrating conceptacle host tissue, giving rise above to the erect sparsely-packed, brush-like tufts of filaments, paraphyses and/or sporangia; erect filaments to 5 (–10) mm long, free, linear,

Fig. 19 *Elachista fucicola*
A–B Habit of plants, epiphytic on *Fucus vesiculosus*. C–D S.P. of young thalli. E. S.P. of thallus showing erect filaments (f) and paraphyses (p). F. Portion of erect filament showing thick walled cells with large numbers of discoid plastids. G. S.P. of thallus showing unilocular sporangia arising at the base of paraphyses. Bar = 36 mm (A), 5 mm (B), 50 μm (C–D, F), 100 μm (E, G).

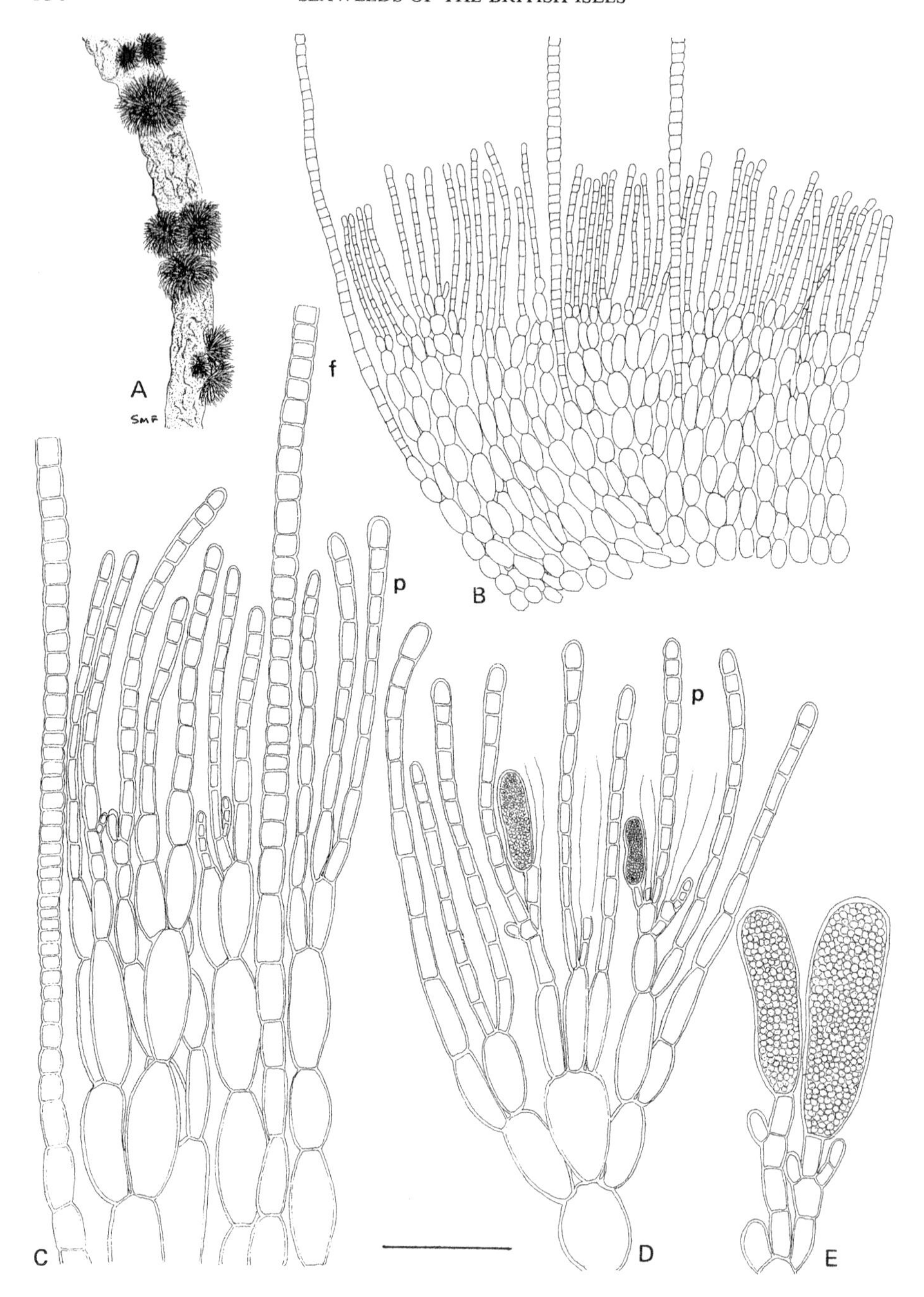

A
SMF
B
f
p
C
D
p
E

uniseriate, comprising thick-walled cells, 1–2 (–3) diameters long, 16–48 × 16–26 µm, each with numerous discoid plastids without pyrenoids; paraphyses abundant, simple, linear to elongate-clavate, multicellular, to 195 (–325) µm long, comprising 10 (–14) cells, 20–47 × 8–21 µm, basal cells elongate-rectangular, 4–6 diameters long, terminal cells rectangular, barrel-shaped or slightly moniliform, 1 (–1·5) diameters long; hairs unknown.

Unilocular sporangia common, borne at the base of the paraphyses, on 1–4 celled pedicels, elongate-cylindrical, 80–120 × 30–55 µm; plurilocular sporangia rare, terminal on modified paraphyses, filiform, uniseriate, simple, to 120 long × *c.* 7–10 µm wide, 5–30 loculi.

Epiphytic on the thongs (receptacles) of *Himanthalia elongata,* emerging from conceptacles, upper to lower eulittoral, exposed or in pools, and shallow sublittoral.

Common and widely distributed all around the British Isles.

Annual, June to October, although infrequently reported throughout the year. Sporangia present in late summer/autumn with the unilocular sporangia occurring first, followed by the plurilocular sporangia. The plurilocular sporangia were first reported by Katpitia & Blackler (1962) and are only known for Fife and Devon.

Thuret (1850), pp. 236–237, pl. 25 (figs 1–3); Newton (1931), pp. 134–135; Hamel (1931–39), pp. 119–120, fig. 28A–B; Katpitia & Blackler (1962), pp. 173–174; Rueness (1977), p. 140, Stegenga & Mol (1983), p. 92, pl. 29, figs 2–4.

Elachista stellaris Areschough (1842), p. 233. Fig. 21

Symphoricoccus stellaris (Areschoug) Kuckuck (1929), p. 34.
Symphoricoccus radians Reinke (1888), p. 17.
Areschougia stellaris (Areschoug) Meneghini (1844), p. 293.
Elachista fracta Gran (1893), p. 28.
Phycophila stellaris (Areschoug) Kützing (1845–71), pl. 97.
Leptonema fasciculatum var. *flagellare* Reinke (1889b), p. 51.

Thallus epiphytic, forming small brush-like, discrete tufts on hosts to 3 (–5) mm long, arising from a small basal, wart-like cushion; basal cushions firm and cartilaginous, or slightly soft and gelatinous, comprising vertical rows of closely packed, di-trichotomously branched filaments of large, thin-walled, colourless, elongate-cylindrical or slightly barrel-shaped cells; lower cells producing downward-growing rhizoidal filaments which penetrate host tissue, terminal cells producing the free erect filaments of the brush-like tufts and sporangia; erect filaments to 5 mm long, linear, simple, tapering slightly at base and apex, with basal meristematic zone, uniseriate, comprising fairly

Fig. 20 *Elachista scutulata*
A. Habit of plant epiphytic on *Himanthalia.* B–C. V.S. of vegetative thallus showing basal cushion of large, thin-walled cells giving rise terminally to exerted filaments (f) and paraphyses (p). D–E. V.S. of fertile thallus showing paraphyses (p) and unilocular sporangia. Bar = 15 mm (A), 200 µm (B), 100 µm (C–D), 50 µm (E).

thick-walled, rectangular to barrel-shaped cells 1–2 (–4) diameters long, 45–104 × 26–50 μm, each with numerous plastids without pyrenoids; hairs unknown.

Unilocular and plurilocular sporangia lateral at the base of the erect free filaments, unaccompanied by obvious paraphyses, more rarely reported in mixed sori on the upper parts of the erect filaments; basally positioned unilocular sporangia pyriform, 85–118 × 28–36 μm, plurilocular sporangia lateral or terminal on small branches, densely crowded, filiform, uniseriate, to 50 μm long (6 loculi) × *c.* 5–8 μm wide; unilocular sporangia on upper parts of erect filaments, globular, lateral, sessile or on one-celled stalks, plurilocular sporangia filiform or conical, lateral.

Epiphytic, on various algae in the lower eulittoral.
Only recorded for Dorset and South Devon.
Spring and summer annual.

Life history investigations of *Elachista stellaris* have been conducted by Kylin (1934, 1937—European isolates) and Wanders *et al.* (1972—Mediterranean isolates). Kylin concluded that this species has an apomeiotic, direct type of life history; zoospores from both unilocular and plurilocular sporangia produced microscopic, fertile (with plurilocular sporangia) filaments from which new macroscopic plants sprouted directly. The culture study by Wanders *et al.* (1972), however, was more significant. They demonstrated a heteromorphic life history (with both a micro- and macrothallus) which was determined by environmental conditions of photoperiod, temperature and nutrients rather than a sexual process. The diploid macroscopic plant (showing the morphology of *E. stellaris*) could reproduce directly by large zoospores from plurilocular sporangia or produce haploid microscopic thalli from smaller zoospores formed (probably by meiosis) in the unilocular sporangia. The microscopic phase could also directly repeat itself via zoospores from plurilocular sporangia, the latter smaller than those developed on the macroscopic plant. New macroscopic plants could then develop directly from the microscopic phase, under certain culture conditions, a phenomenon attributed to the process of 'spontaneous diploidisation'. Their culture results suggest that *E. stellaris* plants would produce plurilocular sporangia throughout the vegetative growth period in the spring and summer, whilst unilocular sporangia would be produced in the autumn. The microthalli produced from the released unispores would then serve as overwintering phases from which would sprout new macrothalli in the following spring.

Reinke (1889a), p. 3, pl. 2, p. 13 (ℵ), pl. 10, figs 10–11; Kuckuck (1929), figs 26–31; Newton (1931), p. 133; Hamel (1931–39), pp. 124–125, fig. 29a–b; Sauvageau (1933b), pp. 179–188; Kylin (1934), pp. 9–13, figs 4–6; Rosenvinge (1935) pp. 24–26; Kylin (1937), pp. 14–15, fig. 5; Kylin (1947), pp. 49–50, fig. 43; Dangeard (1968a), pp. 87–94, pl. 1, figs 1–9; Hoek *et al.* (1972), pp. 57–63, figs 3–5; Pankow (1971), p. 180, figs 219–221; Wanders *et al.* (1972), pp. 458–491, pls I–VI; Rueness (1977), pp. 141–142, fig. 69; Stegenga & Mol (1983), p. 92, pl. 30, fig. 1.

Fig. 21 *Elachista stellaris*
A. Habit of plant epiphytic on *Arthrocladia*. B–G. S.P. of fertile thalli showing a single erect filament with basal unilocular sporangia. Bar = 8 mm (A), 100 μm (B–D), 50 μm (E–G).

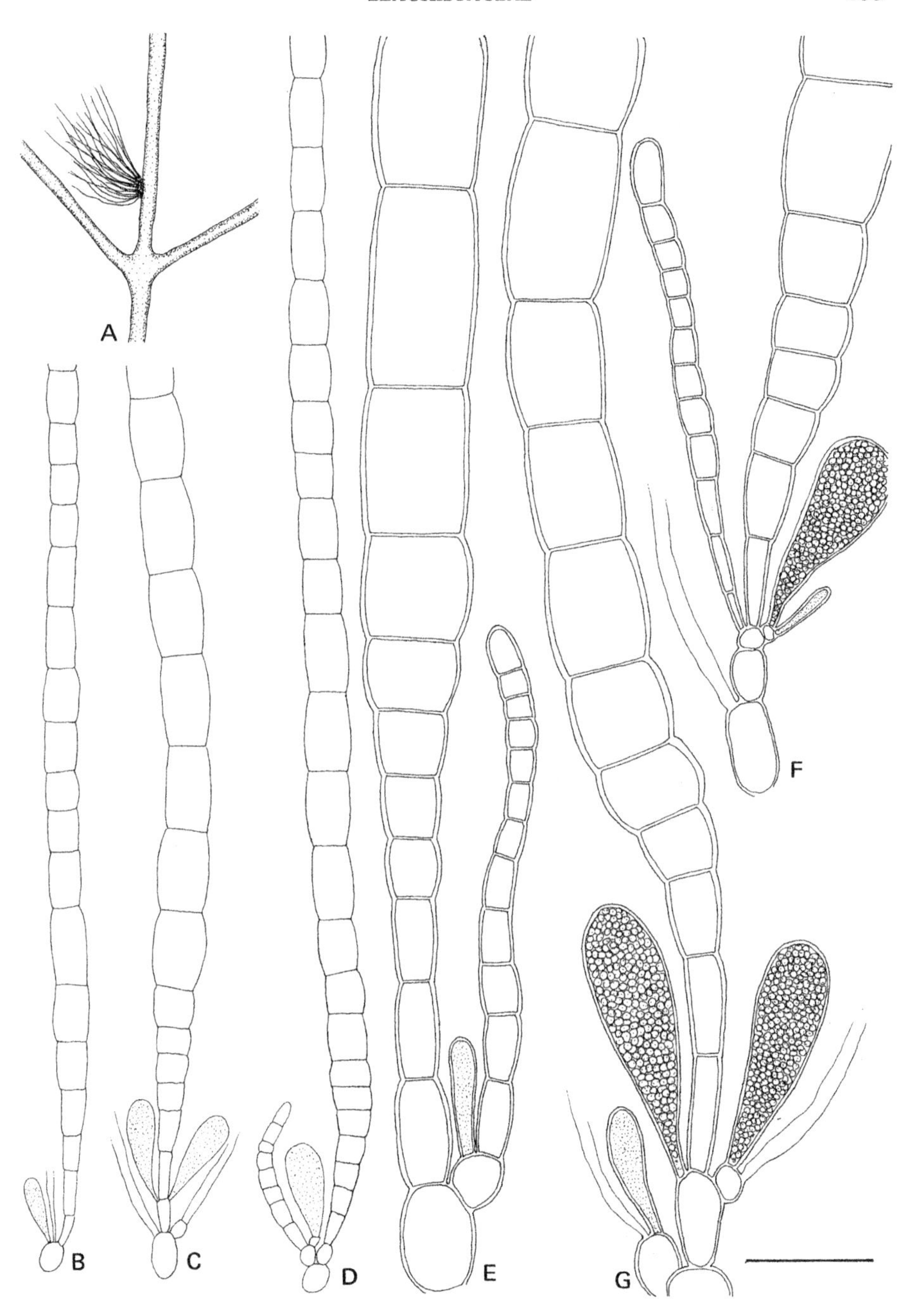
A
B
C
D
E
F
G

HALOTHRIX Reinke

HALOTHRIX Reinke (1888), p. 19.

Type species: *H. lumbricalis* (Kützing) Reinke (1888), p. 19.

Thallus epiphytic, forming small brush-like tufts on host, consisting of erect, simple, or rarely branched, linear, uniseriate filaments arising from a basal system of compacted, branched, rhizoids; growth by an intercalary meristem, usually basally positioned; cells with numerous discoid plastids without pyrenoids.

Plurilocular sporangia intercalary or terminal on erect filaments, formed by subdivision of vegetative cells, densely packed, multiseriate, clustered; unilocular sporangia unknown.

One species in the British Isles:

Halothrix lumbricalis (Kützing) Reinke (1888), p. 19. Fig. 22; Pl. 3

Ectocarpus lumbricalis Kützing (1845), p. 233.
Elachista lumbricalis (Kützing) Hauck (1883–85), p. 354.

Thallus epiphytic, forming small (1–2 mm high), solitary, brush-like tufts on hosts, just visible to the naked eye; consisting of a basal fibrous network of densely packed, irregularly branched, thin, colourless, rhizoidal filaments, 6–8 μm in diameter, penetrating host tissue below, giving rise above to numerous erect filaments; erect filaments predominantly linear throughout, narrowing towards base, simple or rarely branched towards base, comprising cells 1–2 (–2½) diameters long, barrel-shaped to rectangular, 20–85 × 25–65 μm, light coloured, thick-walled, with numerous, well scattered, small discoid plastids without pyrenoids; growth zone intercalary, usually near base, recognisable by a series of distinctly subquadrate, almost lenticular cells; hairs unknown.

Plurilocular sporangia common, formed by internal division of erect filament cells, solitary or more commonly in 1–6 cell chains, slightly enlarged to bulbous, dark coloured, to 88 μm wide, comprising compacted loculi in surface view, 3–5 × 3–4 μm, becoming multiseriate, clustered and projecting later, rounded in surface view; unilocular sporangia unknown.

Epiphytic on *Zostera* leaves in the lower eulittoral and shallow sublittoral.
Rare but widely distributed around the British Isles (Argyll, Isle of Man, Dorset).
Summer annual, recorded April to August.

Culture studies by Pedersen (1979) revealed *Halothrix lumbricalis,* collected from Denmark, to have a 'direct' type of life history. Zoospores from the plurilocular

---

Fig. 22 *Halothrix lumbricalis*
A. Habit of plant showing erect filaments and intercalary plurilocular sporangia. B. Basal region of erect filament. C. Portion of erect filament. D. Portions of erect filament showing plurilocular sporangia. Bar = 200 μm (A), 50 μm (B–C), 100 μm (D).

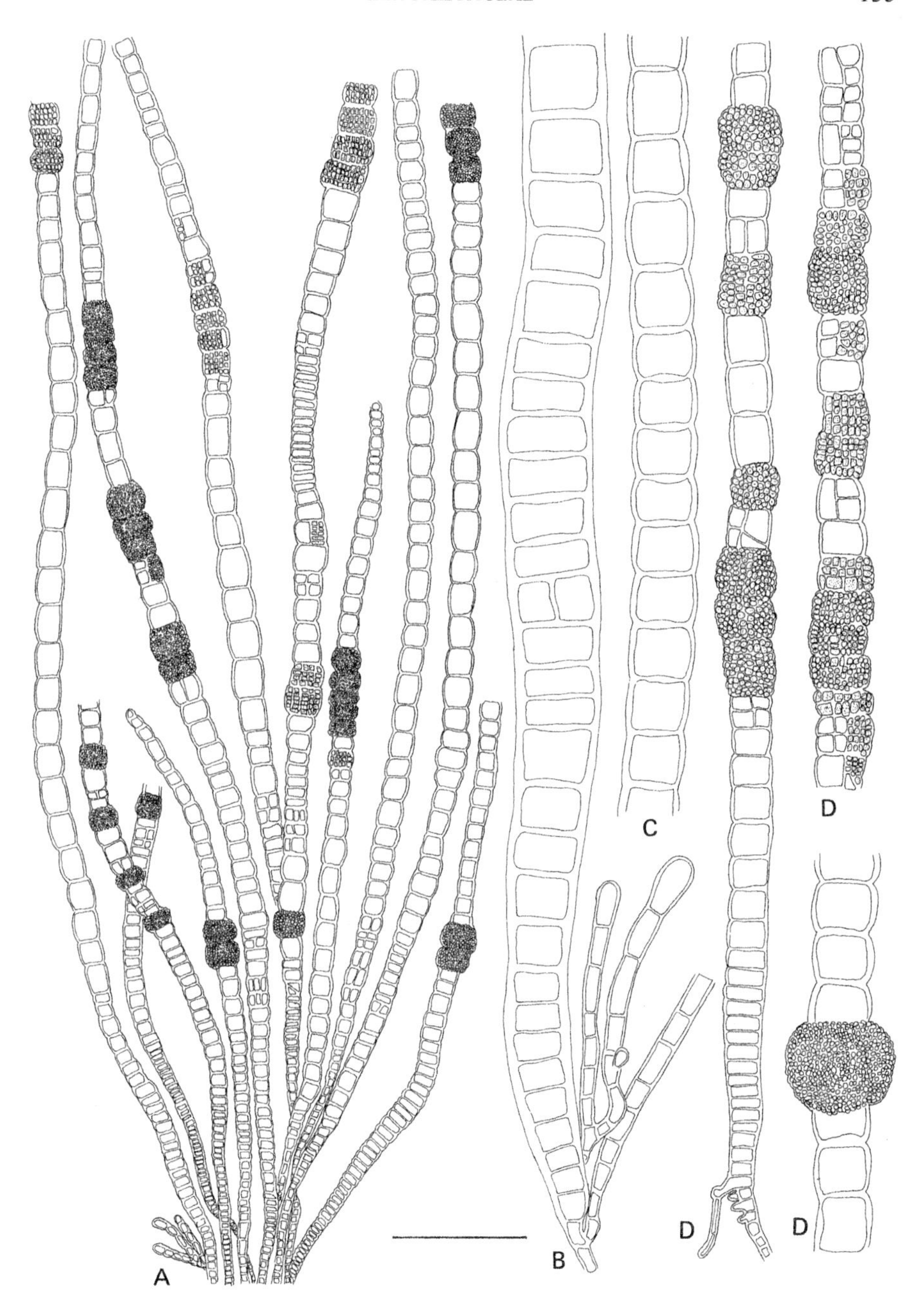
A
B
C
D
D
D
D

sporangia germinated directly, without sexual fusion, to form prostrate systems from which sprouted new erect filaments bearing plurilocular sporangia. Unilocular sporangia were not observed either on field or cultured material.

Reinke (1889a), p. 1, pl. 1; Batters (1892a), p. 174. Kuckuck (1929), pp. 26–28, figs 15–18; Newton (1931), pp. 132–133, fig. 79; Hamel (1931–39), p. 126, fig. 29c–d; Rosenvinge (1935), pp. 37–38, figs. 36–37; Taylor (1957), p. 143; Pankow (1971), p. 180, figs. 223–225; Rueness (1977), p. 141, fig. 68; Pedersen (1979), pp. 151–159, figs. 1–2.

LEPTONEMATELLA Silva

LEPTONEMATELLA Silva (1959), p. 63.

Type species: *L. fasciculata* (Reinke) Silva (1959) p. 63.

Leptonema Reinke (1888), p. 19, non *Leptonema* A. de Jussieu.

Plants epiphytic, epilithic or epizoic, forming small dense tufts of erect filaments arising from a pseudodiscoid base of outwardly radiating branched filaments; erect filaments simple, or little branched at the base, uniseriate, acuminate, with basal intercalary meristem; cells with several plate-like plastids, without pyrenoids in the North Atlantic; hairs unknown.

Unilocular sporangia ovate or elongate-pyriform, sessile or stalked, lateral at the base of the erect filaments; plurilocular sporangia intercalary or terminal on erect filaments, in series, formed by simple subdivision of vegetative cells.

One species in the British Isles:

Leptonematella fasciculata (Reinke) Silva (1959), p. 63. Fig. 23

Leptonema fasciculatum Reinke (1888), p. 19.
L. fasciculatum var. *majus* Reinke (1889b), p. 51; (1889a), p. 13.
L. fasciculatum var. *flagellare* Reinke (1889a), p. 13.
L. fasciculatum var. *uncinatum* Reinke (1889b), p. 51; (1889a), p. 13.
Elachista fasciculata (Reinke) Gran (1893), p. 29.
Leptonema fasciculatum var. *subcylindrica* Rosenvinge (1893), p. 879.

Fig. 23 *Leptonematella fasciculata*
A. Habit of plant. Note basal unilocular sporangia. B. Portions of erect filament showing cells with discoid plastids. C. Portions of fertile erect filaments with mature and dehisced plurilocular sporangia. D–E. Basal regions of erect filaments showing unilocular sporangia. Bar = 0·5 mm (A), 50 μm (B–C, E), 100 μm (D).

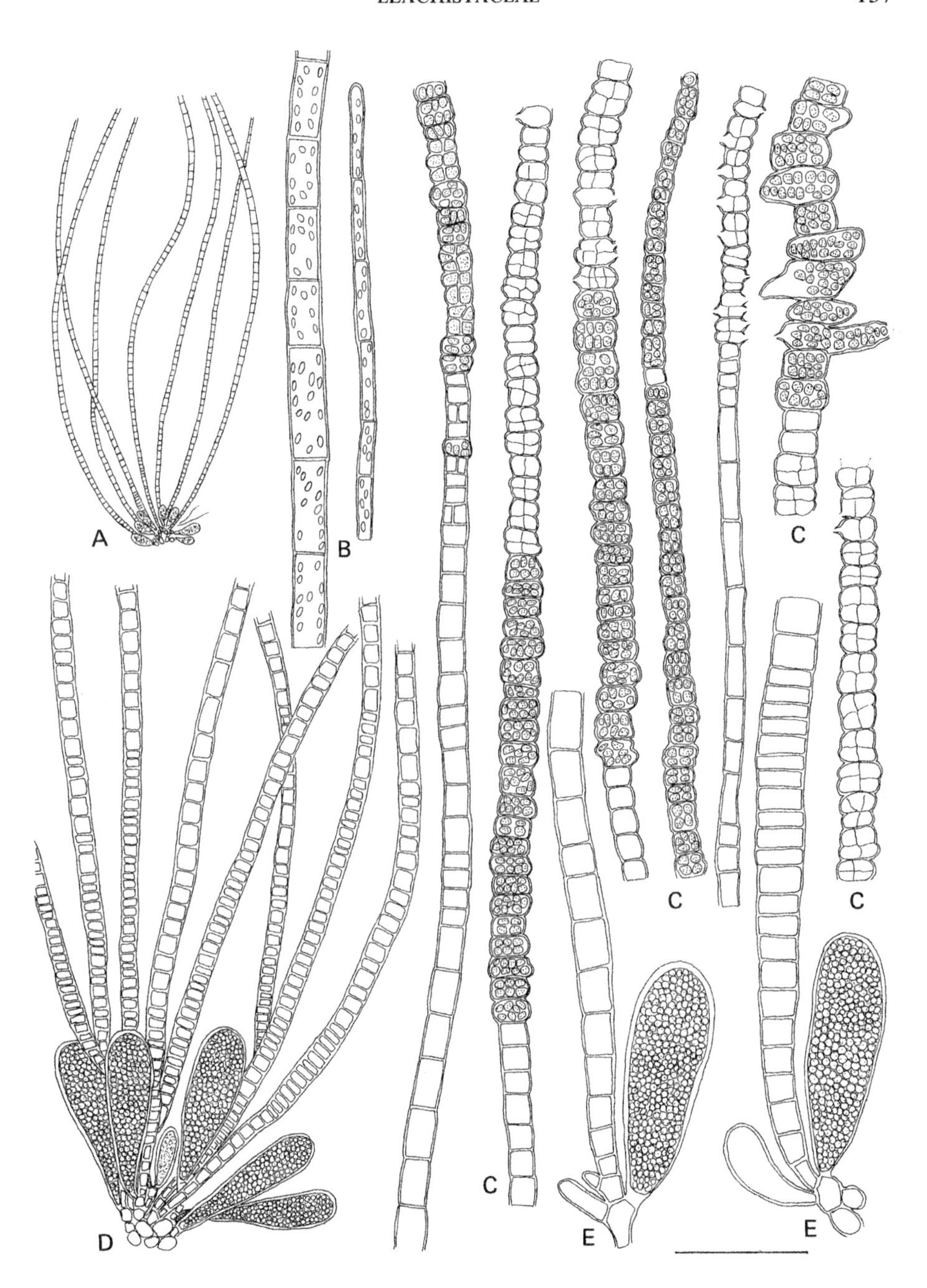
A
B
C
C
C
C
C
D
E
E

Plants usually epiphytic, less commonly epizoic and epilithic, forming erect, dense, light brown, flaccid tufts, solitary, not confluent, arising from the centre of a small pseudodiscoid base; disc comprising outwardly radiating, branched filaments, one-layered or with superimposed layers in parts, either free or closely compacted and pseudodiscoid in structure, cells variable in shape, frequently irregularly distended, 1–4 diameters long, 6–18 × 4–6 µm, with branches often arising from mid cell wall region only; erect tufts to 12 mm long, uniseriate throughout, linear, acuminate, simple or branched towards base, with basal, intercalary meristem, comprising cells largely quadrate or barrel shaped below, rectangular above, 1–3 (–4) diameters long, 11–53 × 7–17 µm, each with several plate-like plastids without pyrenoids in the North Atlantic; hairs unknown.

Unilocular and plurilocular sporangia borne on the same plants; unilocular sporangia rare, borne laterally at the base of erect filaments, sessile or more commonly on one to several celled stalks, ovate or elongate-pyriform, 80–120 × 26–40 µm; plurilocular sporangia common, intercalary or terminal on erect filaments, formed by subdivisions of vegetative cells, usually 4 sporangia per cell produced, 7–9 × 8–18 µm, sometimes laterally protruding to 50 µm in diameter, 4–6 (–8) loculi per sporangium seen in surface view, loculi 3–5 × 3–4 µm.

Epiphytic, epizoic on shells, lower eulittoral pools and sublittoral to 10 m.

Recorded for scattered localities around the British Isles (Shetland Isles, Argyll, Isle of Man, Devon); probably more widely distributed.

Annual, spring and summer.

Leptonematella fasciculata has been investigated in culture by Dangeard (1966b, 1968d, France isolate), Wynne (1969, Washington State isolate) and Pedersen (1978a, West Greenland isolate). All three authors reported the zoospores from the plurilocular sporangia to germinate directly, without sexual fusion, into more or less disc-shaped prostrate systems of uniseriate branched filaments, which developed lateral plurilocular sporangia and sprouted new erect filaments. Plurilocular sporangia, similar to those reported on field collected material, later developed on the erect filaments. Spores from the plurilocular sporangia, on both the erect and creeping filaments, repeated the above sequence of development. Only Pedersen observed unilocular sporangia in culture which developed only under short day conditions. Released unispores behaved in an identical manner to the plurispores. The life history appears, therefore, to be asexual and of the 'direct' type. However, detailed chromosomal studies on material in culture is obviously essential before the exact relationship between the micro- and macrothalli can be determined.

Reinke (1889a), p. 12, pl. 9–10 (figs 1–9); Batters (1892a), p. 174; Batters (1892b), p. 20; Batters (1894), p. 91; Printz (1926), p. 159, pl. 3, fig. 28; Knight & Parke (1931), p. 66, 110–111, pl. 13, figs 35–36, 39; Newton (1931), p. 131, fig. 78A, D, not 78B, C; Hamel (1931–39), pp. 127–128; Taylor (1957), p. 143; Lund (1959), pp. 113–116, fig. 24; Silva (1959), p. 63; Dangeard (1966b), pp. 1692–1694, pls I–II; Dangeard (1968d), pp. 117–130, pls I–III; Dangeard (1969), pp. 86–87; Wynne (1969), pp. 11–13, fig. 4; Pankow (1971), pp. 181–182, figs 226–227; Pedersen (1976), pp. 40–41; Kornmann & Sahling (1977), p. 212, fig. 63; Pedersen (1978a), pp. 61–68, figs 12–16; Stegenga & Mol (1983), p. 83, pl. 25, figs 3–6?; Pedersen (1984), pp. 43–44.

CORYNOPHLAEACEAE Oltmanns

CORYNOPHLAEACEAE Oltmanns (1922), p. 23.

Thallus epiphytic or epilithic, forming small, solitary or confluent globose, hemispherical or finger-like cushions, solid or hollow, simple or shortly forked/branched, usually gelatinous and soft; comprising a basal system of dichotomously branched filaments, free or laterally adjoined and disc-like, radiating outwards on substratum surface, producing erect, branched filaments, either remaining short or more commonly extensive in development, with filaments di-trichotomously divided and closely packed, pseudoparenchymatous, fleshy and mucilaginous in texture, quite easily separable under pressure; comprising large, thin walled cells, elongate-cylindrical or irregularly swollen below, and often producing downwardly extending rhizoidal filaments, gradually becoming smaller, more globose above and terminating in one or more, shorter or longer, simple, pigmented, free or closely compacted filaments (paraphyses), hairs and/or sporangia; plastids plate-like and single or discoid, several to a cell, with pyrenoids; hairs with basal meristem with or without obvious sheath; unilocular and/or plurilocular sporangia present; unilocular sporangia pyriform, elongate-cylindrical or deformed; plurilocular sporangia filiform, simple, uniseriate rarely multiseriate.

Five genera, *Corynophlaea, Leathesia, Microcoryne, Myriactula* and *Petrospongium* occur in the British Isles. Morphologically they comprise small globular, hemispherical or finger-like cushions which are almost exclusively epiphytic on other algae; only *Petrospongium* is commonly reported as a lithophyte. In the British Isles they are also generally either host specific or have been reported epiphytic on only a limited range of algae. For example, *Corynophlaea* is apparently confined to the host species *Chondrus crispus*, epiphytic *Petrospongium* is confined to the brown crustose alga *Ralfsia verrucosa* whilst 3 species of *Myriactula* are each confined to a single host species or genus.

In many respects they resemble, both morphologically and anatomically, the basal, cushion-like pads formed in most species of *Elachista*. In structure they comprise a pseudoparenchymatous thallus of erect, closely-packed, branched filaments of large, colourless cells (sometimes referred to as the 'medulla'), giving rise peripherally to short, multicellular, usually simple, pigmented filaments (termed paraphyses in view of their similarity to these structures in *Elachista*), sporangia and hairs (sometimes referred to as the 'cortex'). Unlike *Elachista*, however, they do not additionally form the long, free, filaments contributing to the brush-like tufts. All thalli are soft and lubricous and easily squashed under light pressure.

Corynophlaea appears to be very closely related to *Leathesia;* indeed it was included as a subgenus of *Leathesia* by Newton (1931). It differs from *Leathesia* in the following points: it forms generally smaller thalli, the erect, colourless, (medullary) cells are less irregularly contorted and not stellate and the paraphyses are much longer and less compact. The single representative species for the British Isles, *C. crispa*, is also host specific (to *Chondrus crispus*) and confined to south west shores of the British Isles, unlike the British Isles representative of *Leathesia* i.e. *L. difformis* which can occur on various host algae and is distributed all around the British Isles.

Petrospongium is represented in the British Isles by a single species *P. berkeleyi* (Greville in Berkeley) Nägeli in Kützing, based on *Chaetophora berkeleyi* described by Greville in Berkeley (1833). It was transferred by the Crouan brothers in 1851 to *Cylindrocarpus* (type species *C. microscopicus* Crouan frat. 1851), a position which has been accepted by most workers (Kuckuck, 1929; Hamel, 1931–39; Feldmann, 1937; Parke & Dixon, 1976). However, based on examination of material of both *Cylindrocarpus berkeleyi* and *C. microscopicus* collected in the British Isles it is concluded that they probably represent unrelated taxa. *C. microscopicus* differs from *C. berkeleyi* in a number of features. It has an extensive network of endophytic filaments in the host red alga *Gracilaria bursa-pastoris*, which emerge as erect microscopic tufts of branched filaments rather than gelatinous cushions; the erect filaments are also not closely packed into a pseudoparenchyma and, although the basal and mid region cells are slightly swollen, there is no obvious differentiation into a medulla-like region of large, thin-walled colourless cells. *C. berkeleyi* is, therefore, transferred to *Petrospongium* where it has been retained by Newton (1931) whilst *C. microcopicus* is removed from the Corynophlaeaceae pending further studies revealing it true taxonomic identity. In this respect it is interesting to note that Batters (1902), Setchell & Gardner (1925) and Newton (1931) linked *C. microscopicus* with the Ectocarpaceae.

Although some workers such as Rosenvinge (1935) and Taylor (1957) recognise a broad concept of the Elachistaceae and include genera such as *Myriactula*, in most studies (Kuckuck, 1929; Newton, 1931; Hamel, 1931–39; Abbot & Hollenberg, 1976; Rueness, 1977) the Corynophlaeaceae is maintained as a separate family to include the small, cushion forming genera. However, it is likely that future studies will reveal the Corynophlaeaceae to be an unnatural group of genera and that the affinity of one or more of them probably lies elsewhere in other families. In this respect it is interesting to note that genera such as *Leathesia* and *Microcoryne*, for example, have often been included in the family Chordariaceae (see Taylor, 1957 for *Leathesia;* see Batters, 1902; Kylin, 1947; Sundene, 1953 for *Microcoryne*).

CORYNOPHLAEA Kützing

CORYNOPHLAEA Kützing (1843), p. 331.

Type species: *C. umbellata* (Agardh) Kützing (1843), p. 331.

Thallus epiphytic, forming small, gelatinous hemispherical cushions on hosts, less than 3 mm in diameter; comprising a prostrate base of outwardly spreading, pseudodichotomously branched, monostromatic filaments, free or irregularly associated and pseudodiscoid, each cell giving rise to erect filaments; erect filaments short, di-trichotomously branched, loosely compacted in a gelatinous matrix, comprising large thin-walled, colourless cells, elongated, cylindrical and narrow below becoming broader and more oval above, terminating in narrow thread-like photosynthetic filaments (paraphyses) hairs and/or sporangia; paraphyses multicellular, gelatinous, simple or branched, slightly recurved with terminal moniliform cells; plastids single, lobed, plate-like or several and discoid, with pyrenoids; hairs with basal meristem but lacking obvious sheath.

Unilocular and plurilocular sporangia borne at the base of the terminal filaments; unilocular sporangia elongate-pyriform; plurilocular sporangia unknown for the British Isles, reported elsewhere to be cylindrical, uniseriate, rarely biseriate.

One species in the British Isles:

Corynophlaea crispa (Harvey) Kuckuck (1929), p. 42. Fig. 24; Pl. 2e

Leathesia crispa Harvey (1857), p. 201.
Leathesia concinna Kuckuck (1897), p. 387.

Thallus forming small light brown, hemispherical spots on host, usually circular and solitary, occasionally irregular and confluent, to 1–3 mm across; consisting of a prostrate system of branched pseudodiscoid filaments giving rise to erect, dichotomously branched filaments loosely compacted together in a gelatinous matrix and easily separable under pressure, erect filament cells, thin-walled, colourless except for 1–4 small plate-like/discoid plastids, elongate below, 75–150 × 10–35 µm, 3–6 (–10) diameters long, becoming more swollen above, 40–120 × 25–50 µm, 1–3 diameters long, finally becoming globular terminally, 18–34 × 15–30 µm, 1–2 diameters wide, and giving rise to numerous paraphyses, hairs and/or unilocular sporangia; paraphyses simple or branched, multicellular, 9–17 cells, 65–140 µm long, curled, comprising cells rectangular or barrel shaped near base, 1–3 diameters long, 6–18 × 5–9 µm, approximately quadrate and crenate in appearance near apex, to 14 µm long × 8–12 µm in diameter; cells usually with one, more rarely two, parietal, lobed, plate-like plastids, with pyrenoids; hairs common, arising from cells at base of paraphyses, 8–9 µm in diameter, with basal meristem, but without obvious sheath.

Unilocular sporangia common, pyriform, 40–70 × 16–25 µm, solitary or in small groups at base of paraphyses; plurilocular sporangia unknown in the British Isles, reported elsewhere to be rare, usually simple and uniseriate, more rarely biseriate, occasionally branched, terminal or lateral on paraphyses.

Epiphytic on blades, especially ones yellow and senescent, of the red alga *Chondrus crispus*, upper eulittoral to shallow sublittoral.

South and west coasts of the British Isles, as far north as Argyll and as far east as Dorset; in Ireland, Mayo, Clare, Wicklow.

Annual, summer and autumn, April to September.

Without microscopic examination this species can easily be confused with young plants of *Leathesia difformis*. However, the latter are usually lighter in colour and more firm and gelatinous with surface cells much more closely packed. Squash preparations of *C. crispa* often reveal endophytic thalli of *Streblonema*-like plants.

Harvey (1857), pp. 201–202, Pl. XII, figs 1A–3A; Batters (1906), p. 2; Cotton (1908a), p. 329; Newton (1931), p. 141; Hamel (1931–39), pp. 142–143, fig. 32F–G; Pybus (1975), pp. 153–155, fig. 1; Pedersen (1983), pp. 3–4, fig. 6a–d.

LEATHESIA Gray

LEATHESIA Gray (1821), p. 301.

Type species: *Leathesia tuberiformis* (J. E. Smith) S. F. Gray (1821), p. 301 (= *Leathesia difformis* (Linneaus) Areschoug (1847), p. 376).

Thallus epiphytic or epilithic, globose, gelatinous and soft, quite easily squashed under pressure, smooth or irregularly convoluted, solid when young, becoming hollow in

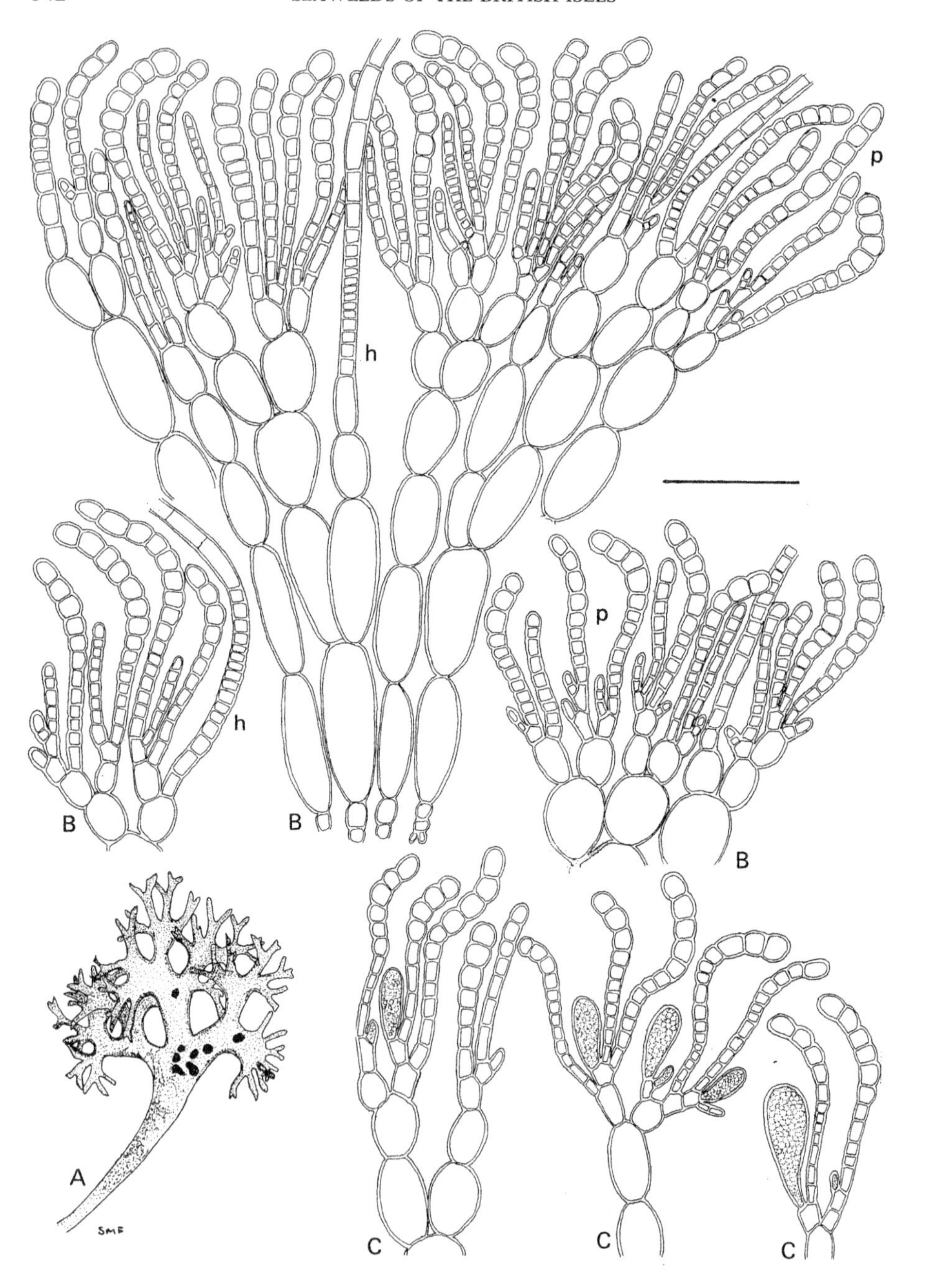
p
h
h
p
B
B
B
A
SMF
C
C
C

mature specimens; comprising erect rows of di-trichotomously branched compacted filaments of large, irregularly shaped and contorted, colourless cells, elongate-cylindrical below, progressively becoming globose towards the periphery and producing short, 2–5 celled, clavate, pigmented, tightly packed, terminal branches (paraphyses); paraphyses' cells with several discoid plastids with pyrenoids; hairs single or in fascicles, arising at base of paraphyses, with basal meristem and sheath.

Plurilocular and unilocular sporangia arising at base of paraphyses; plurilocular sporangia simple, filiform and uniseriate; unilocular sporangia globose or pyriform.

One species in the British Isles:

Leathesia difformis (Linnaeus) Areschoug (1847), p. 376. Fig. 25

Tremella difformis Linnaeus (1755), p. 429.
Nostoc marinum C. A. Agardh (1810–12), p. 45.
Chaetophora marina (C. A. Agardh) Lyngbye (1819), p. 193.
Leathesia tuberiformis Gray (1821), p. 301.
Nostoc mesentericum C. A. Agardh (1824), p. 21.
Corynephora marina C. A. Agardh (1824), p. 24.
Leathesia marina (C. A. Agardh) J. G. Agardh (1848–1876), p. 52.
Corynephora baltica Kützing (1845–71), pl. 211.

Thallus epiphytic, more rarely epilithic, light to yellow brown in colour, firm fleshy and mucilaginous in texture, solid, globose and smooth when young, hollow, irregular and convoluted later, either solitary and discrete or confluent and expanding to 5 (−8) cm in diameter; in squash preparation thalli constructed of upward and outwardly radiating, di-trichotomously branched filaments of cells, loosely packed together and fairly easily separable under pressure, solid throughout at first becoming irregularly hollow later; comprising large, elongate-cylindrical or irregularly contorted and almost stellate, colourless cells, becoming smaller and more globose towards the periphery, terminating in short 2–5 celled clavate, simple, densely crowded, pigmented filaments (paraphyses) hairs, unilocular and/or plurilocular sporangia; cells of paraphyses 2–3 diameters long, elongate-cylindrical below, globose terminally, with 1–3 plate-like plastids with pyrenoids; hairs and sporangia arising at the base of the paraphyses; hairs single or in fascicles with basal meristem and sheath.

Unilocular sporangia globose or pyriform, 20–35 × 17–23 μm; plurilocular sporangia filiform, simple, uniseriate, 15–33 × 4–7 μm, to 5–7 loculi.

Epiphytic on diverse algae, especially the red alga *Corallina officinalis*, upper to lower eulittoral, in pools or emergent.

Commonly distributed around the British Isles.

Annual; spring and summer.

Fig. 24 *Corynophlaea crispa*
A. Habit of plant epiphytic on *Chondrus crispus*. B. S.P. of vegetative thalli showing terminal hairs (h), and paraphyses (p). C. Portions of fertile thalli showing paraphyses and unilocular sporangia. Bar = 30 mm (A), 50 μm (B–C).

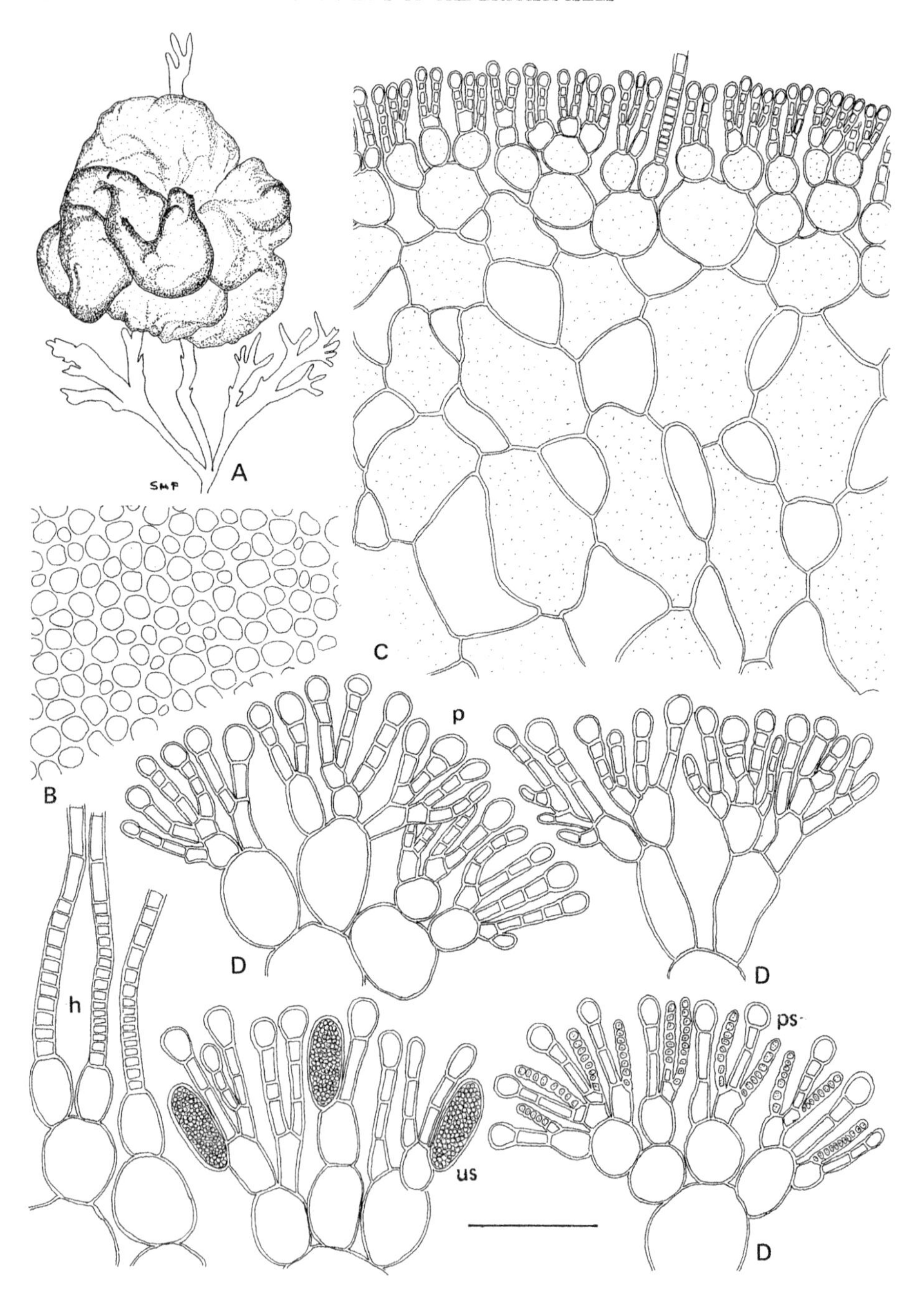
SMF
A
B
C
p
D
D
h
ps
us
D

European isolates of *Leathesia difformis* have been studied in culture by Sauvageau (1925, 1928, 1929), Dammann (1930), Kylin (1933) and Dangeard (1965c, 1969). It is generally considered that the life history is heteromorphic and basically comprises a diploid, sporophytic macrothallus which alternates with a haploid gametophytic microthallus. The life history is further complicated, however, by a direct recycling of the diploid sporophyte via plurispores from plurilocular sporangia on the macrothallus. A similar direct recycling of the haploid gametophyte via plurispores from plurilocular sporangia on the microthallus (by parthenogenesis) also occurs which is in addition to their reported gametic sexual role in returning the sporophyte. Dangeard additionally reported a 'short circuit' life history with the development of the diploid macrothallus via sexual fusion of unispores from the unilocular sporangia (but note Müller 1975 generally casts doubt on reports of unispores fusing). A noteworthy feature of the culture studies (commented upon by Sauvageau, 1925) was the morphological resemblance between the microthalli of *Leathesia difformis* and *Myrionema*. This adds further support to Pedersen's (1984) hypothesis that this genus does not represent a natural phylogenetic unit.

Sauvageau (1925), pp. 1632–1635; Sauvageau (1928), p. 268; Sauvageau (1929), p. 403; Setchell & Gardner (1925), pp. 511–512, pl. 40, fig. 52, pl. 43, figs 65, 66; Dammann (1930), p. 11, fig. 3, pl. 1 (1); Newton (1931), p. 141, fig. 87; Hamel (1931–39), pp. 138–140, fig. 32A–D; Kylin (1933), pp. 64–66, fig. 30; Taylor (1957), p. 149, pl. 12, fig. 5, pl. 14, fig. 8; Rosenvinge & Lund (1947), pp. 8–11, fig. 1; Kylin (1947), pp. 53–54, fig. 46; Lindauer *et al.* (1961), pp. 220–221, fig. 44; Dangeard (1965c), pp. 5–43, pls 1–5; Dangeard (1969), pp. 79–81, fig. 4; Abbott & Hollenberg (1976), pp. 176–177, fig. 142; Rueness (1977), pp. 142–143, fig. 70; Stegenga & Mol (1983), pp. 92–93, pl. 31.

MICROCORYNE Strömfelt

MICROCORYNE Strömfelt (1888), p. 382.

Type species: *M. ocellata* Strömfelt (1888), p. 382.

Thallus epiphytic, erect, minute, globose, cylindrical or clavate, simple or shortly forked/branched, solid, soft and gelatinous, protruding from hosts; constructed of vertical and outwardly radiating rows of closely-packed gelatinous, di-trichotomously branched filaments, easily separable under pressure, comprising colourless cells, predominantly elongate-cylindrical, becoming more globose towards the periphery, terminating in paraphyses, plurilocular sporangia and hairs; paraphyses multicellular, simple, curved comprising cells with several discoid plastids and pyrenoids; hairs with basal meristem without obvious sheath.

Plurilocular sporangia in dense fascicles at base of paraphyses, cylindrical, simple, uniseriate or partly biseriate; unilocular sporangia unknown.

One species in the British Isles:

Fig. 25 *Leathesia difformis*
A. Habit of plant. B. Surface view of thallus showing loosely associated cells in gelatinous matrix. C. V.S. of thallus showing large, irregularly shaped central cells giving rise terminally to short filaments (paraphyses) and a single hair. D. Portions of peripheral thallus region showing paraphyses (p), hairs (h), unilocular (u.s.) and plurilocular sporangia (p.s.). Bar = 17 mm (A), 50 µm (B, D), 100 µm (C).

Microcoryne ocellata Strömfelt (1888), p. 382. Fig. 26; Pl. 4a

Thallus epiphytic, forming small, erect, solitary and discrete extensions from host surface, globose, cylindrical or more commonly clavate, simple or once forked, more rarely with short, radiating branches, solid, soft and gelatinous, easily squashed, to 4 (–5) mm long × 1 mm wide; in squash preparation consisting of upward and outwardly radiating, di-trichotomously branched filaments, loosely packed together and easily separable under pressure, comprising elongate-cylindrical, colourless cells up to 350 μm or more in length × c. 18–25 μm in diameter, becoming shorter, more globose towards the periphery, 32–112 × 17–35 μm, terminating in paraphyses, plurilocular sporangia and/or hairs; paraphyses quite densely crowded, multicellular, simple, curved, to 12(–16) cells, (155–200 μm) long, pigmented, comprising cells rectangular below, becoming barrel-shaped and/or moniliform above, 7–21 × 6–12 μm, each with several discoid plastids and pyrenoids; hairs with distinct basal meristem, enclosing sheath not observed.

Plurilocular sporangia common, in dense fascicles at the base of the paraphyses, arising directly from large colourless cells, either sessile or on branched 1–2 celled stalks, to 15 loculi long, 55 × c. 6–9 μm, simple, cylindrical uniseriate, often partly biseriate, with straight, sometimes oblique cross walls, comprising quadrate, more commonly subquadrate or lenticular loculi, 3–5 × 6–8 μm; unilocular sporangia unknown.

Epiphytic on various algae including *Chorda filum, Dasya hutchinsiae, Palmaria palmata* and *Sauvageaugloia griffithsiana,* in lower eulittoral pools and shallow sublittoral.

Recorded only for Dorset on the south coast of England and Mayo, Ireland; probably more widely distributed around the British Isles.

Summer annual; May to September.

Microcoryne ocellata was originally described as an epiphyte on *Chorda filum* in Norway by Strömfelt (1888). It was subsequently reported for the British Isles, at Weymouth, Dorset, by E. A. L. Batters (Batters, 1892c) and later at Portland by E. M. Holmes (reported in Batters, 1893b), epiphytic on various host algae. It was also reported for Clare Island, County Mayo, Ireland by Cotton (1912). No subsequent collections of this species appear to have been made in the British Isles and it has not been detected in the field by the present author despite numerous searches. The present description and ecological notes are based on the examination of material deposited in the British Museum (Natural History) by E. A. L. Batters, E. M. Holmes and A. D. Cotton.

Strömfelt (1888), p. 382, pl. 3 (2–3); Batters (1892c), pp. 51–52; Batters (1893b), p. 51; Kylin (1907), pp. 81–83, figs 18–19; Kuckuck (1929), pp. 45–46, fig. 53; Newton (1931), pp. 139–140, fig. 85; Hamel (1931–39), pp. 143–144; Rosenvinge & Lund (1947), p. 11; Kylin (1947), pp. 52–53, fig. 45; Rueness (1977), p. 143, fig. 71; Stegenga & Mol (1983), pp. 92–93.

Fig. 26 *Microcoryne ocellata*
A. Habit of plant, epiphytic on *Sauvageaugloia griffithsiana*. B. Habit of young plant. C. S.P. of portion of young plant showing elongated cells of central filaments. D–E. Terminal thallus portions showing paraphyses (p). F–G. Portions of fertile thalli showing plurilocular sporangia arising at base of paraphyses. Bar = 4 mm (A), 0·5 mm (B–C), 50 μm (D–G).

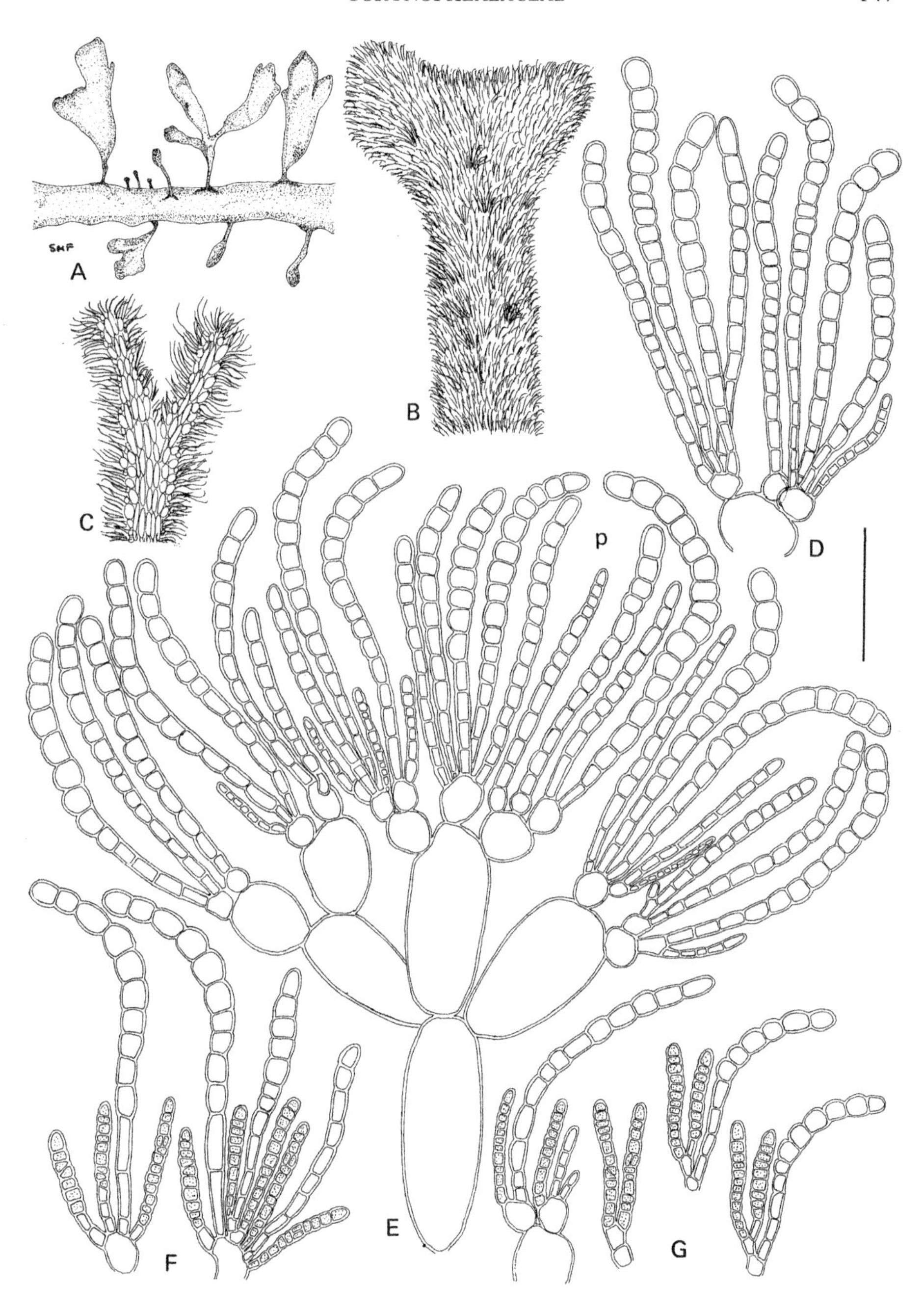
SMF
A
B
C
D
p
E
F
G

MYRIACTULA Kuntze

MYRIACTULA Kuntze (1891–98), p. 415.

Type species: *M. pulvinata* (Kützing) O. Kuntze (1891–98), p. 415. (=*M. rivulariae* (Suhr in Areschoug) J. Feldmann).
Myriactis Kützing (1843), p. 330.
Gonodia Nieuwland (1917), p. 30.

Thallus epiphytic, sometimes partly endophytic, forming minute pustules, cushions or tufts on hosts, solitary or gregarious, discrete or sometimes confluent and spreading; comprising a basal system of branched filaments, with or without downwardly extending endophytic/rhizoidal filaments, giving rise to erect, di-trichotomously branched filaments terminating in paraphyses, hairs and/or sporangia; erect filaments remaining short and little differentiated or vertically extensive and becoming closely packed to form a soft and gelatinous, somewhat cartilaginous pseudoparenchyma of large, thin-walled, colourless cells, elongate-cylindrical below giving rise to downwardly growing rhizoidal filaments, becoming more globose above; paraphyses loosely associated, gelatinous, multicellular, simple, linear, clavate or fusiform, comprising cells with several discoid plastids and pyrenoids; hairs with basal meristem without obvious sheath.

Unilocular and plurilocular sporangia borne at the base of the paraphyses, sessile or shortly stalked; unilocular sporangia pyriform; plurilocular sporangia clustered, cylindrical, simple, uniseriate.

KEY TO SPECIES

1 Thallus epiphytic on the thongs of the brown alga *Himanthalia elongata;* paraphyses commonly less than 14 cells in length . . . . *M. areschougii*
Thallus not epiphytic on *Himanthalia elongata;* paraphyses commonly greater than 14 cells in length 2
2 Thallus epiphytic on *Fucus* spp; paraphyses up to 250 μm in length; cells of paraphyses less than 10 μm in diameter . . . . *M. clandestina*
Thallus not epiphytic on *Fucus* spp; paraphyses commonly greater than 250 μm in length; cells of paraphyses greater than 10 μm in diameter . 3
3 Paraphyses distinctly fusiform; cells of paraphyses larger than 25 μm in diameter; thallus epiphytic on *Cystoseira* and *Sargassum*, more rarely on *Halidrys* and other algae . . . . *M. rivulariae*
Paraphyses clavate, linear or slighty fusiform; cells of paraphyses less than 25 μm in diameter; thallus not reported on *Cystoseira, Sargassum* and *Halidrys* 4
4 Cells of paraphyses less than 16 μm in diameter; thallus epiphytic on *Scytosiphon lomentaria* only . . . . *M. haydenii*

Fig. 27 *Myriactula areschougii*
A. Habit of plant on *Himanthalia*. B. V.S. of thallus embedded within host tissue. C. Portions of thalli showing terminal paraphyses (p), hairs (h) and unilocular sporangia. Bar = 2 mm (A), 200 μm (B), 50 μm (C).

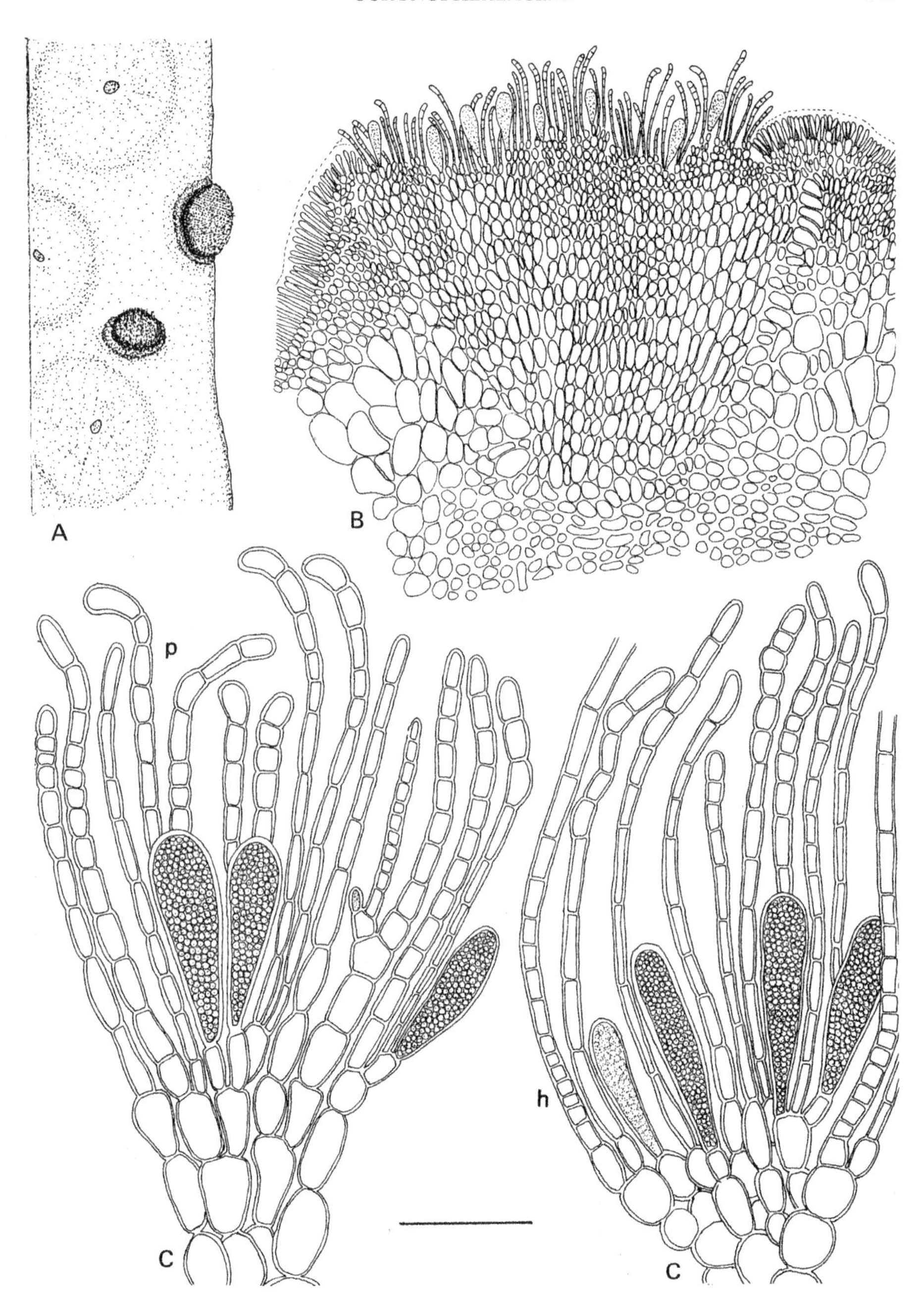
A
B
p
h
C
C

Cells of paraphyses commony greater than 16 µm in diameter; thallus recorded on several hosts, only occasionally on *Scytosiphon lomentaria* 5

5 Thallus epiphytic on *Dictyota dichotoma* only; paraphyses up to 285 µm in length . . . . *M. stellulata*

Thallus not reported on *Dictyota dichotoma;* paraphyses up to 470 µm in length . . . . *M. chordae*

Myriactula areschougii (Crouan frat.) Hamel (1931–39), p. xxxii. Fig. 27

Elachistea areschougii Crouan frat. (1867), p. 160.
Myriactis areschougii (Crouan frat.) Batters (1892a), p. 173.
Gonodia areschougii (Crouan frat.) Hamel (1931–39), p. 132.

Plants epiphytic, forming small, slightly raised pustules on host, circular or occasionally oblong, solitary or rarely grouped, to 2 mm in diameter and just visible to the naked eye; in surface view plants turf-like, similar to a carpet pile, slightly projecting above, and surrounded by raised rim of host material; comprising a soft, gelatinous, somewhat cartilaginous cushion occupying host conceptacle, consisting of upwardly growing, closely packed, dichotomously branched filaments of large, thin-walled, colourless cells, elongate-cylindrical below and occasionally giving rise to branched, multicellular rhizoidal filaments, ramifying down into host tissue, gradually becoming smaller, more globose, above and terminating in paraphyses, unilocular sporangia and/or hairs; paraphyses erect, simple, multicellular, slightly elongate-clavate, curved, usually 8–14 cells, 180–250 µm long, comprising predominantly elongate-rectangular, thin-walled cells throughout, 2–5 diameters long, 13–40 × 5–10 µm becoming less elongate towards the apex, thick-walled, rectangular, barrel-shaped or somewhat irregularly swollen, 1–2 diameters long, 10–21 × 7–13 µm often with terminal cells ruptured, all cells with numerous discoid plastids, with pyrenoids; hairs uncommon, with basal meristem and sheath, 8–10 µm in diameter.

Unilocular sporangia borne at base of paraphyses, elongate-pyriform, 65–105 × 18–28 µm; plurilocular sporangia unknown.

Epiphytic, occupying conceptacles of the large brown alga *Himanthalia elongata,* in lower eulittoral pools.

Only recorded from a few widely scattered localities around the British Isles (Devon, Bute, East Lothian, Berwickshire, Shetland Isles; in Ireland, Antrim, Mayo and Galway).

Plants annual, June to October.

Batters (1885), p. 537: Batters (1892a), p. 173; Sauvageau (1892), pp. 36–41: Kuckuck (1929), p. 37; Newton (1931), p. 142; Hamel (1931–39), pp. 132–134, fig. 30B–C.

Fig. 28 *Myriactula chordae*
A. Habit of plant on *Sauvageaugloia griffithsiana.* B. S.P. of thalli showing erect paraphyses arising from basal cushion of cells. C. Portion of thallus showing paraphyses (p), a hair (h) and plurilocular sporangia. D. Portion of thallus showing paraphyses and a unilocular sporangium. Bar = 2 mm (A), 50 µm (B–D).

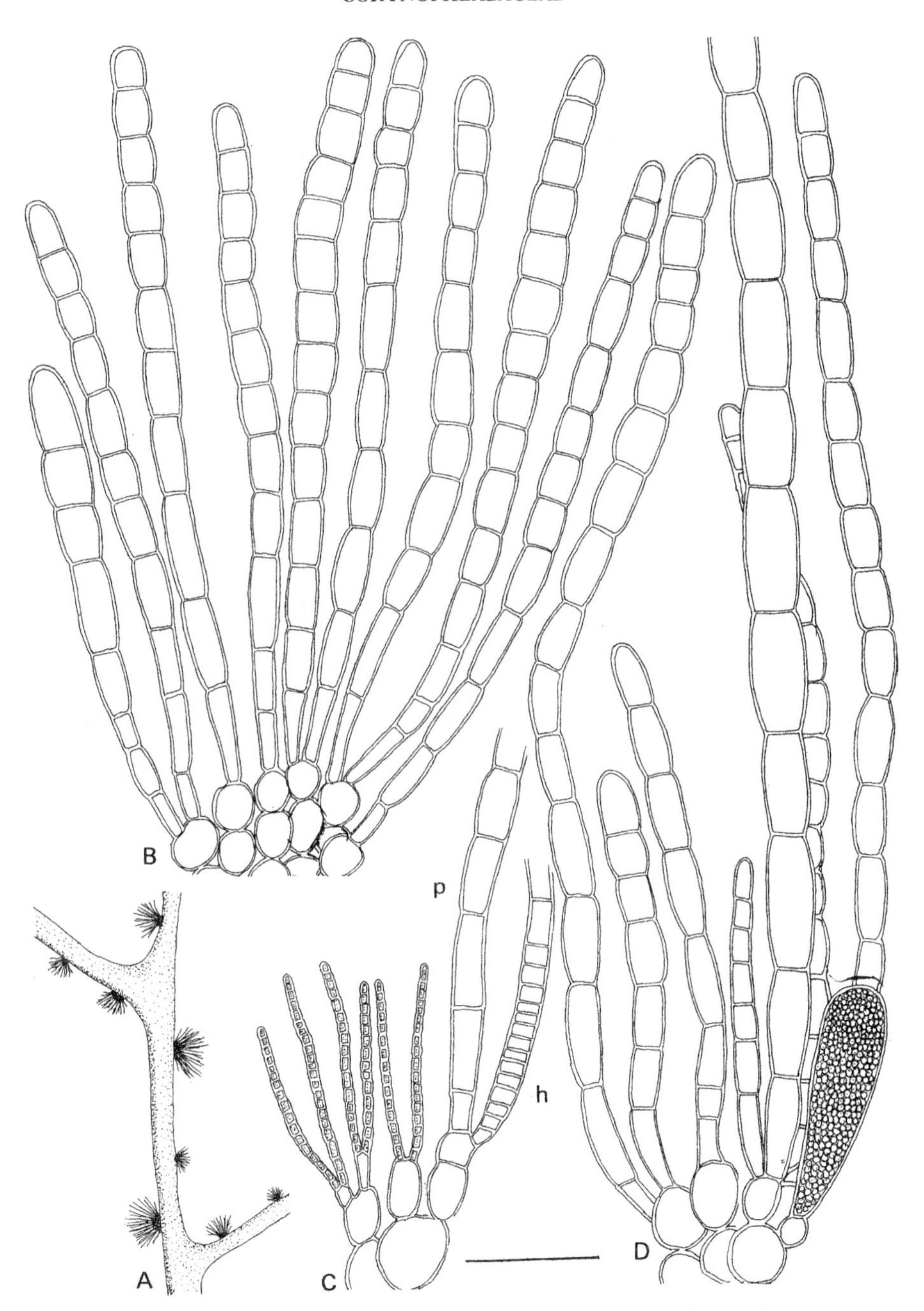

B
p
h
A
C
D

Myriactula chordae (Areschoug) Levring (1937), p. 57. Fig. 28

Elachista stellaris var. *chordae* Areschoug (1875), p. 18.
Elachista chordae (Areschoug) Kylin (1907), p. 61.
Gonodia chordae (Areschoug) Hamel (1931–39), p. 137.
Gonodia pulvinata (Kützing) Nieuwland *f. chordae* (Areschoug) Rosenvinge (1935), p. 31.

Plants epiphytic, forming small, hemispherical cushions on hosts, to 0·5 (–1) mm in diameter, soft and lubricous, easily squashed under pressure; consisting of a small, gelatinous, pseudoparenchymatous base of short, erect, closely packed, di-trichotomously branched filaments of large, thin-walled, colourless cells, becoming smaller above and giving rise terminally to paraphyses, sporangia and/or hairs; paraphyses erect, multicellular, elongate-clavate, becoming linear, simple, to 800 μm long (–22 cells), comprising elongate-cylindrical or rectangular cells 28–52 × 14–29 μm, each with numerous discoid plastids with pyrenoids; hairs common, borne at base of paraphyses, with basal meristem, without obvious sheath, 10–16 μm in diameter.

Unilocular and plurilocular sporangia recorded on the same or different plants, borne at the base of the paraphyses; unilocular sporangia pyriform, 70–94 × 23–33 μm; plurilocular sporangia in dense clusters, sessile or on 1–3 celled stalks, simple, cylindrical, uniserate, to 91 μm long (–19 loculi) × c. 6–8 μm wide.

Epiphytic on various algae, in lower eulittoral pools, and shallow sublittoral.

Recorded for a few scattered localities around the British Ises (Orkneys, Hampshire, Dorset).

Summer annual; May to September.

Hamel (1931–39), pp. 137–138, fig. 30 F; Newton (1931), p. 133; Rosenvinge (1935), pp. 31–37, figs 29–35.

Myriactula clandestina (Crouan frat.) J. Feldmann (1943), p. 223. Fig. 29

Elachista clandestina Crouan frat. (1867), p. 160.
Ectocarpus clandestinus (Crouan frat.) Sauvageau (1892), p. 13.
Gonodia clandestina (Crouan frat.) Hamel (1931–39), p. 134.
Entonema clandestinum (Crouan frat.) Hamel (1931–39), p. xxvi.

Plants primarily endophytic, emerging from host tissue as small rounded pustules, 0·5–1·0 mm in diameter, solitary or more often gregarious; comprising a network of branched, multicellular, filaments, 3–6 μm in diameter ramifying through host tissue, particularly just below surface cells, emerging to the surface, at first forming small swellings or raised bumps, later rupturing through the surface as small pustules; central region turf-like, similar to a carpet pile, spongy in texture, consisting of a basal layer of emergent, branched, endophytic filaments, the cells of which give rise to dense, erect tufts of paraphyses, unilocular sporangia and hairs hardly extending above raised rim of host material; paraphyses gelatinous, easily separable under pressure, simple, multicellular,

Fig. 29 *Myriactula clandestina*
A–B. Habit of plants, forming pustules on *Fucus serratus*. C. S.P. of thalli showing paraphyses (p), hairs (h) and unilocular sporangia (some dehisced) Bar = 4 mm (A), 1 mm (B), 50 μm (C).

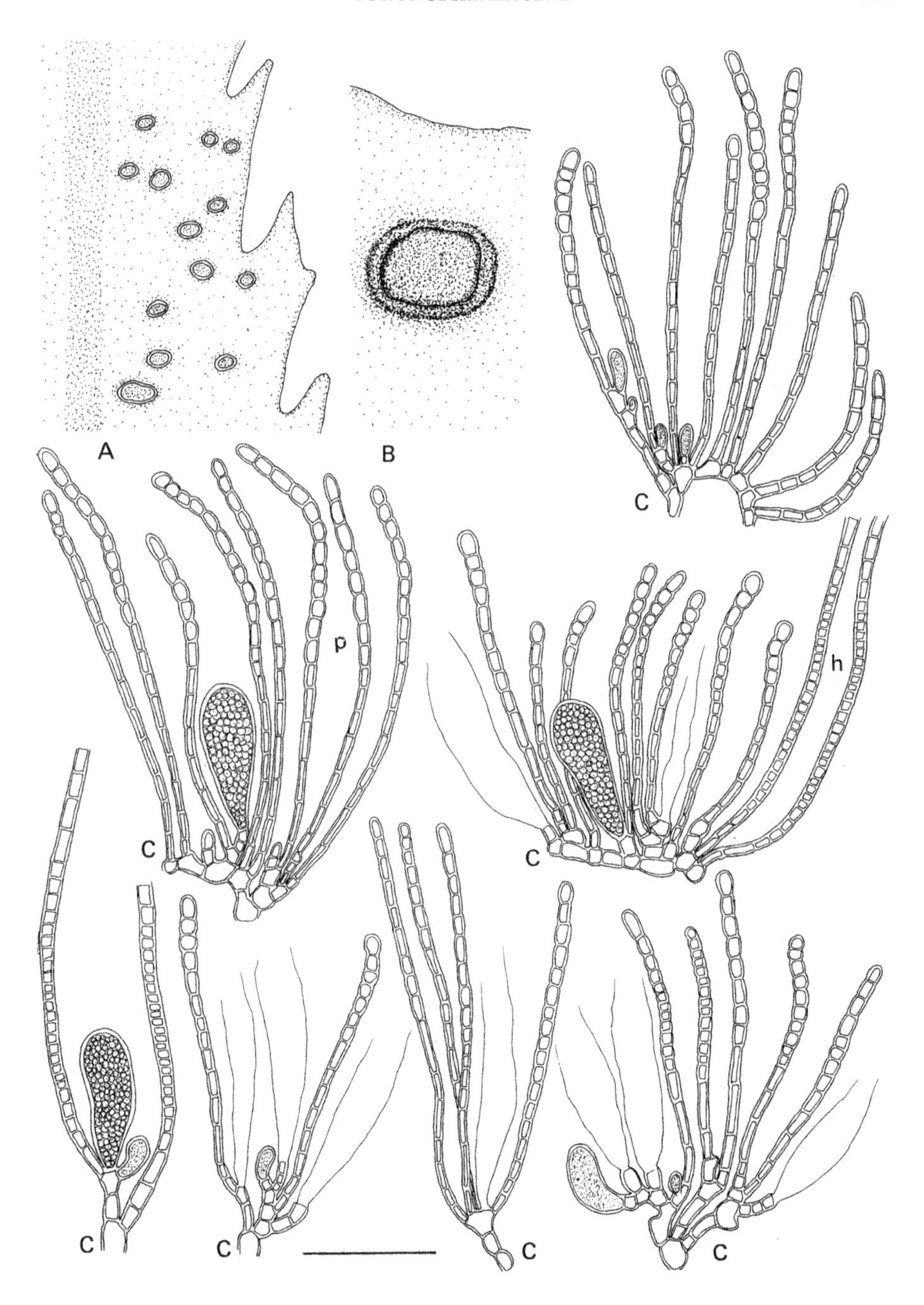
A
B
C
p
h
C
C
C
C
C
C

linear or slightly elongate-clavate, straight or slightly curved, to 180 (−250) μm long, 13–22 (−30) cells, comprising colourless or only faintly pigmented, thick-walled cells throughout, ½–1 (−2) diameters long, rectangular, rarely moniliform, 5–9 (−18) × 6–10 μm, in upper and mid regions, 3–5 diameters long, elongate-rectangular, 12–20 × 3–5 μm towards the base; cells with 2–3 small discoid plastids, without obvious pyrenoids; hairs common, produced from basal filaments, sometimes terminal on paraphyses, with basal meristem, without obvious sheath.

Unilocular sporangia formed directly from basal cells, emerging at base of paraphyses, pear-shaped to elongate-cylindrical, 55–90 × 14–32 μm, often on a single-celled pedicel; plurilocular sporangia unknown.

Epiphytic on *Fucus* spp., particularly *Fucus serratus*, over whole length of thallus, although more commonly near base, in the upper to lower eulittoral, in pools or emergent.

Recorded for widely scattered localities all around the British Isles (Shetland Isles, Fife, Berwick, Yorkshire, Kent, Sussex, Hampshire, Dorset, Devon, Cornwall, Pembroke, Isle of Man; in Ireland, Mayo); probably more common and widespread.

No obvious seasonal distribution pattern; recorded throughout the year with unilocular sporangia.

Crouan frat. (1867), p. 160, pl. 24, fig. 157 (1–2); Sauvageau (1892), pp. 38–41; Batters (1895a), p. 276; Newton (1931), p. 115; Hamel (1931–39), p. 134, fig. 30D–E; Dangeard (1965b), pp. 16–18. Dangeard (1968b), pp. 81–86, pl. 1.

Myriactula haydenii (Gatty) Levring (1937), p. 57. Fig. 30

Elachista haydenii Gatty (1863), p. 162.
Elachista moniliformis Foslie (1894), p. 120.
Myriactis haydenii (Gatty) Batters (1902), p. 36.
Myriactis moniliformis (Foslie) Kylin (1910), p. 13.

Thallus epiphytic, forming minute tufts on host, to 0·5 mm in height, usually confluent and spreading to 2–3 mm in extent; comprising a basal system of outwardly spreading, fairly tightly packed, branched filaments on host surface, comprising cells irregular in shape, 7–13 × 4–7 μm, from which arise short, erect, branched filaments bearing paraphyses, sporangia and/or hairs; erect filaments to 6 cells long, cells rectangular, 6–16 × 4–9 μm with 1–2 (−3) plate-like plastids; paraphyses erect, simple, multicellular, to 400 μm (−23 cells) long, linear or slightly clavate, tapering gradually below, comprising quadrate, barrel-shaped or more commonly rectangular cells, 1–2 (−2½) diameters long, 8–30 × 10–16 μm, each with several discoid plastids without obvious pyrenoids and sometimes with central, large, hyaline vacuole/oil vesicle present; hairs abundant, arising directly from erect filaments or terminal on paraphyses, with basal meristem but without obvious sheath, 9–12 μm in diameter.

Fig. 30 *Myriactula haydenii*
S.P. of thalli showing paraphyses (p), hairs (h) and plurilocular sporangia (some dehisced) arising from the basal layer. Bar = 50 μm.

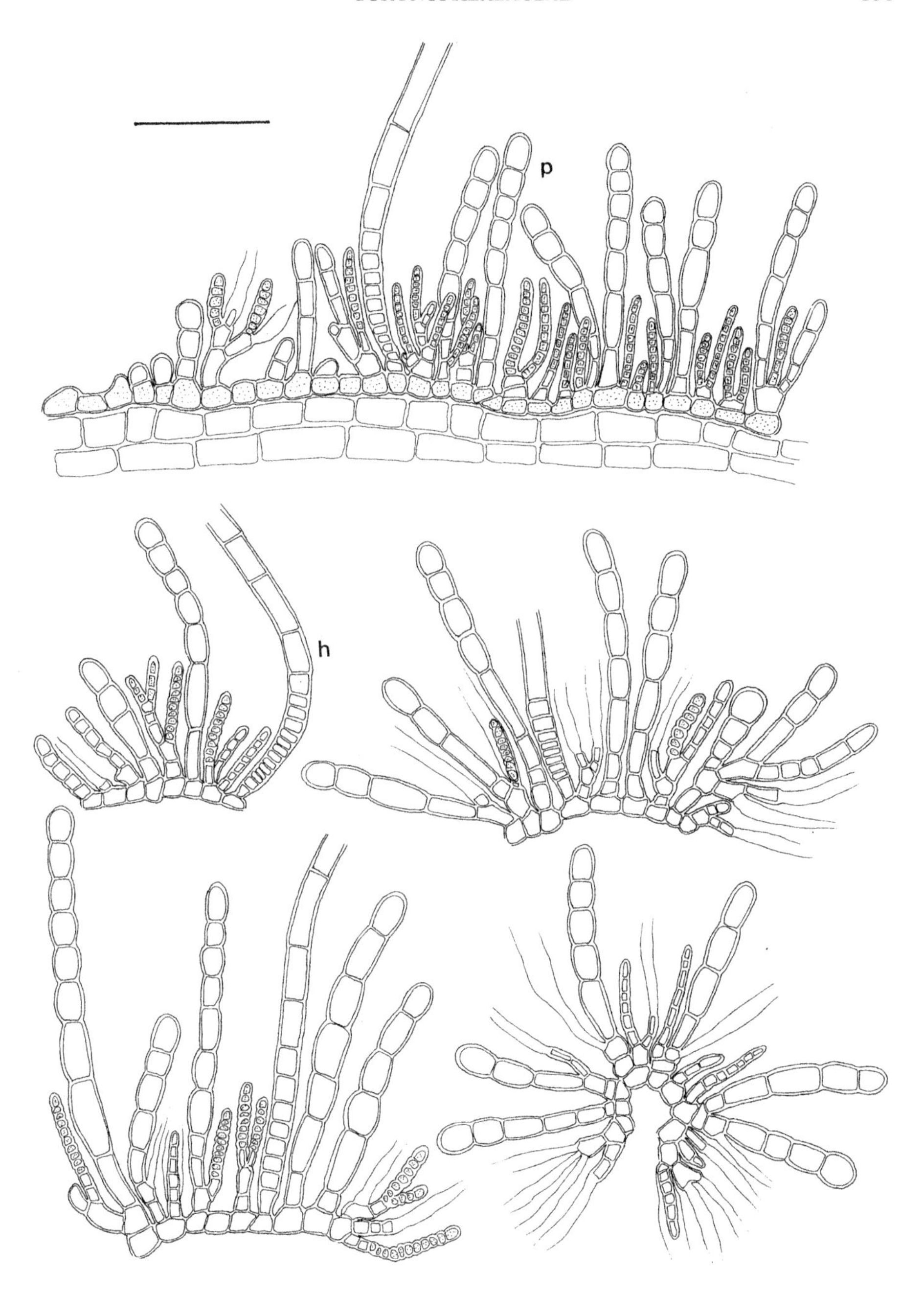
p
h

Unilocular and plurilocular sporangia arising at base of paraphyses, terminal or lateral on erect filaments; plurilocular sporangia abundant, densely crowded, sessile or stalked, simple or rarely branched, cylindrical, uniseriate, to 70 µm long (–11 loculi) × c. 5–7 µm in diameter, often arising within old sporangial wall husks, comprising subquadrate loculi, 3–5 × 5–6 µm with straight, rarely oblique cross walls; unilocular sporangia pyriform, not observed for the British Isles, rarely reported elsewhere in Europe.

Epiphytic on *Scytosiphon lomentaria*, in eulittoral pools.

Only recorded for a few scattered localities around the British Isles (Northumberland, Hampshire, Cheshire, Isle of Man; in Ireland, Cork and Mayo); probably more widespread and common.

Summer annual, June to September.

Foslie (1894), pp. 120–123, pl. 1, figs 3–6; Kuckuck (1929), pp. 37–38, figs 35–36; Newton (1931), p. 143; Levring (1937), p. 57; Kylin (1947), p. 48, fig. 42B; Rueness (1977), p. 145, fig. 74A.

Myriactula rivulariae (Suhr in Areschoug) J. Feldmann (1937), p. 274. Fig. 31; Pl. 2f

Elachista rivulariae Suhr in Areschoug (1842), p. 235.
Myriactis pulvinata Kützing (1843), p. 330.
Elachista attenuata Harvey (1846–51), pl. 28A.
Myriactula pulvinata (Kützing) Kuntze (1891–98).
Gonodia pulvinata (Kützing) Nieuwland (1917), p. 30.
Elachista pulvinata (Kützing) Harvey (1846–51), pl .17.

Plants epiphytic, forming small, hemispherical cushions on hosts, to 2 mm in diameter, 1 mm in height, just visible to the naked eye, soft and lubricous easily squashed under pressure; consisting of a soft and gelatinous, somewhat cartilaginous, pseudoparenchymatous lower region of erect, closely packed, di-trichotomously branched filaments, emergent from host tissue, comprising large, thin-walled, colourless cells, containing only a few discoid plastids, which gradually become smaller above and give rise terminally to paraphyses, sporangia and/or hairs; paraphyses erect, closely packed but gelatinous and easily separated under pressure, fusiform, multicellular, simple, slightly curved, to 620 µm long (–31 cells), tapering gradually towards apex, more sharply towards base, comprising thick-walled cells, mainly barrel-shaped throughout, less commonly rectangular and quadrate, especially at base, later becoming enlarged and swollen in mid-lower region, $\frac{1}{2}$–$1\frac{1}{2}$ diameters long, 16–50 × 25–43 µm remaining much smaller and less obviously swollen in mid-terminal region, 1–$1\frac{1}{2}$ diameters long, 14–32 × 16–26 µm, all cells with numerous discoid plastids with pyrenoids; hairs common, borne at base of paraphyses, with basal meristem, without obvious sheath, c. 9 µm in diameter.

Unilocular and plurilocular sporangia recorded on the same or different plants, borne at the base of the paraphyses; unilocular sporangia sessile or on 1-celled stalks, pyriform,

Fig. 31 *Myriactula rivulariae*
A. Habit of plants epiphytic on an air bladder of *Sargassum muticum*. B. S.P. of vegetative thallus showing erect paraphyses. C–F. S.P. of fertile thalli showing unilocular (u.s.) and plurilocular sporangia (p.s.) arising at base of paraphyses. Note hair (h). Bar = 2·5 mm (A), 100 µm (B–C), 50 µm (D–F).

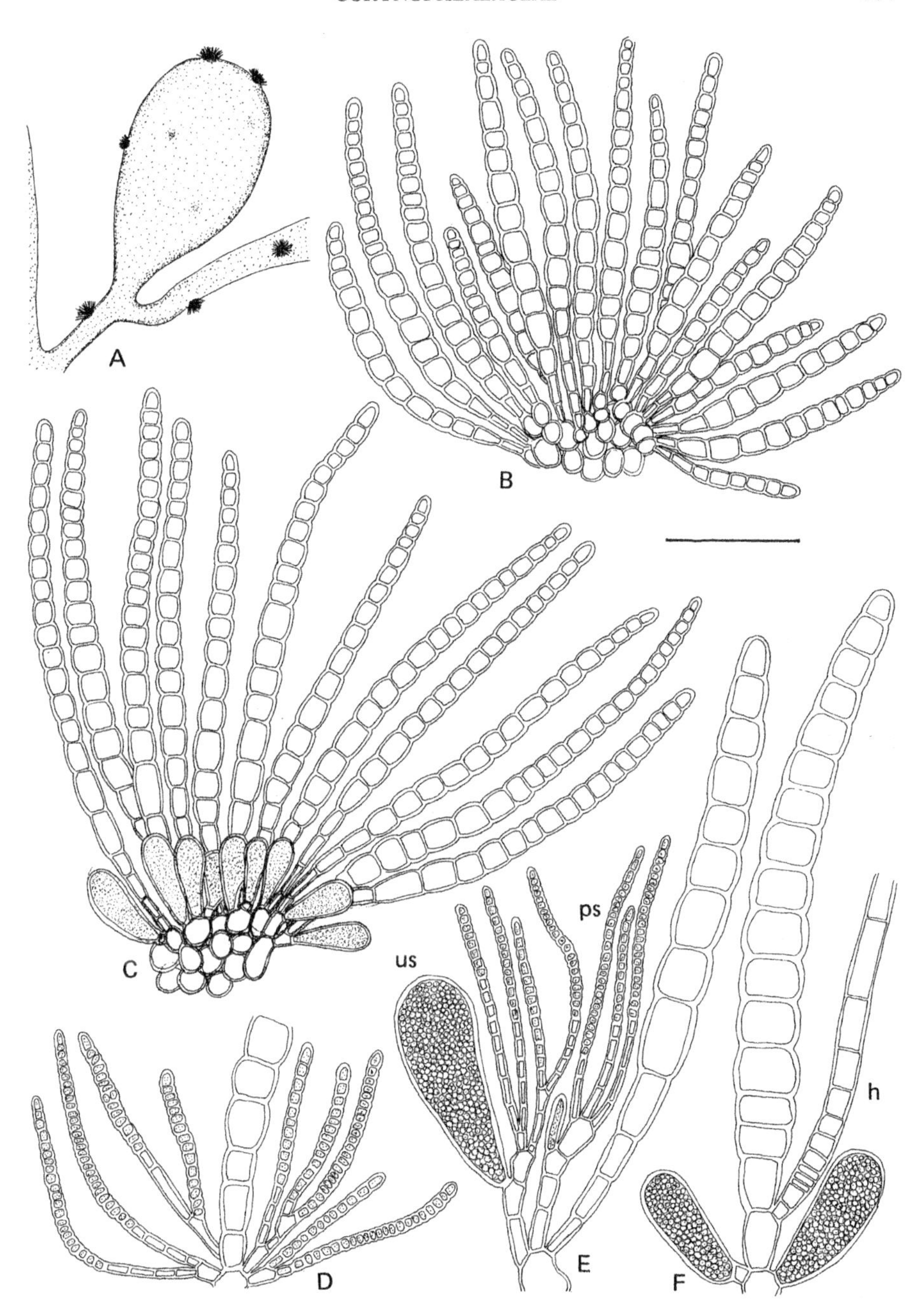
A
B
C
us
ps
h
D
E
F

54–90 × 22–40 μm; plurilocular sporangia on 1–2 celled stalks, simple, cylindrical, uniseriate, to 60 μm long (–10 loculi) × c. 7–9 μm wide.

Epiphytic on *Cystoseira* spp and *Sargassum muticum,* less commonly on *Halidrys siliquosa,* usually emergent from conceptacles or cryptostomata. Also more rarely recorded on various other algae including *Scytosiphon lomentaria* and *Laminaria* spp. upper and lower eulittoral pools and shallow sublittoral.

Common on *Cystoseira* spp. on south west shores of the British Isles (Hampshire, Channel Isles, Devon, Cornwall, Isles of Scilly; in Ireland, Wexford, Waterford, Cork, Kerry, Clare, Galway and Mayo). Recorded on *Halidrys,* from widely scattered localities (Argyll, Isle of Man, Yorkshire), and on other algae from Dorset; probably more widely distributed on these hosts.

Summer annual, July to September.

Kuckuck (1929), pp. 39–40, figs 40–41; Newton (1931), p. 142; Hamel (1931–39), pp. 135–136, fig. 31 (1–2); Rosenvinge (1935), p. 28, figs 27–28; Feldmann (1937), p. 274, fig. 46A–C; Stegenga & Mol (1983), p. 94, pl. 30, figs 2–3.

Myriactula stellulata (Harvey) Levring (1937), p. 57. Fig. 32

Conferva stellulata Harvey (1841), p. 132.
Elachista stellulata (Harvey) Harvey (1846–51), pl. 261.
Myriactis stellulata (Harvey) Batters (1892a), p. 174.
Gonodia stellulata (Harvey) Hamel (1931–39), p. 132.
Myriactula stellulata (Harvey) Feldmann (1937), p. 272.

Plants epiphytic, forming minute, solitary, brush-like tufts on host, 0·5–1 mm in diameter; composed of endophytic, branched filaments of largely colourless or faintly pigmented cells ramifying mainly between the cortical and medullary cell layers of host, occasionally spreading down between medullary cells, which break through to the surface forming small pustules; emergent filaments remaining fairly short, erect, di-trichotomously branched, closely packed and cushion forming, although gelatinous and easily separated under pressure, comprising large, somewhat swollen, thin-walled, colourless cells, becoming smaller above and terminating in paraphyses, sporangia and/or hairs; paraphyses erect, gelatinous, multicellular, to 12 (–17) cells, 155–285 μm long, simple, rarely branched, at first linear or elongate-clavate later becoming slightly fusiform comprising cells elongate-rectangular below, 1–3 diameters long, 16–25 × 7–12 μm becoming more rectangular, barrel-shaped or slightly moniliform above, $\frac{1}{2}$–$2\frac{1}{2}$ diameters long, 7–27 × 10–24 μm each containing several discoid plastids with pyrenoids; hairs common, borne at base of paraphyses, with basal or sub-basal meristem, without obvious sheath, c. 9–11 μm in diameter.

Unilocular and plurilocular sporangia recorded on the same or different plants, borne at the base of the paraphyses; unilocular sporangia common, sessile or shortly stalked,

Fig. 32 *Myriactula stellulata*
A–B. Habit of plants epiphytic on *Dictyota dichotoma.* C. V.S. of fertile thallus emerging from host tissue. D–E. S.P. of thalli showing paraphyses, hairs and unilocular sporangia. Bar = 2 mm (A), 0·5 mm (B), 100 μm (C–D), 50 μm (E).

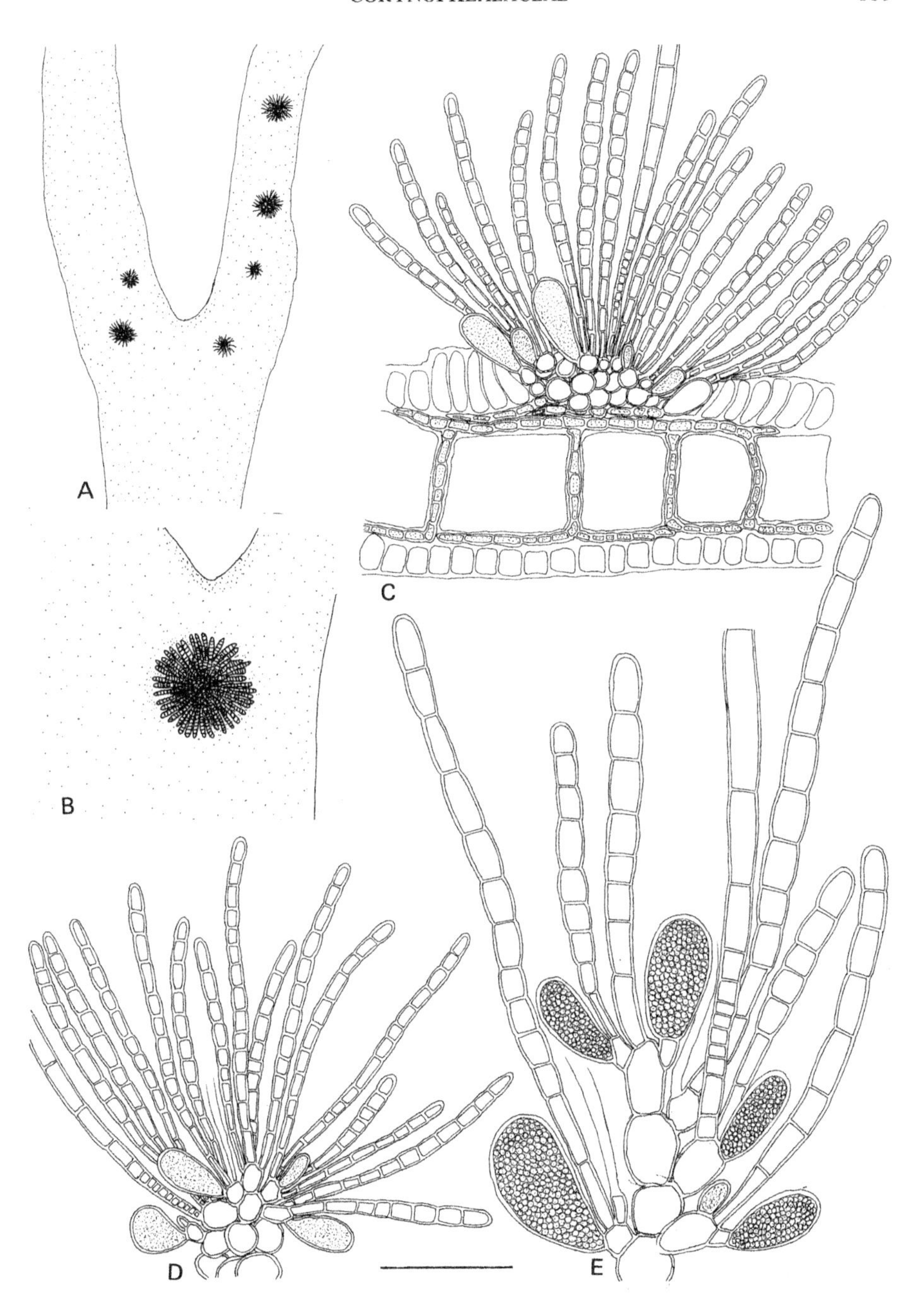
A
B
C
D
E

elongate-pyriform, 36–72 × 20–33 μm; plurilocular sporangia rare, in dense clusters, cylindrical, simple, uniseriate, to 80 μm long (13 loculi) × c. 8–10 μm wide.

Epiphytic, on *Dictyota dichotoma,* usually near the base, in pools in the eulittoral, and shallow sublittoral.

Southern and western shores of the British Isles; Channel Islands and extending eastwards to Hampshire and northwards to Ayr; in Ireland eastwards to Wexford and northwards to Mayo.

Plants annual, June to September and usually fertile; plurilocular sporangia only recorded for the Isle of Man but probably more widely distributed.

Sauvageau (1892), pp. 6–10, pl. 1, figs 1–2; Kuckuck (1929), p. 34, figs 32–34; Newton (1931), p. 142, fig. 88; Hamel (1931–39), p. 132, fig. 30A; Feldmann (1937), pp. 272–274, fig. 45A–C; Rueness (1977), p. 146, fig. 76.

PETROSPONGIUM Nägeli in Kützing

PETROSPONGIUM Nägeli in Kützing (1845–71), p. 2.

Type species: *P. berkeleyi* (Greville in Berkeley) Nägeli in Kützing (1845–71), p. 2.

Thallus epiphytic or epilithic, forming solid, hemispherical, gelatinous cushions; comprising a prostrate system of outwardly spreading, branched filaments giving rise to erect, loosely compacted, di-trichotomously branched filaments, easily separable under pressure; filaments infrequently branched below, comprising large, elongate-cylindrical to globular colourless cells, often with downwardly growing host-penetrating (if epiphytic), multicellular, branched, rhizoidal filaments, becoming more frequently branched and compacted above, comprising quadrate, rectangular to barrel-shaped smaller, pigmented cells; hairs and sporangia associated with terminal, pigmented filaments; hairs with basal meristem and sheath.

Unilocular sporangia laterally inserted at the base of the terminal, pigmented filaments, sessile or shortly stalked, elongate-cylindrical becoming deformed; plurilocular sporangia reported to be terminal on pigmented filaments, multiseriate.

One species in the British Isles;

Petrospongium berkeleyi (Greville in Berkeley) Nägeli in Kützing (1845–71), p. 2. Fig. 33

Chaetophora berkeleyi Greville in Berkeley (1833), p. 5.
Leathesia berkeleyi (Greville in Berkeley) Harvey (1846–51), p. 176.
Cylindrocarpus berkeleyi (Greville *in* Berkeley) Crouan frat (1851, p. 363).

Thallus epiphytic or epilithic, forming small, light to dark brown, hemispherical cushions, discrete and circular, or confluent and irregularly spreading to 10 (–20) mm in

Fig. 33 *Petrospongium berkeleyi*
A. Habit of plant epiphytic on *Ralfsia verrucosa*. B. Surface view of thallus showing loosely associated cells. C. Portion of vegetative thallus. D–E. Portions of fertile thalli showing unilocular sporangia. Bar = 6 mm (A), 50 μm (others).

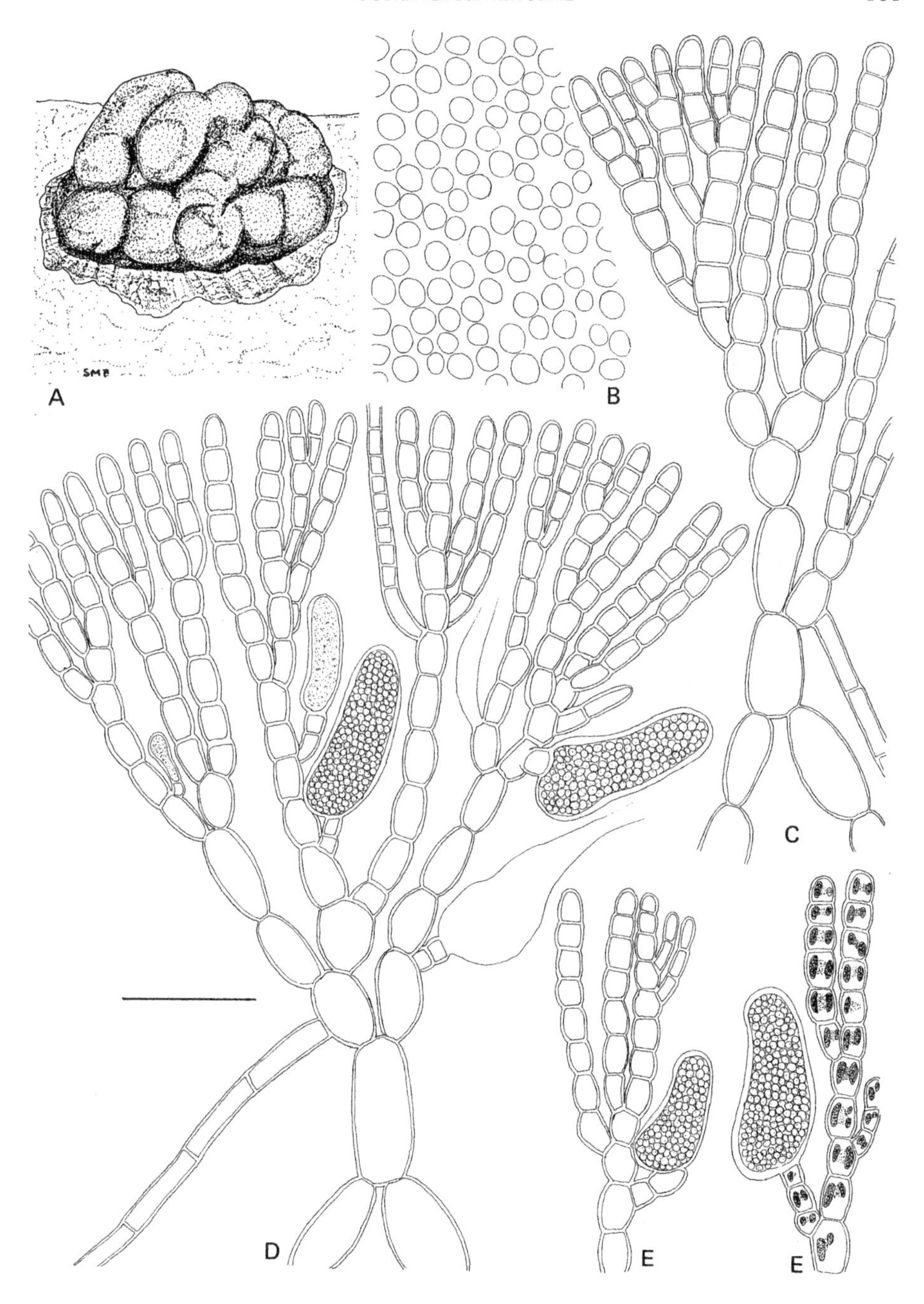
SMB
A
B
C
D
E
E

diameter, 2–3 (−4) mm high; in squash preparation consisting of erect, multicellular, branched filaments, loosely compacted in a mucilaginous matrix; lower and mid regions sparsely dichotomously branched, comprising large, colourless cells, elongate-cylindrical below, becoming more globose above, 1–4 diameters long, 25–65 × 15–30 μm, frequently producing from the basal region downwardly growing, branched, cylindrical or irregularly contorted, colourless rhizoidal filaments; rhizoid cells 28–50 × 6–12 μm, reputed to penetrate below into host tissue; upper regions of thallus more frequently di-trichotomously branched forming closely compacted pigmented, straight, thread-like filaments, comprising cells quadrate, rectangular, or more frequently barrel-shaped in appearance, 1 ($-1\frac{1}{2}$) diameters long, 9–16 × 8–13 μm, each with a single, plate-like plastid with pyrenoid occupying upper cell region; hairs produced at base of terminal thread-like filaments, with basal meristem and sheath, 8–10 μm in diameter.

Unilocular sporangia common, lateral on thread-like filaments, sessile or shortly stalked, elongate-cylindrical or slightly deformed and characteristically wider below, basally, sometimes laterally, attached, 50–81 × 17–32 μm; plurilocular sporangia rare, reported to be terminal on the thread-like filaments, pod-like, multiseriate, 46–113 × c. 11 μm.

Epilithic on bedrock, less obviously epiphytic on crusts of the brown alga *Ralfsia verrucosa*, in the upper to lower eulittoral, especially in shaded situations.

Southern and western shores of the British Isles, extending northwards to Orkney and eastwards to Dorset; in Ireland northwards to Donegal, eastwards to Wexford, Antrim.

Plants annual, summer and autumn.

Caram (1957) reported *P. berkeleyi* (as *Cylindrocarpus berkeleyi*) to have an isomorphic type of life history. Meiosis was assumed to occur in the unilocular sporangia to form haploid zoospores which developed into the plurilocular sporangia-bearing haploid plants. She also reported a 'short-circuit' life history involving direct development of the diploid plant from zygotes produced by sexual fusion of the unispores.

Crouan frat. (1851), p. 364, pl. 17, figs 12–13; Crouan frat. (1852), pl. 25, fig. 159; Hanna (1899), pp. 461–464; Kuckuck (1929), p. 46, fig. 54; Miranda (1931), p. 26, fig. 1; Newton (1931), p. 140, fig. 86; Hamel (1931–39), pp. 147–148, fig. 33a–b; Caram (1957), pp. 440–443; Dangeard (1969), pp. 81–82.

POGOTRICHACEAE Pedersen

POGOTRICHACEAE Pedersen (1978a), p. 66.

Plants erect, filamentous becoming parenchymatous, unbranched, either terete, solid or tubular, or foliose and membranaceous, one or two cells thick, without hairs; plurilocular and unilocular sporangia scattered or grouped, developed directly from surface vegetative cells by simple transformation.

This small family was created by Pedersen (1978a) to include the two genera *Pogotrichum* Reinke and *Omphalophyllum* Rosenvinge. Two species of *Pogotrichum* were recognised, *P. filiforme* Reinke (usually referred to as *Litosiphon filiformis* (Reinke) Batters) and *P. setiforme* (Rosenvinge) Pedersen (usually referred to as *Litosiphon filiformis* (Rosenvinge) Rosenvinge) whilst a single species, *O. ulvaceum* Rosenvinge, was included in the monotypic genus *Omphalophyllum*. In the British flora only one species of *Pogotrichum, P. filiforme,* is represented; *O. ulvaceum* is an arctic deep water species originally described from Greenland and *P. setiforme* has been described only from Greenland and Denmark. In addition, *P. setiforme* is doubtfully distinct as it appears to be included within the circumscription of *P. filiforme* (see below, also see Pedersen, 1978a, p. 65).

According to Pedersen the characteristic features of the genus *Pogotrichum* which distinguish it from the genus *Litosiphon* as typified with *L. pusillus* (Carm. ex Hook.) Harvey include (a) the absence of hairs; (b) the extensive confluent development of sporangia which can leave greater parts of the thallus empty following spore release and (c) plurilocular sporangia are formed directly in surface vegetative cells, following simple subdivision, without preceding vegetative divisions.

Examination of British material has confirmed these differences and, therefore, support is given to Pedersen's (1978a) resurrection of Reinke's (1892) old name *Pogotrichum* to include the entity usually referred to as *Litosiphon filiforme* (Reinke) Batters.

POGOTRICHUM Reinke

POGOTRICHUM Reinke (1892), p. 61.

Type species: *Pogotrichum filiforme* Reinke (1892), p. 62.

Plants epiphytic or epilithic, forming tufts of erect thalli attached by basally produced rhizoidal filaments, sometimes arising from a small basal disc; erect thalli simple, filiform, cylindrical or slighty elongate-clavate, uniseriate becoming pluriseriate and parenchymatous above; in section outer cells usually small and pigmented, inner cells usually large and colourless, sometimes without size differentiation; cells with several discoid plastids and associated pyrenoids; hairs unknown.

Unilocular and plurilocular sporangia borne on the same or different thalli, developing from surface vegetative cells, single or more commonly clumped; unilocular sporangia ovate, or elliptical, formed directly in vegetative cells; plurilocular sporangia formed directly in vegetative cells by subdivision.

One species in the British Isles:

Pogotrichum filiforme Reinke (1892), p. 62. Fig. 34

Litosiphon filiformis (Reinke) Batters (1902), p. 25.
Pogotrichum filiformis f. gracilis Batters (1892b), p. 18.

Plants epiphytic, forming erect, light to dark brown tufts on hosts; tufts discrete, not confluent, commonly gregarious and densely fringing host surface, each comprising large numbers of erect, densely packed shoots arising from the centre of a discoid base; basal disc one cell layered, comprising outwardly radiating, branched, closely packed filaments of cells, not firmly attached to host surface and often transversely divided with straight and oblique cross walls to form plurilocular sporangia; erect shoots 10–20 (–50) mm long, 16–60 (–115) μm wide, filiform, terete, solid, cylindrical, uniseriate, often becoming pluriseriate and parenchymatous above by frequent longitudinal and transverse divisions; cells subquadrate, quadrate or rectangular in uniseriate regions, 0·5–1 (–2) diameters long, 14–55 × 18–35 μm, more irregularly shaped in polysiphonous regions, 13–34 × 12–27 μm, in transverse but not longitudinal rows below, irregularly placed above, each with several discoid plastids with pyrenoids; in section inner cells colourless, outer cells pigmented, approximately equal in size or slightly differentiated into medulla and cortex; hairs not present.

Unilocular and plurilocular sporangia borne on different, less commonly similar shoots, single, or more commonly grouped, formed from surface cells; unilocular sporangia rare, ovate, elliptical, bulging slightly, 19–29 × 16–21 μm; plurilocular sporangia common formed directly from single cortical cells following subdivisions, often in continuous sori, 14–26 × 12–16 μm.

Epiphytic on various hosts, in particular *Laminaria saccharina* and *L. digitata,* lower eulittoral pools and shallow sublittoral.

Not uncommon and widely distributed around the British Isles.

Annual, spring and summer (April to September).

Recent culture studies on Greenland material by Pedersen (1978a), using a wide variety of culture conditions, indicate that a direct life history is in operation, without a sexual process; swarmers from the plurilocular sporangia on both the prostrate and erect parts of the thalli germinated without copulation to repeat the fertile parental phase. A similar 'direct' life history was reported by Kuckuck (1917) and Kylin (1937).

Reinke (1892), p. 62, pl. 41, figs 12–25; Batters (1892b), pp. 18–19; Batters (1902), p. 25; Kuckuck (1917), pp. 557–566, figs 1–3; Newton (1931), p. 181; Hamel (1931–39), pp. 219–220, fig. 43 (8–10); Taylor (1957), pp. 162–163; Kylin (1947), p. 75, fig. 60B–D; Rosenvinge & Lund (1947), pp. 20–21, fig. 4; Lund (1959), pp. 150–152, fig. 33; Jaasund (1965), pp. 85–87, figs 24–25; Kornmann & Sahling (1977), p. 128, fig. 67; Rueness (1977), p. 162, fig. 94; Pedersen (1978a), pp. 61–68, figs 1–11; Stegenga & Mol (1983), p. 97, pl. 23, figs 2–6.

Fig. 34 *Pogotrichum filiforme*
A. Habit of plants epiphytic on *Laminaria*. B. Portions of erect thalli showing different stages of development. C. Transverse sections (T.S) of three thalli at different stages of development. D. Portion of erect thallus with unilocular sporangia. E. Portions of erect thalli at different stages of development, with plurilocular sporangia. Note dehisced sporangia with released contents. Bar = 12 mm (A), 50 μm (others).

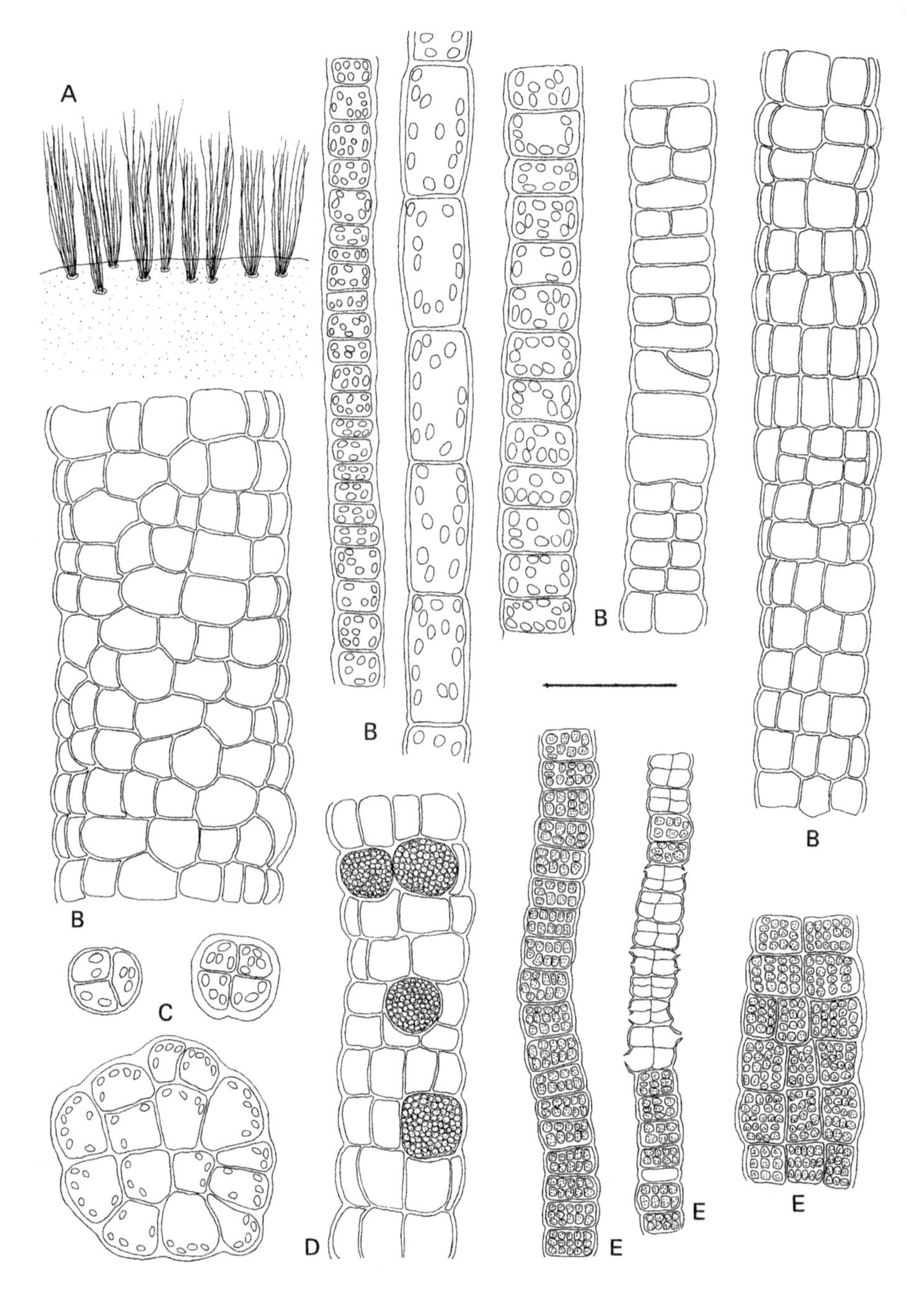
A
B
B
B
B
B
C
D
E
E
E

MYRIOTRICHIACEAE Kjellman

MYRIOTRICHIACEAE Kjellman (1890), p. 46

Plants endophytic or epiphytic; endophytic plants microscopic, filamentous and branched; epiphytic plants usually macroscopic, erect, tufted, uniseriate and filamentous, or multiseriate and thalloid, arising from either a creeping, filamentous or discoid base; thalloid shoots solid, parenchymatous, terete, filiform/cylindrical or elongate-clavate, flaccid, with or without internal differentiation into an outer cortex of small, pigmented cells and a central medulla of long, colourless cells, simple or radially branched with branches uniseriate or multiseriate; hairs with basal meristem, lacking sheath; unilocular and plurilocular sporangia occurring on the same or different plants, borne on basal filaments or erect shoots and/or branches, single or grouped, either formed directly (or indirectly following divisions) in vegetative cells, or arising laterally, more rarely terminally (plurilocular sporangia only), sessile or stalked; unilocular sporangia spherical, ovate or elliptical, plurilocular sporangia ovate, cylindrical, conical, uniseriate or biseriate.

Three genera are presently included in the Myriotrichiaceae for the British Isles; *Litosiphon* Harvey (1846–51) and the two closely related genera *Myriotrichia* Harvey (1834) and *Leblondiella* Hamel (1931–39). *Leblondiella,* a monotypic genus, based on *Myriotrichia densa* Batters (1895b), has received fairly widespread recognition (see Newton, 1931; Parke & Dixon, 1976), although more recently rejected by Rueness (1977). It differs from *Myriotrichia* in the following features: it has a discoid rather than a filamentous base; the mature, branched, erect shoots are filiform and linear, sharply attenuate at base and dark brown/black in colour rather than elongate-clavate, with a long tapering base and light/mid brown in colour; the short radial branches are dichotomously and/or unilaterally branched with branches strongly recurved, rather than irregularly or radially branched with divergent branches; the constituent cells are more moniliform in *Leblondiella* than in *Myriotrichia* and remain uniseriate and filamentous rather than becoming multiseriate and parenchymatous; hairs are also uncommon in *Leblondiella* rather than abundant as in *Myriotrichia.* It is quite possible that the above differences will prove insufficient to maintain *Leblondiella* as an autonomous genus. However, it is retained here pending the results of future culture studies.

The removal of the genus *Litosiphon* from the Punctariaceae (a position accepted by authors such as Parke & Dixon (1976), Rueness (1977) and Kornmann & Sahling (1977) to the Myriotrichiaceae follows the proposal of Pedersen (1984). By comparing *Litosiphon pusillus* and *Punctaria plantaginea* (both the type species of their respective genera), Pedersen concludes that they differ in a number of important characteristics. These include: the occurrence of sympodial branching in *L. pusillus* but not *P. plantaginea*; sheathed hairs are absent in *L. pusillus* but present in *P. plantaginea*; the embryospore occasionally shows immediate differentiation in *L. pusillus* but not in *P. plantaginea*; the microthalli are *Streblonema*-like in *L. pusillus* but *Hecatonema*-like in *P. plantaginea.* In particular the occurrence of the two characteristics, sympodial branching and immediate differentation of the embryospore (which are shared with *Myriotrichia,* type genus of the Myriotrichiaceae) persuaded Pedersen to transfer *Litosiphon* from the Punctariaceae to this family.

LEBLONDIELLA Hamel

LEBLONDIELLA Hamel (1931–39), p. xl.

Type species: *L. densa* (Batters) Hamel (1931–39), p. XL.

Plants epiphytic, forming small, dark brown/black, flaccid tufts on host, solitary or gregarious, arising from a small, discoid, attachment base; comprising up to several erect, filiform shoots, uniseriate, simple and sharply tapering below, multiseriate and radially branched above with obtuse apex; branches either sparse and tufted or confluent and expansive, clothing central parenchymatous axis, short, dichotomously or unilaterally branched, strongly recurved, comprising cylindrical to moniliform cells, each with several discoid plastids; hairs uncommon with basal meristem, lacking sheath.

Unilocular and plurilocular sporangia borne on the same or different shoots; unilocular sporangia spherical or ovoid, sessile or stalked, at base of radial branches; plurilocular sporangia lateral on primary axis below, lateral and terminal on radial branches above, solitary or clustered, sessile or stalked, cylindrical to conical, uniseriate or biseriate.

One species in the British Isles:

Leblondiella densa (Batters) Hamel (1931–39), p. xl. Fig. 35

Myriotrichia densa Batters (1895b), p. 313.

Plants forming erect tufts on host, single or gregarious, each comprising 2–5 shoots arising from a disc-like base; erect shoots to 25 (–40) mm long, 200 (–300) μm wide, simple, dark brown/black, filiform, more or less linear, soft and flaccid in texture, obtuse above, sharply tapering to base; basal disc to 300 (–600) μm in diameter, comprising an outward spreading network of dichotomously/irregularly branched filaments of thin-walled, colourless, often irregularly shaped cells, weakly joined together in a mucilaginous matrix; erect shoots uniseriate and simple below comprising cells $\frac{1}{2}$ (–1) diameters long, 16–31 × 27–36 μm becoming multiseriate above by longitudinal and transverse divisions to form a narrow terete thallus, with quadrate or irregularly shaped surface cells which produce short, radial branches, either sparsely to form irregular tufts or more often abundantly to clothe the primary axis; lateral filaments dichotomously or unilaterally branched, with branches strongly recurved and slightly attenuate, to 130 μm long (8 cells) with cells moniliform, subcylindrical to cylindrical, $\frac{1}{2}$–1 (–2) diameters long, 8–34 × 8–34μm, each with several discoid plastids without obvious pyrenoids; hairs uncommon, with basal meristem and lacking sheath, 10–16 μm in diameter.

Unilocular and plurilocular sporangia more commonly reported on different shoots, sometimes mixed; unilocular sporangia less common, sessile or shortly stalked, at base of branched filaments, spherical or ovoid 34–60 μm in diameter; plurilocular sporangia common arising directly from primary axis, more commonly on lateral branches, single or clustered, sessile or stalked, cylindrical, subcylindrical, lanceolate or conical, to 10 loculi long, 25–48 (–60) × 8–13 μm, uniseriate, more commonly biseriate with straight or oblique cross walls.

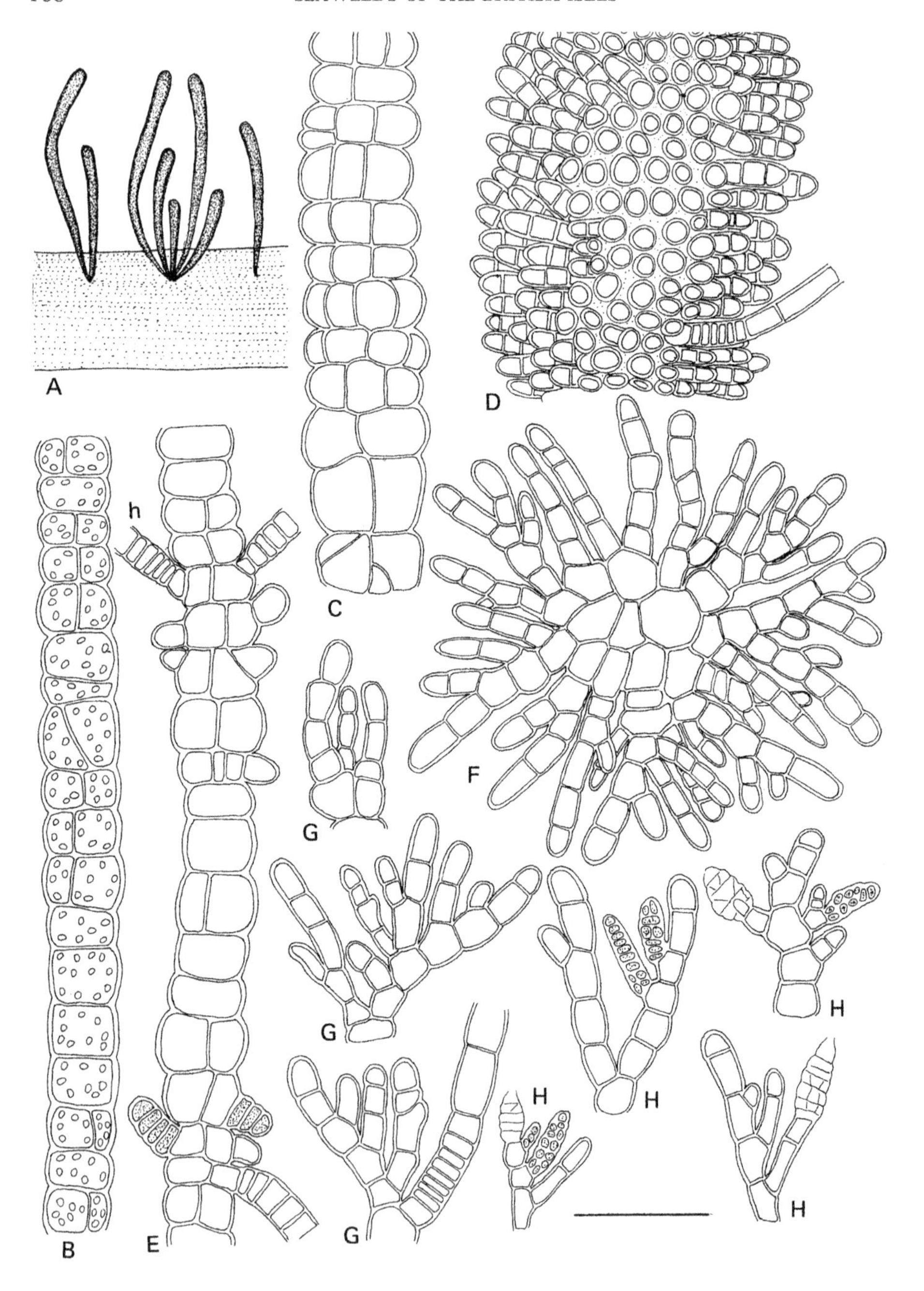
A
B
C
D
E
h
F
G
G
G
G
H
H
H
H
H

Epiphytic on decaying leaves of *Zostera*, lower eulittoral pools and shallow sublittoral. Only recorded for a few scattered localities around the British Isles (Hampshire, Dorset, S. Devon, Cornwall, Isle of Man, Argyll).

Annual; summer (June to August).

Buffham (1891), pp. 322–323, figs 4–12; Batters (1895c), p. 169; Batters (1895b), pp. 311–313, pl. xi, figs 11–13; Kuckuck (1899a), p. 75; Sauvageau (1931), pp. 76–77; Newton (1931), p. 174; Hamel (1931–39), p. 239, xl–xli, fig. 46V.

LITOSIPHON Harvey

LITOSIPHON Harvey (1846–51), p. 43.

Type species: *L. pusillus* (Carmichael ex Hooker) Harvey (1849), p. 43 (= *L. laminariae* (Lyngbye) Harvey (1849), p. 43).

Plants epiphytic, gregarious, forming tufts of erect shoots on hosts, attached by downwardly penetrating rhizoidal filaments; shoots simple, solid, parenchymatous, filiform/cylindrical or elongate-clavate; in section comprising an outer cortex of 1–2 small, pigmented cells enclosing a central medulla of large, elongate, colourless cells; cells with several discoid plastids with pyrenoids; hairs common with basal meristem, lacking sheath.

Unilocular and plurilocular sporangia borne on the same or different shoots, developing from surface cells, single or grouped; unilocular sporangia ovate or elliptical, formed directly from a single vegetative cell; plurilocular sporangia formed indirectly from vegetative cells following vegetative division.

Of the three species of *Litosiphon* recognised for the British Isles by Parke & Dixon (1976) viz. *L. filiformis* (Reinke) Batters, *L. pusillus* (Carmichael ex Hooker) Harvey and *L. laminariae* (Lyngbye) Harvey only the latter (*L. laminariae*) is accepted in the present text. *L. filiformis* is removed to the genus *Pogotrichum* Reinke, following the proposal of Pedersen (1978a), whilst *L. pusillus* is included within the synonomy of *L. laminariae*. Certainly there appears to be little justification for maintaining a distinction between *L. laminariae* and *L. pusillus* based on previously accepted criteria such as host specificity (*L. laminariae* on *Alaria esculenta*, *L. pusillus* on *Chorda filum*), thallus colour (dark brown thalli in *L. laminariae*, light brown thalli in *L. pusillus*), thallus size (longer, broader thalli in *L. pusillus* compared to *L. laminariae*) and basal attachment system (pustule formation on host surface by *L. laminariae* only). The discovery of a wider range of host plants as well as a range of intermediate forms of these two species indicate their probably conspecificity.

One species in the British Isles:

Fig. 35 *Leblondiella densa*
A. Habit of plants epiphytic on *Zostera*. B–D. Portions of erect thalli showing increased anatomical development. E. Portion of thallus showing hairs (h) and short, lateral plurilocular sporangia. F. T.S. of well developed thallus showing branched lateral filaments. G. S.P. of three portions of vegetative thallus showing branched lateral filaments. H. S.P. of four portions of fertile thallus showing plurilocular sporangia. Bar = 4 mm (A), 50 μm (others).

Litosiphon laminariae (Lyngbye) Harvey (1849), p. 43. Figs 36, 37; Pl. 4b

Bangia laminariae Lyngbye (1819), p. 84.
Asperococcus laminariae (Lyngbye) Crouan frat. (1852), no. 64.
Punctaria laminariae (Lyngbye) Crouan frat. (1867), p. 167.
Pogotrichum hibernicum Johnson (1892), p. 6.
Litosiphon hibernicum (Johnson) Batters (1902), p. 25.
Litosiphon pusillus (Carmichael ex Hooker) Harvey (1849), p. 43.
Asperococcus pusillus Carmichael in Hooker (1833), p. 277.
Dictyosiphon pusillus (Carmichael) Areschoug (1847), p. 149.
Punctaria pusilla Crouan frat. (1867), p. 167.

Plants epiphytic forming erect, light to dark brown, commonly gregarious, dense tufts on hosts, each tuft solitary, not confluent, comprising up to 40 (−70) erect shoots, closely associated and attached at the base by an intricate mat of downwardly growing secondary rhizoids; rhizoids sometimes penetrating and producing small (usually less than 1 mm) swellings on host surface; erect shoots 10–35 (−70) mm long, 50–180 (−280) μm wide (−22 cells), simple, filiform, terete, firm, solid, linear or more commonly, slightly elongate-clavate, fairly smooth, occasionally tortuous in parts, parenchymatous throughout except at extreme base of young shoots, with obtuse apex; comprising surface cells in straight, more commonly spirally twisted, longitudinal, less obviously, transverse rows, cells 1–2 (−4) diameters long, 8–32 × 8–21 μm, quadrate or rectangular in shape, each with several discoid plastids with pyrenoids; in transverse section shoots comprising an outer cortex of 1–2 small, pigmented cells enclosing a central medulla of large, slightly elongated, colourless cells; hairs abundant, arising singly from surface cells, 13–16 μm in diameter, multicellular with short basal meristem and lacking sheath.

Unilocular and plurilocular sporangia usually borne on different, occasionally the same shoots, discrete or confluent, occurring over greater length of shoot, formed in surface cells; unilocular sporangia variable in shape, usually oval or elliptical, 16–35 × 15–27 μm; plurilocular sporangia not formed in subdivided surface cells but, in daughter cells following transverse and longitudinal vegetative divisions, variable in shape, 11–14 × 6–11 μm in surface view, slightly protruding, each comprising 2–4 (−6) loculi.

Epiphytic on various hosts, especially *Chorda filum* and *Alaria esculenta* but also commonly occurring on *Scytosiphon, Sacchoriza* and *Laminaria* spp., lower eulittoral pools and shallow sublittoral.

Common and widely distributed around the British Isles.

Annual; summer and autumn.

Culture studies by Sauvageau (1929, 1933a) and Dangeard (1965a) (Atlantic coast of France), Kylin (1934) and Nygren (1975a, b, 1979) (Sweden) and Pedersen (1978a, 1981c, 1984) (Denmark) seem to indicate that a 'direct' life history is operating, with the released swarmers directly repeating, without copulation, plants of similar morphology.

Fig. 36 *Litosiphon laminariae*
A. Habit of plants epiphytic on *Alaria esculenta*. B. Habit of plants epiphytic on *Chorda filum*. C. Habit of young plant. Bar = 34 mm (A), 16 mm (B), 1·2 mm (C).

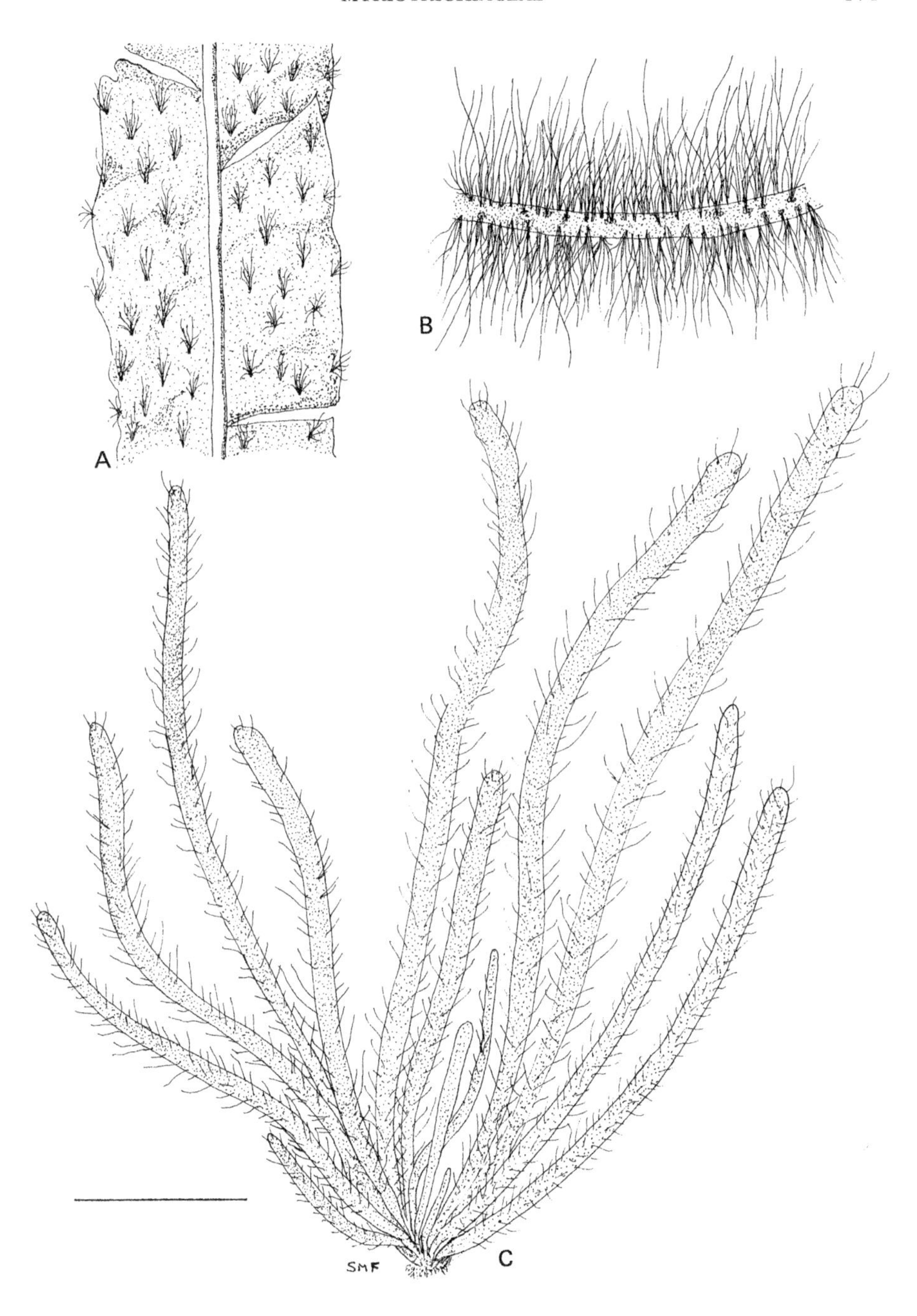
A
B
C
SMF

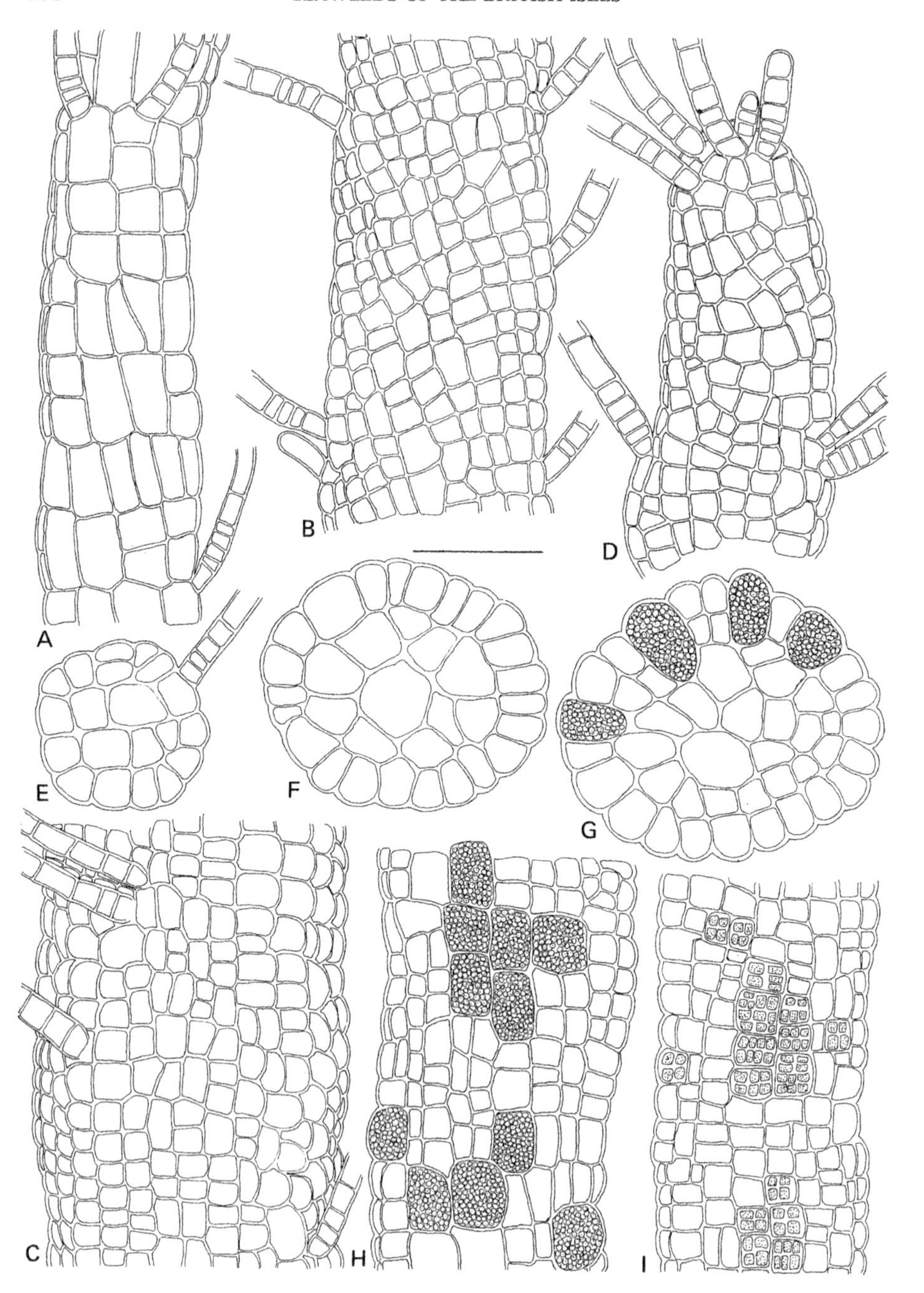

A
B
D
E
F
G
C
H
I

Certainly there is no basis for assuming, as did Kylin, that an alternation of generations takes place in this species. Fertile prostrate systems (microthalli) are also commonly reported (Nygren, 1975a, Sauvageau, 1933a; Pedersen, 1978a, 1981c, 1984) which may exist alone under various suboptional conditions of reduced salinity and light intensity (Pedersen, 1981c, 1984). Their morphology has been described as streblonemoid and similarities have been drawn between the *Litosiphon* microthalli and species of *Streblonema,* such as *S. thuretii* Sauvageau, *S. volubile* (Crouan frat.) Thuret in Le Jolis, *S. danicum* Kylin and *S. oligosporum* Strömfelt.

Harvey (1849), p. 43, pl. 8, fig. D; Batters (1892b), p. 19; Johnson (1892), pp. 5–6; Sauvageau (1929), pp. 350–362; Newton (1931), p. 180, fig. 113; Hamel (1931–39), pp. 217–219, fig. 43 (4–6); Kylin (1933), pp. 25–33, figs 5–7; Sauvageau (1933a), pp. 5–22, figs 1–4; Levring (1937), pp. 66–68; Kylin (1947), p. 74, fig. 60A; Rosenvinge & Lund (1947), pp. 15–19, fig. 3; Jaasund (1957), pp. 218–222, fig. 7 c–g; (1965), pp. 83–85; Dangeard (1969), pp. 72–74, fig. 2B; Nygren (1975a), pp. 131–141, figs 16–21; Nygren (1975b), pp. 143–147, fig. 4; Rueness (1977), p. 162, fig. 95; Pedersen (1978a), pp. 61–68, figs 17–19; Pedersen (1981c), pp. 194–217; Stegenga & Mol (1983), p. 99; Pedersen (1984), pp. 26–32, figs 14–19.

MYRIOTRICHIA Harvey

MYRIOTRICHIA Harvey (1834), p. 299.

Type species: *M. clavaeformis* Harvey (1834), p. 300.

Plants epiphytic forming small, erect tufts on hosts arising from an endophytic, or more commonly, epiphytic base of branched, rhizoidal filaments; tufts comprising several erect, flaccid shoots, uniseriate, becoming multiseriate; cells with several discoid plastids without obvious pyrenoids; hairs abundant, with basal meristem but lacking sheath.

Unilocular and plurilocular sporangia occurring on the same or different shoots, borne on basal filaments, or more commonly on erect shoots; unilocular sporangia sessile or stalked, solitary or grouped, lateral on primary axis or at base of radial branches, spherical; plurilocular sporangia sessile or stalked, solitary or grouped, lateral on primary axes or more commonly lateral and terminal on radial branches, simple or branched, uniseriate to biseriate, cylindrical to conical.

One species in the British Isles:

Fig. 37 *Litosiphon laminariae*
A–C. Portions of erect thalli with lateral hairs. D. Terminal portion of thallus. E–F. T.S. of vegetative thalli. G. T.S. of fertile thallus with unilocular sporangia. H. Portion of fertile thallus with unilocular sporangia. I. Portion of fertile thallus with plurilocular sporangia. Bar = 50 μm.

Myriotrichia clavaeformis Harvey (1834), p. 300. Figs 38, 39; Pl. 4c–e

Myriotrichia filiformis Harvey (1841), p. 44.
Myriotrichia repens Hauck (1879), p. 242.
Dichosporangium repens (Hauck) Hauck (1883–1885), p. 339.
Myriotrichia harveyana Nägeli (1847), p. 149.
?*Streblonema sphaericum* (Derbès et Solier in Castagne) Thuret in Le Jolis (1863), p. 73.
M. clavaeformis var. *subcylindrica* Batters (1895b), p. 312.

Plants forming erect tufts on host, usually visible to the naked eye, solitary, rarely confluent, comprising several erect shoots attached at the base by spreading, filamentous, branched, rhizoidal filaments; erect shoots filiform, linear, often slighty nodose, becoming elongate-clavate, to 6–30 (–40) mm long, 0·2–0·5 mm wide, simple or with short radial branches, soft and flaccid in texture; shoots uniseriate and filamentous throughout, or more commonly uniseriate below, multiseriate and parenchymatous above; parenchymatous regions discrete and shoots nodose, sometimes becoming confluent and expansive; surface cells of parenchymatous regions giving rise to lateral branches, hairs and/or sporangia; lateral branches radially produced either irregular and discrete or confluent and closely-packed, sheathing the primary axis, short, uniseriate, becoming multiseriate, simple or irregularly and alternatively branched, usually longer in terminal regions contributing to clavate appearance of shoots, later terminated by plurilocular sporangia; cells quadrate, cylindrical or barrel-shaped, $\frac{1}{2}$–1 (–2) diameters long, 8–36 × 13–20 (–40) μm, each with several discoid plastids without obvious pyrenoids; hairs abundant, 10–13 μm in diameter, colourless except for short, 2–4 celled, pigmented basal meristem, without enclosing sheath.

Unilocular and plurilocular sporangia commonly borne on different plants, occasionally mixed; unilocular sporangia sessile or stalked on primary axis or lateral branches, single or more commonly clumped in sori, spherical, 20–40 (–52) μm diameter; plurilocular sporangia terminal or lateral on lateral branches, single or more commonly clustered in dense sori, simple or branched, uniseriate to biseriate with straight and oblique cross walls, cylindrical, subcylindrical or conical, 4–12 loculi, 20–58 × 7–13 μm.

Endophytic or more commonly epiphytic on various algae and *Zostera;* much more commonly recorded on brown algae, in particular species of *Scytosiphon, Asperococcus, Punctaria* and *Stictyosiphon;* in eulittoral pools and shallow sublittoral.

Not uncommon and widely distributed around the British Isles.

Annual; summer.

In the present treatment, agreement is expressed with Pedersen (1978b) that the three entities described as *M. repens, M. filiformis* and *M. clavaeformis* represent progressive stages in development of the same species, for which *M. clavaeformis* has priority. The form described under *M. repens* is a microscopic epi/endophyte, occurring on various

---

Fig. 38 *Myriotrichia clavaeformis*
A. Short erect filament with terminal hair. B. Terminal portion of erect filament with hair. C. Three short filaments with terminal clusters of plurilocular sporangia. D. Terminal portions of two erect filaments with plurilocular sporangia. E. Portion of more differentiated erect thallus. F. Portion of erect thallus with lateral plurilocular sporangia. G–I. Portions of erect thalli with lateral branches and unilocular sporangia. Bar = 50 μm.

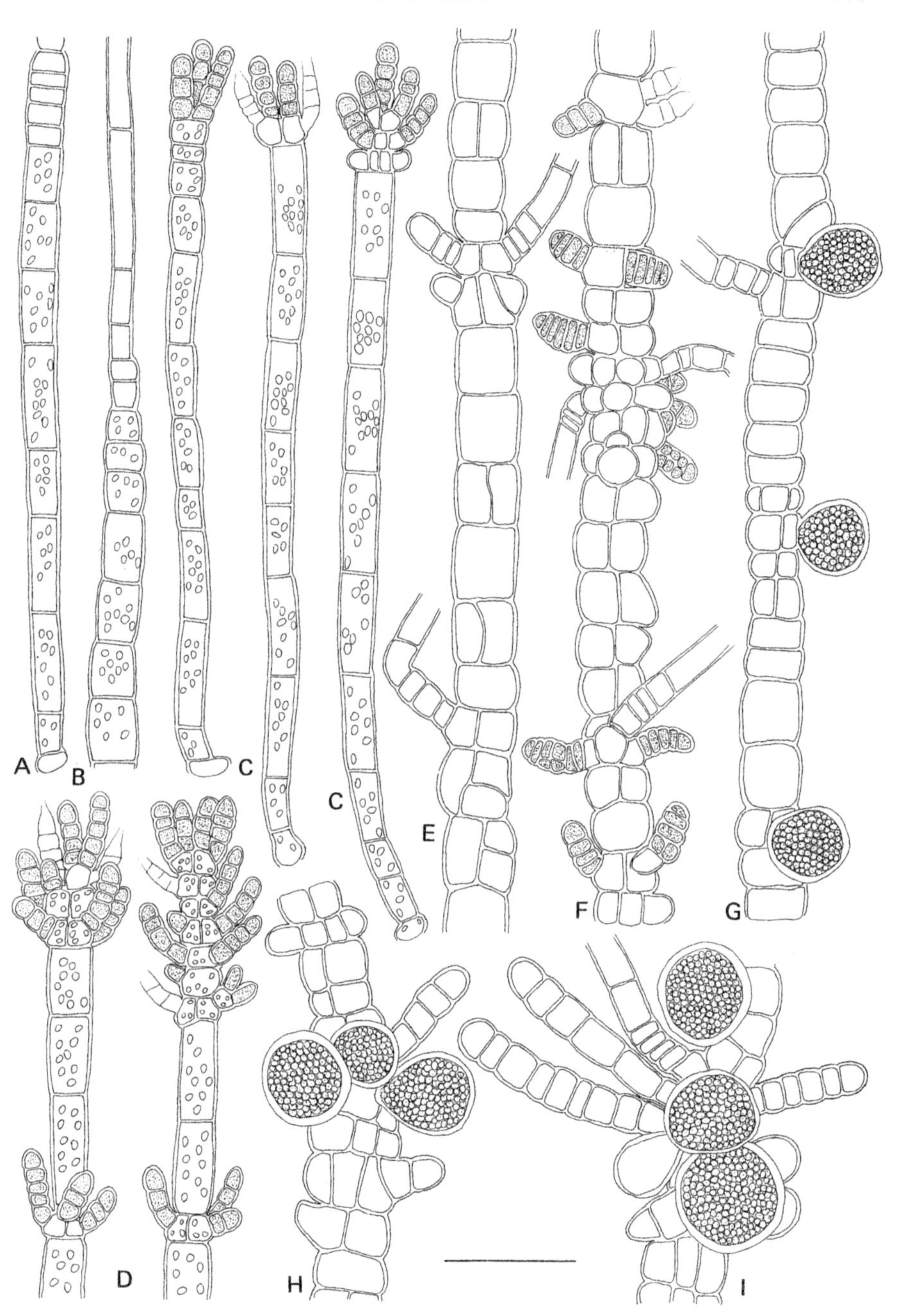
A
B
C
C
E
F
G
D
H
I

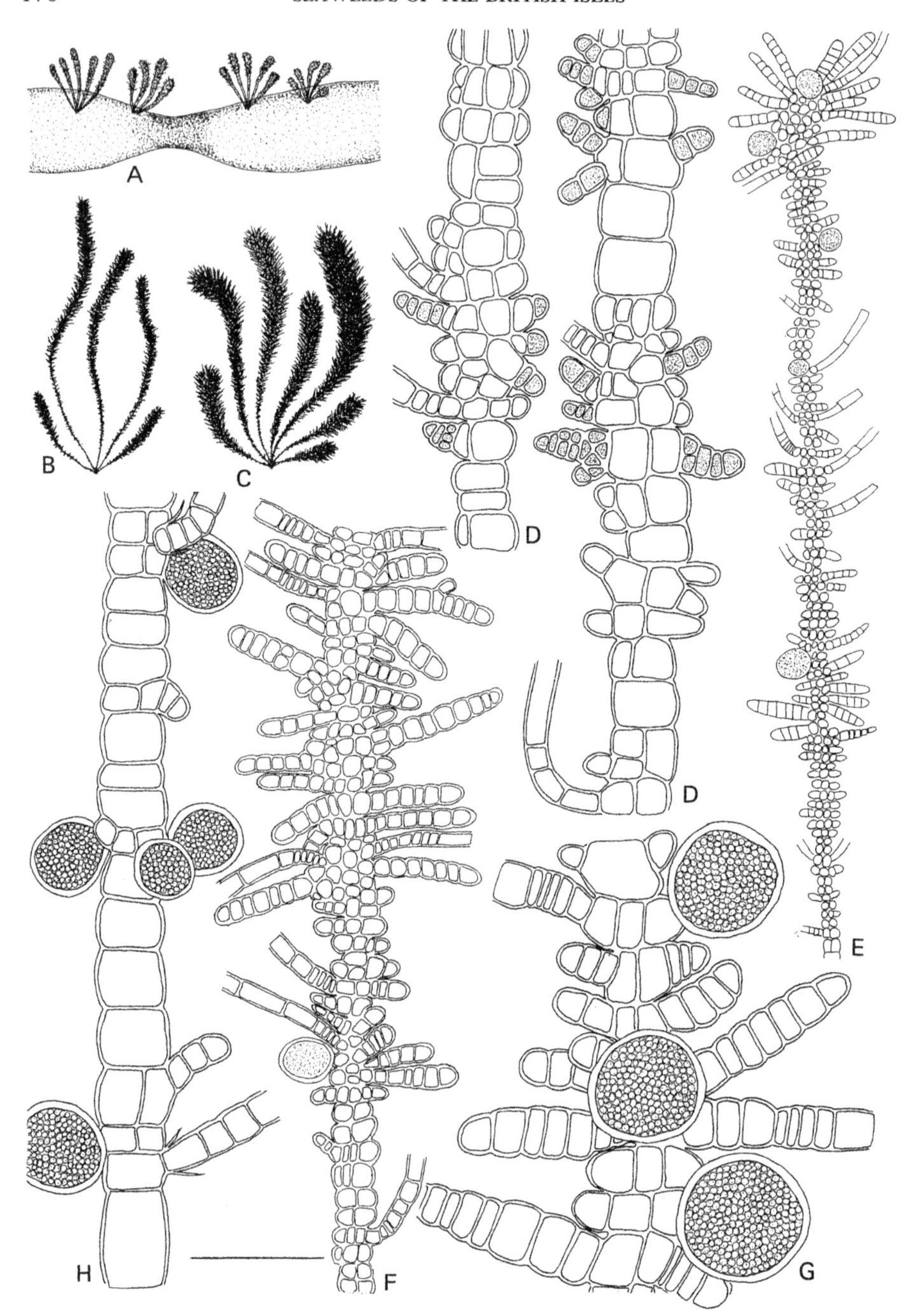
A
B
C
D
D
E
F
G
H

hosts particularly under suboptimal conditions of light and salinity. It comprises a basal system of outwardly spreading and penetrating rhizoidal filaments giving rise to short simple, erect filaments which are predominantly uniseriate, partly biseriate, bearing hairs and single, or more usually, clustered, unilocular and/or plurilocular sporangia. Occasionally both sporangia can be observed on the basal filaments and in many respects this growth form has been likened to *Streblonema sphaericum* (see Pedersen, 1978b for discussion).

With increasing length of the erect shoots and more regular development of parenchymatous knots of tissue along the terminal region of the uniseriate filament, the form described under *M. filiformis* is produced. Very often short branches, unilocular and plurilocular sporangia are associated with these node-like regions on the erect shoot. Finally, with the greater part of the erect shoot parenchymatous and clothed in the lateral branches and sporangia, the form described under *M. clavaeformis* is produced. All the above described forms may be observed either separately or in association.

Although *M. clavaeformis* has been investigated in laboratory culture by various authors such as Karsakoff (1892), Sauvageau (1931), Kylin (1933), Dangeard (1965d) and Pedersen (1978b) (as *M. clavaeformis, M. filiformis* and/or *M. repens*) most of these studies were often incomplete and the life history is still not fully understood. In general the evidence points towards the occurrence of a direct, asexual, life history with streblonemoid microthalli. Loiseaux's (1969) report of a *Myriotrichia* sp (provisionally identified as *M. clavaeformis*) being connected in life history with a *Hecatonema maculans* microthallus was undoubtedly incorrect (see Pedersen, 1978b, p. 290 and Fletcher, 1984, p. 193).

Harvey (1834), p. 300, pl. 138; Harvey (1841), p. 44; Harvey (1846–51), pl. 101; Hauck (1879), p. 242, pl. 4, figs 1–2; Karsakoff (1892), pp. 433–444, figs I–II, pl. xiii, figs 1–10; Batters (1893a), p. 23; Batters (1895b), pp. 312–313; Sauvageau (1897b), p. 36, fig. 1; Kuckuck (1899a), pp. 21–30, 37–41, pl. 3, figs 1–10; Sauvageau (1931), pp. 51–77, figs 10–12; Newton (1931), pp. 174–175, fig. 109; Hamel (1931–39), pp. 233–239, fig. 46, III, IV & VI; Knight & Parke (1931), p. 64, pl. xi, figs 21–22; Kylin (1933), pp. 33–36, figs 8–10; Taylor (1957), pp. 161–162, pl. 10, figs 4–7; Kylin (1947), pp. 70–72, fig. 58; Rosenvinge & Lund (1947), pp. 48–55, figs 16–18; Waern (1952), pp. 159–162, figs 71–73; Dangeard (1965d), pp. 79–98, pls XVI–XIX; Dangeard (1969), pp. 78–79; Rueness (1977), pp. 155–157, figs 89–90; Pedersen (1978b), pp. 281–291, figs 1–34.

PUNCTARIACEAE (Thuret) Kjellman

PUNCTARIACEAE (Thuret) Kjellman (1880), p. 9.
Punctarieae Thuret in Le Jolis (1863), p. 14

Plants epiphytic, less commonly epilithic, forming erect, macroscopic shoots or microscopic spots or tufts; erect shoots initially filamentous, becoming parenchymatous, filiform, cylindrical or dorsiventrally flattened, simple or proliferous around margin, solid or hollow, arising from a discoid holdfast; structure of equal sized cells or more

---

Fig. 39 *Myriotrichia clavaeformis*
A. Habit of plants on *Scytosiphon lomentaria*. B–C. Habit of plants. D. Portions of erect thalli with plurilocular sporangia. E–H. Four portions of erect thalli with unilocular sporangia. Bar = 6 mm (A), 2 mm (B–C), 50 μm (D, F, H), 200 μm (E), 100 μm (G).

commonly with inner medulla of large, elongate, colourless cells and outer cortex of smaller, pigmented cells; cells with several discoid plastids and pyrenoids; hairs solitary, or grouped, arising superficially or immersed, with basal meristem and enclosing sheath; plurilocular and unilocular sporangia solitary or grouped in sori, produced from surface cells, with or without associated, multicellular, paraphysis-like filaments.

Microscopic thalli comprising a basal layer of outwardly radiating, branched filaments, free or laterally united and pseudodiscoid/discoid, monostromatic or in part distromatic, giving rise to erect filaments, hairs, ascocysts and/or plurilocular sporangia; erect filaments short, simple, or little branched, comprising cells with one to several plate-like plastids with pyrenoids; hairs with basal meristem and sheath; plurilocular sporangia on basal layer, sessile or stalked, or lateral and terminal on erect filaments, multiseriate; unilocular sporangia unknown.

Out of the four genera recognised for this family by Parke & Dixon (1976) only *Asperococcus* and *Punctaria* are included here; *Litosiphon* has been transferred to the Myriotrichiaceae following Pedersen (1984) whilst *Desmotrichum* has been synonymised with *Punctaria*. Both *Asperococcus* and *Punctaria* are predominantly macrophytes and easily distinguished on morphological features; *Asperococcus* forms inflated or compressed tubular thalli whilst *Punctaria* forms dorsiventrally flattened, solid thalli. Also included are the microscopic genera *Chilionema* and *Hecatonema* removed from the Myrionemataceae because of their possession of a part parenchymatous thallus, hairs with basal meristem and enclosing sheath, and life history studies which have shown some species to be connected in life history with species of *Asperococcus* and *Punctaria*.

ASPEROCOCCUS Lamouroux

ASPEROCOCCUS Lamouroux (1813), p. 277.

Type species: *non designatus*.

Plants consisting of erect tubular fronds arising from a small stipe and basal disc or holdfast; fronds simple, solid or hollow, cylindrical, filiform, or elongate clavate, inflated or dorsiventrally compressed, with or without dark brown punctations; comprising an inner medulla of large, colourless medullary cells enclosed by an outer cortex of smaller pigmented cells; cells with several discoid plastids and pyrenoids; hairs solitary or grouped arising superficially, with basal meristem and sheath.

Plurilocular and/or unilocular sporangia present, in discrete or continuous sori, arising from surface cells, associated with multicellular, paraphysis-like filaments; unilocular sporangia spherical or pyriform; plurilocular sporangia cylindrical or conical, protruding, uniseriate or multiseriate.

The genus *Asperococcus* is characterised by an erect, simple, usually hollow thallus internally differentiated into an inner medulla of large, colourless cells and an outer cortex of small, pigmented cells; additional characters include cells with several discoid plastids with pyrenoids, hairs with a basal meristem and sheath and sporangia arising from surface cells and usually associated with short, multicellular, paraphysis-like filaments. The four species present in the British Isles are quite easily distinguished on morphological characters.

Although European isolates of three of these species (*A. compressus*, *A. fistulosus* and *A. turneri*) have been investigated in culture, their life histories are still not fully understood and often conflicting interpretations have been made by various research workers. In general it seems likely that a direct type of life history is operating, which is heteromorphic (comprising both macro- and microthalli) and monophasic; reported observations of copulation between plurispores derived from plurilocular sporangia on the microthalli (Knight *et al.*, 1935 for *A. fistulosus* and *A. turneri*) and between unispores derived from unilocular sporangia on the macrothalli (Knight *et al.*, 1935 and Dangeard, 1968c for *A. fistulosus*) must be treated with caution (see Pedersen, 1984, pp. 19–20).

An interesting aspect of the life history studies was the resemblance between the microthalli produced in the cultures of *Asperococcus* and the genus *Hecatonema*. Similar characteristics included very often a discoid/pseudodiscoid thallus, short upright simple or little branched filaments, siliquose multiseriate plurilocular sporangia borne on the basal filaments or lateral/terminal on the erect filaments, cells with several discoid plastids and pyrenoids and hairs with basal meristem and enclosing sheath (see especially Kylin (1933), fig. 14; Dangeard (1968c) pl. 5A; Pedersen (1984) figs 8–9). Pedersen identified the microthalli produced in his cultures of *A. fistulosus* with one or more of four species of *Hecatonema* depending on the degree of illuminance under which they were grown (see p. 185). As an adjunct to these studies it is interesting to note the reports by Loiseaux (1969) and Fletcher (1984) of *Asperococcus*-like macrothalli developing in cultures of field collected plants of *Hectatonema maculans* (see p. 185).

KEY TO SPECIES

1	Thallus solid and less than 3 mm long and 0·5 mm wide	*A. scaber*
	Thallus hollow and more than 5 cm long and 2 mm wide	2
2	Thallus distinctly dorsiventrally flattened	*A. compressus*
	Thallus distinctly inflated	3
3	Thallus cylindrical and linear, gradually attenuate towards base, flaccid but fairly strong, up to 20 mm wide; paraphyses to 190 μm and 7 cells long	*A. fistulosus*
	Thallus inflated and bulbous, sharply attenuate towards base, soft and delicate, easily torn, up to 40 mm wide; paraphyses to 120 μm and 4 cells long	*A. turneri*

Asperococcus compressus Griffiths ex Hooker (1833) p. 278. Figs 40A, 41

Haloglossum compressum (Griffiths ex Hooker) Hamel (1931–39), p. 225.
Haloglossum griffithsianum Kützing (1843), pl. 52 I.

Plants forming erect, gregarious, olive green becoming light brown, thalli arising from a discoid holdfast; thalli simple, flaccid, hollow, but dorsiventrally compressed, elongate-clavate, to 40 (– 80) cm long, 8–20 mm wide, covered throughout with dark brown punctations; in section, thalli distinctly hollow with an inner medulla of 1–3 large, colourless,

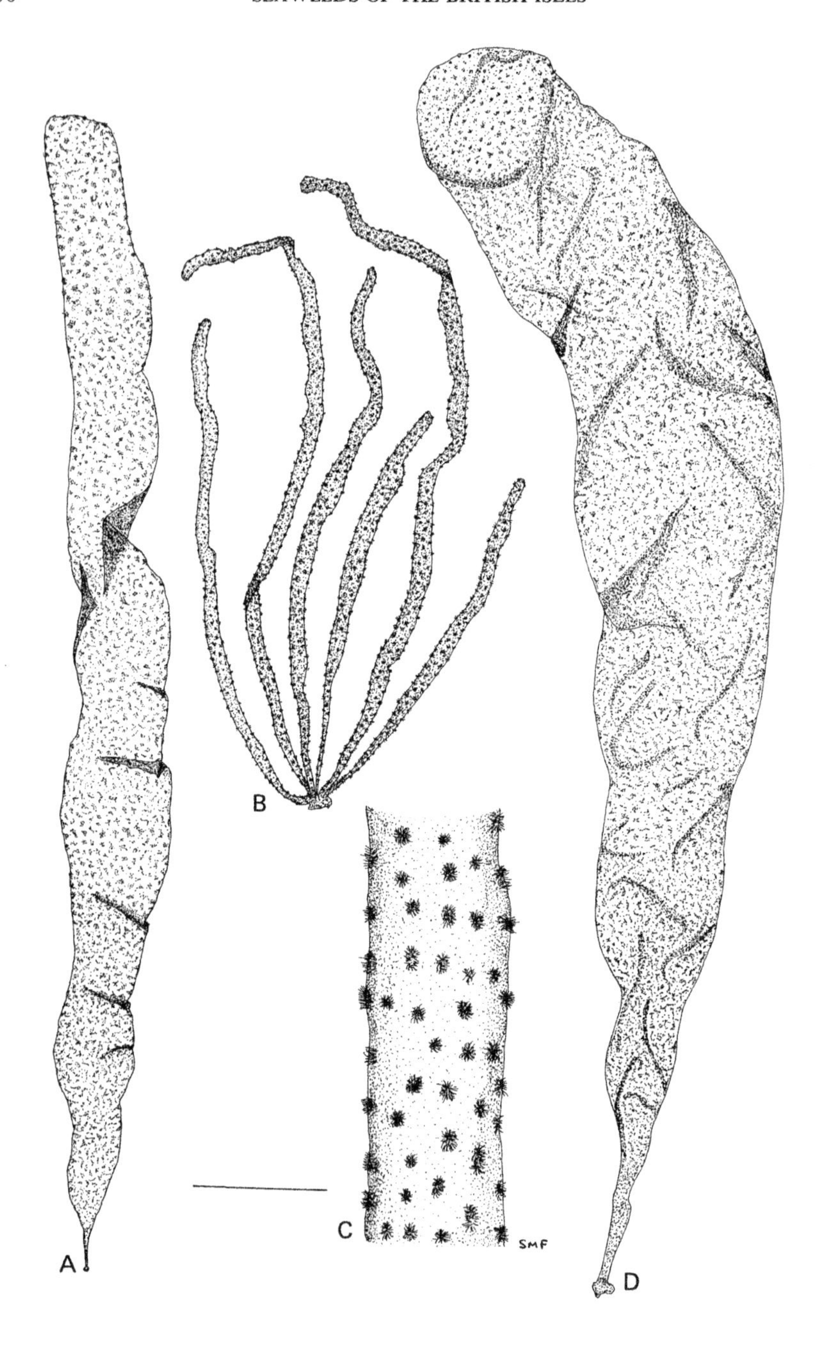
A
B
C
D
SMF

sometimes ruptured cells, enclosed by an outer cortex of 1–2 smaller pigmented cells; surface cells quadrate or rectangular, usually in rows, 20–30 (–40) × 13–20 (–30) μm, each with several discoid plastids and pyrenoids; hairs common, arising from surface cells and grouped in discrete punctate sori, with basal meristem and enclosing sheath, 11–14 μm in diameter.

Unilocular sporangia common, crowded in discrete, irregularly shaped, punctiferous sori, produced from surface cells, with associated hairs and paraphysis-like filaments; unilocular sporangia rounded in surface view, 27–46 μm in diameter, globose or pyriform in section, usually sessile occasionally stalked 40–60 × 26–43 μm; paraphyses erect, simple, colourless, linear or clavate, 1–3 (–4) celled, 58–87 × 13–23 μm, arising from surface cells; plurilocular sporangia rare, not observed on British material, reputed to be on separate plants, in small sori, cylindrical, conical, sessile or stalked, to 35 μm long × 9–15 μm wide.

Epiphytic, epilithic in eulittoral pools and sublittoral to 10 m.
Rare, widely distributed around the British Isles.
Annual, spring and summer, April to September.

Reinke (1878a) and Sauvageau (1929) reported a 'direct' type of life history in this species with unispores from the unilocular sporangia germinating to produce creeping, filamentous, sometimes plurilocular sporangia-bearing microthalli from which new, erect macrothalli developed. Fertile microthalli, described as hecatonemoid were also reported in cultures of this species by Pedersen (1984).

Reinke (1878a), pp. 268–269; Sauvageau (1895), pp. 336–338, fig. 1; Kuckuck (1912), p. 178, pl. 8, fig. 11; Sauvageau (1929), pp. 366–381, figs 15–17; Newton (1931), p. 172; Knight *et al.* (1935), pp. 86–87; Hamel (1931–39), pp. 225–226; Rosenvinge & Lund (1947), pp. 47–48; Rueness (1977), p. 158; Pedersen (1984), p. 57, fig. 36F.

Asperococcus fistulosus (Hudson) Hooker (1833), p. 277. Figs 40B, C, 42; Pl. 5

Ulva fistulosa Hudson (1778), p. 569.
Conferva echinata Mertens in Roth (1806), p. 1701.
Asperococcus echinatus (Mertens in Roth) Greville (1830), p. 50.
Asperococcus rugosus Lamouroux (1813), p. 62.
Asperococcus echinatus var. *filiformis* Reinke (1888), p. 19.
Asperococcus fistulosus var. *vermicularis* Griffiths ex Harvey (1846–51), pl. 194.
Asperococcus echinatus f. *villosa* Kylin (1907), p. 77.

Plants forming erect, gregarious, light olive to dark brown thalli arising from a discoid holdfast; thalli simple, flaccid, hollow, cylindrical, linear or slightly elongate-clavate, 5–40 cm long × 2–5 (–20) mm wide, gradually attenuate below, more sharply attenuate

Fig. 40 *Asperococcus compressus*
A. Habit of plant.
Asperococcus fistulosus
B. Habit of plant. C. Portion of erect thallus.
Asperococcus turneri
D. Habit of plant. Bar = 3 cm (A–B), 1·5 cm (D), 1·2 mm (C).

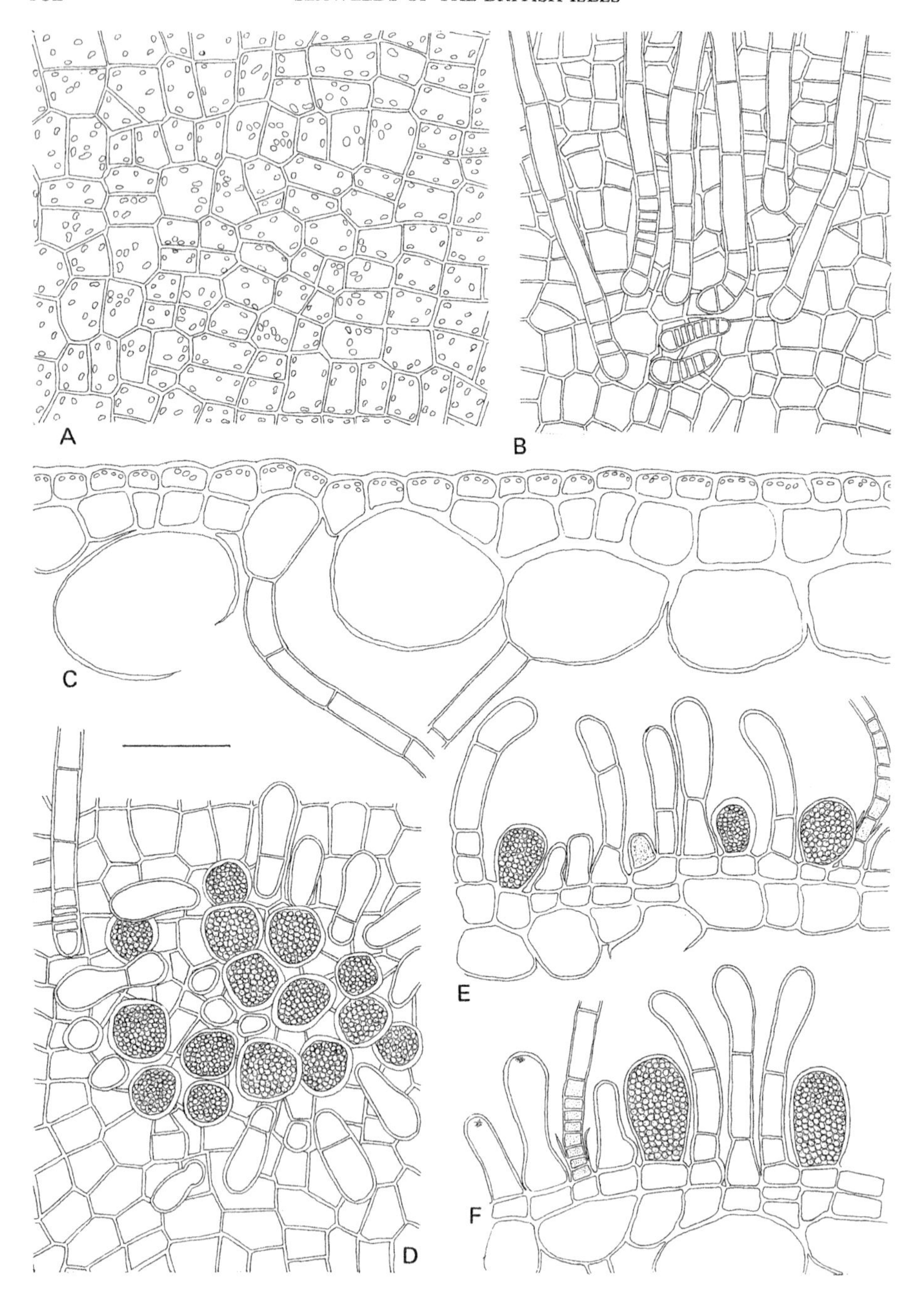
A
B
C
D
E
F

and obtuse above, surface unevenly inflated and contoured, slightly roughened in texture, covered throughout with dark brown punctations; surface cells quadrate or rectangular, usually in rows, 10–21 (–44) × 8–19 (–26) μm, each with large numbers of discoid plastids with associated pyrenoids; in section, thalli distinctly hollow with an outer cortex of 1 (–2) small plastid-bearing cells enclosing an inner medulla of 1–3 large, longitudinally elongate, colourless, sometimes ruptured cells; hairs abundant, grouped in discrete, punctate sori, with dark brown basal meristem and enclosing sheath, 13–18 μm in diameter.

Plurilocular and unilocular sporangia borne on the same or different thalli, crowded in discrete, irregular shaped, abundant and extensive, punctiferous sori, produced from cortical cells, with associated hairs and paraphysis-like filaments; unilocular sporangia abundant, rounded in surface view, 40–70 μm diameter, globose or pyriform in section, 30–85 × 20–70 μm, sessile; plurilocular sporangia rare, only reported for the Isle of Man, dimensions uncertain; paraphyses erect, simple, or rarely branched, rigid, thick-walled, darkly pigmented, especially terminal cell, linear or slightly elongate-clavate, multicellular 2–5 (–7) cells, 80–120 (–190) μm long, arising directly from surface cells, cells subquadrate, quadrate or rectangular, 18–47 × 18–36 μm.

Epiphytic on various algae, especially *Fucus*, more rarely epizoic, in upper to lower eulittoral pools and sublittoral to 18 m.

Common and widely distributed around the British Isles.

Annual, summer (May to September) less commonly reported throughout the year.

European isolates of this species have been the subject of laboratory culture studies by Sauvageau (1929), Knight & Parke (1931), Kylin (1934), Knight *et al.* (1935), Rosenvinge & Lund (1947), Dangeard (1968c) and Pedersen (1984). In the majority of these studies the life history has been generally interpreted to conform to the haplodiplophasic, heteromorphic type, with unispores derived from the unilocular sporangia on the diploid macroscopic phase producing creeping, microscopic, filaments bearing plurilocular sporangia which are considered to represent the gametophyte phase. However, copulation of the plurispores from the plurilocular sporangia was only reported by Knight & Parke (1931) and Knight *et al.* (1935); usually the zoospores behaved asexually to produce several generations of fertile microscopic filaments. An additional feature of the life history reported by Knight *et al.* and Dangeard was the copulation of zoospores from the unilocular sporangia, thus apparently directly repeating the diploid macroscopic phase (diplophasic life history).

From the results of a more recent laboratory culture study, Pedersen (1984) proposed that a 'direct' type of life history was probably operating in this species. He questioned the validity of reports of spore fusions and thought it unlikely that the fertile microthalli represented gametophytes. He, therefore, interpreted the life history as monophasic and

---

Fig. 41 *Asperococcus compressus*
A. Surface view of thallus showing cells with numerous discoid plastids. B. Surface view of thallus with hairs. C. Section of vegetative thallus showing outer 1–2 layers of small cortical cells and inner layers of large, medullary cells enclosing central cavity. Note rhizoidal extensions of medullary cells. D. Surface view of fertile thallus showing unilocular sporangia and associated 1–2 celled paraphyses. E–F. Sections of fertile thalli showing unilocular sporangia and paraphyses arising from surface cells. Bar = 50 μm.

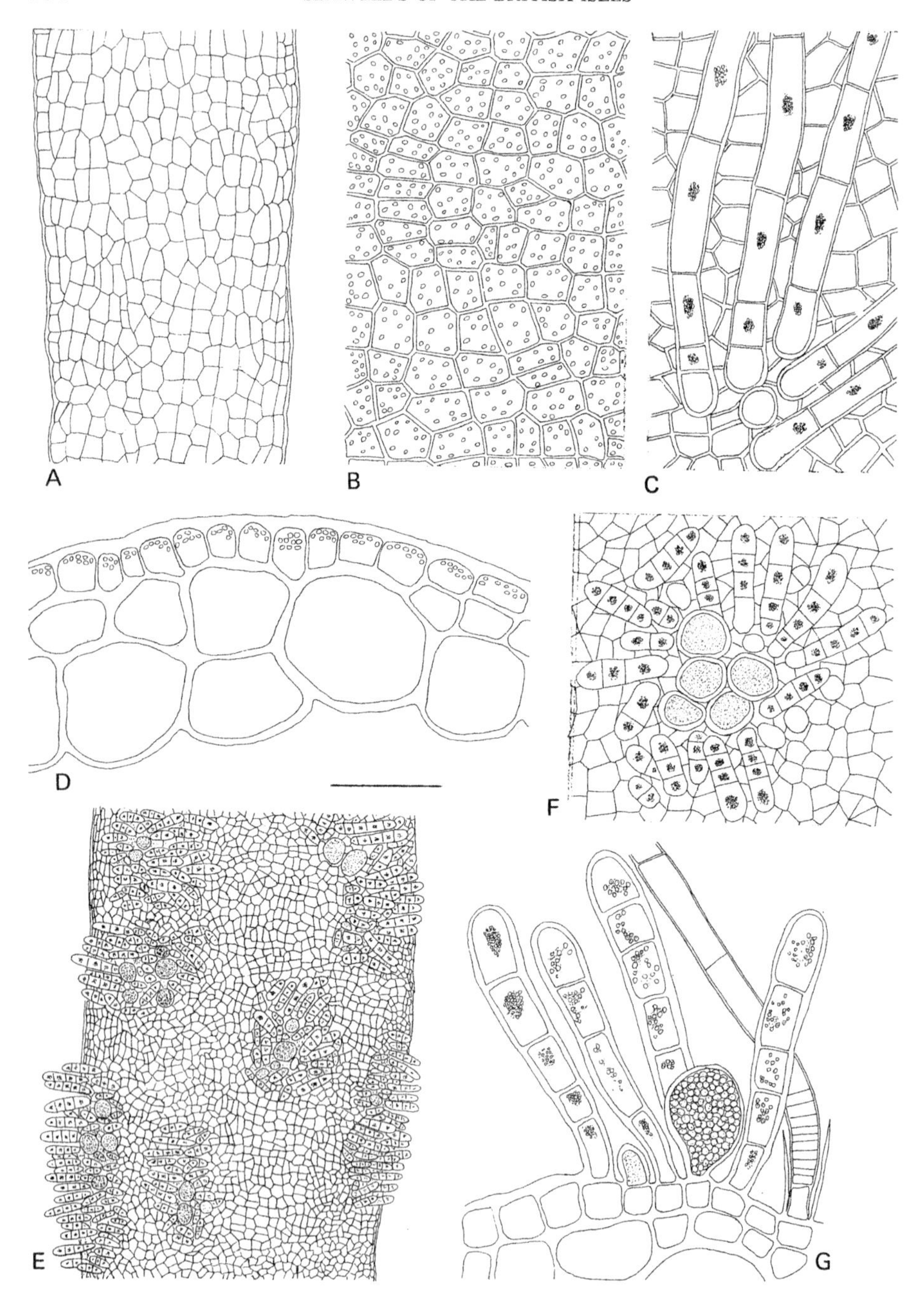
A
B
C
D
F
E
G

heteromorphic with the microthalli serving a useful function as a resting phase during unfavourable environmental conditions. Agreement is expressed with Pedersen in this interpretation of the life history.

A noteworthy feature of the life history studies on this species reported by Kylin (1934), Dangeard (1969) and Pedersen (1984) was the hecatonemoid growth form of the fertile microthalli. Kylin noted the resemblance of the microthalli to *Hecatonema maculans* (Collins) Sauvageau and *Ectocarpus terminalis* Kützing (probably = *H. maculans*) whilst Pedersen identified the microthallus as either *Hecatonema reptans* Sauvageau and *Ectocarpus repens* Reinke or *Hecatonema maculans* (Collins) Sauvageau and *Hecatonema terminalis* (Kützing) Kylin depending on the culture conditions of illumination at 4°C.

Hecatonema maculans was also linked in life history with an *Asperococcus* species (probably *A. fistulosus*) by Loiseaux (1979) and Fletcher (1984). Both authors reported a facultative relationship between the *Hecatonema*-like microthalli and the thalloid macrothalli which was governed by environmental conditions, particularly temperature and photoperiod. All released spores behaved asexually and no evidence was obtained of microthalli functioning as gametophytes or of an obligate alternation between the two expressions due to a difference in ploidy level. These results add support to Pedersen's (1984) interpretation of the life history of European isolates of *A. fistulosus*.

Reinke (1889a), p. 7, pl. 4; Reinke (1889b), p. 53; Sauvageau (1929), pp. 381–385, fig. 18; Knight & Parke (1931), pp. 59, 105–110; Newton (1931), p. 172; Hamel (1931–39), pp. 222–223, fig. 43, VII; Kylin (1934), pp. 13–15, figs 7–8; Knight *et al.* (1935), pp. 87–90; Taylor (1957), pp. 169–170, pl. 15, fig. 7, pl. 16, figs 5–6; Rosenvinge & Lund (1947), pp. 43–47, fig. 15; Dangeard (1969), pp. 75–76, fig. 3A, pls. XIV–XV; Rueness (1977), p. 159, fig. 92; Stegenga & Mol (1983), p. 99, pl. 34; Pedersen (1984), pp. 18–20, figs 8, 9, 36A, 37.

Asperococcus scaber Kuckuck (1899a), p. 18. Fig. 43

Plants forming erect, solitary or gregarious thalli, up to 13 mm long and 0·25 mm wide arising from a small pseudo-discoid base of outwardly spreading, compacted rhizoids; thalli simple, filiform, cylindrical, linear or elongate-clavate, gradually attenuating above and below, rarely obtuse above, solid, parenchymatous; surface cells quadrate or rectangular, in longitudinal less obviously transverse rows, 10–27 × 10–21 μm, each with numerous discoid plastids with pyrenoids; in section thalli solid with 4–6 large, colourless central cells, with or without a small central cavity, enclosed by 1–2 smaller, pigmented cortical cells; hairs common, single, arising from surface cells, with basal meristem and sheath.

Fig. 42 *Asperococcus fistulosus*
A. Portion of erect vegetative thallus. B. Surface view of vegetative thallus showing cells with several discoid plastids. C. Surface view of vegetative thallus with hairs. D. Section of vegetative thallus showing outer small cortical cells and inner large medullary cells enclosing central cavity. E. Portion of fertile thallus showing unilocular sporangial sori. F. Surface view of fertile thallus showing unilocular sporangia and associated paraphyses. G. Section of fertile thallus showing paraphyses, a single hair with basal sheath and unilocular sporangia. Bar = 50 μm (A–D, G), 100 μm (F), 650 μm (E).

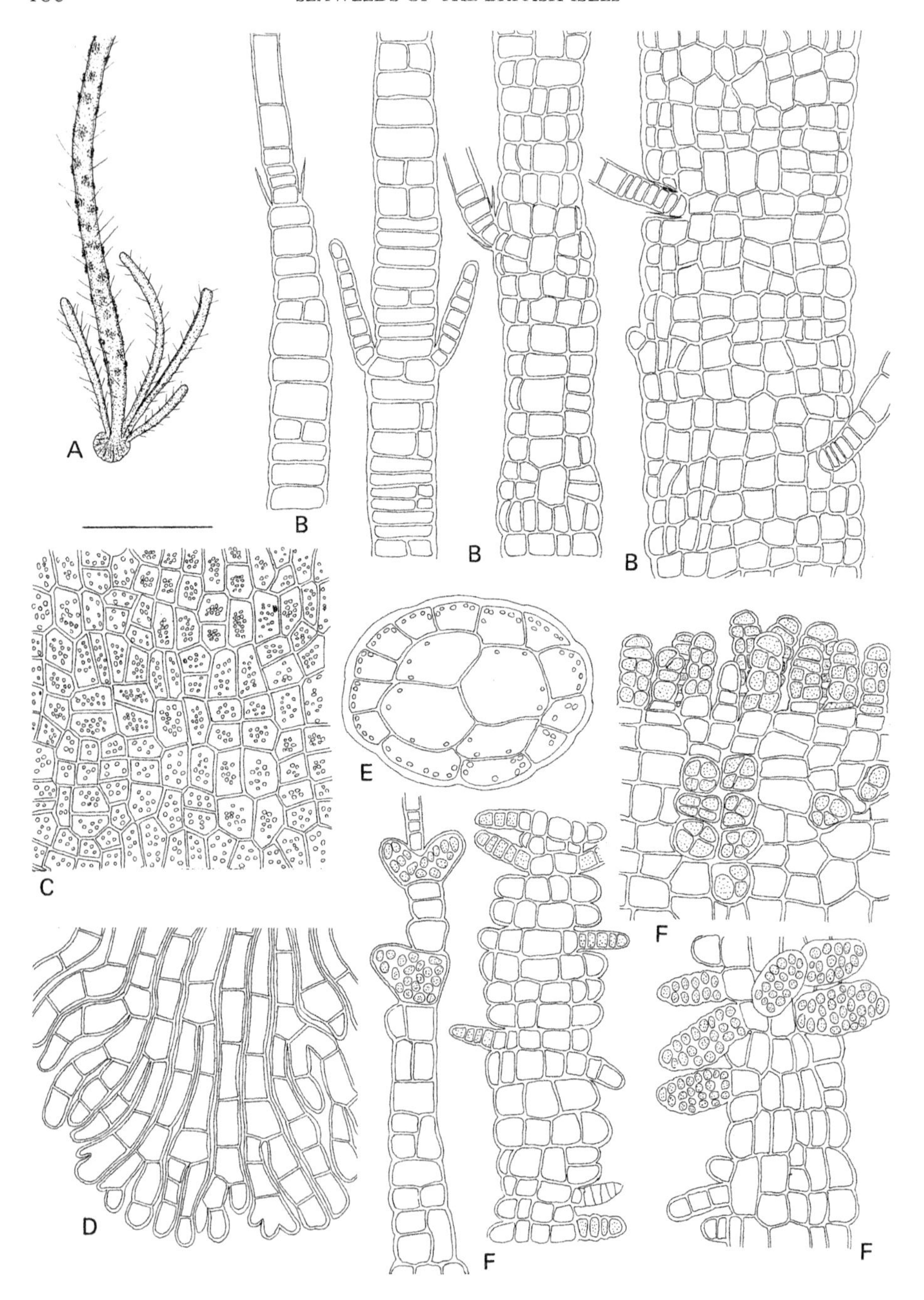
A
B
B
B
C
E
F
D
F
F

Plurilocular sporangia common, borne in crowded extensive sori often occupying greater length of erect shoot, arising and protruding laterally from surface cells, cylindrical or more commonly conical, sessile or stalked, uni- to multiseriate, 22–55 × 10–18 μm, rarely with associated short, 2–3 celled, paraphysis-like filaments; unilocular sporangia not observed on British Isles material, reputed to be on the same or separate plants, uncommon, spherical or pyriform, to 45 μm long × 30–35 μm wide.

Epilithic on stones, in pools, in the lower eulittoral.
Rare, recorded only from Dorset and Argyll.
Collections made in April; information inadequate for comment on seasonal distribution.

The records from the British Isles are based on collections made by E. A. L. Batters in April 1892 and 1897. Slides of this material are deposited in the British Museum (Natural History). No collections have been made by the present author.

Batters (1902), p. 28; Kuckuck (1899a), pp. 14–20, figs 1–4, pl. II; Newton (1931), pp. 171–172; Hamel (1931–39), pp. 224–225; Knight *et al.* (1935), pp. 90–91.

Asperococcus turneri (Smith) Hooker (1833), p. 277. Figs 40D, 44

Ulva turneri Smith in Smith & Sowerby (1790–1814), p. 2570.
Asperococcus bullosus Lamouroux (1813), p. 62.
Encoelium bullosum C. Agardh (1820–28), p. 146.

Plants forming erect, solitary, or more commonly gregarious, olive to light brown thalli arising from a discoid holdfast; thalli simple, membranous, very soft and flaccid, easily torn, hollow, inflated and bulbous, cylindrical, 10–30 cm long × 5–40 mm wide, attenuate sharply below to a narrow stipe-like region, obtuse above, surface with or without small, dark-coloured punctations; surface cells quadrate or rectangular, in rows, 23–62 × 14–50 μm, each with numerous discoid plastids and pyrenoids; in section thalli distinctly hollow with an outer cortex of 1, rarely 2 small, plastid-bearing cells enclosing an inner medulla of 1–2 (–3) large, longitudinally elongate, colourless, often ruptured cells; hairs abundant, arising from surface cells, solitary, later often grouped and associated with reproductive sori, with dark brown basal meristem and enclosing sheath, 19–23 (–28) μm in diameter.

Plurilocular and unilocular sporangia borne on the same or different thalli, crowded in discrete, irregularly shaped, punctiferous sori, widely scattered over blade surface, produced from the surface cells, with associated hairs and paraphysis-like filaments; unilocular sporangia abundant, rounded in surface view 28–55 μm in diameter, pyriform and sessile in transverse section, 60–82 × 52–72 μm; plurilocular sporangia rare, ellipsoidal in section, sessile or on 1–3 celled stalks, multiseriate, 34–56 × 13–25 μm; paraphyses usually uncommon, erect, multicellular, simple, rigid, thick-walled, darkly pigmented (especially the terminal cell), linear or slightly elongate-clavate, to 1–3 cells

Fig. 43 *Asperococcus scaber*
A. Habit of plant. B. Portions of erect thalli at various stages of development. Note lateral hairs. C. Surface view of thallus showing cells with numerous discoid plastids. D. Peripheral region of basal attachment disc. E. T.S. of vegetative thallus. F. Portions of fertile thalli showing plurilocular sporangia. Bar = 2 mm (A), 50 μm (others).

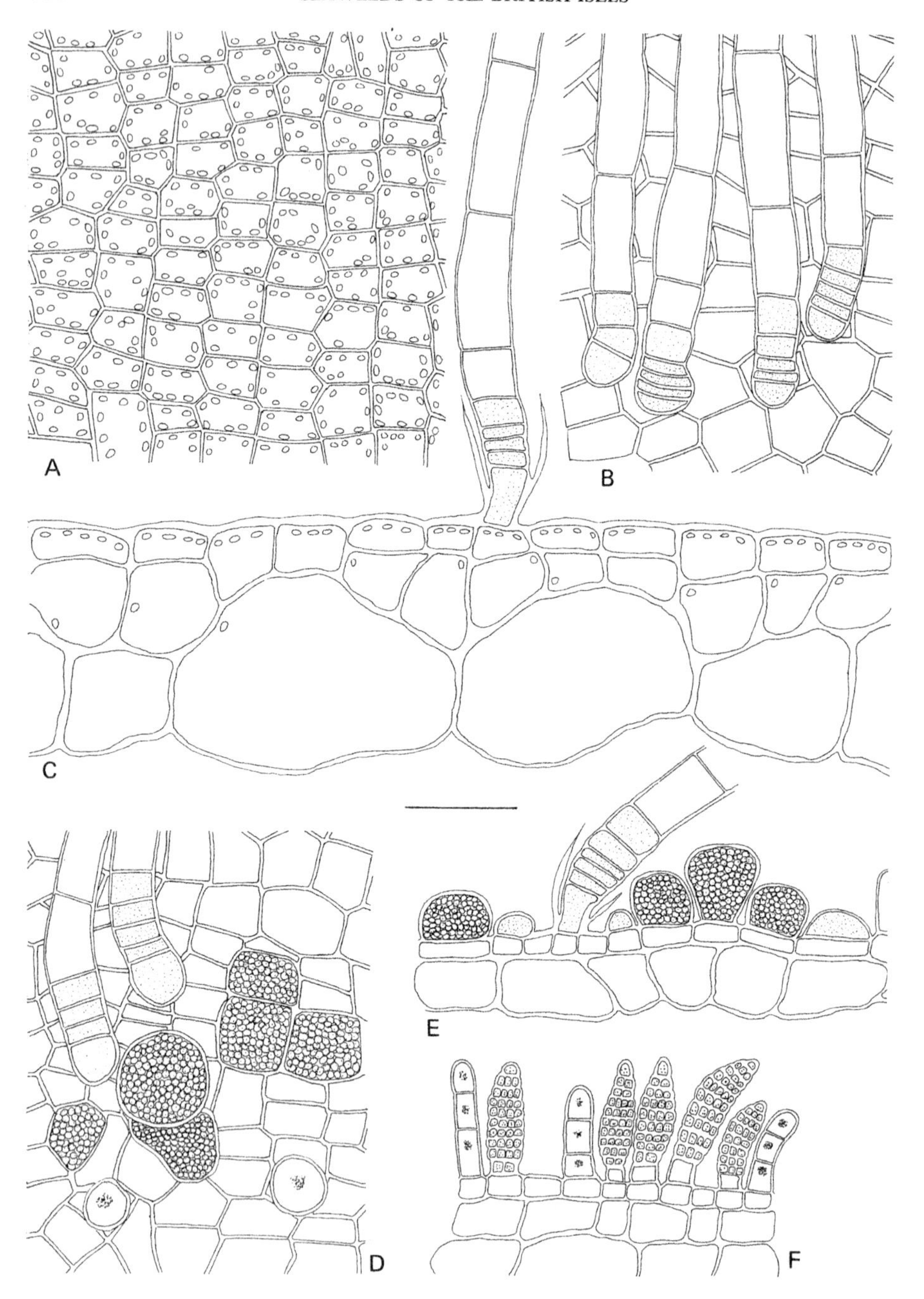
A
B
C
D
E
F

(120 μm) long, comprising quadrate to rectangular cells 1–2 diameters long, 32–55 × 26–32 μm.

Epilithic, less commonly epiphytic on various algae, upper to lower eulittoral pools and sublittoral to 12 m.

Not uncommon and widely distributed around the British Isles.

Annual, summer (April to October).

Culture studies have been carried out on this species by Sauvageau (1929), Knight *et al.* (1935), Kylin (1918, 1933) (all as *A. bullosus*), and Pedersen (1984). A classical heteromorphic, haplodiplontic life history was reported by Knight *et al.*, with zoospores from the unilocular sporangia on the macroscopic phase germinating directly to form microscopic, creeping, filamentous gametophytes bearing plurilocular sporangia, the gametes of which copulated to reform the *Asperococcus* plants. A 'direct' diplophasic life history, however, was reported by Sauvageau (1929) and Kylin (1933) in which zoospores from either unilocular or plurilocular sporangia formed creeping microscopic filaments with or without plurilocular sporangia, which could directly form new *Asperococcus* plants. Successive generations of this microscopic phase could also be produced by the direct development, without copulation, of the zoospores from the plurilocular sporangia. The culture results of Pedersen (1984) are in approximate agreement with those of Sauvageau and Kylin. The life history appeared to be of the 'direct' type with macrothalli developing directly from the microthalli. Copulation between spores was not observed and the life history was interpreted as apparently monophasic and heteromorphic. I would agree with this proposed life history for *A. turneri* and express doubt on the haplodiplontic, heteromorphic cycle reported by Knight *et al.*

As in *A. fistulosus* it is interesting to note the similarity between the plurilocular sporangia-bearing microthalli produced in the cultures of *A. turneri* by Kylin and Pedersen and the genus *Hecatonema* (see especially fig. 14 in Kylin and fig. 36E in Pedersen).

Lamouroux (1813), p. 62, pl. 6, fig. 5; Buffham (1891), pp. 321–323, pl. 314, figs 1–3; Sauvageau (1929), pp. 385–396, figs 19–20; Newton (1931), p. 172, fig. 107; Hamel (1931–39), pp. 223–225; Kylin (1933), pp. 38–44, figs 12–16; Knight *et al.* (1935), pp. 91–97, figs 1–3; Kylin (1947), pp. 75–76, fig. 60e–f, pl. 11, fig. 38; Rosenvinge & Lund (1947), pp. 41–43, fig. 14; Rueness (1977), pp. 159–160, pl. 21, fig. 3; Pedersen (1984), pp. 56–57, figs 35–36E.

CHILIONEMA Sauvageau

CHILIONEMA Sauvageau (1897a), p. 263.

Type species: *non designatus.*

Fig. 44 *Asperococcus turneri*
A. Surface view of thallus showing cells with numerous plastids. B. Surface view of thallus with hairs. C. Section of vegetative thallus showing outer 1–2 layers of small cortical cells (and emergent hair with basal sheath) and inner 1–2 layers of larger medullary cells enclosing central cavity. D. Surface view of fertile thallus showing unilocular sporangia. E. Section of fertile thallus with unilocular sporangia and hair arising from surface cells. F. Section of fertile thallus with plurilocular sporangia and associated paraphyses arising from surface cells. Bar = 50 μm.

Thallus epiphytic, forming small spots on hosts, consisting of a layer of prostrate filaments, laterally adjoined and disc-like, with or without prominent, peripheral apical cells, monostromatic or distromatic in central parts, giving rise, densely or irregularly, to erect filaments, hairs, ascocysts and/or sporangia; erect filaments usually of equal height, remaining short, simple, rarely branched, comprising cells with 1–3 plate-like plastids with pyrenoids; hairs common, usually terminal or erect filaments, rarely arising from basal layer, with basal meristem and sheath; ascocysts in some species usually borne on basal layer, more rarely terminal on erect filaments.

Plurilocular sporangia borne on basal layer, sessile or on many celled stalks, more rarely terminal or lateral on erect filaments; unilocular sporangia unknown.

Chilionema is an epiphytic genus characterised by a mainly distromatic basal disc which gives rise, somewhat irregularly, to short, erect filaments of approximately equal size which are simple or rarely branched near the base. Hairs (with basal meristem and enclosing sheath) and multiseriate, plurilocular sporangia are also borne on the disc and are sessile or shortly stalked. Unilocular sporangia are unknown. It is closely related to (and possibly synonymous with) *Hecatonema*, type species *H. maculans* (Collins) Sauvageau, which is based on *Phycocelis maculans* described by Collins (1896) on the North East American coast. In his original paper Sauvageau described and figured three 'formes' of *H. maculans* which represented progressive stages of increasing development of the erect filaments. The 'première' forme (form one) was described with short, simple, erect filaments and plurilocular sporangia borne on the basal disc whilst the 'deuxième' and 'troisième' forms (forms 2 and 3 respectively) had well developed, erect, branched filaments on which the plurilocular sporangia were mainly borne, either terminally or laterally. I would prefer, for the time being, to understand *H. maculans* (and hence *Hecatonema*) in terms of forms 2 and 3 described by Sauvageau. It is pertinent here to point out that the description of form 2 was based upon examination of material sent by Collins (no. 274 Phycotheca Boreali—America). This will, therefore, exclude form 1 which can be conveniently transferred to the genus *Chilionema*, in particular to the species *C. ocellatum* (Kützing) Kuckuck.

In view of the above discussion, two species of *Hecatonema* which have been reported for the British Isles, viz. *H. foecundum* (Strömfelt) Loiseaux and *H. hispanicum* (Sauvageau) Loiseaux would also appear to be more suitably placed in *Chilionema*. They form small, epiphytic, discoid basal thalli bearing both short, erect, predominantly simple, filaments and sessile or stalked, multiseriate, plurilocular sporangia. The new combinations are, therefore, proposed:

Chilionema foecundum (Strömfelt) Fletcher comb. nov.
Basionym: *Phycocelis foecunda* Strömfelt (1888), pl. 3, fig. 5.
Chilionema hispanicum (Sauvageau) Fletcher Comb. nov.
Basionym: *Ascocyclus hispanicus* Sauvageau (1897a), p. 274, figs 26–27.

Four species are currently recognised for the British Isles; *C. foecundum, C. hispanicum, C. ocellatum* and *C. reptans*. Differences between the latter two species are very small and they may well be conspecific. *C. foecundum* and *C. hispanicum* are characterised by the frequent presence of ascocyst-like cells arising from the basal layer. *C. hispanicum* differs from the other species of this genus in having a monostromatic basal layer and cells with a single, plate-like plastid. In some respects it shows resemblance to *Symphyocarpus strangulans* Rosenvinge, family Lithodermataceae (see p. 92). However, it would seem better at this time for the species to be retained in *Chilionema* pending further investigation.

Any proposals that *Chilionema* comprises a rather disparate group of entities would find support in the life history variations which have been reported (usually under the genus *Hecatonema*). For example, *Chilionema foecundum* and *C. hispanicum* (as *Hecatonema foecundum* and *H. hispanicum*, respectively) were both reported by Loiseaux (1967b and 1966/1967b respectively) to have a haplodiplontic, heteromorphic life cycle with a haploid gametophytic microthallus and a diploid sporophytic 'macrothallus'. An additional 'short-circuit' life history with a fusion of unispores was reported in *C. hispanicum*. *Chilionema*-like thalli also appear to have been reported as microthalli in the life histories of species of the macroalgal genus *Asperococcus*. For example material referable to the 'premiere' form of *Hecatonema maculans* described by Sauvageau (1897a) and transferred to the species *Chilionema ocellatum* in the present treatment was reported to be a phase in the life history of an *Asperococcus* sp. by Loiseaux (1969) and Fletcher (1984). Also Pedersen (1984) reported *Hecatonema reptans* Sauvageau and *Ectocarpus repens* Reinke, both of which are included in the synonomy of *Chilionema* species in the present treatment, as phases in the life history of *Asperococcus fistulosus*. Clearly the situation is unsatisfactory and more extensive studies, particularly involving culture work, are necessary to ascertain fully the taxonomic relationship between *Chilionema* and *Hecatonema* and their respective involvement as prostrate stages in the life histories of *Asperococcus* and *Punctaria* species.

KEY TO SPECIES

1	Thalli without ascocyst-like cells.	2
	Thalli with ascocyst-like cells	3
2	Epiphytic on the red alga *Palmaria;* plurilocular sporangia to 65 μm long, subconical or ovate-lanceolate in shape, hairs common	*C. ocellatum*
	Epiphytic on various algal hosts, particularly fucoids; plurilocular sporangia to 90 μm long, elongate-lanceolate in shape; hairs uncommon	*C. reptans*
3	Basal cells clearly distromatic in parts; plastids multilobed	*C. foecundum*
	Basal cells monostromatic throughout; plastids platelike	*C. hispanicum*

Chilionema foecundum (Strömfelt) Fletcher nov. comb. Figs 45A, B, 46

Phycocelis foecunda Strömfelt (1888), p. 383.
Myrionema foecundum (Strömfelt) Sauvageau (1897a), p. 170.
Ascocyclus sphaerophorus Sauvageau (1897a), p. 280.
Ascocyclus islandicus Jónsson (1903), p. 149.
Ascocyclus foecundus (Strömfelt) Cotton (1912), p. 122.
Ascocyclus saccharinae Cotton (1912), p. 122.
Chilionema børgensenii Printz (1926), p. 137.
Ascocyclus distromaticus Taylor (1937b), p. 228.
Hecatonema foecundum (Strömfelt) Loiseaux (1967a), p. 338.

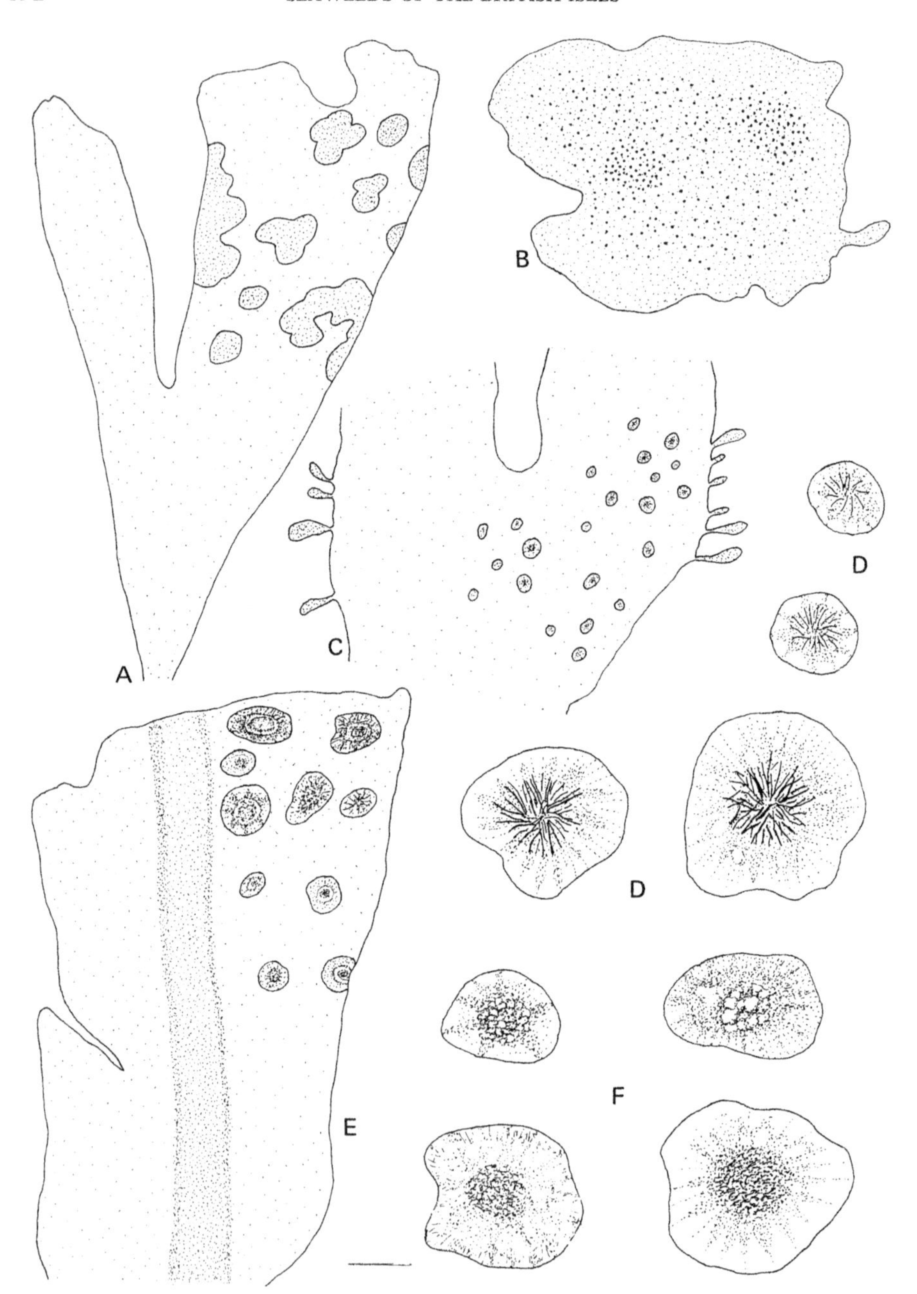

A
B
C
D
D
E
F

Thallus epiphytic, forming loosely adherent light to dark brown spots on hosts, usually solitary and circular 1–5 mm in diameter, occasionally irregular and confluent; central region dark brown, turf-like, similar to a thick pile, composed of densely packed vertical filaments, peripheral region of light-coloured, branched, outwardly radiating filaments, composed of rectangular cells, 1–3 diameters long, 9–18 × 5–8 μm, with a single multilobed plastid; terminal apical cells not noticeably coalesced; in section or squash preparation, consisting of loosely united, outwardly radiating, monostromatic, frequently distromatic filaments, easily separable under pressure, giving rise, in central regions to erect, slightly gelatinous, loosely associated, occasionally branched, erect filaments of 6–9 (−13) cells, 80–134 μm high; cells cylindrical, 1–2 diameters long, 8–21 × 8–13 μm, with a single large multilobed plastid with 1–3 pyrenoids; ascocyst-like cells common, either produced from basal cells or terminal on erect filaments; hairs frequent, arising from basal layer, with basal meristem and sheath.

Plurilocular sporangia usually abundant, either on basal cells or terminal on erect filaments, biseriate or (more rarely) triseriate, subcylindrical, attenuate towards apex, 26–59 × 9–18 μm, to 12 loculi long, frequently formed within empty sporangial husks; unilocular sporangia unknown.

Epiphytic, on *Palmaria palmata* and *Laminaria saccharina* blades, lower eulittoral and shallow sublittoral.

Recorded for scattered localities around the British Isles (Shetland Isles, Kent, Hampshire, Dorset, Devon, Anglesey, Argyll; in Ireland, Wexford and Mayo.

Summer annual, commonly recorded in August and September.

Chilionema foecundum has been investigated in laboratory culture by Loiseaux (1964, 1967b) (as *Ascocyclus sphaerophorus* Sauvageau). The life history was reported to be heteromorphic and haplodiplophasic. Plurispores released from plurilocular sporangia on the diploid, discoid thallus germinated directly and, by the process of heteroblasty, formed two morphologically different thalli: ectocarpoid, filamentous growths and stellate, later discoid growths. Both growth forms produced plurilocular sporangia, the plurispores from which directly repeated the diploid ectocarpoid and/or discoid growths. In addition a sexual cycle was observed with unilocular sporangia developing on the diploid ectocarpoid growths, the unispores from which germinated directly to form filamentous ectocarpoid, gametophytes. Released plurispores from 'gametangia' on the plants were observed to fuse and the zygote reformed the ectocarpoid, diploid thalli.

Batters (1892a), p. 174; Batters (1892b), p. 20; Sauvageau (1897a), pp. 280–285, figs 28–29; Børgesen (1902), p. 427, fig. 82; Cotton (1912), pp. 122–123, pl. X, figs 4–9; Printz (1926), pp. 137–139, pl. 2, figs 12–17; Newton (1931), p. 159; Hamel (1931–39), p. 100, fig. 25 VII; Levring (1937), pp. 51–52, fig. 7C–E; Jaasund (1957), pp. 228–230, figs 9–10; Taylor (1957), pp. 156–157, pl. 11, figs 7–10, 12; Loiseaux (1967a), pp. 329–347, figs 3, 5D, 6, pl.1; Rueness (1977), p. 133.

Fig. 45 *Chilionema foecundum*
A–B. Habit of plants on *Palmaria palmata.*
Chilionema ocellatum
C–D. Habit of plants on *Palmaria palmata.*
Chilionema reptans
E–F. Habit of plants on *Fucus serratus,* Bar = 10 mm (A), 0·4 mm (B, D, F), 30 mm (C), 5 mm (E).

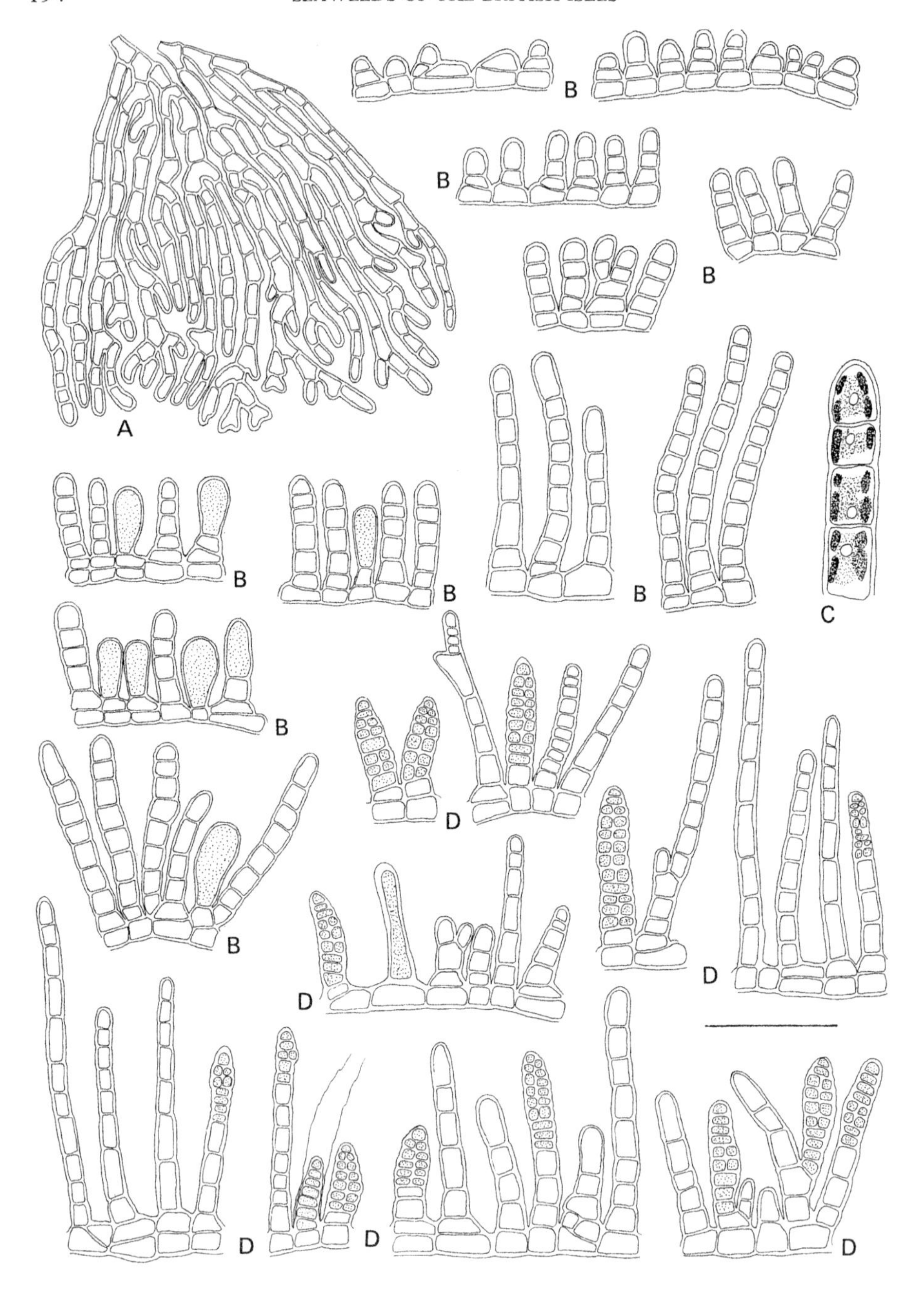
A
B
B
B
B
B
B
B
B
B
C
D
D
D
D
D
D

Chilionema hispanicum (Sauvageau) Fletcher nov. comb. Fig. 47

Ascocylus hispanicus Sauvageau (1897a), p. 274.
Hecatonema hispanicum (Sauvageau) Loiseaux (1967a), p. 338.

Thallus epiphytic, forming dark brown/black spots on hosts, solitary and circular, to 5 or more millimetres in diameter, becoming confluent and irregular, several centimetres across; comprising a monostromatic base of laterally adjoined, firmly attached, outwardly spreading filaments of cells, giving rise to erect filaments, plurilocular sporangia, ascocysts and/or hairs; basal cells rounded, quadrate or rectangular in section, usually longer than wide, 8–21 × 10–16 µm; erect filaments loosely associated, rarely branched, to 10 cells (–91 µm), cells subquadrate, quadrate or rectangular, $\frac{1}{2}$–2 diameters long, 5–14 × 5–11 µm, each with a single terminal, plate-like plastid; ascocysts common, usually sessile on basal layer or on 1–2 celled stalks, conical in shape, 27–50 × 13–20 µm; hairs infrequent, arising from basal layer, with basal meristem and sheath, c. 8–9 µm in diameter.

Plurilocular sporangia common, terminal on erect filaments, or on 1–3 celled stalks, biseriate, frequently triseriate, subcylindrical, 40–50 × 11–17 µm; unilocular sporangia unknown.

Epiphytic on the brown algae *Saccorhiza polyschides* and *Laminaria saccharina* probably present on other hosts, lower eulittoral and shallow sublittoral.

Only recorded for Dorset by E. A. L. Batters but probably more widely distributed. Recorded in September; insufficient data to comment on seasonal distribution.

Chilionema hispanicum has been investigated in culture by Loiseaux (1966; 1967b) (as *Ascocyclus hispanicus* Sauvageau). The life history was reported as heteromorphic and haplodiplophasic. A 'short-circuit' life cycle was also reported to occur by the fusion of unispores from the unilocular sporangia of the sporophyte.

Sauvageau (1897a), pp. 274–280, figs 26–27; Hamel (1931–39), p. 99, fig. 25 VIII; Loiseaux (1966), pp. 68–71, figs 1–2; Loiseaux (1967a), p. 338; Newton (1931), p. 159.

Chilionema ocellatum (Kützing) Kuckuck (1953), p. 325. Figs 45C, D, 48

Phyllactidium ocellatum Kützing (1843), p. 295.
Myrionema leclancheri Harvey (1846–51), pl. 41A.
Myrionema ocellatum (Kützing) Kützing (1849), p. 540.
Ascocyclus ocellatus (Kützing) Reinke (1889b), p. 44.
Chilionema nathaliae Sauvageau (1897a), p. 263.
Hecatonema maculans (Collins) Sauvageau (1897a), p. 248 'Première forme'.
Chilionema ocellatum (Reinke) Sauvageau (1897a), p. 273 but see Pedersen (1984), p. 60.

Fig. 46 *Chilionema foecundum*
A. Peripheral region of thallus showing outwardly spreading branched vegetative filaments. B. S.P. of vegetative thallus showing erect filaments arising from a monostromatic/distromatic base. Note enlarged ascocysts. C. Portion of erect filament showing lobed plate-like plastid. D. S.P. of fertile thallus showing erect filaments, plurilocular sporangia and ascocysts. Bar = 50 µm (A–B, D), 20 µm (C).

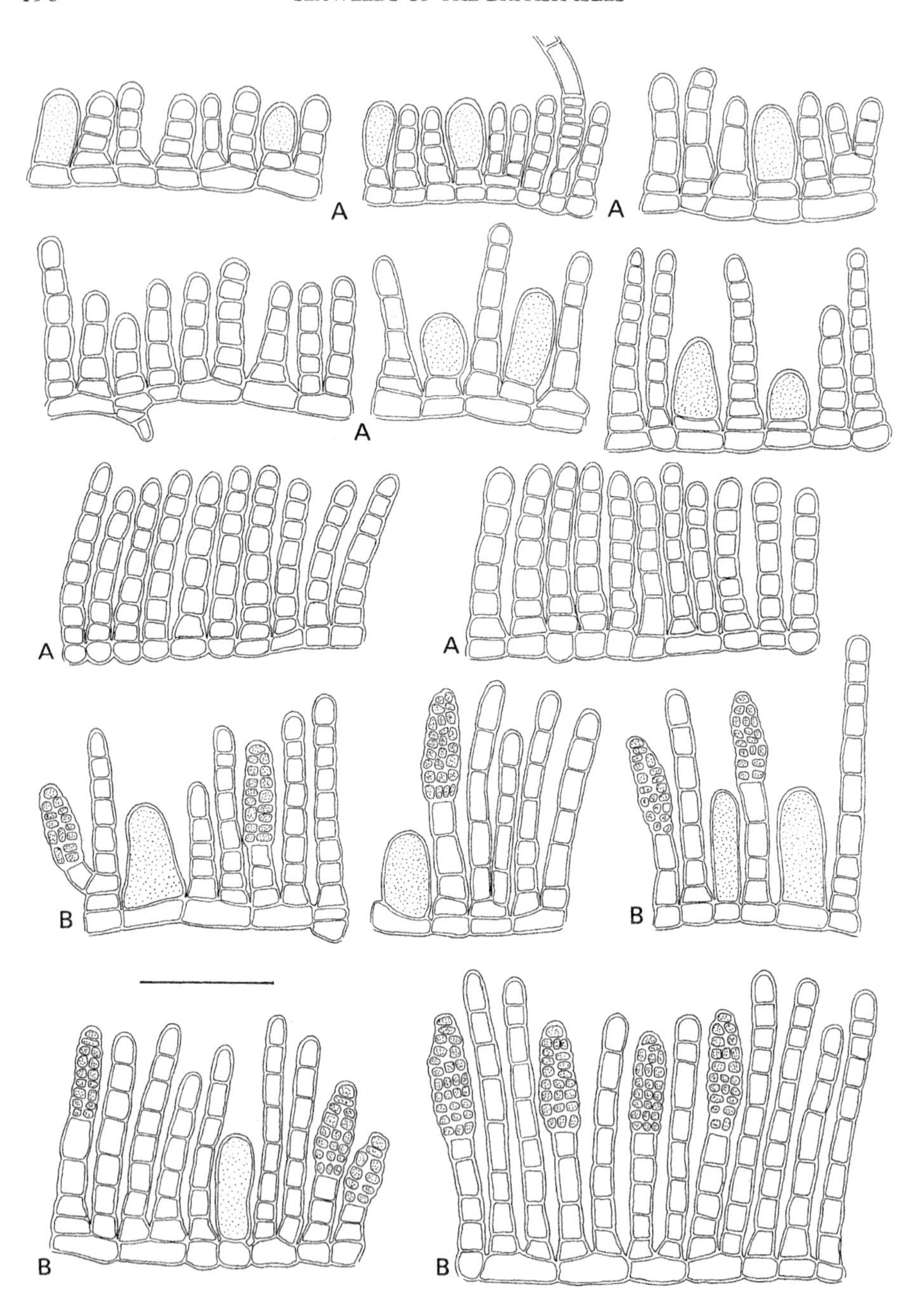
A
A
A
A
A
B
B
B
B

Thallus epiphytic, forming light brown spots on hosts, usually solitary and circular, 1–3 mm in diameter, occasionally confluent and irregular; in surface view, comprising a light brown, peripheral region of branched, outwardly radiating, monostromatic filaments, quite firmly united and discoid in structure, with peripheral row of dark-coloured apical cells enclosed within outer cuticle, cells quadrate or rectangular in shape, 1–2 diameters long, 12–30 × 9–20 µm with 3–5 plate-like plastids with pyrenoids; outer central region of scattered, rounded, thick-walled cells, 13–21 µm in diameter, inner central region turf-like, similar to a sparse carpet pile; in squash preparation consisting of an outer monostromatic, inner distromatic base, from most cells in the central region of which arise erect filaments, hairs and/or plurilocular sporangia; basal cells thick-walled, 1–3 diameters long, 12–26 × 7–13 µm; erect filaments, short, linear, simple, slightly mucilaginous, to 11 (–17) cells, 180 (–280) µm high, comprising thick-walled cells, quadrate or more usually rectangular, 1–2 (–3½) diameters long, 9–26 (–42) × 9–13 µm with 2–3 plate-like plastids with pyrenoids; hairs common, stalked and arising from the basal cells, or more usually, terminal on the erect filaments, 13–16 µm in diameter with basal meristem and sheath; rhizoids not observed.

Plurilocular sporangia common, usually on 1–6 celled stalks arising from the basal layer, biseriate or more commonly multiseriate, simple, ovate-lanceolate or subconical, 40–65 × 14–23 µm, to 10 loculi; unilocular sporangia unknown.

Epiphytic on the red alga *Palmaria palmata,* in the lower eulittoral and shallow sublittoral.

Recorded for scattered localities – Channel Isles, Dorset, Devon, Pembroke, Anglesey, Isle of Man; in Ireland, Wexford and Mayo. Probably more common and widely distributed.

Summer annual; June to September.

Culture studies of *Hecatonema maculans* 'première form' (Sauvageau, 1897a, p. 248) (transferred to *Chilionema ocellatum* in the present treatment) have been reported by Loiseaux (1969) and Fletcher (1984). In both investigations the alga was shown to be connected in life history with a species of *Asperococcus* (probably *A. fistulosus*) although the macroalga was misidentified as a *Myriotrichia* sp. by Loiseaux. The life history was heteromorphic and monophasic and lacked a sexual cycle. The two morphological expressions (*Hecatonema*-like microthalli and *Asperococcus*-like macrothalli) did not undergo an obligate alternation but were produced in response to environmental conditions. *Hecatonema*-like thalli were produced under conditions of 18 or 20°C, (Loiseaux) and 15/20°C, 16–8 h light/dark photoregime (Fletcher) whilst *Asperococcus*-like thalli were formed at 12°C (Loiseaux) and 10°C, 8–16 h light/dark photoregime (Fletcher).

Reinke (1889b), p. 44; (1889a), p. 19, pl. 15 (1–2), p. 22, pl. 19 (5–6) Batters (1893a), p. 23; Sauvageau (1897a), 'Première forme', pp. 248–256, figs 18–19; Batters (1900), pp. 371–372; Newton (1931), p. 156, fig. 97 (as *Hecatonema maculans* Sauvageau); Hamel (1931–39), pp. 96–98,

---

Fig. 47 *Chilionema hispanicum*
A. V.S. of vegetative thallus showing erect filaments, hairs and ascocysts arising from a monostromatic base. B. V.S. of fertile thallus showing erect filaments, ascocysts and plurilocular sporangia. Bar = 50 µm.

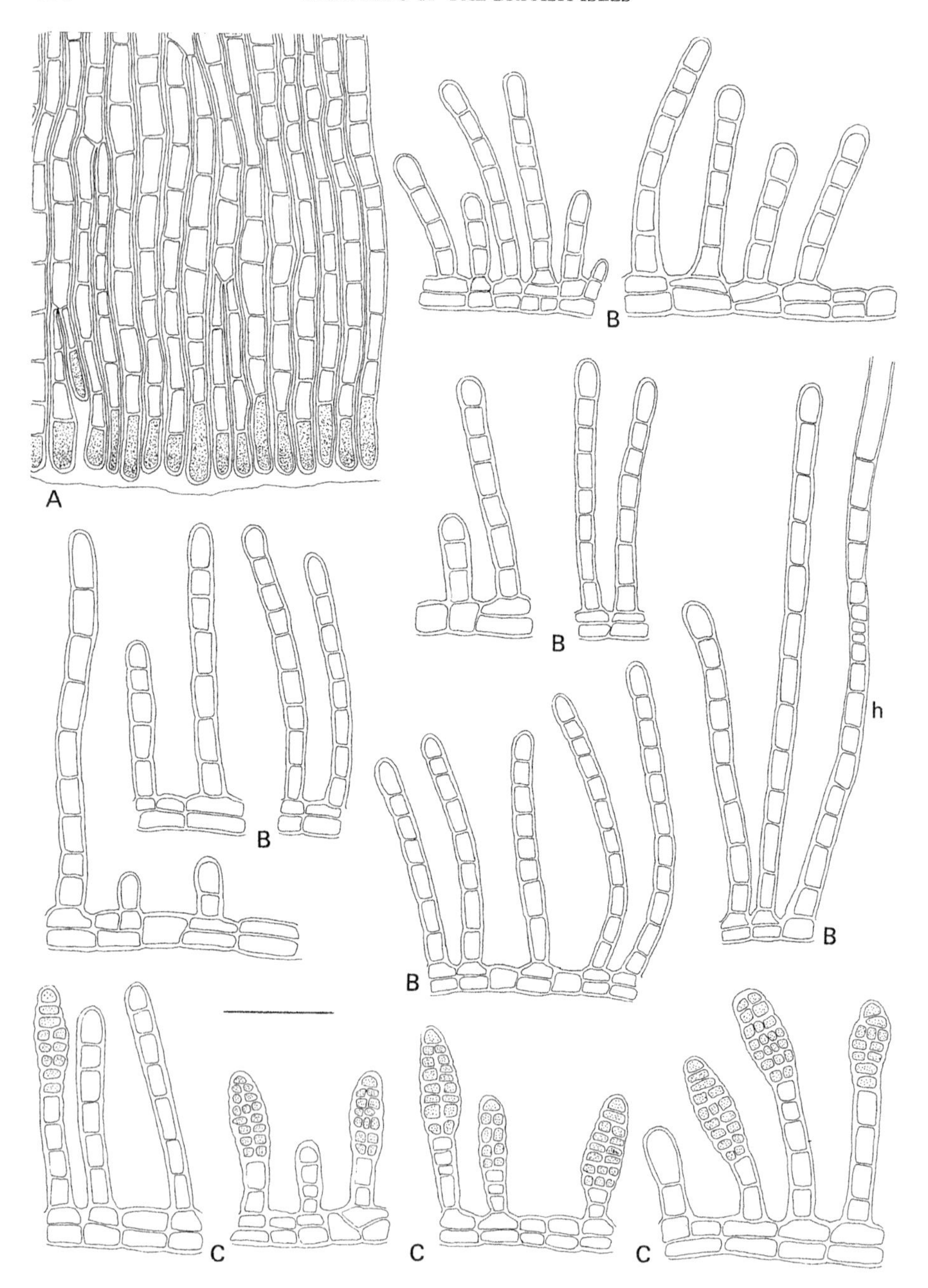
A
B
B
B
h
B
B
C
C
C

fig. 25 III-V (as *Hecatonema maculans*, *Chilionema nathaliae* and *C. ocellatum*); Kuckuck (1953), pp. 325–334, figs 4–8; Taylor (1957), p. 129 (as *Hecatonema maculans* f. *Sauvageauii*); Loiseaux (1969), pp. 11–15, figs 1–5 (as *Hecatonema maculans* Sauvageau); Rueness (1977), p. 131, fig. 56.

Chilionema reptans (Crouan frat.) Sauvageau (1897a), p. 268. Figs 45E, F, 49

Ectocarpus reptans Crouan frat. (1867), p. 161.
Ectocarpus repens Reinke (1889a), p. 21; (1889b), p. 42.
Ascocyclus reptans (Crouan frat.) Reinke (1889b), p. 44.
Phycocelis reptans (Crouan frat.) Kjellman (1890), p. 81.
Hecatonema reptans (Reinke) Sauvageau (1897a), p. 273 but see Pedersen (1984), p. 60.
Hecatonema reptans (Kjellman) Kylin (1907), p. 41 but see Pedersen (1984), p. 60.
Hecatonema fucicola Kylin (1907), p. 42.

Thallus epiphytic, forming thin, closely adherent spots on hosts, usually solitary and circular, 1–4 mm in diameter, occasionally confluent and irregular; in surface view comprising a monostromatic peripheral region of outwardly spreading, laterally adjoined (discoid), branched filaments of quadrate to rectangular cells, 1–2½ diameters long, 6–21 × 8–16 μm, each with 1–4 plate-like plastids with pyrenoids, an outer central region of irregularly spaced, thick-walled, rounded cells, 5–8 μm in diameter and an inner central region which is raised, turf-like, similar to a thin carpet pile; in section or squash preparation consisting of a monostromatic, later distromatic, basal layer of cells giving rise, in central region, to erect filaments, hairs and/or plurilocular sporangia; erect filaments loosely associated, simple, linear, slightly mucilaginous, 10–15 (–28) cells, 220 (–390) μm long, occasionally arising within sporangial husks, comprising cells quadrate or rectangular, ½–2 (–3) diameters long, 7–31 × 8–12 μm each with 1–2 lobed, plate-like plastids; hairs common, terminal on erect filaments with basal meristem and sheath, *c.* 19 μm in diameter; rhizoids not observed.

Plurilocular sporangia common, usually sessile on basal layer or on 1–3 celled stalks, more rarely terminal or lateral on erect filaments, simple, cylindrical, biseriate later often triseriate, 58–91 × 14–22 μm, to 15 loculi, with straight, occasionally oblique cross walls; unilocular sporangia unknown.

Epiphytic on various hosts, in particular *Fucus serratus*, lower eulittoral, in pools and shallow sublittoral.

Rare, but widely distributed around the British Isles (Shetland Isles, Essex, Kent, Dorset, Argyll).

Recorded June to November.

Pedersen (1984, p. 20) identified *Hecatonema reptans* Sauvageau and *Ectocarpus repens* Reinke, both of which are included here in the synonomy of *Chilionema reptans*, with the microthallus produced in cultures of *Asperococcus fistulosus*. In view of the results of culture studies of *Hectonema maculans*—première forme (Collins) Sauvageau (see p. 204) the situation is clearly unsatisfactory and more work is required to delimit *Chilionema* and *Hecatonema* and their species.

Fig. 48 *Chilionema ocellatum*
A. Peripheral region of thallus showing outwardly spreading, laterally united, branched filaments. B. S.P. of vegetative thallus showing erect filaments, and hair (h) arising from a distromatic base. C. S.P. of fertile thallus showing plurilocular sporangia. Bar = 50 μm.

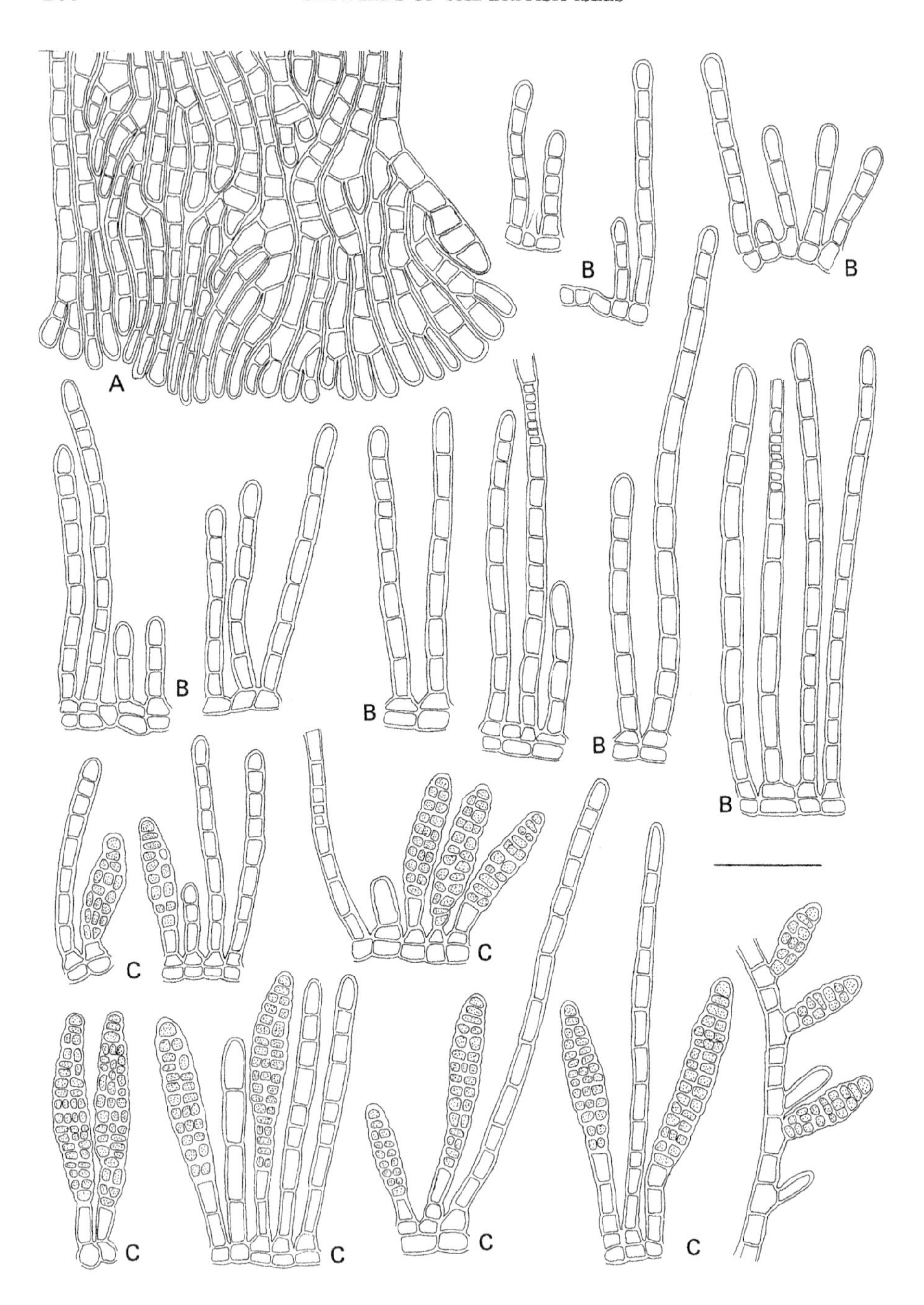
A
B
B
B
B
B
B
B
B
C
C
C
C
C
C

Crouan frat. (1867), p. 161, pl. 24, fig. 158 (3–4); Sauvageau (1897a), pp. 268–273, fig. 25; Reinke (1889b), p. 44; (1889a), p. 19, pl. 15 (3–6) (as *Ascocyclus reptans*); Cotton (1912), p. 21; Newton (1913), p. 157, fig. 98; Hamel (1931–39), p. 29, fig. 25 VI: Kylin (1947), pp. 15–16, fig. 12; Rueness (1977), p. 132, fig. 57.

HECATONEMA Sauvageau

HECATONEMA Sauvageau (1897a), p. 256.

Type species: *H. maculans* (Collins) Sauvageau (1897a), p. 256.

Thallus epiphytic or epilithic, forming small solitary or confluent tufts up to 1 mm high, consisting of a basal layer of outwardly spreading, dichotomously branched filaments, either free or laterally associated and pseudodiscoid, monostromatic, sometimes partly distromatic, giving rise to short, erect, free, unbranched or little branched filaments (usually all of equal height), plurilocular sporangia and/or hairs; cells with several discoid or 1–3 plate-like plastids with pyrenoids; hairs common, lateral or terminal on erect filaments with basal meristem and sheath.

Plurilocular sporangia common, borne on basal layer or terminal, more usually lateral on erect filaments, multiseriate; unilocular sporangia unknown on British material.

In the present treatment *Hecatonema* is characterised by the 'deuxième' and 'troisième forme' of *Hecatonema maculans* described by Sauvageau (1897a, pp. 256 & 259, figs 20–21) and characterised by a filamentous or discoid basal layer, which is monostromatic at first, later becoming distromatic and which gives rise to short, erect, branched filaments, hairs with a basal meristem and sheath and multiseriate, plurilocular sporangia. The sporangia are usually lateral or terminal on the erect filaments, rather than arising directly from the basal layer. It very closely resembles and is not easily distinguished from the genus *Chilionema,* and the two genera might well be synonymous (see p. 190). However, they are here maintained separately pending the results of further studies.

Particular interest has been given over recent years to reports of the involvement of *Hecatonema* species and *Hecatonema*-like microthalli in the life histories of species of the macroscopic genera *Asperococcus, Desmotrichum* and *Punctaria.* These include investigations by Sauvageau (1929), Kylin (1933, 1937), Dangeard (1963a), Loiseaux (1969), Clayton & Ducker (1970), Rhodes (1970), Clayton (1974), Rietema & van den Hoek (1981), Lockhart (1982), Fletcher (1984) and Pedersen (1984). In general there appears to be a facultative relationship between the *Hecatonema*-like microthalli and the thalloid macrothalli rather than one based on differences in ploidy level. Indeed, the two types of morphological expression were often shown to be produced in response to environmental conditions.

It is partly because of these life history connections that *Hecatonema* was removed from the Myrionemataceae and placed in the Punctariaceae in this treatment, thus

---

Fig. 49 *Chilionema reptans*
A. Peripheral region of thallus showing outwardly spreading filaments. B. S.P. of vegetative thallus showing erect filaments and hairs arising from a distromatic base. C. S.P. of fertile thallus showing erect filaments and plurilocular sporangia. Bar = 50 µm.

following the procedure of Christensen (1980). However, much more work is required, particularly culture studies, to elucidate fully the extent of involvement and relationship of *Hecatonema*-like thalli and described taxa of *Hecatonema* (and *Chilionema*) with macroscopic members of the Punctariaceae. In this respect it is pertinent to note culture studies which have revealed *Hecatonema* plants to behave independently and do not develop macrothalli in culture (see Edelstein *et al.*, 1971; Clayton, 1974; Pedersen, 1984).

Four species of the genus *Hecatonema* are listed for the British Isles by Parke & Dixon (1976). Of these *H. foecundum* (Strömf.) Loiseaux and *H. hispanicum* (Sauvageau) Loiseaux have been transferred to *Chilionema,* q.v. whilst *H. liechtensternii* (Hauck) Batters has been transferred to *Myrionema*. Therefore, this genus is represented in the British flora only by *Hecatonema maculans*.

One species in the British Isles:

Hecatonema maculans (Collins) Sauvageau (1897a), p. 256. Fig. 50

Phycocelis maculans Collins (1896), p. 459.
Hecatonema terminalis (Kützing) Kylin (1937), p. 8.
Hecatonema maculans (Collins) Sauvageau (1897a), Deuxième forme and Troisième forme.
?*Ectocarpus terminalis* Kützing (1845–71), p. 236.

Thallus forming small tufts of filaments, up to 1 mm high, solitary or confluent and spreading, similar to a sparse carpet pile, to several centimetres in diameter; comprising a basal network of outwardly spreading dichotomously branched filaments giving rise to erect filaments, plurilocular sporangia and/or hairs; in squash preparation basal filaments free or sometimes irregularly and laterally associated and pseudodiscoid in appearance, monostromatic, or in parts distromatic, comprising cells variable in shape, although usually longer than wide 13–25 × 13–17 μm; erect filaments to 30 (–60) cells long, 0·5–1 mm, linear, free, unbranched or more commonly sparingly and laterally branched with branches characteristically widely divergent at first, later recurving towards parental filament, comprising cells 1–2 diameters long, 13–34 × 12–17 μm, with thick hyaline walls, containing several discoid or 1–3 plate-like, lobed plastids with pyrenoids; hairs common, terminal on erect filaments or branches, 9–12 μm diameter, with basal meristem and outer sheath.

Plurilocular sporangia common, sessile or stalked on basal layer, or more commonly terminal or lateral on erect filaments; laterally positioned sporangia widely divergent at first, later recurved towards parental filament, sessile or stalked, solitary or more commonly in unilateral groups; sporangia biseriate, or more commonly multiseriate, elongate, ovate-lanceolate in shape, occasionally bifurcate at tip, 65–90 (–135) × 20–28 μm, to 30 loculi; unilocular sporangia unknown.

Fig. 50 *Hecatonema maculans*
A–C. Erect filaments arising from basal layer. D–E. Portions of branched, erect filaments. F. S.P. of fertile thallus showing erect filaments arising from a monostromatic/distromatic base with lateral/terminal plurilocular sporangia. Bar = 50 μm.

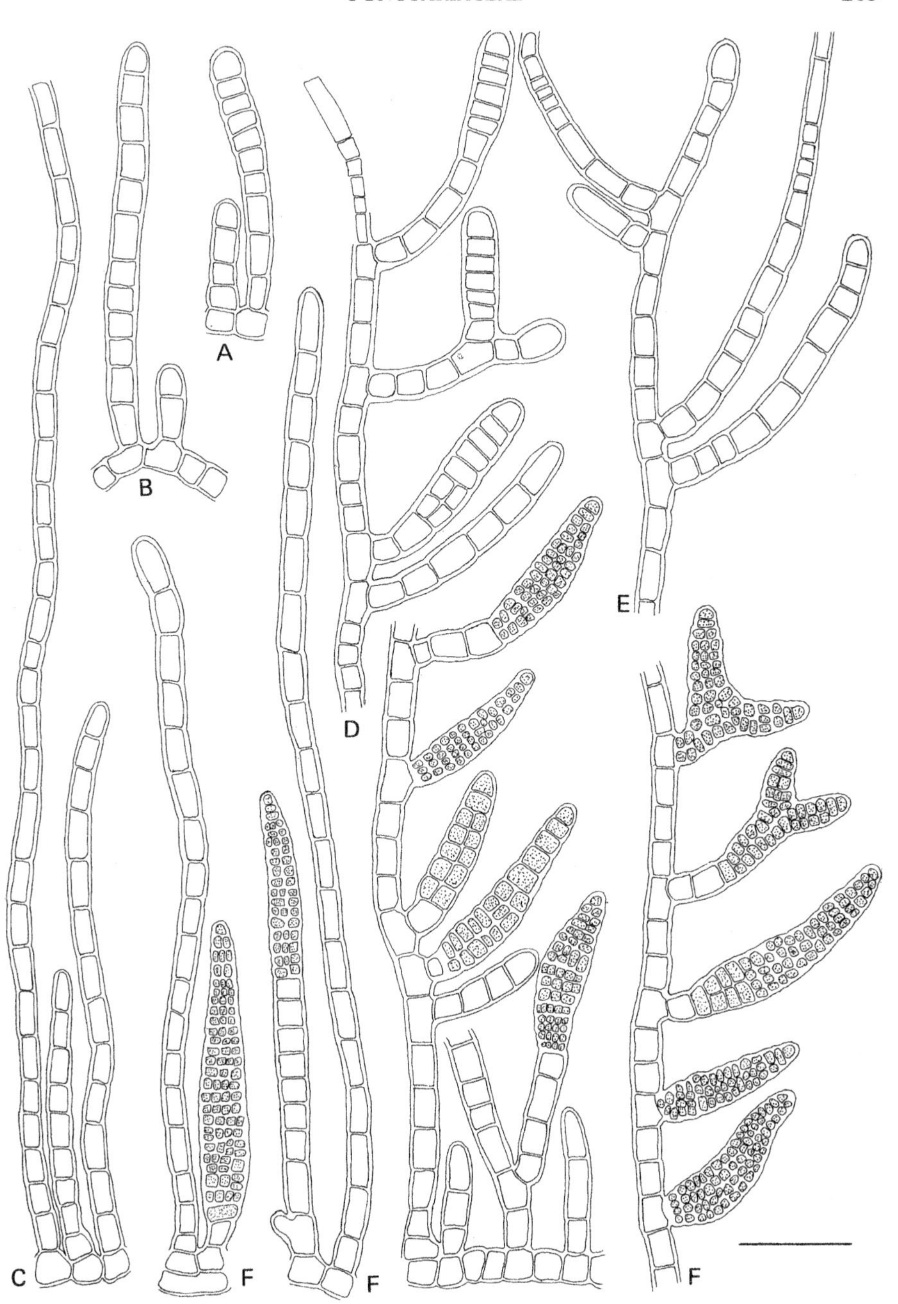
A
B
C
D
E
F
F
F
F

Epiphytic on various algae, more rarely epilithic, in the eulittoral and shallow sublittoral.

Rare and widely distributed around the British Isles.

Recorded throughout the year.

A small number of culture studies have linked *Hecatonema maculans* in life history with macroscopic genera in the Punctariaceae. In three studies the *H. maculans* plants were connected with *Punctaria tenuissima* (as *Desmotrichum undulatum*) and *Punctaria latifolia* (Clayton & Ducker, 1970; Clayton, 1974; Fletcher, 1984—Sauvageau's 'deuxième forme' only) whilst in one study (Pedersen, 1984) it was linked with *Asperococcus fistulosus*. Support is thus given to the proposal by Clayton (1974) that *H. maculans* probably represents a mixture of genotypes which are morphologically indistinguishable. This is further exemplified by reports of isolates of *H. maculans* possessing an independent life history (Edelstein *et al.*, 1971; Clayton, 1974).

Collins (1896), pp. 459–460, pl. 278, figs 1–5; Batters (1900), p. 371; Børgesen (1926), pp. 52–55, figs 27–28; Hamel (1931–39), pp. 51–52, fig. 17A; Kylin (1937), p. 8; Kylin (1947), p. 15, figs 10–11; Taylor (1957), pp. 153–154; Cardinal (1964), pp. 76–77, fig. 40; Edelstein *et al* (1971), pp. 1248–1249, figs 4, 11–12; Clayton (1974), pp. 790–793, figs 28–29; Rueness (1977), pp. 132–133, fig. 58.

PUNCTARIA Greville

PUNCTARIA Greville (1830), p. 52.

Type species: *P. plantaginea* (Roth) Greville (1830), p. 53.

Diplostomium Kützing (1843), p. 298.
Phycolapathum Kützing (1843), p. 299.
Desmotrichum Kützing (1845), p. 244.
Homoeostroma J. Agardh (1896), p. 7.
Nematophlaea J. Agardh (1896), p. 12.
Rhadinocladia Schuh (1900), p. 3.

Thallus consisting of an erect flattened blade arising from a small stipe and discoid holdfast; blade simple, entire or with marginal proliferations present, solid, flaccid or subcoriaceous, 2–4 (–9) cells thick, central medullary cells if present, large, thick-walled, colourless, longitudinally elongate, transversely rounded, outer cortical cells smaller, pigmented; in surface view cells large, in rows, each with numerous discoid plastids and associated pyrenoids; hairs single or grouped, with basal meristem and enclosing sheath.

Unilocular and plurilocular sporangia on the same or different blades, extensive on both surfaces, solitary or grouped, formed in surface cells; unilocular sporangia uncommon or rare, ovate; plurilocular sporangia common, quadrate or polygonal in surface view, conical, sometimes bifurcate, protruding above surfaces in transverse sections.

The genus *Punctaria* is principally characterised by an erect, simple, flattened blade, variable in dimension and shape, usually becoming several cells thick and with only slight internal differentiation into an inner medulla of large colourless cells and an outer cortex of 1–3 small photosynthetic cells. The cells contain several discoid plastids with pyrenoids, hairs have a basal meristem and sheath and the multiseriate sporangia develop within the cortical cells. These characteristics should enable the genus to be distinguished from other morphologically similar genera such as *Petalonia* and *Laminaria*.

The early stage of development comprises an erect uniseriate filament terminated by one or several hairs. Transverse and longitudinal divisions then produce a laterally expanded blade mainly two cells thick. These cells then divide parallel to the blade surface to produce a four-celled thick thallus, the outer cells of which then assume the main photosynthetic role and become heavily pigmented. Further transverse and longitudinal divisions of these cells are also responsible for the increased thickness of the blades seen in some species and the formation of the often projecting sporangia.

The extent of the development of the erect blades is an important character used in the recognition of the species. In *P. tenuissima* (which includes *Desmotrichum undulatum* in the present treatment) lateral development of the blades appears to be limited (to less than 12 mm) and they remain ribbon-shaped. The thickness of the blades also rarely exceeds 60 μm and usually comprises two, sometimes four cell layers. Some very reduced forms of this species have also been described (formerly under *Desmotrichum*) with very narrow blades, which can be uniseriate in parts. Much wider and thicker blades, on the other hand, are characteristic of the other *Punctaria* species found in the British Isles. However, some doubt must be expressed on the validity of these characters in species' identification. Considerable overlap does seem to occur amongst British material. The usefulness of such characters as blade dimension, shape, colour and thickness, the presence or absence of hairs, whether hairs are solitary or grouped, present on the blade surface or in the margin, the dimensions and extent of protrusion of the sporangia, is undoubtedly questionable and may well have to be abandoned. A particularly noteworthy paper by Rietema and van den Hoek (1981) has revealed many of these characters to be modified by environmental conditions of temperature, day length and light intensity.

Life history studies of a number of species of *Punctaria* have indicated that a 'direct' monophasic life history is operating with unispores and/or plurispores producing small microthalli, from which budded plants identical with the parents, i.e. with apparently apomeiotic unilocular sporangia. This has been shown for *P. latifolia* in both the North Atlantic and Australia (Sauvageau, 1929, 1933a; Dangeard, 1963a, 1966a (as *P. crouanii*); Clayton & Ducker, 1970; Pedersen, 1984), for *P. plantaginea* and *P. orbiculata* Jao collected in Newfoundland and Washington respectively (South, 1980) and for *P. tenuissima* (as *Desmotrichum undulatum*) on the west coast of Sweden (Kylin, 1933), the north east coast of North America (Rhodes, 1970; Lockhart, 1982), the Netherlands (Rietema and van den Hoek, 1981) and Denmark (Pedersen, 1984). Not uncommonly, fertile microthalli (with plurilocular sporangia) were observed in the cultures which Pedersen (1984) identified with both the genus *Hecatonema* and the species *Streblonema effusum* Kylin. It is also interesting to note that cultures of *Hecatonema maculans* were reported by Clayton (1974) and Fletcher (1984) to give rise to macrothalli resembling *Punctaria* and *Desmotrichum*. The relationship between these two morphological expressions (i.e. microthallus and macrothallus) appeared to be facultative and influenced by enviromental conditions (particularly temperature, day length and light intensity) rather than differences in ploidy level. Indeed the only report of a sexual life history in *Punctaria* was given by Knight (1929) in which she claimed to have observed copulation of unispores in *P. plantaginea:* in view of the results of more recent publications this now needs confirmatory studies.

KEY TO SPECIES

1 Thallus dark brown, 6–8 cells in thickness, firm and subcoriaceous, plurilocular sporangia rare or absent 2
Thallus light/olive brown, 2–4 (rarely 6) cells in thickness, soft and flaccid, plurilocular sporangia common 3
2 Thallus lanceolate becoming oblong or orbicular, to 70 × 24 cm, with ruffled margins; restricted to the Isles of Scilly . . . . . . *P. crispata*
Thallus lanceolate becoming obovate, to 30 × 5 cm with smooth margins; widely recorded around the British Isles . . . . *P. plantaginea*
3 Thallus to 8 cm in width, ovate, linear-lanceolate or oblong in shape; 2–4 (rarely 6) cells in thickness . . . . *P. latifolia*
Thallus to 2 cm in width, elongate-lanceolate to ribbon shaped; 1–2 (rarely 4) cells in thickness . . . . *P. tenuissima*

Punctaria crispata (Kützing Batters (1900), p. 372. Fig. 51

Phycolapathum crispatum Kützing (1843), p. 299.
Punctaria laminarioides Crouan frat. (1867), p. 167.

Plants forming erect, dark brown, dorsiventrally flattened blades arising from a short, cylindrical stipe and discoid holdfast; erect blades solitary or gregarious, simple, often irregularly split, solid, firm and subcoriaceous, oblong, orbicular or lanceolate, to 30 (–70) cm long, 24 cm broad, with ruffled margins, tapering sharply below; surface cells either in rows or irregularly placed, markedly thick-walled, quadrate, rectangular or irregular in shape, 15–39 × 13–26 μm, each with numerous discoid plastids with pyrenoids; in transverse section blades to 180–(220) μm thick, 6–8 cells, slightly differentiated into inner, large, colourless cells and outer (1–2 cells) smaller, pigmented cells; hairs not observed.

Unilocular sporangia scattered on both sides of blade, slightly immersed, formed from within surface cells, orbicular or ovate in surface view, 26–35 × 22–30 μm, hemispherical or quadilateral, usually slightly protruding, more rarely sunken in transverse section, 28–40 × 23–30 μm; plurilocular sporangia unknown.

Epiphytic on *Zostera* leaves, lower eulittoral pools.
Only known for the Isles of Scilly.
Annual; summer (May to October)

The distinguishing features of this species include the large, dark brown, subcoriaceous, peripherally ruffled blades, easily mistaken for *Laminaria saccharina,* the apparent absence of hairs, the relatively small, inconspicuous unilocular sporangia which are sometimes sunken and the absence of plurilocular sporangia.

Batters (1900), pp. 372–372; Newton (1931), p. 186; Hamel (1931–39), p. 216–217, fig. 44, 6.

Fig. 51 *Punctaria crispata*
A. Habit of plant. B. Surface view of vegetative thallus showing cells with several discoid plastids. C. V.S. of vegetative thallus. D. Surface view of fertile thallus showing unilocular sporangia. E–F. V.S. of fertile thalli showing unilocular sporangia. Bar = 4 cm (A), 50 μm (B, D, F), 100 μm (C, E).

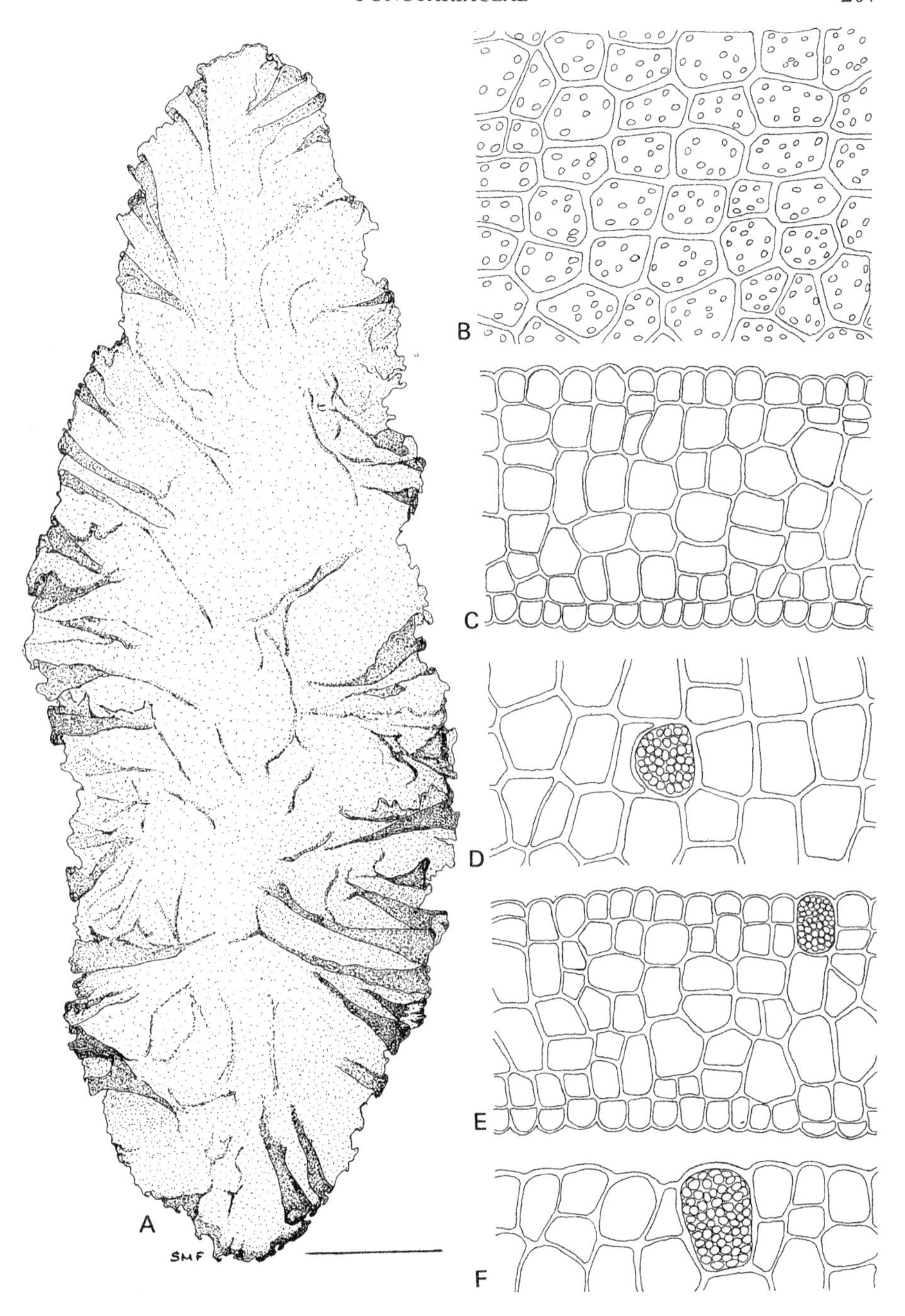
A
B
C
D
E
F
SMF

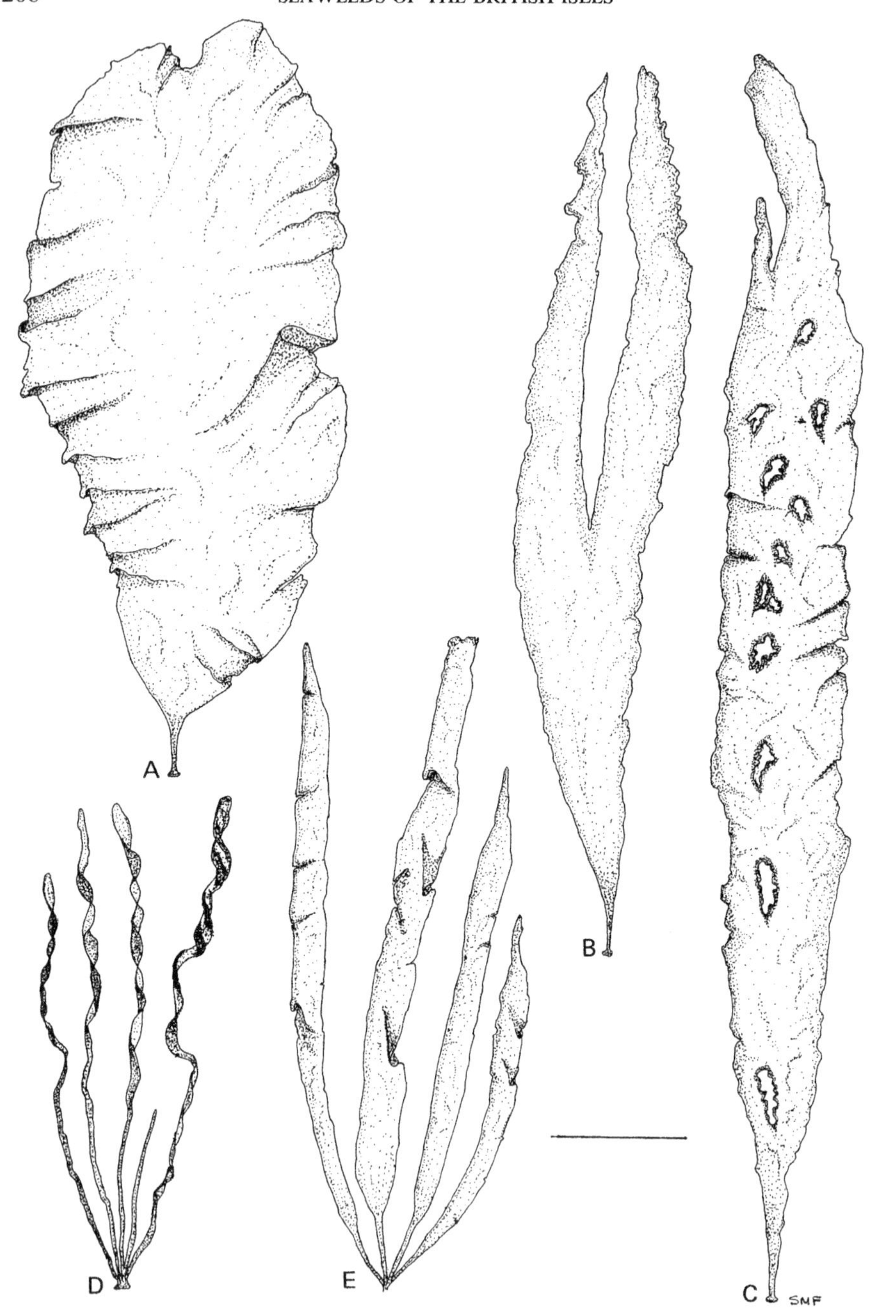
A
B
C
D
E
SMF

Punctaria latifolia Greville (1830), p. 52. Figs 52A, 53; Pl. 6a

Punctaria plantaginea var. *crouani* Thuret in Le Jolis (1863), p. 70.
Punctaria crouanii (Thuret) Bornet et Thuret (1876–80), p. 15.
Punctaria hiemalis Kylin (1907), p. 70.

Plants forming erect, light or greenish brown, membranous, dorsi-ventrally flattened blades arising from a small discoid holdfast; erect blades simple, sometimes proliferous around margins, solitary or more commonly gregarious, flaccid, slightly lubricous, solid, elongate-ovate, linear-lanceolate or oblong, to 30 (−45) × 2–4 (−8) cm, with flat or ruffled margins, tapering below, fairly sharply to a short, narrow stipe, 2–4 (−8) mm in length; surface cells in rows, quadrate to rectangular, 17–30 (−50) × 11–26 (−35) µm, each with large numbers (up to 30) of discoid plastids and associated pyrenoids; in transverse section blades 60–120 (−160) µm thick, 2–4 (−6) cells, either with quadrate or subquadrate cells (in 2–4 celled thick thalli) or with large, colourless, thick-walled, rounded, central cells (usually elongate-rectangular in longitudinal section) and small, quadrate, subquadrate to vertically elongate, pigmented, surface cells (in 4–6 celled thick thalli); hairs abundant, arising from surface or subsurface cells, single or more commonly grouped, 10–16 µm in diameter, fairly flaccid, with short basal meristem and enclosing sheath.

Plurilocular and unilocular sporangia borne on the same or different blades, widely distributed on both surfaces, formed within surface cells; plurilocular sporangia common, solitary and scattered, or grouped in extensive patches, formed by direct subdivision of surface cells, more commonly formed in 2–4 daughter cells following vegetative divisions, quadrate to rectangular, 14–42 (−59) × 12–26 (−42) µm in surface view, subquadrate or more commonly conical and projecting, sometimes bifurcate, superficial or immersed, 25–46 × 13–26 µm in transverse section; unilocular sporangia rare, orbicular or ovate, 40–50 × 36–43 µm in surface view, ovate or vertically elongate in transverse section.

Epiphytic on a wide variety of algae, less commonly epilithic, upper to lower eulittoral pools and sublittoral to 15 m.

Common and widely distributed around the British Isles.

Annual; summer.

Included within the synonymy of *P. latifolia* in the present treatment is the morphologically and anatomically similar *P. crouanii* which, according to Hamel (1931–39), is primarily distinguished by the occurrence of elongate plurilocular sporangia. However, such erect, protruding, plurilocular sporangia have been observed on British material of three species and, therefore, the taxonomic importance of this character is questionable. Certainly they were very common on *P. latifolia* in agreement with French material

---

Fig. 52 *Punctaria latifolia*
A. Habit of plant.
Punctaria plantaginea
B–C. Habit of plant.
Punctaria tenuissima
D–E. Habit of plant. Bar = 22 mm (A), 21 mm (B), 19 mm (C), 30 mm (D–E).

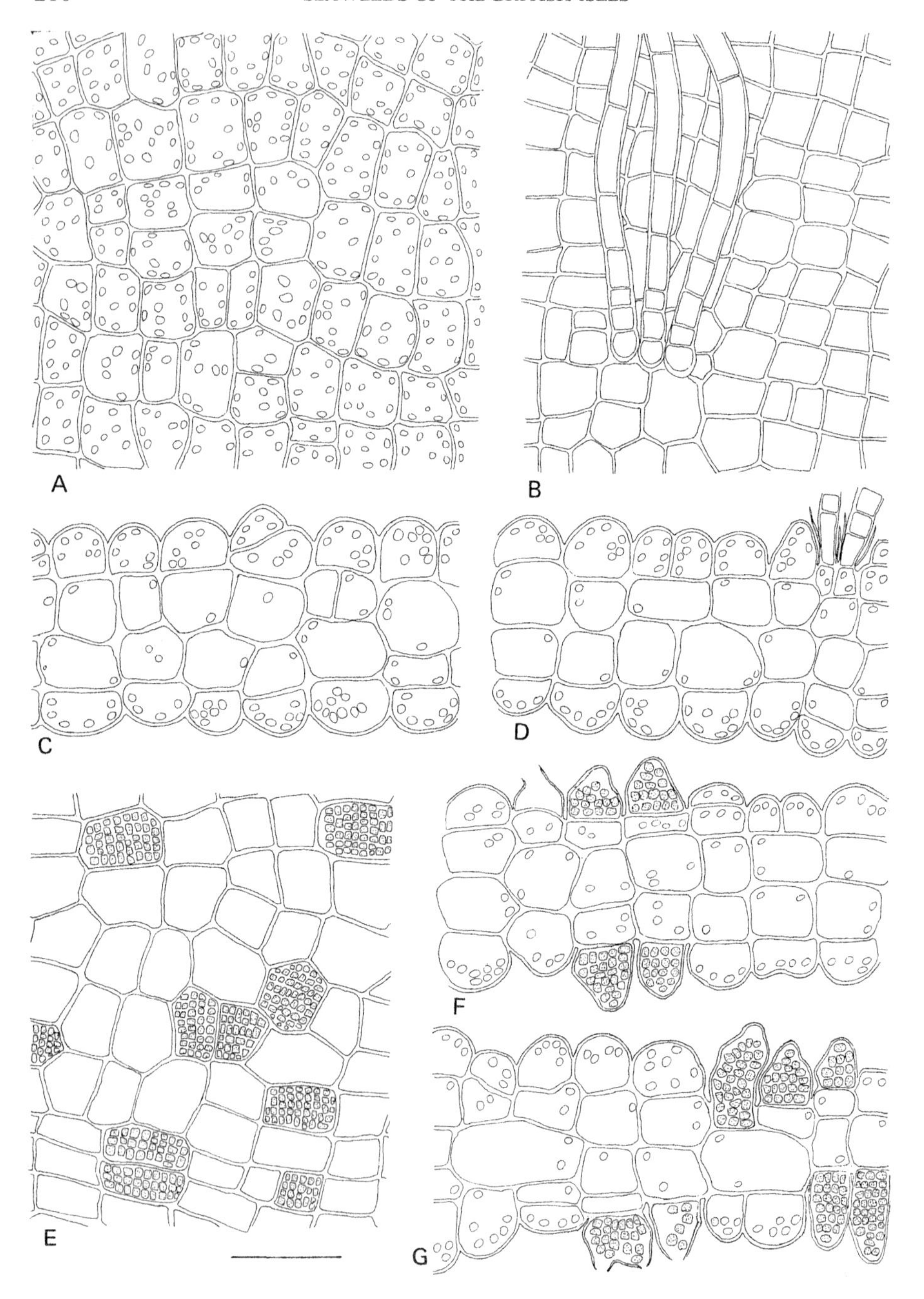
A
B
C
D
E
F
G

of this species described by Dangeard (1963a) and Australian material described by Clayton & Ducker (1970).

Punctaria latifolia has been investigated in laboratory culture by Sauvageau (1929, 1933a), Dangeard (1963a, 1966a – as *P. crouanii*), Clayton & Ducker (1970) and Pedersen (1984). The life history of all isolates was shown to be of the 'direct' type with unispores and/or plurispores from the macrothallial blades germinating directly to form microthalli from which sprouted a new generation of erect blades. Copulation between spores has not been observed and the life history can additionally be described as monophasic and heteromorphic. In general the microthalli showed close resemblance to *Hecatonema*. A life history connection between *Hecatonema* and *Punctaria latifolia* was also reported in Australia by Clayton (1974).

Kylin (1907), pp. 70–72, fig. 17; Sauvageau (1929), pp. 334–349, figs 11–14; Newton (1931), pp. 184–185, fig. 116; Hamel (1931–39), pp. 211–213, fig. 44 1–2; Sauvageau (1933a), pp. 1–4; Taylor (1957), p. 166, pl. 15, fig. 5; Dangeard (1963a), pp. 205–224, pls 1–3; Jaasund (1965), p. 91; Dangeard (1966a), pp. 157–167, pls 1–3; Dangeard (1969), pp. 70–72, fig. 2A; Rhodes (1970), pp. 312–314, figs 1–11; Clayton & Ducker (1970), p. 293–300, figs 1–10; Cabioch (1976), pp. 23–26, fig. 1; Kornmann & Sahling (1977), p. 278, fig. 16; Rueness (1977), p. 164, fig. 96A, B; Stegenga & Mol (1983), pp. 99–103, pls 35–36; Pedersen (1984), pp. 21–22, fig. 10.

Punctaria plantaginea (Roth) Greville (1830), p. 53. Figs 52B, C, 54; Pl. 6b

Ulva plantaginea Roth (1800), p. 243.
Phycolapathum plantagineum (Roth) Kützing (1849), p. 483.
Laminaria plantaginea (Roth) C. Agardh (1817) p. 20.
Zonaria plantaginea (Roth) C. Agardh (1820) p. 138.

Plants forming erect, light to dark brown, dorsiventrally flattened blades, arising from a small discoid holdfast; erect blades solitary or more commonly gregarious, simple, solid, subcoriaceous, entire or often irregularly or longitudinally split, lanceolate or obovate, gradually tapering below to a short stipe, to 15 (–30) × 1–3 (–5) cm; surface cells quadrate to rectangular, in longitudinal less obvious transverse rows, 20–50 × 14–39 μm each with large numbers of discoid plastids and associated pyrenoids; in section blades 120–180 (–210) μm thick, 4–6 (–9) cells, inner cells large, colourless, thick-walled, quadrate or elongate-rectangular in longitudinal section, rounded in transverse section, outer cells smaller, pigmented, quadrate to vertically elongate; hairs abundant, arising from surface cells, single or more commonly grouped, thick-walled, fairly stiff, 15–20 (–25) μm in diameter, with extensive basal meristem and enclosing sheath.

Plurilocular and unilocular sporangia borne on the same or different blades, widely distributed on both surfaces, formed within surface cells; plurilocular sporangia solitary

---

Fig. 53 *Punctaria latifolia*
A. Surface view of thallus showing cells with numerous discoid plastids. B. Surface view of thallus showing hairs. C–D. V.S. of vegetative thalli. E. Surface view of fertile thallus showing plurilocular sporangia. F–G. V.S. of fertile thalli showing plurilocular sporangia. Bar = 50 μm.

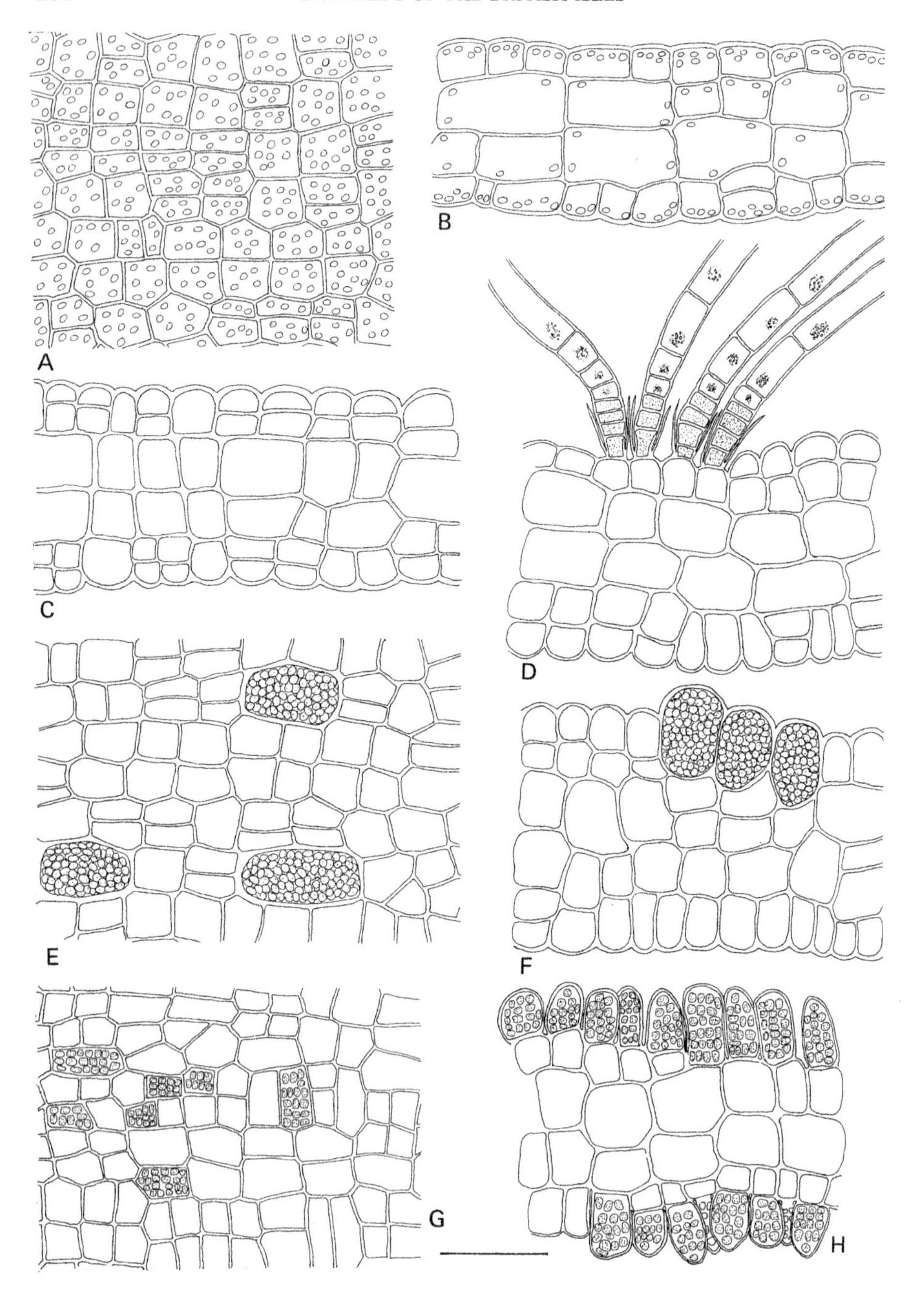
A
B
C
D
E
F
G
H

or grouped, formed by direct subdivision of surface cells, or following 2–4 vegetative divisions, quadrate to rectangular in surface view, 20–40 × 11–32 μm, conical, often bifurcate, projecting from the surface in section, 29–52 × 14–29 μm; unilocular sporangia orbicular in surface view 45–85 × 31–40 μm, ovate in section, slightly projecting, 32–52 × 29–43 μm.

Epilithic, epizoic, upper to lower eulittoral pools, sublittoral to 18 m.
Not uncommon; reported for scattered localities around the British Isles.
Annual; spring and summer.

North Atlantic isolates of *P. plantaginea* have been studied in culture by Knight (1929), South (1980) and Pedersen (1984). Both South and Pedersen reported a 'direct' asexual life history with unispores germinating into sterile microthalli from which further generations of unilocular sporangia-bearing erect plants were produced. In the culture study by Knight a sexual life history was reported, with unispores fusing together. Very few details, however, were given in Knight's paper and her claim must remain suspect. Also noteworthy and requiring further investigation is the suggestion by Pedersen (1984) that plants of *P. plantaginea* reported with plurilocular sporangia may represent a cold-water form of *P. latifolia*.

Hamel (1931–39), pp. 213–215, fig. 44(3); Taylor (1957), pp. 166–167, pl. 15, fig. 4, pl. 16, fig. 4; Jaasund (1965), pp. 89–91, fig. 27; Pankow (1971), p. 194, pl. 21, fig. 2; Kornmann & Sahling (1977), p. 128, fig. 68A–C; Rueness (1977), p. 64, pl. 21, fig. 4, fig. 97; South (1980), pp. 266–272, figs 1–5; Stegenga & Mol (1983), p. 103; Pedersen (1984), pp. 20–22.

Punctaria tenuissima (C. Agardh) Greville (1830), p. 54. Figs 52D, E, 55

Zonaria tenuissima C. Agardh (1824), p. 268.
Punctaria undulata J. Agardh (1836), p. 15.
Diplostomium tenuissimum Kützing (1849), p. 483.
Desmotrichum balticum Kützing (1849), p. 470.
Desmotrichum scopulorum Reinke (1888), p. 18.
Desmotrichum undulatum (J. Agardh) Reinke (1889b), p. 57.
Desmotrichum balticum f. *paradoxum* Gran (1897), p. 37.
Punctaria baltica (Kützing) Batters (1902), p. 26.
Desmotrichum scopulorum f. *fennicum* Skottsberg (1911), p. 5.

Plants forming erect, olive to light brown, dorsiventrally flattened blades, arising from a small discoid holdfast and short stalk; erect blades solitary or more commonly

Fig. 54 *Punctaria plantaginea*
A. Surface view of vegetative thallus showing cells with numerous plastids. B–D. V.S. of vegetative thalli with lateral hairs (D). E. Surface view of fertile thalli with unilocular sporangia. F. V.S. of fertile thallus with unilocular sporangia. G. Surface view of fertile thallus with plurilocular sporangia. H. V.S. of fertile thallus with plurilocular sporangia. Bar = 50 μm.

gregarious, simple, solid, linear, elongate-lanceolate, tapering to base and apex, occasionally obtuse at apex, soft and flaccid, frequently spirally twisted with crisp margins, entire or with small serrations, to 10 (–20) cm × 2–6 (–12) mm; surface cells quadrate or rectangular, in rows, 14–48 × 13–27 μm, each with numerous, small discoid plastids with associated pyrenoids; in transverse section blades 32–66 μm thick, 1–2 (–4) cells, cells rectangular, higher than wide, 30–38 × 13–17 μm, in one cell thick regions, quadrate, subquadrate 15–21 × 15–20 μm in two cells thick regions and with large, colourless, longitudinally elongate, transversely rounded, central cells and smaller, pigmented, quadrate or wider than high surface cells in four cells thick regions; hairs abundant, single or grouped, arising from blade surface or margin, 10–12 μm in diameter with small, basal meristem and enclosing sheath.

Plurilocular and unilocular sporangia borne on the same or different blades, single or grouped on both surfaces, formed from surface cells; plurilocular sporangia common, quadrate or rectangular in surface view, 24–36 × 20–27 μm, conical and sometimes bifurcate and projecting above thallus surface in section, 26–42 × 14–23 μm; unilocular sporangia rare, ovoid or quadrate in surface view, 30–40 μm in diameter.

Epiphytic on various algae, upper to lower eulittoral pools and shallow sublittoral.
Widely distributed around the British Isles.
Annual; spring and summer.

Culture studies by Kylin (1933), Rhodes (1970), Lockhart (1982), Rietema & van den Hoek (1981) and Pedersen (1984) on *Punctaria tenuissima* (as *Desmotrichum undulatum*) reveal isolates of this species to have an asexual 'direct' life history. In all cultures spores developed directly into a microthallus, from which new erect blades were produced. The production of erect blades from many isolates was notably influenced by environmental factors including temperature, light intensity, light quality, photo-period and nutrient content of medium. Environmental conditions also controlled fertility of the microthalli (reported by Pedersen to resemble *Hecatonema* and *Streblonema effusum* Kylin). Isolates from different geographical regions did, however, vary in their response to various environmental conditions suggesting this species is represented by various genotypes. *Punctaria tenuissima* is possibly identical to *Punctaria latifolia*.

Harvey (1846–51), pl. 248; Reinke (1889a), p. 15, pls 11–13; Batters (1892b), pp. 17–18; Kylin (1907), pp. 66–70, fig. 16; Newton (1931), pp. 185–186; Kylin (1933), pp. 36–38, figs 11–12; Taylor (1957), p. 165, pl. 15, fig. 6, pl. 16, figs 1–2; Kylin (1947), pp. 72–73, fig. 59A, B; Rosenvinge & Lund (1947), pp. 6–11, fig. 1; Jaasund (1957), pp. 216–218, fig. 6, (1965), pp. 87–89, fig. 26; Rhodes (1970), pp. 321–314, figs 1–11; Pankow (1971), pp. 192–194, fig. 249, pl. 22, fig. 1; Rueness (1977), p. 160, fig. 93, p. 164; Rietema & van den Hoek (1981), pp. 321–335, figs 1–10.

Fig. 55 *Punctaria tenuissima*
A. Portions of erect thallus showing enlargement of blade. B. V.S. of thallus margin. C–D. Portions of narrow thalli with lateral mature and dehisced plurilocular sporangia. E–F. Edge portions of erect thalli showing protruding plurilocular sporangia. Bar = 100 μm (third A from left only, E), 50 μm (others).

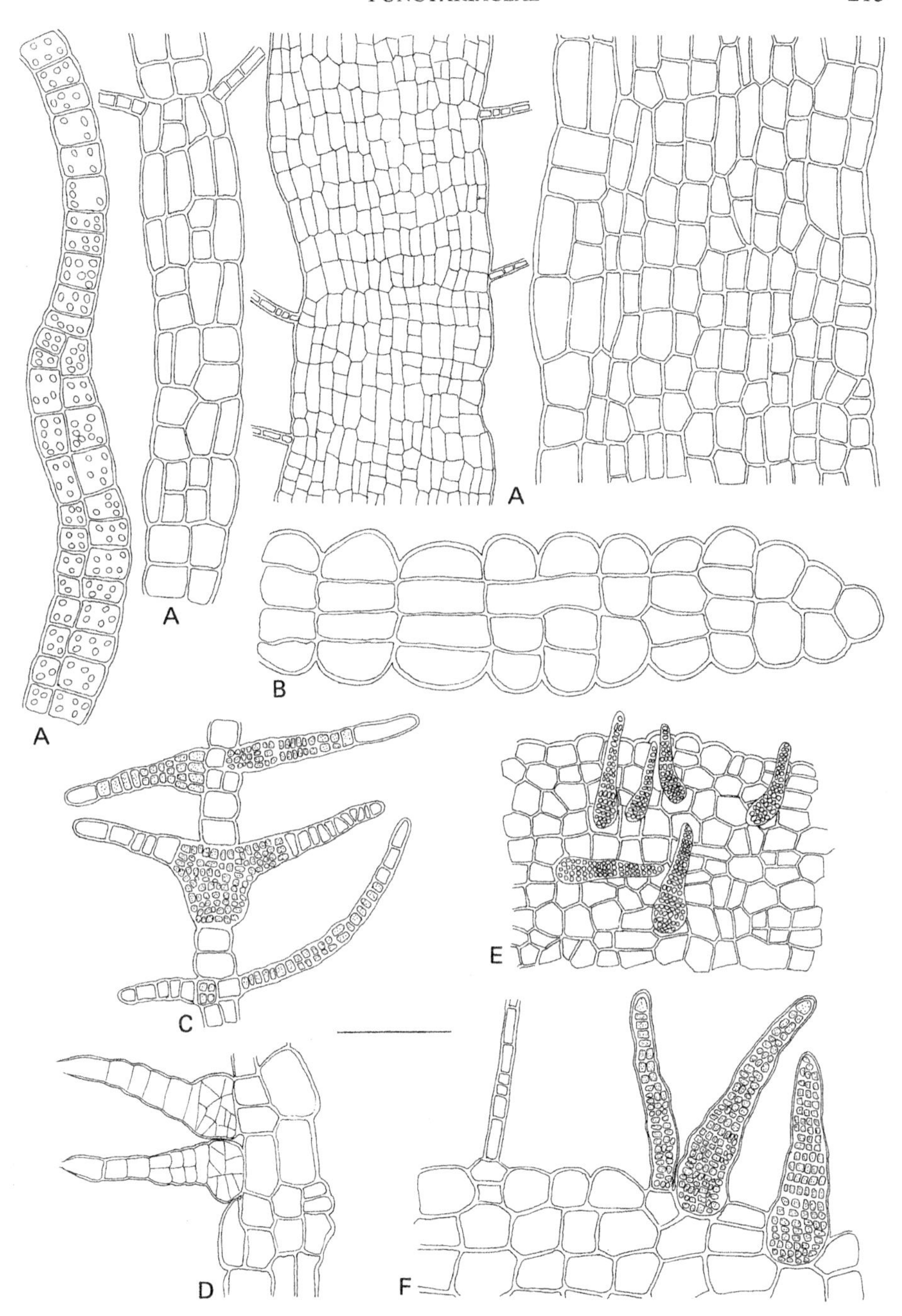
A
A
A
A
B
C
D
E
F

SCYTOSIPHONACEAE Farlow

SCYTOSIPHONACEAE Farlow (1881), pp. 15, 62 [as Scytosiphoneae].

Thallus erect or prostrate, epilithic, epizoic or epiphytic; erect thalli macroscopic, thalloid, saccate, cylindrical or dorsiventrally flattened, simple, hollow or entire; in structure parenchymatous, with inner medulla of large, colourless cells and outer cortex of small, pigmented cells; cells with a single, large, plate-like plastid with 1–2 pyrenoids; hairs with basal meristem with or without obvious sheath; plurilocular sporangia in local or extensive sori in terminal thallus region, in closely packed, vertical columns, uni- to multiseriate, with or without associated ascocysts; unilocular sporangia unknown on blade surface but in some species reported on additional prostrate thalli.

Prostrate thalli microscopic or macroscopic, densely tufted, resembling a carpet pile, slightly pulvinate or crustose; comprising outwardly spreading, branched filaments either free, with or without longitudinally divided cells, or compacted and pseudodiscoid/ discoid in structure, giving rise to erect, short, simple or little-branched filaments, loosely aggregated and gelatinous or laterally adjoined, weakly or quite firmly and pseudoparenchymatous in structure; cells with a single, plate-like plastid with or without obvious pyrenoid; hairs solitary or grouped, arising superficially or immersed, with basal meristem but lacking obvious sheath; unilocular sporangia common, irregularly distributed on basal filaments, sessile or stalked, more commonly grouped in immersed or raised mucilaginous sori, terminal on compacted erect filaments, with or without associated, further projecting, multicellular paraphyses arising from the supporting cell; plurilocular sporangia rare, terminal on compacted erect filaments, usually grouped in sori without accompanying paraphyses, uniseriate to multiseriate, with or without terminal sterile cells.

This family is represented in the British Isles by five genera: two pseudoparenchymatous (syntagmatic) crustose genera (*Ralfsia* and *Stragularia*) and three parenchymatous thalloid genera (*Colpomenia, Petalonia* and *Scytosiphon*). Also included here is the small microscopic tufted alga *Compsonema saxicolum* and the unilocular sporangia bearing material of the crustose alga *Microspongium gelatinosum*. The inclusion of *C. saxicolum* and the crustose algae in the Scytosiphonaceae is based on cytological and life history features, and essentially follows the proposal of Pedersen (1981c).

Circumscription of the family is based on the parenchymatous erect thalli, in particular the following two criteria: a single plate-like plastid with pyrenoid in each cell and the occurrence of plurilocular sporangia only on the erect thalli, which are in closely packed, mainly uniseriate, vertical columns, with or without accompanying paraphysis-like cells. These features were even considered by Feldmann (1949) to be sufficiently important to erect the order Scytosiphonales. Reaction to this order has been mixed; it was, for example, accepted by Jaasund (1965), Wynne (1969), Wynne & Loiseaux (1976), Nakamura (1972), Abbott & Hollenberg (1976), Rueness (1977), Bold & Wynne (1978) and Christensen (1980) but rejected by Russell & Fletcher (1975), Parke & Dixon (1976), Kornmann & Sahling (1977) and South & Hooper (1980). Indeed it is pertinent here to note that even the family Scytosiphonaceae has not received full support, with genera such as *Petalonia* and *Scytosiphon* variously included in the Punctariaceae or Asperococcaceae (Taylor 1957; Lindauer *et al.*, 1961; Pankow 1971).

As Bold & Wynne (1978) point out, an additional, more recent and significant criterion for the delimitation of the group, is the involvement of prostrate, principally crustose, expressions bearing unilocular sporangia in the life histories of many of the members. This heteromorphic life history is in contrast to the isomorphic life history originally suggested by Feldmann (1949). The relationship between the two morphological expressions is not fully understood, but an extensive study of several genera in Japan (Nakamura & Tatewaki 1975) suggests that the basic life history probably involves a heteromorphic alternation between an erect bladed gametophytic (i.e. sexual) phase and a prostrate, essentially crustose, sporophytic (i.e. asexual) phase. The sexual plants are isomorphic and heterothallic releasing either iso- or anisogametes from the plurilocular sporangia. Fusion results in the development of the prostrate phase bearing unilocular sporangia in which meiosis occurs. Release and germination of the unispores completes the life cycle with the formation of the erect gametophytic thallus. In addition to this sexual, biphasic, life cycle, an abbreviated asexual life history was reported, via parthenogenetic development of the gametes. Depending on the culture conditions of temperature and daylength used, these parthenogametes could either repeat the parental expression or form unilocular sporangia bearing crustose thalli. The unilocular sporangia in these haploid crusts do not undergo meiosis and the released unispores behave asexually and develop similarly to the parthenogametes. Both sexual and abbreviated asexual life histories were reported by Nakamura & Tatewaki in five species of the Scytosiphonaceae: *Petalonia fascia* differed in possessing only an abbreviated life history. This abbreviated asexual life history appears to be in accord with most other studies on Scytosiphonaceae members throughout the world, including those on isolates collected in the North Atlantic. It is possible, therefore, that the erect-bladed thalli represent potential gametophytes, in which sexuality is exceptional or requires specific environmental conditions for its fulfilment.

A notable feature of the heteromorphic life histories revealed in this family is the range of prostrate expressions which have been described. These have frequently been linked in identity with thalli of such genera as *Streblonema, Compsonema, Myrionema, Microspongium* and *Stragularia* (usually identified as *Ralfsia*). In the North Atlantic particular connections have been established between parenchymatous erect bladed genera and *Stragularia* species as well as material identified as *Compsonema saxicolum* and *Microspongium gelatinosum*. For this reason the genus *Stragularia,* and by necessity the closely related genus *Ralfsia,* are included in the Scytosiphonaceae. Also included are the two above mentioned species which are separated from their respective genera in the Myrionemataceae (in which they were unsuitably placed) and provisionally interpreted as 'phases' in the life histories of *Scytosiphon* and/or *Petalonia* spp. Indeed the possession of a single, plate-like plastid in each cell would alone justify the inclusion of the above prostrate algae in the family Scytosiphonaceae (Christensen 1980; Pedersen 1981c, 1984).

The prostrate thalli can be distinguished using a combination of morphological and anatomical features. *C. saxicolum* forms confluent tufts whilst the *M. gelatinosum* phase, *Ralfsia* and *Stragularia* form crustose thalli. Unlike the encrusting *Ralfsia* and *Stragularia, M. gelatinosum* forms slightly pulvinate thalli comprising erect filaments which are only loosely associated and easily glide apart under slight pressure. *Stragularia* and *Ralfsia* are distinguished by the curvature of the erect filaments (erect or slightly curved in *Stragularia,* markedly curved, almost prostrate at first, in *Ralfsia*), the cohesion of the erect filaments (quite firmly adjoined but separable under pressure in *Stragularia,* tightly adjoined and not separable in *Ralfsia*), and the distribution of the reproductive sori on the crust surface (indefinite in *Stragularia,* discrete in *Ralfsia*).

The three thalloid genera included here are more easily distinguished: *Colpomenia* forms inflated saccate thalli, *Petalonia* forms dorsiventrally flattened thalli and *Scytosiphon* forms tubular thalli.

COLPOMENIA (Endlicher) Derbès et Solier

COLPOMENIA (Endlicher) Derbès et Solier (1851), p. 95.

Type species: *C. sinuosa* (Mertens ex Roth) Derbès et Solier (1851), p. 95.

Thallus spherical or saccate, membranous, hollow, furrowed, firm and entire or more frequently collapsed and irregularly split, attached at the base by rhizoidal filaments; surface cells irregularly placed, each with a single plate-like, parietal plastid and pyrenoid; comprising an inner medulla of large, thick-walled, colourless cells and an outer cortex of smaller, pigmented cells; hairs tufted, grouped, arising from depressions, with basal meristem, with or without obvious sheath.

Plurilocular sporangia crowded, discrete or extensive on thallus surface, in closely packed vertical columns, uni- to multiseriate, with ascocyst-like cells; unilocular sporangia unknown on thallus surface, reported to be associated with a ralfsioid phase in the life histories of Pacific isolates only.

Two species of *Colpomenia* have been recorded in the North Atlantic: *C. sinuosa* (Roth) Derbès et Solier which is restricted to warm-temperate waters and only extends as far north as the Spanish coast in the Bay of Biscay (Hamel, 1928; Miranda, 1931; Ardré, 1970) and *C. peregrina* (Sauvageau) Hamel which is mainly cool-temperate in distribution and extends from Morocco to Norway. The two species are distinguished mainly on features such as surface configuration, hair origin, wall thickness and sorus shape. However, many of these were rejected by Clayton (1975) in her comparative study of populations of these two species in Australia. Only the shape of the sori (irregular/extensive in *C. peregrina,* punctate in *C. sinuosa*) and the presence of a cuticle on the plurilocular sporangia (in *C. sinuosa* only) were considered by her to be effective.

A number of *Colpomenia* species have been investigated in laboratory culture throughout the world; these include *C. bullosa* Yamada (Nakamura & Tatewaki, 1975; Blackler, 1981), *C. peregrina* (Sauvageau, 1927a; Dangeard, 1936b; Clayton, 1979; Blackler, 1981) and *C. sinuosa* (Kunieda & Suto, 1938; Wynne, 1972a—as *f. deformans* Setchell & Gardner; Blackler, 1981). In the investigations carried out by Blackler, Dangeard, Sauvageau and Wynne (isolates from Europe and the Pacific coast) direct asexual life histories were reported with the plurispores from the macrothalli germinating directly to form filamentous microthalli from which sprouted a new generation of erect thalli. Usually the microthalli were fertile with plurilocular sporangia, although their significance or possible function was not discussed. However, in the investigations of Kunieda & Suto (1938), Nakamura & Tatewaki (1975) (Japanese isolates) and Clayton (1979) (Australian isolates) a sexual life history was reported; this was additionally shown by Nakamura & Tatewaki and Clayton to be heteromorphic with a gametophytic plurilocular sporangia-bearing *Colpomenia* phase alternating with a sporophytic unilocular sporangia-bearing filamentous or ralfsioid phase. It would appear, therefore, that geographical isolates of each species have different life histories.

One species in the British Isles:

Colpomenia peregrina (Sauvageau) Hamel (1931–39), p. 201. Figs 56, 57; Pl. 7

Colpomenia sinuosa (Mertens ex Roth) Derbès et Solier var. *peregrina* Sauvageau (1927a), p. 321.

Plants consisting of erect, spherical or saccate, hollow thalli, 3–7 (−9) cm across, entire and fleshy when young, frequently furrowed, collapsed and irregularly torn later, broadly attached at the base by localised patches of rhizoidal filaments produced from the outer cortical cells; saccate portion membranous, smooth and slightly lubricous outside, roughened inside, olive brown, drying to green, surface cells polygonal, irregularly displaced, 8–16 × 5–12 μm, each with a single, parietal, plate-like plastid and pyrenoid; hairs frequent, immersed in pits, with basal meristem, with or without obvious sheath; in section, thalli with an outer cortex of 1–3 small, pigmented cells and an inner medulla of 1–4 layers of large, thick-walled, frequently ruptured and collapsed colourless cells.

Plurilocular sporangia crowded, in extensive or discrete dark patches on thallus surface, in closely packed vertical columns arising from outer cortical cells, to 12 loculi (−55 μm) long, uni- to multiseriate, 5–11 μm in diameter, usually with abundant scattered ascocyst-like cells present, 17–35 × 7–13 μm; unilocular sporangia unknown on thallus surface and not reported in the life history of Atlantic isolates.

Epiphytic, rarely epilithic, upper to lower eulittoral pools and damp rock, and sublittoral to 3 m.

Generally distributed around the British Isles, but more common on south western shores.

Thalli annual, appearing in early summer and dying back during late autumn and early winter, rarely recorded in winter and spring and probably lying dormant as basal filaments. At Bembridge on the Isle of Wight, plants are abundant in May and June, less frequent during July and August with a noticeable increase in numbers again in September and October.

Colpomenia peregrina represents an introduced species to the North Atlantic, probably originating from the Pacific (see Jones, 1974 and Farnham, 1980 for discussion). It was first reported at the beginning of the century (1906) in the Gulf of Morbihan on the

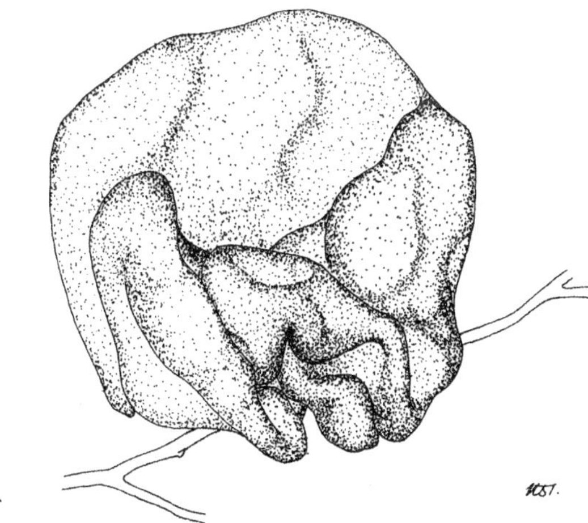

Fig. 56 *Colpomenia peregrina*
Habit of plant. Bar = 20 mm.

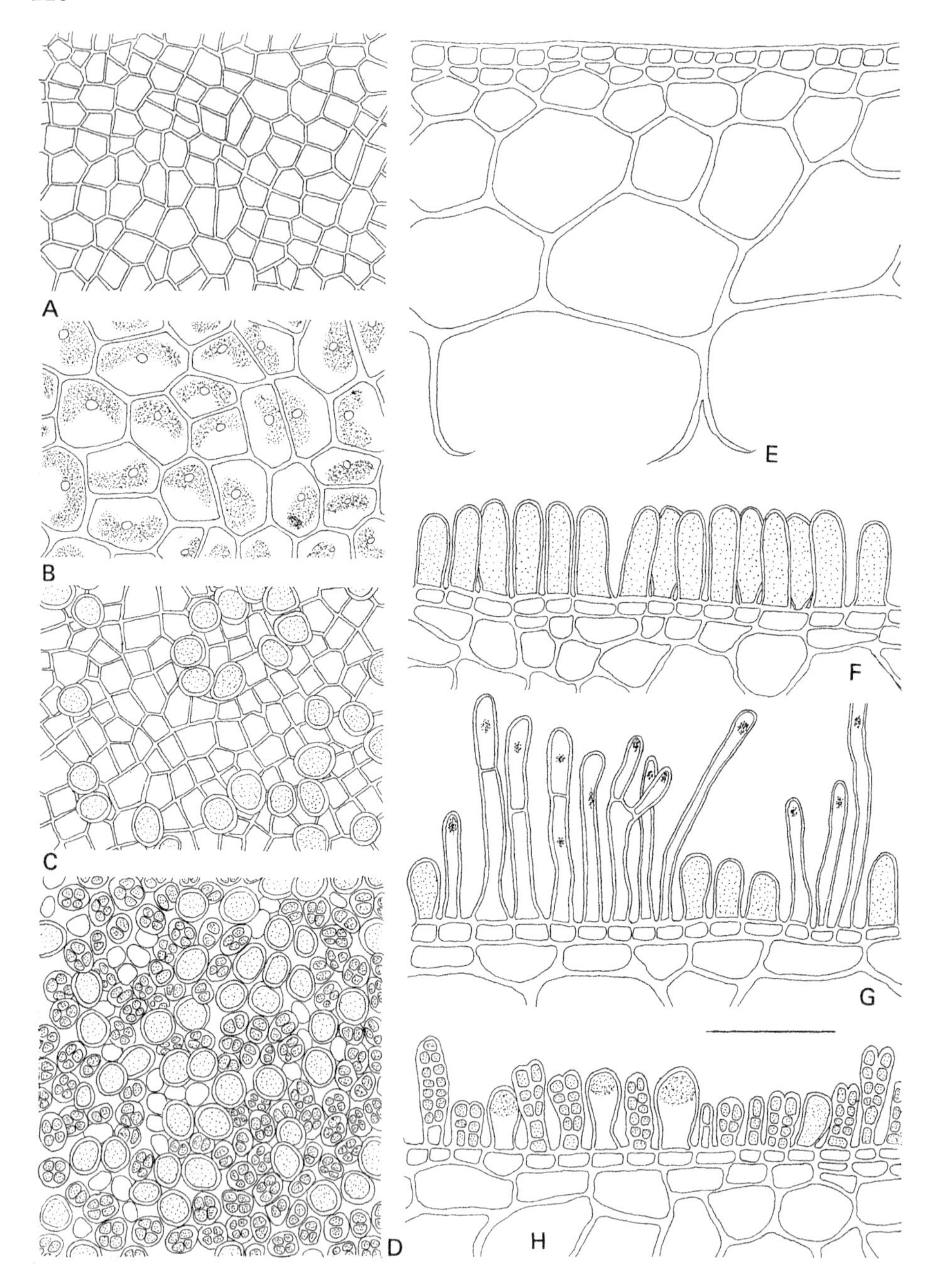
A
B
C
D
E
F
G
H

west coast of France (Sauvageau, 1918) and has since spread as far north as Norway (Rueness, 1977). It was first observed in the British Isles by Cotton (1908b, c; 1911) and surprisingly was not included in Newton (1931).

Culture studies on Atlantic isolates by Sauvageau (1927a), Dangeard (1963b) and Blackler (1981) reveal the occurrence of a direct, asexual life history. Plurispores from plurilocular sporangia on the saccate blades germinated directly to form filamentous microthalli bearing plurilocular sporangia from which directly sprouted a further generation of saccate macrothalli. There was no evidence of a sexual cycle and the life history can therefore be described as heteromorphic and monophasic. This is in contrast to the heteromorphic, biphasic life history described for this species in Australia (Clayton, 1979). Clayton revealed that the saccate macrothalli are potential gametophytes which alternate with filamentous or ralfsioid, unilocular sporangia bearing, sporophytes. Sexual reproduction was described as anisogamous, with large female gametes and small male gametes. However, functional gametophytes were described as rare, with a restricted winter occurrence; usually the life history is 'direct' with the plurispores (which could be either zoospores or gametes) behaving asexually and germinating directly into similar thalli, not unlike the sequence described for European isolates by Sauvageau and Dangeard. More culture studies on European isolates of this species throughout its seasonal occurrence, are therefore required to determine if a sexual life history, with a unilocular sporangia-bearing ralfsioid phase, is present, similar to that reported in Australia.

Cotton (1980b), pp. 73–77; Cotton (1908c), pp. 82–83; Cotton (1911), pp. 153–157; Sauvageau (1918); Sauvageau (1927a), pp. 308–353, figs 1–8: Hamel (1931–39), pp. 201–202; Feldmann (1949), pp. 103–115, fig. 2; Blackler (1964), p. 50; Dangeard (1963b), pp. 66–73, pl. XII–XIII, XX–XXI; Gayral (1966), p. 254, pl. XXXIII; Blackler (1967), pp. 5–8; Jones (1974), p. 107; Clayton (1975), pp. 187–195, figs 1–14; Abbott & Hollenberg (1976), pp. 1–10, figs 1–10; Rueness (1977), p. 167, pl. XXII, 5; Clayton (1979), pp. 1–10, figs 1–10; Farnham (1980), p. 877; Clayton (1980b), pp. 113–118, figs 1–3; Blackler (1981), p. 133; Parsons (1982), pp. 295–297, figs 6–7, 13; Stegenga & Mol (1983), p. 103, pl. 37, figs 1–2; Vandermeulen *et al.* (1984), pp. 325–326, figs 5–8.

COMPSONEMA SAXICOLUM Phase

'Compsonema saxicolum' phase of *Petalonia/Scytosiphon* Fig 58; Pl. 8a, b

?Myrionema saxicola Kuckuck (1897), p. 381
Compsonema saxicola (Kuckuck) Kuckuck (1953), p. 343.

Thallus epilithic or epizoic, forming small, light brown tufts not visible to the naked eye, discrete and circular or more commonly confluent (resembling a carpet pile) and spreading to approximately 1 mm in diameter; in squash preparation consisting of a basal layer

Fig. 57 *Colpomenia peregrina*
A–B. Surface view of thallus. Note single plate-like plastid with pyrenoid. C. Surface view of thallus showing ascocysts. D. Surface view of thallus showing ascocysts and plurilocular sporangia with 1–4 loculi. E. V.S. of vegetative thallus showing 1–2 layers of small, cortical cells and 3–4 layers of large, medullary cells. Note inner cells, enclosing central cavity, are ruptured. F–G. V.S. of thalli showing variously shaped ascocysts. H. V.S. of fertile thallus showing ascocysts and plurilocular sporangia arising from surface cells. Bar = 20 μm (B), 50 μm (others).

of branched, outwardly spreading, irregularly contorted, loosely associated filaments (pseudodiscoid) comprising mainly rectangular cells, 7–13 × 5–8 μm, frequently longitudinally divided (biseriate), many of which, especially in the central regions give rise to erect filaments, hairs and/or unilocular sporangia; erect filaments free and not laterally adjoined, loosely associated in a gelatinous matrix and easily separable under pressure, simple or rarely branched, to 280 μm long, 14–25 (–35) cells, slightly tapering above, comprising cells mainly quadrate, sometimes rectangular, 5–11 × 5–7 μm, each with a single, plate-like and parietal plastid occupying the upper cell region, with pyrenoid; hairs not uncommon arising from basal layer or terminal on erect filaments, with basal meristem and sheath.

Unilocular sporangia common, oval to pyriform, 40–70 × 19–35 μm, usually arising directly from basal cells, sessile or on 1–3 celled stalks, less frequently lateral or terminal on the erect filaments; plurilocular sporangia unknown.

Epilithic on bedrock, or epizoic on limpets, in the littoral fringe;

Only recorded for the south coast of England (Kent, Dorset, Devon) and west coast of Ireland (Mayo); probably more widely distributed.

Probably a summer annual; recorded from April to October.

This species is probably common and widely distributed around the British Isles. It appears to favour colonisation of limpet shells in the littoral fringe, forming small confluent tufts just visible under a low power binocular microscope. Algae frequently associated with *C. saxicolum* include *Blidingia* spp., *Petroderma maculiforme, Hecatonema* sp., *Ralfsia verrucosa, Enteromorpha* spp., young fucoid germlings and blue-green algae. It is particularly characterised by the unbranched erect filaments, cells with a single, plate-like plastid, basally positioned unilocular sporangia and a parenchymatous prostrate system with frequent biseriate cells.

C. saxicolum has been investigated in laboratory culture by Pedersen (1981c) and the present author (unpublished information). Perdersen revealed the occurrence of a direct, monophasic life history for Danish material; the unispores behaved asexually and germinated directly into fertile thalli essentially similar to the parental field material. Culture studies by the present author, using material collected from Dorset on the south coast of England, revealed a very similar sequence of development; however, an additional erect thallus sprouted from the basal filaments. These erect thalli were poorly developed in culture but were recognisable as like *Petalonia* or *Scytosiphon*. The involvement of *C. saxicolum* in the life history of a member of the Scytosiphonaceae finds support in the investigations of Loiseaux (1970b) and Fletcher (1981a). Loiseaux established a life history connection between the Californian species *Compsonema sporangiiferum* Setchell & Gardner (which she noted to be very similar to *C. saxicolum*) and a minute *Scytosiphon,* close to *S. pygmaeus* Reinke (usually included within the synonomy of *Scytosiphon lomentaria*) whilst Fletcher observed fertile microthalli in cultures of *Petalonia filiformis,* established from material collected on the south east coast of England, which showed a marked similarity to *C. saxicolum* (see p. 236).

Fig. 58 *Compsonema saxicolum*
A. S.P. of vegetative thalli showing erect filaments arising from a distromatic base. B. S.P. of fertile thalli showing unilocular sporangia. Bar = 50 μm.

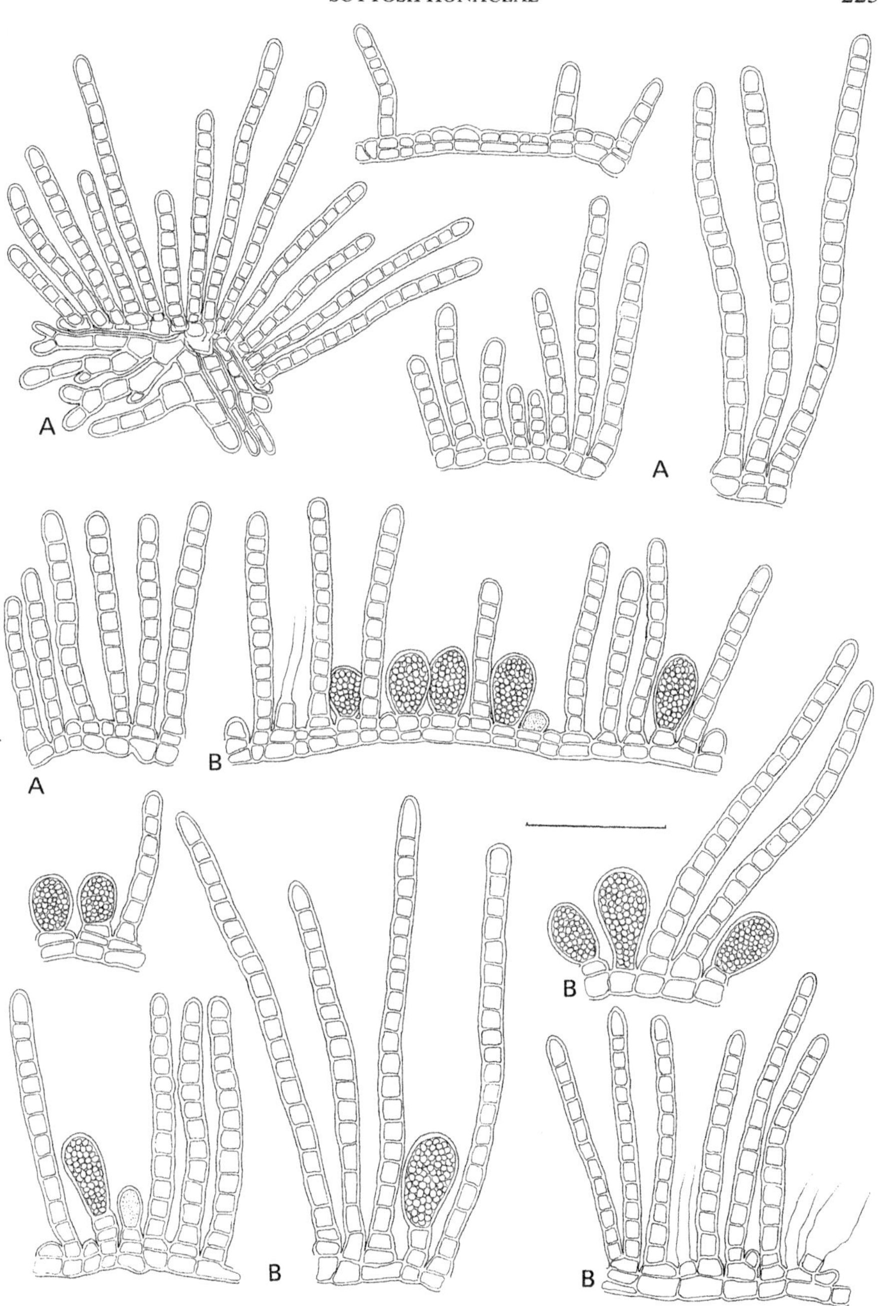
A
A
A
B
B
B
B

An additional interesting aspect of the above described heteromorphic life histories was the role played by environmental conditions, particularly temperature and photoregime, in the relationship between the *Compsonema* microthalli and the erect macrothalli. For example Fletcher (unpublished) revealed erect thallus production to be enhanced in low temperature/short day conditions, whilst fertile *Compsonema*-like thalli showed maximum development in warm temperature/long day conditions. These results would agree with the winter/spring seasonal occurrence of *Petalonia* and *Scytosiphon* spp. around the British Isles and the generally summer/autumn field occurrence of the *C. saxicolum* plants. Certainly more culture studies are required to establish the extent of occurrence of *Compsonema*-like thalli in the life histories of genera and species of the Scytosiphonaceae and their exact relationship.

Compsonema saxicolum was originally described from Helgoland by Kuckuck (1897) as (?) *Myrionema saxicola*. It was subsequently reported for the British Isles at Swanage, Dorset by Batters (1900, p. 371) and described in Newton (1931, p. 151). Kuckuck (1953) later transferred the species to the genus *Compsonema* Kuckuck (1899a), removing the latter from its original place in the Myrionemataceae to the Ectocarpaceae. The placement of *Compsonema* into the Ectocarpaceae was accepted by Parke (1953) and Parke & Dixon (1964) although later (Parke & Dixon, 1968, 1976) the genus was transferred back to the Myrionemataceae, probably as a result of the work by Loiseaux (1967a). On the basis of the culture results it is here included in the Scytosiphonaceae following the proposal of Pedersen (1981c). Other features consistent with its inclusion in this family include the frequent occurrence of longitudinal divisions in the basal cells and the presence of a single, plate-like plastid with pyrenoid in each cell.

Kuckuck (1897), pp. 381–382, fig. 8; Batters (1900), p. 371; Cotton (1912), p. 121; Kuckuck (1953), pp. 343–347, fig. 14; Pedersen (1981c), pp. 213–214, fig. 55; Fletcher (1981a), pp. 103–104; non *Myrionema saxicola* in Knight & Parke (1931) p. 68, pl. XII, figs 27–29.

MICROSPONGIUM GELATINOSUM Phase

'Microspongium gelatinosum' phase of *Scytosiphon lomentaria* Figs 59A, 60; Pl. 8c

Microspongium gelatinosum Reinke (1889b), p. 46 (unilocular sporangial material only).

Thallus crustose or slightly pulvinate and spongy, medium to dark brown, to approximately 0·5 mm thick, discrete and circular or more commonly confluent and spreading, to 30 mm or more in extent, firmly attached to substratum by undersurface, usually without rhizoids; in surface view central vegetative cells rounded, quite closely packed, 7–11 µm in diameter, edge region cells rectangular or quadrate 8–18 × 5–10 µm; consisting of a monostromatic base of coherently spreading, laterally united filaments with a marginal row of apical cells giving rise, over the greater part, to simple, or little branched, erect, slightly gelatinous filaments only loosely adjoined and quite easily separable under pressure, except basal 1–3 cells; filaments to 25 cells (–430 µm) long, comprising mainly rectangular cells, 5–10 µm in diameter each with a single, pariental, plate-like plastid occupying upper cell region, with pyrenoid.

Unilocular sporangial sori extensive over central thallus region, not visible in surface view; unilocular sporangia elongate-pyriform or elongate-cylindrical, 48–100 × 16–27 µm, sessile or more usually on 1–3 celled pedicels, borne at the base of

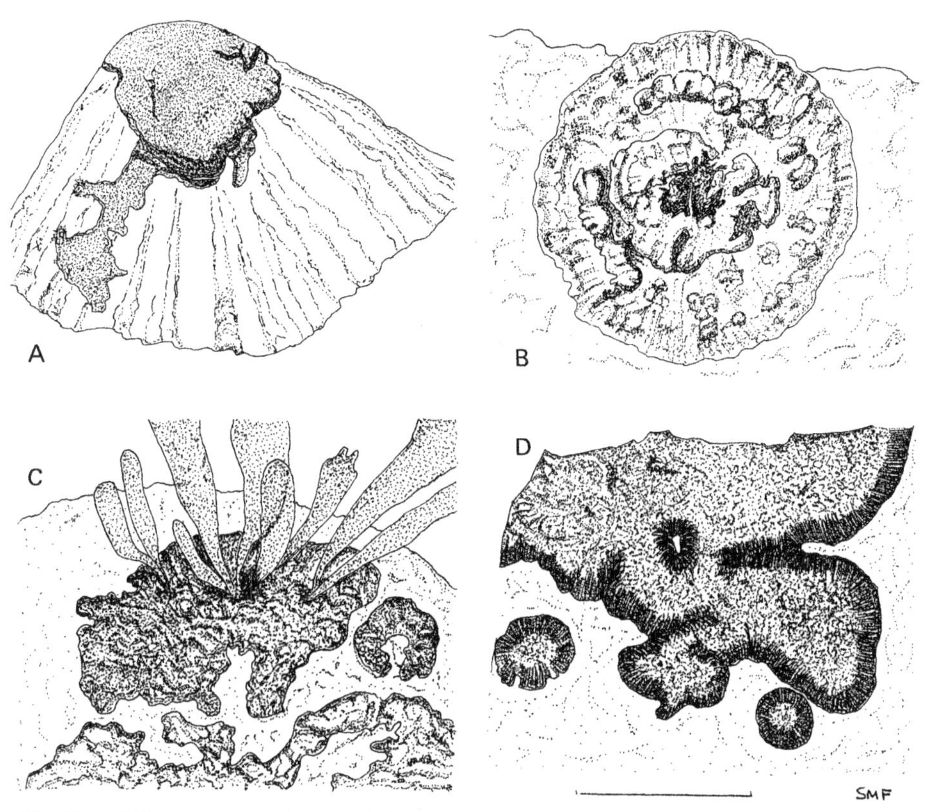

Fig. 59 *'Microspongium gelatinosum'* phase
A. Habit of plant on limpet.
Ralfsia verrucosa
B. Habit of plant.
Stragularia clavata
C. Habit of plants. Note emergent *Petalonia fascia* blades.
Stragularia spongiocarpa
D. Habit of fertile plants. Bar = 17 mm (A), 15 mm (B–C), 18 mm (D).

loosely associated, gelatinous, multicellular paraphyses; paraphyses linear or elongate-clavate, 95–170 × 5–7 μm, mostly of 6–8 cells which are 1–2½ diameters long above, narrower and up to 4(–9) diameters long below; plurilocular sporangia not recorded on crustose thalli in the British Isles, only reported on Danish specimens by Lund (1966, fig. 3N) forming terminal, uniseriate clusters on branches of the erect filaments; plurilocular sporangia usually confined to the erect-bladed *Scytosiphon lomentaria* phase.

Epilithic and epizoic, in pools and emergent, upper to lower eulittoral, usually in association with other crustose brown algae such as *Petroderma maculiforme, Stragularia clavata* and *Ralfsia verrucosa.*

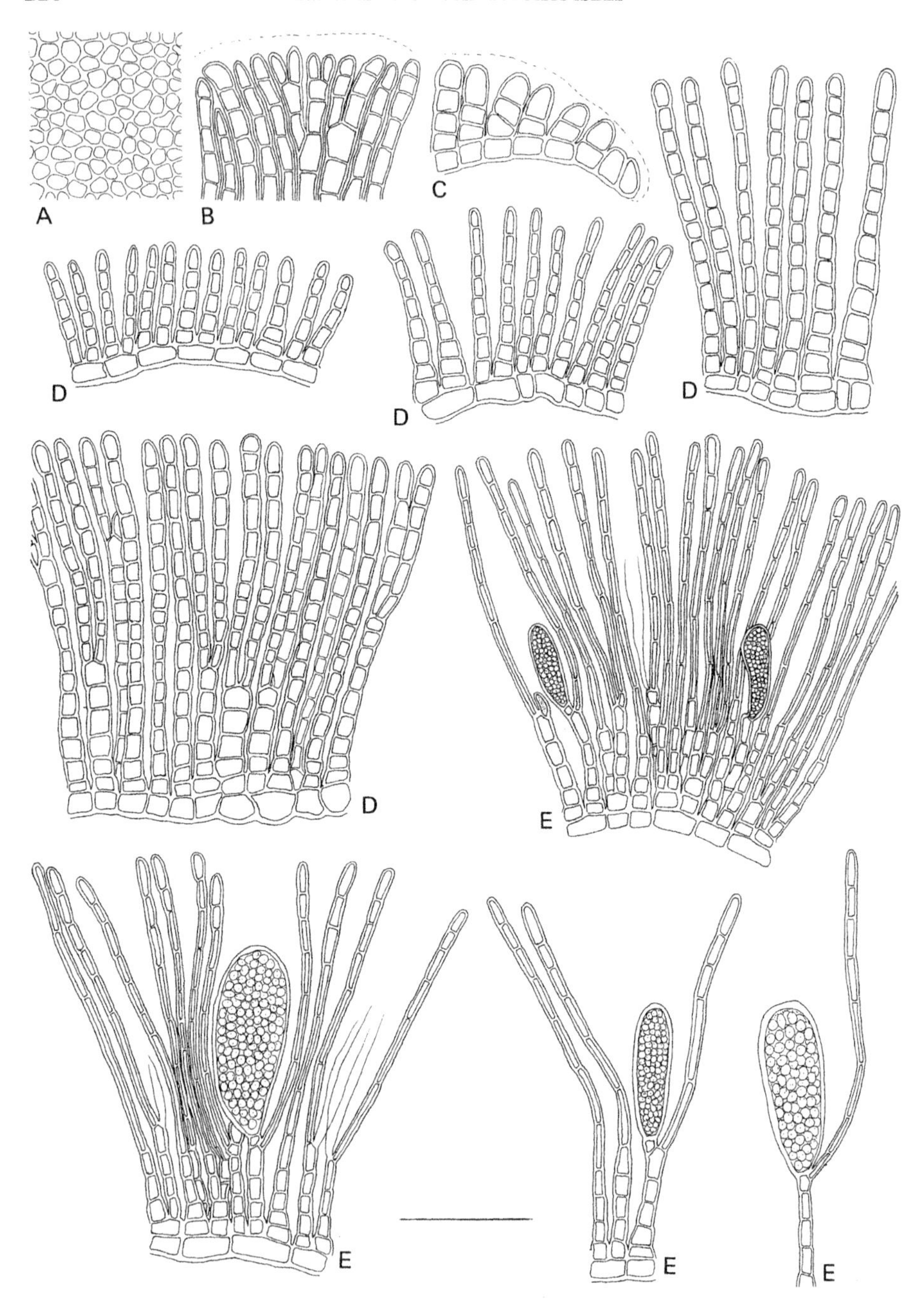
A
B
C
D
D
D
D
E
E
E
E

Uncommon, recorded for widely scattered localities around the British Isles (Fife, East Lothian, Berwick, Northumberland, Yorkshire, Kent, Sussex, Hampshire, Dorset, Devon, Pembroke, Argyll; in Ireland, Dublin).

Vegetative thalli recorded throughout the year and probably perennial; unilocular sporangia recorded September to April.

The above described crustose phase of *S. lomentaria* represents the unilocular sporangia bearing plants of *Microspongium gelatinosum* originally discovered by Reinke (1889b) in the Baltic. It is now recognised that these plants represent a different taxon to the plants of *M. gelatinosum* he had earlier described with plurilocular sporangia (1888) and they must consequently be separated (see p. 100). The unilocular sporangia bearing material (herein termed the *M. gelatinosum* 'phase') is very similar in structure to the genus *Stragularia,* from which it differs in having slightly more pulvinate, spongy crusts, unilocular sporangial sori not clearly distinguishable from the vegetative thallus in surface view and erect vegetative filaments which are loosely associated (except near the base) and easily separated, forming almost fan-shaped squash preparations. In addition the paraphyses in the *M. gelatinosum* phase are less clearly discernible from the erect vegetative filaments. Using the above characteristics field material of the *M. gelatinosum* 'phase' can quite easily be distinguished from *Stragularia* spp., even the closely similar *S. clavata.* This does negate the use of the descriptive term *'Ralfsia'*-like for the unilocular sporangial thalli of the *M. gelatinosum* phase as proposed by Kristiansen & Pedersen (1979), despite the nomenclatural benefits that would accrue.

Field and laboratory culture studies of the *M. gelatinosum* phase in the North Atlantic have revealed it to occur as a phase in the life history of the macroscopic alga *Scytosiphon lomentaria* (see Lund, 1966–Denmark; McLachlan *et al.,* 1971—Nova Scotia; Fletcher, 1978—south coast of England). Unispores from the unilocular sporangia on the crusts germinated directly, without sexual fusion, to form uniseriate filamentous, knot-filamentous and/or *'Microspongium'*-like prostrate systems. In addition to repeating the parental phase and developing unilocular sporangia, these thalli gave rise to erect, parenchymatous, tubular fronds identified as *S. lomentaria.* The evidence indicates that both the crustose and erect thalli probably have the same ploidy level and the life history can, therefore, be described as heteromorphic and monophasic. Fletcher revealed a facultative relationship between the erect fronds and the 'basal' growths which was influenced by environmental conditions. For example, frond initiation appeared to be enhanced in low temperature/short day conditions whilst basal growths, particulary crustose development was enhanced in warm temperature/long day conditions. The involvement of crustose thalli in the life history of *S. lomentaria* is supported by pluri-spore culture studies of the latter alga throughout the world although there is some conflicting evidence from widely different geographical isolates as to whether these are *'Microspongium'*-like or *'Ralfsia'*-like (see p. 251).

Fig. 60 *'Microspongium gelatinosum'* phase
A. Surface view of central crust region. B. Surface view of peripheral crust region. C. S.P. of crust edge showing terminal apical cell. D. S.P. of vegetative thalli showing loosely adjoined erect filaments. E. S.P. of fertile thalli showing unilocular sporangia arising at base of paraphyses. Bar = 50 μm.

On the basis of the culture studies of North Atlantic isolates and the occurrence of a single, plate-like plastid in each cell, *M. gelatinosum* Reinke (unilocular sporangia material only) is removed from the Myrionemataceae and here included in the Scytosiphonaceae.

Reinke (1889a), pp. 11–12, pl. 8; Reinke (1889b), pp. 46–47 (in both references, unilocular sporangial thalli only); Lund (1966), pp. 70–78, figs 1–4; McLachlan *et al.*, (1971), pp. 82–87, figs 1–10; Fletcher (1974a), pp. 78–91, figs 13–16, pls 23–25; Fletcher (1978), pp. 371–398, various figs; Kristiansen & Pedersen (1979), pp. 31–56, various figs; Tanaka & Chihara (1980c), pp. 339–341, fig. 1B (as *Ralfsia bornetii*)

PETALONIA Derbès et Solier nom. cons.

PETALONIA Derbès et Solier (1850), p. 265.

Type species: *P. debilis* (C. Agardh) Derbès et Solier (1850), p. 266 (=*P. fascia* (O. F. Müller) Kuntze (1891–98), p. 419).

Ilea Fries (1835), p. 321.
Phyllitis Kützing (1843), p. 342.

Plants consisting of erect, light to dark brown, bilaterally flattened blades arising from an expanded encrusting base, small discoid holdfast or fibrous mat of rhizoidal filaments; blades single or gregarious, simple, solid or with small central cavities, extremely variable in shape, although usually linear to broadly lanceolate, tapering sharply or slowly to base, less obviously to apex, sometimes spirally twisted in narrow specimens, entire and without proliferations, rarely longitudinally indented/split, with or without ruffled, delicate margins, delicate and flaccid to subcoriaceous in texture, often terminally eroded on mature specimens; in surface view cells small, rectangular to polygonal, irregularly arranged or in longitudinal, less obviously transverse rows, each containing a single, parietal, plate-like plastid with 1–2 pyrenoids; in section comprising an inner medulla of large, thick-walled, longitudinally elongate, transversely rounded, colourless cells, solid throughout or with small intercellular cavities, surrounded by an outer cortex of 1–3 smaller, pigmented cells; hairs single or tufted arising from surface or depressions, with or without obvious basal meristem and sheath.

Plurilocular sporangia in crowded, closely-packed sori, extensive in terminal blade region, in vertical columns, arising by division of cortical cells, uniseriate or biseriate with subquadrate to rectangular loculi; unilocular sporangia not known on blade surface but in some species reported on crustose *Ralfsia*- or *Compsonema*-like thalli.

The genus *Petalonia* is notoriously variable in its morphology with numerous dubious species and varieties described in the past. The distinction between the three species, reported here for the British Isles, is not entirely satisfactory and further detailed population studies are required. The principal characters used are the shape and width of the blades, and unpublished culture studies by the present author indicate that these might be genotypically determined. Other useful characters include the type of basal system present (expanding crust or holdfast in *P. fascia*, fibrous filamentous rhizoids in *P. filiformis* and *P. zosterifolia*) and both the local and geographical distribution patterns (*P. filiformis* and *P. zosterifolia* occur exposed in the upper eulittoral and littoral fringe, and are mainly distributed along the east and west coasts of the British Isles respectively, whilst *P. fascia* is a tide pool species with an unrestricted distribution.)

Petalonia blades may be confused with those of species of *Punctaria* Greville and young *Laminaria* Lamouroux. However, the latter two genera may be distinguished from *Petalonia* by the following anatomical and cytological details: in surface view, the cells of *Punctaria* and *Laminaria* blades are much larger and contain several discoid plastids; in section *Punctaria* shows no pronounced differentiation into cortical and medullary cells, the cells often being approximately equal in size, whilst in *Laminaria* the blades are three-layered with an outer, epidermis-like meristoderm of small pigmented cells, enclosing a cortical zone of large, thin-walled, colourless cells and a central medullary region; whilst *Petalonia* is attenuate towards the base, *Laminaria* and sometimes *Punctaria* have a distinct stipe.

In the genus the majority of life history studies have been carried out on *P. fascia* (Sauvageau, 1929; Kylin, 1933; Kunieda & Arasaki, 1947; Dangeard, 1962b, 1963b, 1969; Caram, 1965; Edwards, 1969; Hsiao, 1969, 1970; Wynne, 1969, 1972b; Rhodes & Connell, 1973; Fletcher, 1974b; Roeleveld *et al.*, 1974; Nakamura & Tatewaki, 1975); less frequently *P. zosterifolia* (Kuckuck, 1912; Dangeard, 1962a; Fletcher, 1974b; Nakamura & Tatewaki, 1975) and *P. filiformis* (Fletcher, 1981a) have been investigated. Usually a 'direct' type life history was reported with the plurispores germinating directly into prostrate systems from which sprouted a further generation of erect blades. The prostrate systems were variously described as filamentous, knot-filamentous and/or crustose and were not uncommonly fertile with plurilocular and/or unilocular sporangia. These fertile basal prostrate growths showed marked similarity to thalli of the genera *Streblonema, Compsonema* and *Ralfsia*. The occurrence of *Ralfsia*-like thalli in the life history of *P. fascia* was further supported by culture studies of field collected material of *Ralfsia* species (Wynne, 1969; Edelstein, *et al.*, 1970; Rhodes & Connell, 1973; Fletcher, 1974b; Roeleveld, *et al.*, 1974; Fletcher, 1978). In general the majority of studies revealed a facultative relationship between the prostrate and erect bladed expressions which was more determined by environmental conditions such as temperature, day length and nutrient levels rather than the two expressions representing rigidly alternating diploid and haploid phases. This asexual type of life history could, therefore, be interpreted as heteromorphic and monophasic. However, there have been reports of copulation by plurispores released from blades of both *P.zosterifolia* (Kuckuck, 1912; Nakamura & Tatewaki, 1975) and *P. fascia* (Kunieda & Arasaki, 1947), suggesting the additional occurrence of a sexual life history in this genus. Nakamura and Tatewaki's study in Japan is particularly noteworthy; in addition to an asexual cycle not unlike that described above, they revealed a heteromorphic alternation between the erect-bladed sexual generation, they plurispores from which behaved as gametes, and a crustose asexual generation bearing unilocular sporangia in which meiosis occurs during the formation of unispores. The life history could, therefore, be described as heteromorphic and haplodiplontic. More culture studies on isolates collected throughout the world including the North Atlantic region are now necessary to determine the extent of occurrence of this sexual cycle in *Petalonia* species.

KEY TO SPECIES

1 Blades 100–200 (−400) μm broad, not markedly dorsiventrally flattened, often oval in cross section and frequently spirally twisted . . . . *P. filiformis*

Blades more than 0·5 mm broad, clearly dorsiventrally flattened and not spirally twisted. 2

2 Blades linear 0·5–2 (–4) mm broad; basal attachment by rhizoidal filaments . . . . *P. zosterifolia*

Blades ovate to linear-lanceolate, more than 5 mm broad; basal attachment by small holdfast frequently overlying a crustose thallus . . . . *P. fascia*

Petalonia fascia (O. F. Müller) Kuntze (1891–1898), p. 419. Figs 61A, 62; Pl. 9

Fucus fascia O. F. Müller (1771–82), pl. 768.
Ilea fascia (O. F. Müller) Fries (1835) p. 321 pro parte.
Phyllitis fascia (O. F. Müller) Kützing (1843), p. 342.
Phycolapathum cuneatum Kützing (1845–71), pl. 49II.
Phyllitis caespitosa Le Jolis (1963), p. 68.
Scytosiphon fascia (O. F. Müller) Crouan frat. (1867), p. 169.
Phyllitis fascia (O. F. Müller) Kützing var *debilis* (C. Agardh) Hauck (1883–85), p. 391.
Petalonia debilis (C. Agardh) Derbès et Solier (1850), p. 266.

Plants forming erect, light to dark brown, single or more commonly gregarious, dorsiventrally flattened blades arising from a thin expanding crust or small discoid holdfast; erect blades simple, solid, thin and flaccid, becoming thicker and subcoriaceous, linear-lanceolate to broadly-lanceolate to 15 (–40) cm × 5–15 (–40) mm, tapering slowly or sharply to base, usually rounded, less frequently tapering above, entire and without proliferations, more rarely longitudinally indented above and often terminally eroded, with flat, in parts occasionally ruffled margins; in surface view cells irregularly placed or in longitudinally less obviously transverse rows, rectangular to polygonal, small, 6–13 × 4–8 μm, each enclosing a single, plate-like, parietal plastid with conspicuous pyrenoid; in section blades to 85 μm (vegetative) and 210 μm (reproductive) thick comprising an inner medulla of 3–6 large, thick-walled, longitudinally elongate, transversely rounded, often collapsed and ruptured colourless cells, enclosed by an outer cortex of 1–3 small, pigmented cells; hairs rarely observed on field collected material, single or grouped arising from surface cells or depressions, with basal meristem, lacking obvious sheath.

Plurilocular sporangia in dark brown sori extensive in terminal blade region, in closely packed vertical columns arising from cortical cells, uniseriate, occasionally biseriate to 60 μm (–14 loculi) long × 3–6 μm in diameter, with subquadrate to rectangular loculi, without paraphyses; unilocular sporangia not known on blade surface but recorded on *Stragularia*-like crustose thallus.

Epilithic, or more rarely epiphytic, upper to lower eulittoral, in open shallow pools and particularly channels in both sheltered and moderately exposed localities; tolerant of some sand cover and some variation in salinity.

Very common and generally distributed around the British Isles.

Blades annual; found throughout the year but especially common during late winter and spring; plants probably persist throughout remainder of year as perennial *Stragularia*-like basal crusts.

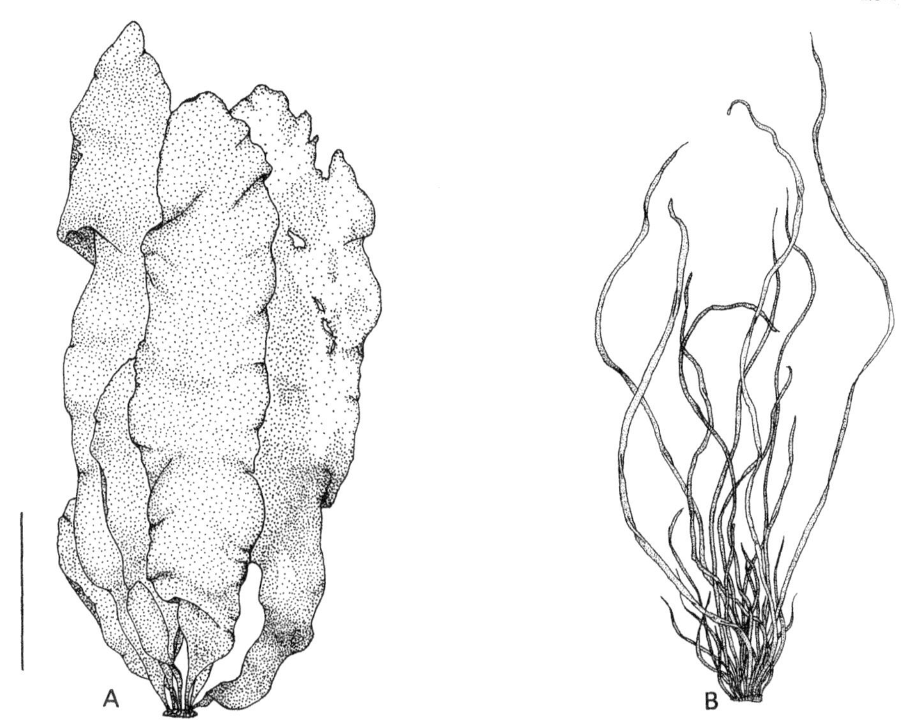

Fig. 61 *Petalonia fascia* A. Habit of plant. *Petalonia filiformis* B. Habit of plant. Bar = 20 mm.

Isolates of *P. fascia* have been investigated in laboratory culture throughout the world including the North Atlantic (Reinke, 1878a; Sauvageau, 1929—as *Phyllitis debilis* Kützing; Kylin, 1933; Dangeard, 1962b, 1963b; Caram, 1965; Edwards, 1969; Hsiao, 1969; Rhodes & Connell, 1973; Fletcher, 1974b; Roeleveld *et al.*, 1974), the Pacific coast of North America (Wynne, 1969; Wynne, 1972a, b) and Japan (Yendo, 1919; Kunieda & Arasaki, 1947; Nakamura & Tatewaki, 1975). In the great majority of these studies a 'direct' type of life history was shown, with the plurispores released from the erect blades germinating directly without sexual fusion to form prostrate systems, from which sprouted further generations of erect blades. Considerable variation was shown in the morphology of these prostrate systems, which were *Streblonema*-like, *Myrionema*-like and/or *Ralfsia*-like. As an adjunct to these studies there have been reports of *Ralfsia*-like crusts collected in the field which, when investigated in laboratory culture, produced erect blades identifiable as *P. fascia* (Wynne, 1969; Edelstein, *et al.*, 1970; Rhodes & Connell, 1973; Fletcher, 1974b, 1978; Roeleveld *et al.*, 1974). Three species of '*Ralfsia*' have to date been implicated in the life history of *P. fascia,* viz. *R. californica* Setchell & Gardner (Wynne, 1969), *R. clavata* (= *Stragularia clavata* here) (Edelstein *et al.*, 1970, Roeleveld, *et al.*, 1974; Fletcher 1974b, 1978) and *R. bornetii* Kuckuck (included within *S. clavata* here) (Edelstein *et al.*, 1970). This would suggest that at least two entities are involved in our concept of *P. fascia* each with its own distinct '*Ralfsia*' phase. However, for the present I would prefer to consider the three above crustose entities as variants

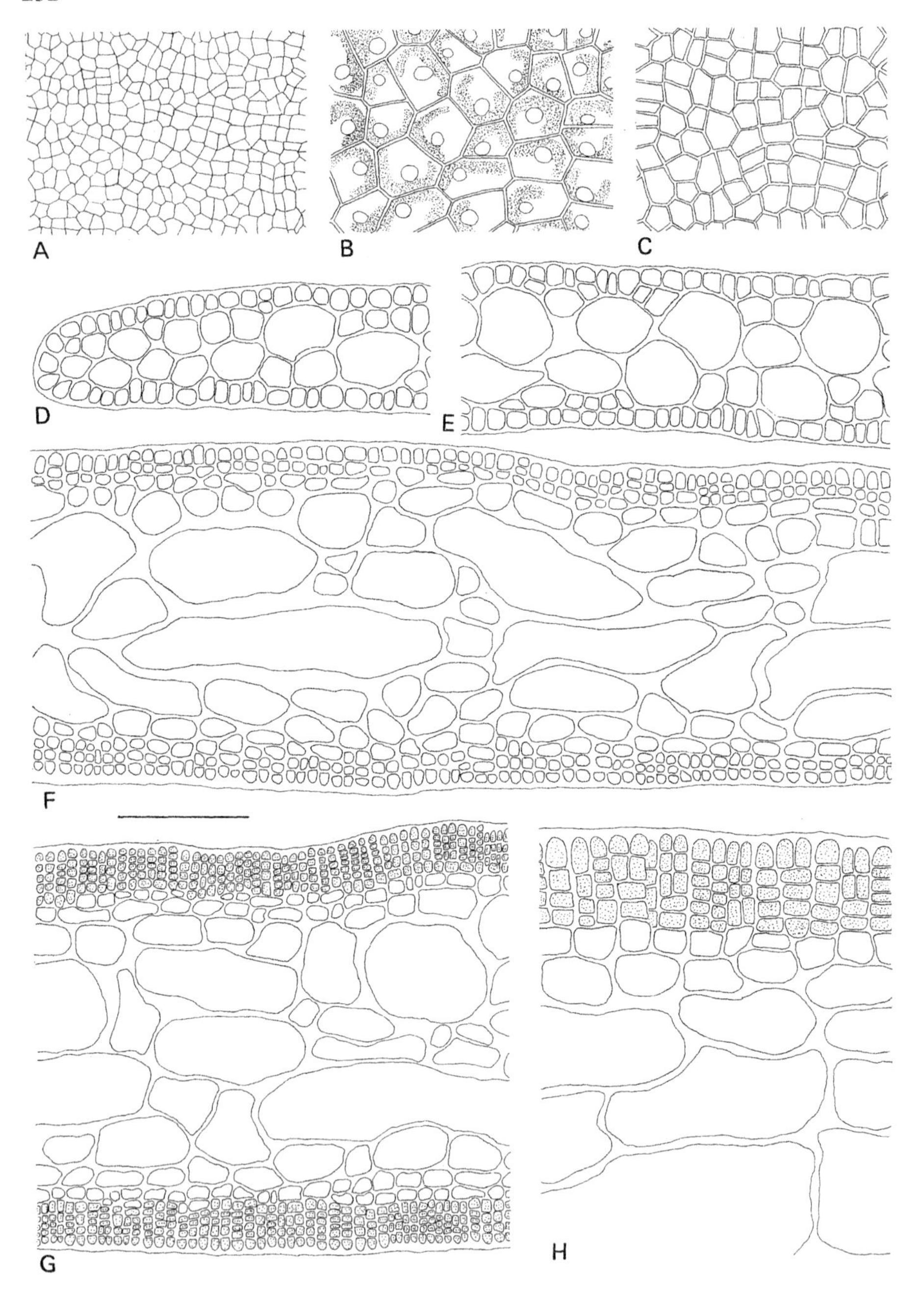
A
B
C
D
E
F
G
H

of a single species for which the name *R. clavata* would have priority (transferred to *Stragularia clavata* in the present treatment). *P. fascia* can, therefore, be considered to represent a range of genotypes with slightly differing crustose bases. Certainly *P. fascia* is polymorphic with a number of 'forma' and 'varieties' previously attributed to it. Work by Wynne (1969) further indicated that these morphological variants are probably genotypically determined. He found that only the narrow, astipitate blades (forma *fascia*) were capable of producing crusts in culture bearing unilocular sporangia. Blades of forma *caespitosa* produced only sterile crusts whereas in cultures of forma *debilis* no thickened crusts were seen at all. Note also that the prostrate systems produced in cultures of *P. fascia* by Edwards (1969) were almost exclusively filamentous.

A significant feature of the above described 'direct' type life histories was the facultative relationship observed between the prostrate and erect bladed expressions. Plurispores derived from the plurilocular sporangia on the erect-bladed expression and unispores derived from unilocular sporangia on the crustose expression could inherently undergo identical processes of germination and development and produce the same range of morphological expressions. Indeed environmental conditions such as the combination of different temperatures and photoperiods (Edwards, 1969; Wynne, 1969; Fletcher, 1974b, 1978; Roeleveld *et al.*, 1974; Nakamura & Tatewaki, 1975) and the nutrient level (Hsiao, 1969) determined the ratio between the erect bladed and crustose thalli. In general low temperatures and short day photoperiods favoured the formation of erect blades whilst high/warm temperatures and long day photoperiods favoured prostrate thallus development. These studies generally lend support to the predominantly winter/spring seasonal occurrence of the erect blades in the North Atlantic and suggest a summer seasonal growth of the basal systems (i.e. crusts). Studies by Wynne (1969) and Nakamura & Tatewaki (1975) further suggest that a period of low temperatures and short days are then necessary for unilocular sporangia development on the crusts; certainly this would agree with the winter fertility observed on crusts of *S. clavata* around the British Isles.

It may well be that this 'direct' life history revealed for *P. fascia* is 'abbreviated' and that the basic life history involves a heteromorphic alternation between an erect bladed (i.e. *Petalonia*-like) gametophyte and a crustose (i.e. *Ralfsia*-like) sporophyte. Such a heteromorphic haplodiplontic life history might have remained undetected because of its rarity (perhaps requiring specific environmental conditions) or geographical isolation. Pertinent to this suggestion has been the report of copulation between plurispores derived from erect blades of *P. fascia* in Japan (Kunieda & Arasaki, 1947). Such a life history with a fairly strict sexual, heteromorphic alternation was also shown for *P. zosterifolia* by Nakamura & Tatewaki (1975) in Japan; however, a 'direct' life history similar to that shown for *P. fascia* was also accomplished by the parthenogenetic development of the plurispores derived from the erect blades, with culture conditions of temperature and photoperiod determining the ratio of formation of the erect bladed or crustose expressions.

Fig. 62 *Petalonia fascia*
A–B. Surface view of vegetative thallus. Note cells with single plate-like plastid and pyrenoid. C. Surface view of fertile thallus showing plurilocular sporangia. D. T.S. of thallus margin. E–F. T.S. of vegetative thalli. Note outer small cortical cells and inner large medullary cells. G–H. T.S. of fertile thallus showing outer vertical rows of plurilocular sporangia. Bar = 20 μm (B–C, H), 50 μm (others).

Reinke (1878a), pp. 262–267, pl. 11, figs 1–2; Batters (1883), p. 113; Yendo (1919), pp. 171–184, pl. II, figs 1–14; Sauvageau (1929), pp. 316–330, figs 8–10—as *Phyllitis debilis* Kützing; Newton (1931), p. 176, fig. 110; Hamel (1931–39), pp. 197–198; Kylin (1933), pp. 44–47, fig. 17; Rosenvinge & Lund (1947), pp. 31–34, fig. 10; Feldmann (1949), pp. 103–115, fig. 1C–G; Gayral (1958), p. 212, fig. 29, pl. XXVI; Taylor (1957), pp. 167–168, pl. 14, fig. 5, pl. 15, fig. 3; Lindauer *et al.*, (1961), pp. 255–256, fig. 64; Dangeard (1962b), pp. 3290–3292, pl. 1; Dangeard (1963b), pp. 48–57, 84, pl. VII–VIV; Caram (1965), pp. 86–98, figs 1–8; Nakamura (1965), pp. 109–110, fig. 5; Gayral (1966), p. 249, pl. XXXV; Dangeard (1969), pp. 60–70, fig. 1B, pl. XI–XII; Edwards (1969), pp. 85–87, figs 38–42; Hsiao (1969), pp. 1611–1616, figs 1–4; Wynne (1969), pp. 17–31, figs 6–8, pls. 6–13; Edelstein *et al.*, (1970); Hsiao (1970), pp. 1359–1361, figs 1–2; Wynne (1972a) pp. 133–136; Wynne (1972b), pp. 137–141, figs 24–27; Rhodes & Connell (1973), pp. 211–215, figs 1–10; Fletcher (1974a), pp. 92–108, figs 17–19, pls. 26–30; Roeleveld *et al.*, (1974), pp. 410–426, figs 1–7; Nakamura & Tatewaki (1975), pp. 72–76, figs 11–14; Abbott & Hollenberg (1976), p. 200, fig. 163; Kornmann & Sahling (1977), pp. 134 & 137, fig. 71A–C, fig. 72A–F; Rueness (1977), p. 168, fig. 100, pl. XXII (2); Fletcher (1978), pp. 371–398, various figs; Fletcher (1980), p. 42, pl. 15, figs 1–4; Stegenga & Mol (1983), p. 105, pl. 38, figs 1–5.

Petalonia filiformis (Batters) O. Kuntze (1891–98), p. 419. Figs 61B, 63; Pl. 10a, b

Phyllitis filiformis Batters (1888), p. 451.

Plants forming erect, light to yellow brown, gregarious, slightly dorsiventrally flattened to oval blades arising from a fibrous mat of rhizoidal filaments; erect blades simple, solid or with small central cavities, thin and flaccid, narrow, filiform and ribbon-like, entire, without proliferations or obviously ruffled margins, frequently spirally twisted, to 4–6 (–8) cm long × 100–350 (–400) µm wide, tapering slowly to base, usually terminally eroded; in surface view cells small, rectangular to polygonal, 4–16 × 4–10 µm arranged in straight or spiral longitudinal, less obviously transverse rows, each enclosing a single, plate-like, parietal plastid with conspicuous pyrenoid; in section blades to 75 µm (vegetative) and 130 µm (reproductive) thick comprising an inner medulla of 2–4 (–6) large, thick-walled, longitudinally elongated, transversely rounded sometimes irregularly collapsed, colourless cells enclosed by an outer cortex of 1–3 layers of small, pigmented cells; hairs not observed on field material.

Plurilocular sporangia in slightly darker coloured sori extensive in terminal blade region, in closely packed vertical columns arising from cortical cells, uniseriate, to 30 µm (–7 loculi) long × 3–6 µm in diameter, with subquadrate to rectangular loculi, without paraphyses; unilocular sporangia not known on blade surface but recorded on *Compsonema*-like thalli.

Epilithic, exposed on the sides of vertical cliff faces, large boulders, etc., in the littoral fringe.

Fig. 63 *Petalonia filiformis*
A. Portion of narrow blade. B. Surface view of vegetative thallus showing single plate-like plastid with pyrenoid. C. Surface view of fertile thallus showing plurilocular sporangia. D–F. T.S. of vegetative thalli showing outer small cortical and inner large medullary cells. Note central cavities. G–H. T.S. of fertile thalli showing outer vertical rows of plurilocular sporangia. Bar = 20 µm (B–C), 50 µm (others).

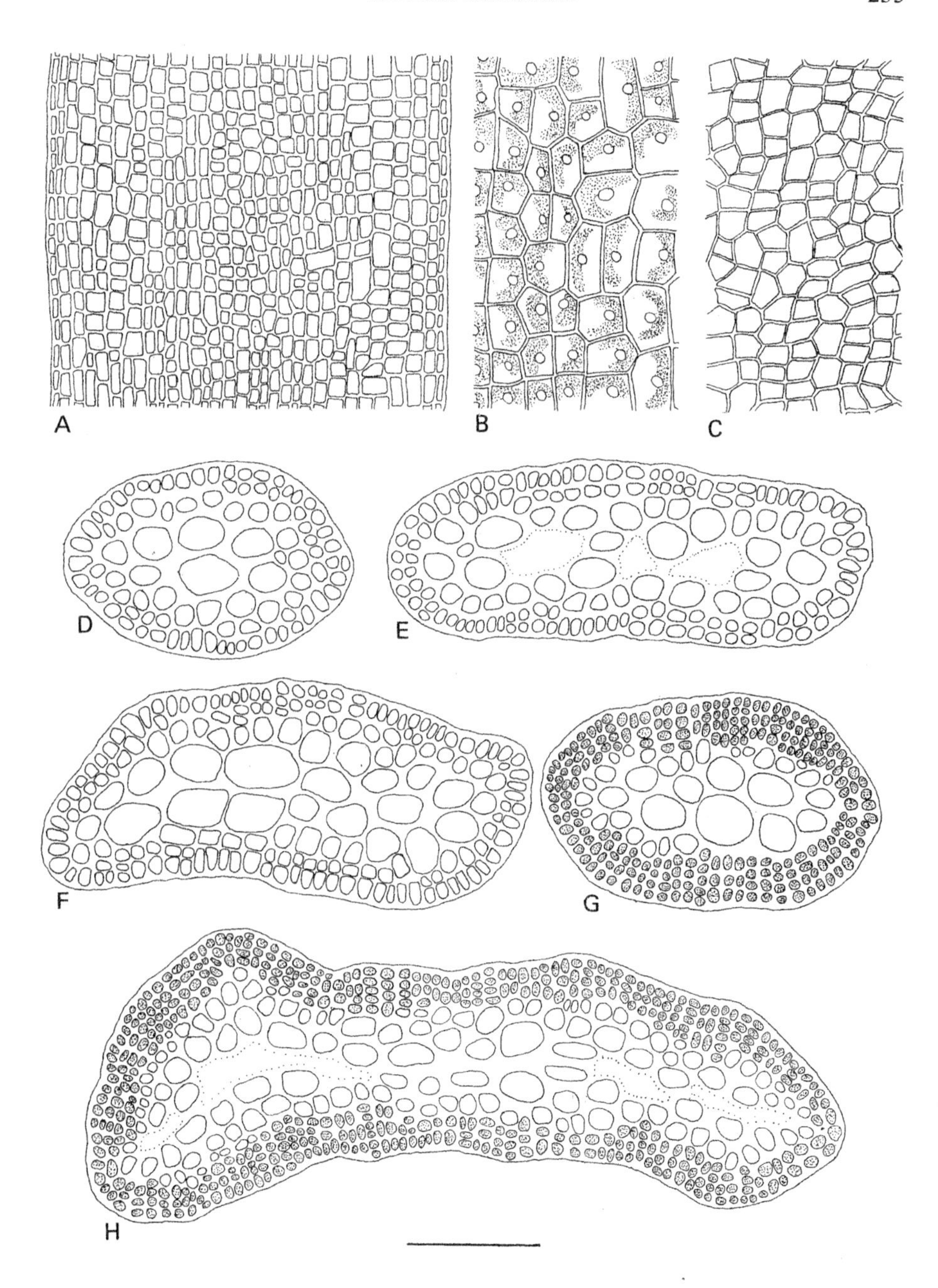
A
B
C
D
E
F
G
H

Apparently confined to south eastern and eastern shores of England and Scotland extending west to Sussex and north to Fife (Fife, Northumberland, Yorkshire, Essex, Kent, Sussex).

Blades annual, January to May, with plurilocular sporangia.

P. filiformis is a fairly uncommon inhabitant of the upper littoral region, where it forms isolated, sometimes dense, stands on the vertical faces of cliffs, harbour walls, breakwaters, large boulders, etc. in moderately exposed areas. It occurs more commonly in north facing positions, and can often be found at the entrance to caves, crevices, etc. Collections have been made from a wide variety of substrata including chalk (Kent), sandstone (Northumberland, Fife) and both concrete and brick-built artificial sea walls (Kent). The gregarious, light to yellow brown thalli contribute to a generally low-lying sward of algae which can include *Ulothrix flacca, Urospora penicilliformis, Blidingia minima, Enteromorpha* spp., *Pilayella littoralis, Callithamnion hookeri, Gelidium pusillum* and *Polysiphonia urceolata,* often with underlying crusts of *Petroderma maculiforme* and *Ralfsia verrucosa* present. The only reports of this species occurring outside the British Isles have been by Munda (1979) for Iceland and Kuckuck (1897) for Helgoland.

Culture studies by Fletcher (1981a) indicated that a 'direct' life history was operating; the plurispores from the erect blades behaved asexually and germinated into branched, filamentous or knot-filamentous bases, from which sprouted a further generation of blades. These prostrate systems further developed unilocular sporangia, which were usually accompanied by erect, multicellular, paraphysis-like filaments; these fertile growths were considered by Fletcher to show a marked similarity to the brown alga *Compsonema saxicolum* (see p. 222). This facultative relationship between the erect-bladed (i.e. *Petalonia*) expression and the microthallial (i.e. *Compsonema*) expression was also revealed by Fletcher (unpublished) to be influenced by environmental conditions. For example blade production and development was more enhanced in low temperature/short day conditions whilst microthallus development was enhanced in warm temperature/long day conditions. These results are in agreement with the observed winter/spring seasonal occurrence of the erect blades and the survival of the plants as *Compsonema*-like microthalli during the more unfavourable summer and autumn periods (Fletcher, 1974a).

The life history of *P. filiformis* can, therefore, be described as asexual, heteromorphic and monophasic with both expressions probably having the same ploidy level. This may well be, however, an abbreviated life history. More culture studies of isolates are required to determine if *Ralfsia*-like thalli are additionally present, as in *P. fascia* and whether a sexual cycle is also operating albeit of rare occurrence.

Batters (1888), pp. 451–452, pl. 18, figs 1–6; Newton (1931), p. 176; Fletcher (1974a), pp. 109–127, figs 20–23; Fletcher (1981a), pp. 103–104.

Petalonia zosterifolia (Reinke) O. Kuntze (1891–98), p. 419. Figs 64, 65; Pl. 10c

Phyllitis zosterifolia Reinke (1889b), p. 61.

Plants forming erect, medium to dark brown, single or more commonly gregarious, dorsiventrally flattened blades arising from an intricate mat of basally produced, branched, rhizoidal filaments; erect blades simple, solid or with central cavities, fairly

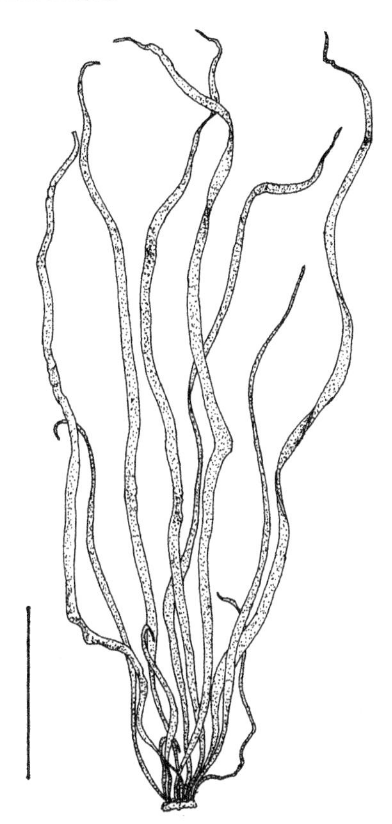

Fig. 64 *Petalonia zosterifolia*
Habit of plant. Bar = 30 mm.

thin and flaccid, more or less linear, entire, without proliferations or ruffled margins, to 15 (–25) cm long × 0·5–2 (–4) mm wide, tapering slowly to base, usually terminally eroded; in surface view cells small, rectangular to polygonal, 6–18 × 4–11 μm, arranged in longitudinal, less obviously transverse rows, each enclosing a single, plate-like parietal plastid with conspicuous pyrenoid; in section blades to 140 μm (vegetative) and 210 μm (reproductive) thick comprising an inner medulla of 4–6 (–8) large, thin walled, longitudinally elongate, transversely rounded, often collapsed, colourless cells enclosed by an outer cortex of 1–3 small, pigmented cells; hairs common, single, arising from surface cells with basal meristem and sheath.

Plurilocular sporangia in dark-brown sori extensive in terminal blade region, in closely packed vertical columns arising from cortical cells, uniseriate, to 40 μm (–10 loculi) long × 3–6 μm wide, with subquadrate to rectangular loculi, without paraphyses; unilocular sporangia not recorded for the British Isles, reported by Dangeard (1962a, 1963b) in France on basal microthalli in culture.

Epilithic, littoral fringe to upper eulittoral, on the sides of vertical cliff faces, platforms, large boulders, etc. in moderately to very exposed areas. Also not uncommon on the upper wave-washed regions of exposed floating structures such as buoys.

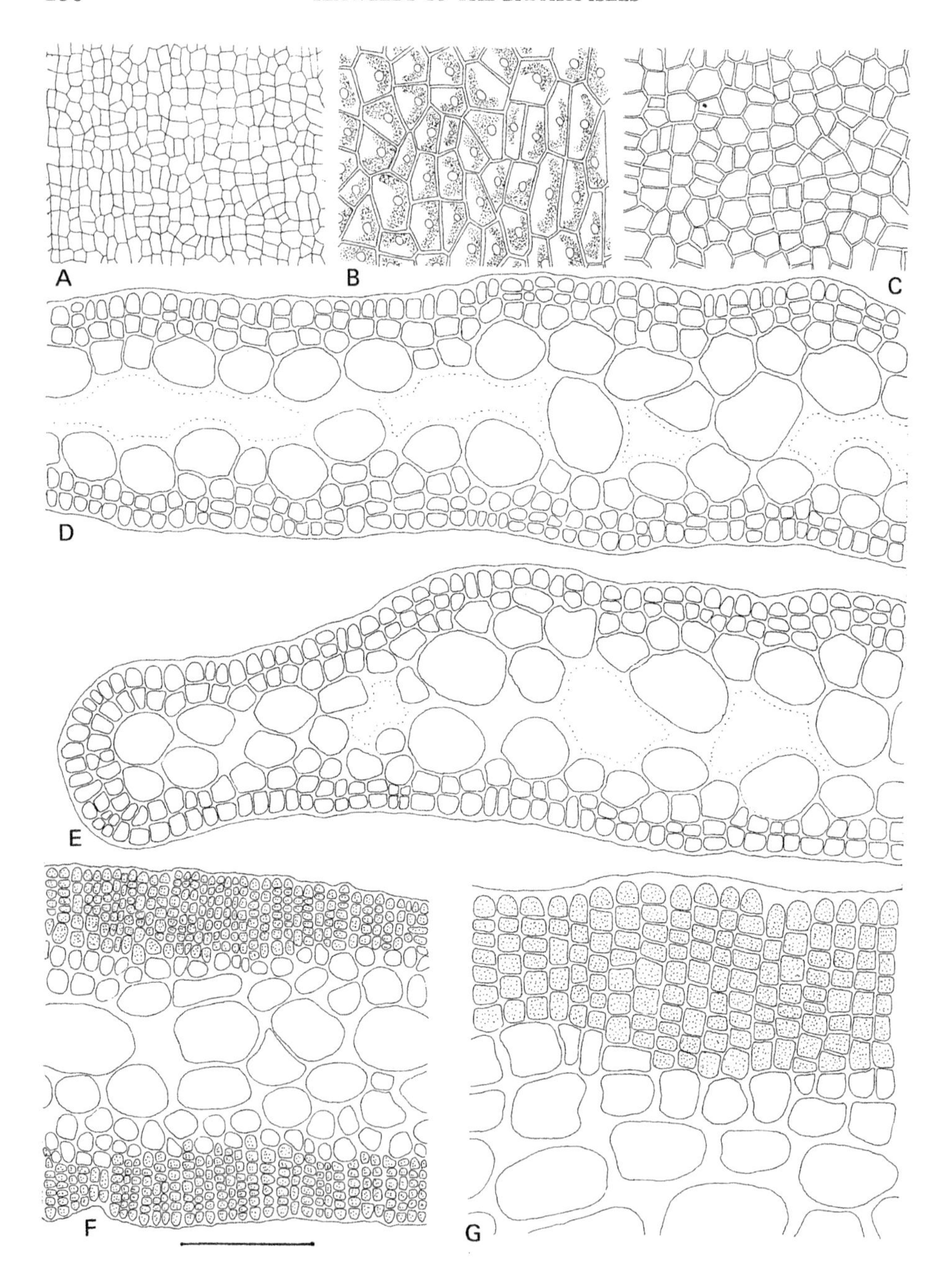
A
B
C
D
E
F
G

Recorded for scattered localities all around the British Isles but more common on the south west and west coasts (Northumberland, Hampshire, Dorset, Devon, Cornwall, Pembroke, Anglesey, Isle of Man).

Blades annual, December to March with plurilocular sporangia.

Cultures studies by Fletcher (1974b) on isolates of this species in Dorset revealed that a 'direct' type life history was operating. The plurispores released from the blades behaved asexually and germinated directly into branched, filamentous microthalli from which sprouted a further generation of erect macrothalli. *Ralfsia*-like crustose thalli similar to those present in the life history of *P. fascia* were not observed. A similar life history was reported for the French North Atlantic coast by Dangeard (1962a, 1963b) although in his cultures the microthalli became fertile with both plurilocular and unilocular sporangia. The life history of these two European isolates can, therefore, be described as heteromorphic and monophasic.

However, in Japan, Nakamura & Tatewaki (1975) reported a sexual life history with an erect bladed (i.e. *Petalonia*) gametophyte bearing plurilocular sporangia, alternating with a *Ralfsia*-like sporophyte bearing unilocular sporangia. Plurispores released from the blades behaved as isogametes and the resultant zygote germinated into *Ralfsia*-like thalli bearing unilocular sporangia in which meiosis occurred. This cycle can, therefore, be described as heteromorphic and haplodiplontic. However, an abbreviated cycle also occurred involving parthenogenetic development of the plurispores. The development of these parthenogametes varied according to the culture conditions of temperature and daylength; under cool, long-day conditions erect blades were produced whilst under warm, long-day conditions crusts or tufts were produced. It is interesting to note that these culture results with parthenogametes are not unlike those obtained for European isolates. More culture studies are, therefore, required on North Atlantic isolates to determine whether an additional sexual life history is operating similar to that reported in the North Pacific.

Newton (1931), p. 176; Hamel (1931–39), pp. 198–199; Rosenvinge & Lund (1947), pp. 34–37, fig. 11; Dangeard (1962a), pp. 1895–1896, pl. 1; Dangeard (1963b), pp. 12–47, pls 14–16; Nakamura & Tatewaki (1975), pp. 65–71, figs 6–10, pl. 2; Dangeard (1969), pp. 60–70, fig. 1A; Kornmann & Sahling (1977), p. 137, fig. 73; Rueness (1977), pp. 168–169; Fletcher (1980), pp. 42–43, pl. 15(5); Stegenga & Mol (1983), p. 105, pl. 38, figs 6–7.

RALFSIA Berkeley

RALFSIA Berkeley in Smith & Sowerby (1843), suppl. III, pl. 2866, 1 Nov. 1841.

Type species: *R.deusta* (C. Agardh) Berkeley (= *R. fungiformis* (Gunnerus) Setchell & Gardner (1924), p. 11).

Fig. 65 *Petalonia zosterifolia*
A–B. Surface view of vegetative thallus. C. Surface view of fertile thallus showing plurilocular sporangia. D–E. T.S. of vegetative thallus showing outer small cortical cells and inner large medullary cells. Note central cavities. F–G. T.S. of fertile thallus showing outer vertical rows of plurilocular sporangia. Bar = 20 μm (B–C, G), 50 μm (others).

Thallus encrusting, discrete and orbicular, becoming confluent and irregular, often superimposed in vertical section, firm, glossy, coriaceous, smooth, becoming bullate and verrucose in older thalli, loosely or moderately attached to the substratum by under-surface, with or without rhizoids; comprising a monostromatic basal layer of outwardly spreading, frequently branched, firmly united filaments, with terminal, synchronously growing, apical cells, giving rise immediately behind to firmly united, branched filaments, at first almost prostrate, giving the appearance of a polystromatic base, later becoming erect, remaining fairly short and covered by a thick surface cuticle; plastids single, plate-like without obvious pyrenoid; hairs infrequent, tufted, in depressions, with basal meristem, lacking obvious sheath.

Unilocular and plurilocular sporangia borne on the same or different thalli, in raised, discrete or spreading, gelatinous sori; unilocular sporangial sori common, discrete, prominent and wart-like, very gelatinous, with sporangia pyriform, sessile or pedicellate at the base of clavate, multicellular, simple paraphyses; plurilocular sporangial sori infrequent, spreading, slightly raised and gelatinous, with sporangia in closely packed uniseriate, vertical columns terminal on erect filaments, without paraphyses, with terminal sterile cell and enclosed within a surface cuticle.

Since the beginning of this century five species of the genus *Ralfsia* have been recognised for the British Isles (Batters, 1902; Newton, 1931; Parke & Dixon, 1976). These are *R. clavata, R. disciformis, R. pusilla, R. spongiocarpa* and *R. verrucosa.* In his investigation of the marine algae of Berwick-on-Tweed Batters (1890, pp. 286–289) identified three species, viz. *R. clavata, R. spongiocarpa* and *R. verrucosa* which he classified within two subgenera: *Euralfsia* which had delimited sori and curved erect filaments and *Stragularia* originally described as a genus by Strömfelt (1886) with *S. adhaerans* (= *S. clavata*) as the type, which had continuous sori and vertically orientated filaments. He placed *R. verrucosa* in *Euralfsia* and *R. clavata* and *R. spongiocarpa* in *Stragularia*. This subdivision of the genus *Ralfsia* has not been widely accepted (see Taylor, 1957; Parke & Dixon, 1976; Abbott & Hollenberg, 1976; Rueness, 1977); indeed Loiseaux (1968, p. 306) concluded on the basis of culture work that there was no justification for maintaining a distinction between *Euralfsia* and *Stragularia*.

The division of *Ralfsia* into two subgenera was, however, accepted by Newton (1931, p. 152) although unlike Batters (1902, p. 42) she included *R. clavata* and *R. spongiocarpa* with *R. verrucosa* in *Euralfsia*; the two remaining species, *R. disciformis* and *R. pusilla,* were placed in *Stragularia*.

In the present work it is concluded that the five species of *Ralfsia* previously described for the British Isles form a rather disparate group of entities which can be usefully separated using the characters originally proposed by Batters in his subgeneric classification. In disagreement with Newton, it is considered that only *R. verrucosa* possesses *Euralfsia* features whilst the remaining four species are better assigned to *Stragularia*. Furthermore it is proposed that these distinguishing characters are sufficient taxonomic criteria to warrant readoption of *Stragularia* as a genus, following the treatments of Kjellman (1893), Weber-van Bosse (1913) and Hamel (1931–39, p. XXXI). Additional evidence for the separation of *Stragularia* spp from *Ralfsia sensu stricto* is provided by the results of some culture studies which indicate that *Stragularia,* unlike *Ralfsia,* comprise a group of crustose entities connected in life history with various erect thalloid members of the Scytosiphonaceae (see pages 253).

One species in the British Isles:

Ralfsia verrucosa (Areschoug) J. Agardh (1848–76), p. 62. Figs 59B, 66

Cruoria verrucosa Areschoug (1843), p. 264.

Thallus epilithic, encrusting, comparatively thick, olive to dark brown, individual and orbicular or more frequently confluent and indefinite in outline to 5 (–10) cm in extent, loose to moderately well attached to substrata by underside, usually without rhizoids; peripheral region olive brown, smooth, coriaceous, quite closely adherent, but with margins usually quite easily raised intact, central regions dark brown, verrucose, bullate, brittle, loosely attached with rust-red coloured underside, frequently with young superimposed thalli; in surface view cells rounded or polygonal, 4–9 μm in diameter, each with a single plate-like plastid without obvious pyrenoid; thallus pseudoparenchymatous in structure, comprising a monostromatic basal layer of outwardly spreading, frequently branched, laterally united filaments, with terminal synchronously growing apical cells, giving rise immediately behind to strongly united and not readily separable branched filaments, at first prostrate, giving the appearance of a polystromatic base, later becoming erect, remaining fairly short, tapering slightly, covered terminally by a thick cuticle; thalli not usually exceeding 14 (–20) cells, 120 (–210) μm thick, superimposed thalli up to 1 mm thick; erect filament cells quadrate to rectangular, 4–11 × 5–8 μm, each with a single, terminal, plate-like plastid without obvious pyrenoid; hairs infrequent, grouped, emerging from depressions, with basal meristem, lacking obvious sheath.

Unilocular and plurilocular sporangia usually borne on separate, rarely on the same plants, terminal on erect filaments in raised, gelatinous sori; unilocular sporangial sori yellow brown, common, discrete, prominent and wart-like, spongy, very gelatinous, often concentrically zoned just behind growing margin, comprising obovate to pyriform unilocular sporangia, 60–105 × 15–37 μm, sessile or on 1–2 celled pedicels, lateral at the base of multicellular paraphyses; paraphyses simple, clavate, 7–11 (–14) cells, to 115 (–150) μm long, comprising rectangular cells 1–2 diameters long 7–12 × 6–7 μm in mid/upper regions usually with an enlarged subcylindrical to subpyriform apical cell, 10–20 × 6–8 μm, much narrower, 1–4 (–5) diameters long, 12–18 × 3–5 μm below; plurilocular sporangial sori infrequent, slightly raised and gelatinous, spreading, with plurilocular sporangia in closely packed vertical columns terminal on erect filaments, without paraphyses, to 135 μm long (–27 loculi) long, uniseriate, rarely biseriate, 6–8 μm in diameter, with subquadrate loculi, and terminal, dark coloured sterile cell embedded in thick cuticular layer.

Epilithic, epizoic, from the littoral fringe to the lower eulittoral, in pools and emergent, especially in damp places.

Abundant and widely distributed all around the British Isles.

Plants perennial; unilocular sporangia widely recorded throughout the year, although more common in the winter, November to March; plurilocular sporangia only recorded from scattered localities (Berwick, Northumberland, Hampshire, Dorset, Devon, Cornwall and Isle of Man), and without any obvious seasonal distribution.

R. verrucosa is the most common crustose brown alga recorded in the littoral region around the British Isles. It occurs on very exposed to very sheltered shores over the whole intertidal range, usually in pools but quite frequently exposed on rock, particularly in damp situations, e.g. under *Fucus* cover, in cracks and crevices, on soft, water-retaining, substrata such as chalk, etc. It occurs on a wide variety of substrata, both natural and

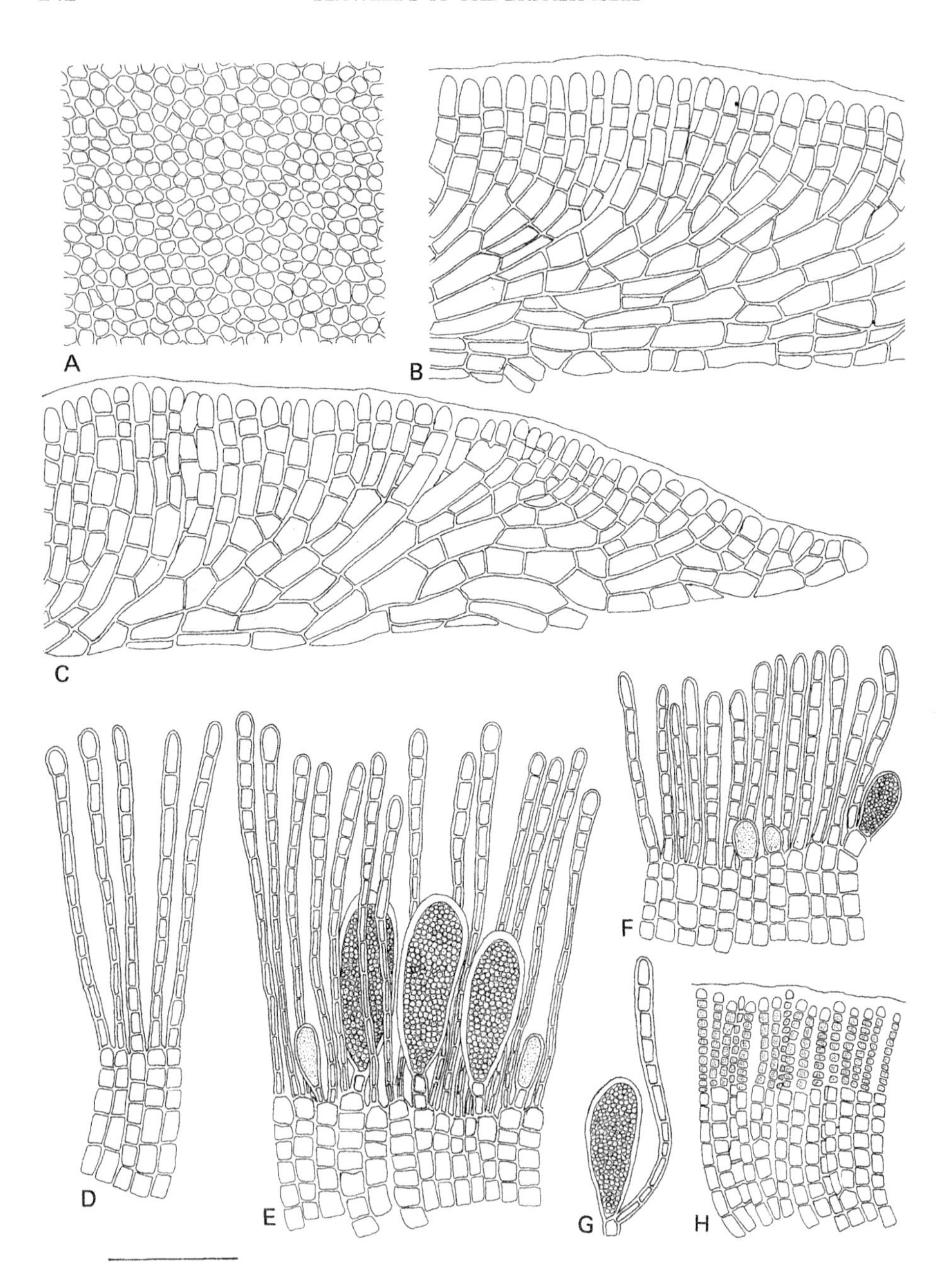
A
B
C
D
E
F
G
H

man-made, and is commonly epizoic on barnacles, limpets, mussels and top shells; it has never, however, been observed as an epiphyte.

It can be distinguished from the other crustose brown algae recorded for the British Isles by the following combination of characteristics: it forms moderately well attached, leathery, verrucose crusts common in the littoral region only, abundantly distributed all around the British Isles; in radial tangential section the erect filaments are markedly curved as they arise from the basal layer, almost prostrate at first, giving the appearance of a polystromatic base; the erect filaments are tightly bound together and can only be squashed apart with difficulty; cells contain a single plate-like plastid without obvious pyrenoid; the unilocular sporangia occur in discrete, raised sori on crust surface, and are lateral at the base of 7–14 celled, clavate paraphyses; the plurilocular sporangia occur in slightly raised, spreading sori on crust surface, and form closely packed, uniseriate, vertical columns with a sterile terminal cell.

Culture studies have only been carried out on North Atlantic isolates of *R. verrucosa* by Loiseaux (1968), Edelstein *et al.* (1971) and Fletcher (1974b). The life history appears to be of the 'direct' type with spores derived from both unilocular and plurilocular sporangia usually directly producing crusts similar to those of the parent. The cycle appears to be apomeiotic and without any indication of a sexual process. There is no evidence that *R. verrucosa* has a heteromorphic life history similar to that shown by some members of the closely related genus *Stragularia*.

Areschoug (1843), p. 264, pl. 9, figs 5–6; J. Agardh (1848–76), p. 62, fig. 212, pl. 40; Reinke (1889a), pp. 9–10, pl. 5, figs 1–13; Setchell & Gardner (1925), pp. 497–498; Newton (1931), p. 153; Hamel (1931–39), pp. 106–108, fig. 26A–B; Tokida (1954), p. 82; Taylor (1957), p. 123, pl. 11, figs 1–2; Lindauer *et al.* (1961), pp. 211–212, fig. 38; Jaasund (1964), pp. 131–133, figs 2–3; Loiseaux (1968), pp. 297–302, figs 1–2; Edelstein *et al.* (1971), pp. 1249–1250; Nakamura (1972), pp. 152–153, fig. 3; Fletcher (1974a), pp. 67–77, figs 11–12, pls 17–22; Rueness (1977), p. 130, fig. 55; Kormann & Sahling (1977), pp. 118–120, fig. 62; Fletcher (1978), pp. 371–373 (various figures); Tanaka & Chihara (1980b), pp. 227–231, fig. 1A, fig. 2A–B; Stegenga & Mol (1983) p. 82, pl. 28, figs 3–6.

Excluded species

Ralfsia disciformis Crouan frat. (1867), p. 166.

This species was described by the Crouan brothers from material dredged at Brest. The description was inadequate and, as Batters (1896) pointed out, comes very near to that of *Ralfsia clavata* (= *Stragularia clavata* in the present text), from which it differs by the shorter, 1–2 celled, less clavate paraphyses which are hardly any longer than the unilocular sporangia. The record for the British Isles is based on material dredged from

Fig. 66 *Ralfsia verrucosa*
A. Surface view of crust cells. B. R.V.S. of thallus showing upwardly curving laterally adjoined filaments. C. R.V.S. of thallus margin. D–G. V.S. of fertile thalli showing erect filaments giving rise terminally to paraphyses and unilocular sporangia. H. V.S. of fertile thallus showing terminal rows of plurilocular sporangia. Note sterile cap cell. Bar = 50 µm.

deep water near the mouth of the river Yealm and in Plymouth Sound (Batters 1896). Examination of material deposited in the British Museum (Nat. Hist.) reveal 2–4 celled paraphyses and unilocular sporangia with size ranges within those given for *S. clavata* in the present work; *R. disciformis* is, therefore, excluded from the British flora.

Ralfsia pusilla (Strömfelt) Batters (1892a), p. 174.

R. pusilla was originally described (as *Stragularia pusilla*) by Strömfelt (1888) for the coast of Norway. It was first reported for the British Isles in Cumbrae, Argyll by Batters (1892a) and subsequently recorded for Dorset, Northumberland and Orkey. Examination of collections of the above material, deposited in the British Museum (Nat. Hist.) confirmed the identity of *Ralfsia*-like thalli (except for the Cumbrae specimen) but they appeared to show a marked similarity to *Stragularia clavata*. The crusts were also much larger than those of *S. pusilla* described by Strömfelt and different in both the number and shape of cells in the paraphyses. For example Strömfelt's (1888) fig. 4 showed paraphyses with 9–13 cells which were c. 1–2 diameters long, compared to only 4–7 cells, which were markedly elongate below, in material identified as this species collected in the British Isles. Clearly the situation is unsatisfactory and it seems best to exclude this species from the British flora pending further studies.

SCYTOSIPHON C. Agardh

SCYTOSIPHON C. Agardh (1820), p. 160 (*nom. cons.*).

Type species: *S. filum* var. *lomentaria* (Lyngbye) C. A. Agardh (= *Chorda lomentaria* Lyngbye; *S. lomentaria* (Lyngbye) Link (1833), p. 233 typ. cons.; See I.C.B.N. (1961), p. 213.

Plants consisting of erect, yellow to dark brown, tubular thalli arising from a slightly expanded encrusting base, small discoid holdfast or fibrous mat of rhizoidal filaments; thalli single or gregarious, simple, hollow, cylindrical, inflated or slightly dorsiventrally flattened, flaccid or slightly rigid, with or without regular constrictions on mature thalli, tapering slowly to base, more sharply to apex; in surface view cells small, rectangular to polygonal, irregularly arranged, each with a single, plate-like parietal plastid with 1–2 pyrenoids; in section thalli parenchymatous, hollow, comprising an inner medulla of 2–5 layers of large, thick-walled, longitudinally elongate, transversely rounded or irregularly shaped colourless cells, enclosed by an outer cortex of 1–3 layers of small, pigmented cells; hairs single or grouped arising from surface cells or depressions.

Plurilocular sporangia in continuous sori, extensive in terminal thallus region, in closely packed vertical columns, arising by division of cortical cells, uni- to biseriate, rarely multiseriate, with subquadrate to rectangular loculi, with or without unicellular ascocyst-like cells; unilocular sporangia not known on thallus surface but reported on crustose *Microspongium* or *Compsonema*-like thalli.

The genus *Scytosiphon* is characterised by an erect, simple, hollow, sometimes constricted thallus with small surface cells, each with a single plate-like plastid with pyrenoids, a parenchymatous mode of construction with an inner medulla of large colourless cells and an outer cortex of small pigmented cells, and plurilocular sporangia only which occur in densely packed, vertical columns extensive in the terminal regions. In the British Isles this genus is similar in habit to *Asperococcus* (in particular *A. fistulosus*), and small plants of *Chorda*, in particular *Chorda filum*. However, it can easily be distinguished from these genera using a combination of the above described characteristics.

Life history studies on *Scytosiphon* throughout the world have generally revealed the occurrence of an asexual 'direct' type life history. This has been reported in *S. dotyi* (Wynne, 1969), *S. complanatus* (Pedersen, 1980) and *S. lomentaria* (by numerous authors but see especially Dangeard, 1963b; Wynne, 1969; Rhodes & Connell, 1973; Fletcher, 1974b; Dring & Lüning, 1975a & b; Kristiansen & Pedersen, 1979; Pedersen, 1980). The prostrate thalli from which the erect thalli sprouted varied considerably in their morphology and could be described as *Streblonema*-like, *Compsonema*-like, *Microspongium*-like and *Ralfsia*-like; they were also often reported to be fertile with plurilocular and/or unilocular sporangia. The life history could, therefore, be described as heteromorphic with both a prostrate and erect expression. Culture conditions, including temperature, photoperiod and light quality, determined the growth pattern of all germinating spores types and, therefore, the final morphological expression which was produced. In apparent contrast to this direct type life history have been the results of studies of *Scytosiphon* in the Pacific by Nakamura & Tatewaki (1975) and Clayton (1980a) in which a sexual cycle was reported, with the gametophytic phase being represented by *Scytosiphon* thalli and the sporophytic phase being represented by ralfsioid crusts. However a 'direct' asexual life history also occurred by the development of parthenogametes liberated from plurilocular sporangia on the erect thalloid phase. They proposed that this parthenogenetic life history would be in agreement with the 'direct' life history reported for isolates in other geographical regions.

Two species are recorded for the British Isles; *S. dotyi* Wynne and *S. lomentaria* (Lyngbye) Link. *S. dotyi* Wynne, a Pacific species, represents a new addition to the marine flora of the British Isles. *S. dotyi* is distinguished from *S. lomentaria* in the following points (see Wynne, 1969, p. 36): the thalli are relatively short and narrow, lacking constrictions; hairs are tufted, emerging from depressions; paraphyses are absent; it occurs in the littoral fringe, exposed on vertical rock faces; the life history is 'direct' without a crustose expression. However, the reliability of these characters is questionable and the two species may well be conspecific.

KEY TO SPECIES

Thalli to 40 cm long, flaccid in texture; mature thalli constricted at intervals, with single hairs and ascocysts present; widely distributed in the littoral, usually in pools; recorded throughout the British Isles . . . . *S. lomentaria*

Thalli not exceeding 14 cm in length, fairly rigid in texture; mature thalli unconstricted at intervals, with grouped hairs and without ascocysts present; recorded in littoral fringe only, exposed on rock; restricted to the east coast of England (Yorkshire and Kent) . . . . *S. dotyi*

Scytosiphon dotyi Wynne (1969), p. 34. Figs 67A, 68; Pl. 11c

S. lomentaria f. *complanatus* minor Setchell & Gardner (1925), p. 534.
S. lomentaria f. *tortilis* Yamada (1935), p. 12.

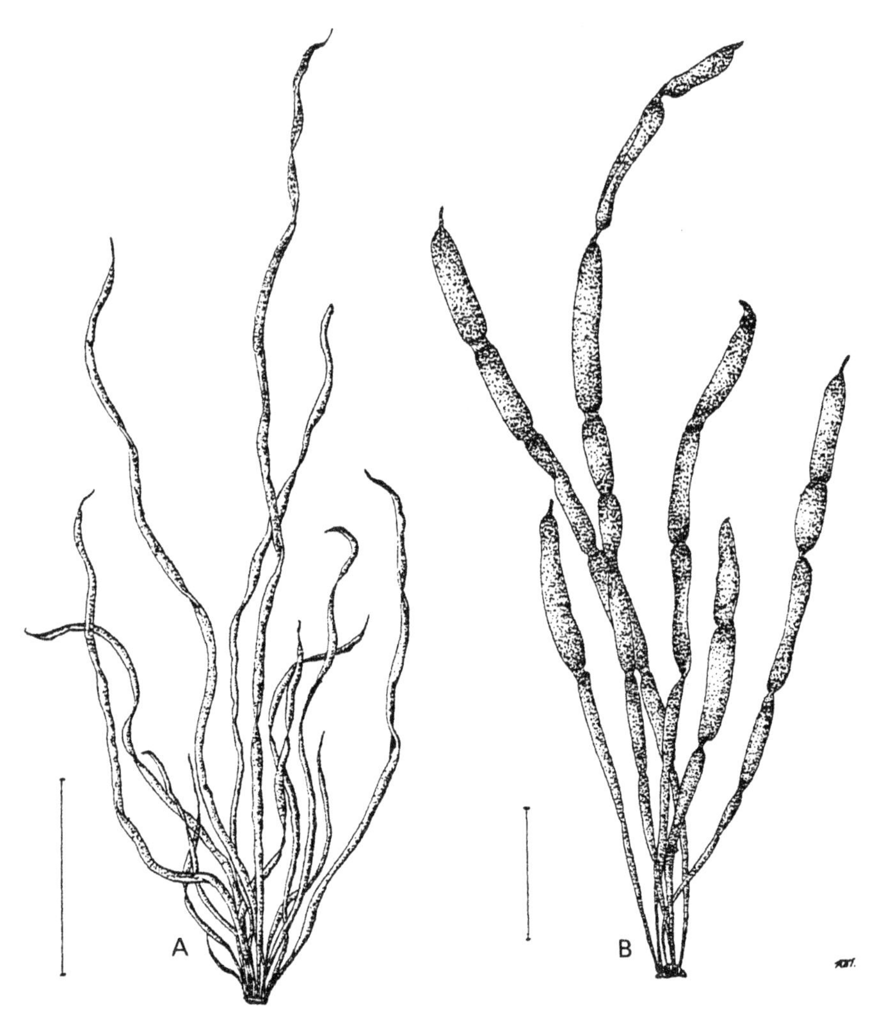

Fig. 67 *Scytosiphon dotyi* A. Habit of plant. *Scytosiphon lomentaria* B. Habit of plant. Bars = 20 mm.

S. lomentaria var. *complanatus* Rosenvinge (1893), p. 863 sensu Smith (1944), p. 130.
S. attenuatus sensu Doty (1947), p. 34 (pro parte) not *Chordaria attenuata* Foslie (1887), p. 176, not *S. attenuatus* Kjellman (1883), p. 259.
S. lomentaria f. *cylindricus nanus* Tokida (1954), p. 105.

Fig. 68 *Scytosiphon dotyi*
A–B. Surface view of vegetative thallus. Note single, plate-like plastid with pyrenoid. C. Surface view of fertile thallus showing plurilocular sporangia. D. T.S. of young thallus showing outer small cortical cells, inner large medullary cells and large central cavity. E. T.S. of small fertile thallus showing terminal rows of plurilocular sporangia. F. Plurilocular sporangia. Bar = 20 μm (B–C, F), 50 μm (others).

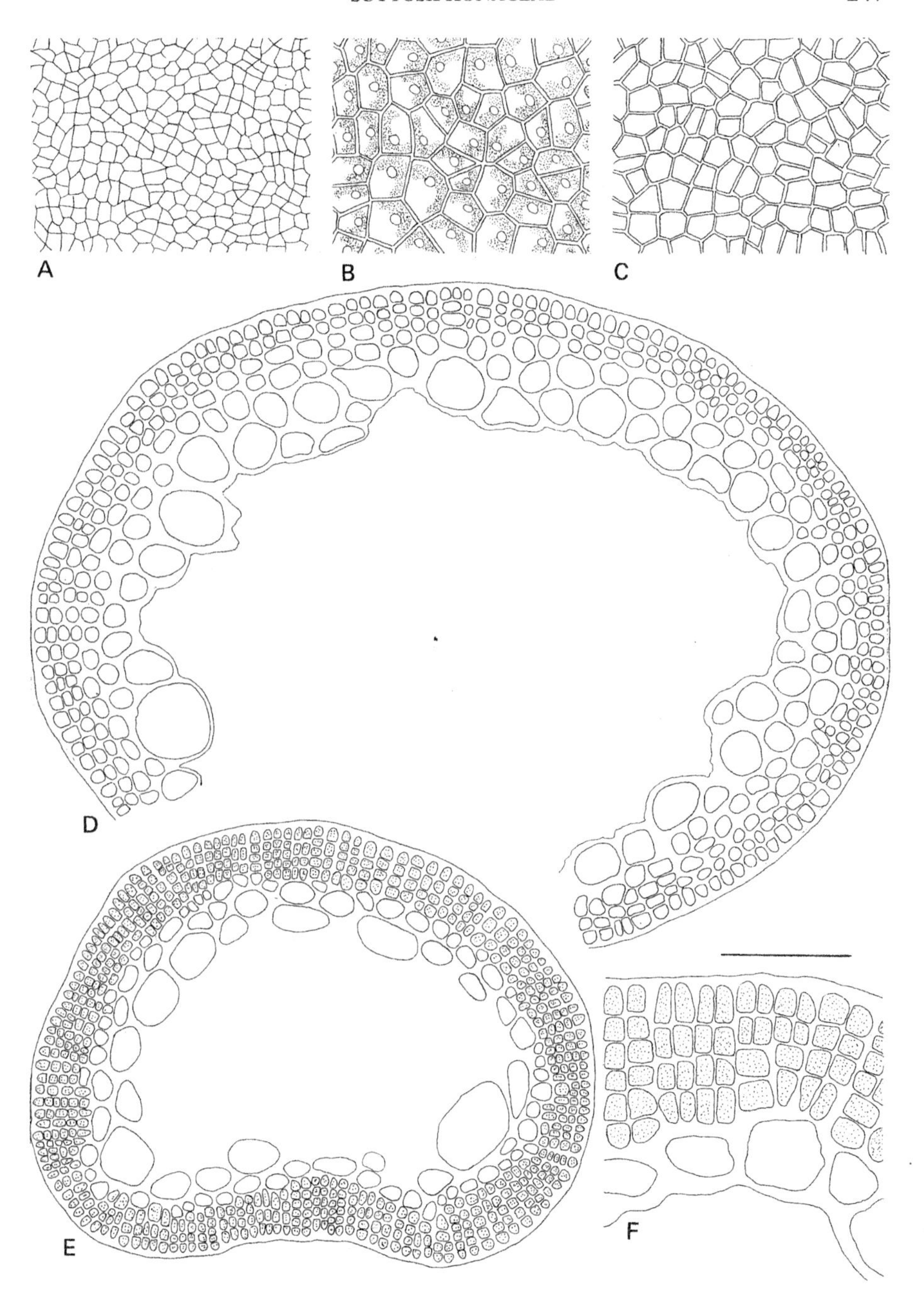
A
B
C
D
E
F

Plants forming erect, gregarious, mid to dark brown, tubular thalli arising from a small discoid holdfast; erect thalli simple, hollow, cylindrical or more usually slightly dorsiventrally flattened, flaccid to slightly rigid in texture, without constrictions although sometimes twisted, to 14 cm long × 0·25–1 mm wide, attenuate above and below, sometimes with curled terminal region; in surface view cells small, irregularly arranged, 4–13 × 4–8 μm each with a single plate-like plastid with pyrenoid; in section thalli parenchymatous, with a large central cavity, enclosed by an inner medulla of 1–5 large, thick-walled, longitudinally elongate, transversely rounded, colourless cells and an outer cortex of 1–3 small, pigmented cells; hairs mainly grouped, arising from cortical cells sunken in fertile regions.

Plurilocular sporangia in dark-coloured continuous sori, extensive in terminal thallus region, in closely packed vertical columns, arising by division of cortical cells, uniseriate, sometimes terminally biseriate, to 35 μm (–7 loculi) long, with subquadrate to rectangular loculi, without accompanying ascocyst-like cells; unilocular sporangia unknown.

Epilithic, littoral fringe, exposed on the sides of vertical cliff faces, large boulders, etc. particularly with a north-facing aspect, in moderately exposed areas.

Restricted to the east coast of England (Yorkshire and Kent); probably more widely distributed.

Plants annual; February to May.

Culture studies by Wynne (1969) revealed *S. dotyi* to have a 'direct', asexual life history. Under a range of culture conditions plurispores from the erect thalli germinated directly to produce small polystromatic crusts of only limited extent, from the centre of which sprouted a further generation of erect thalli. A distinct heteromorphic development of the spores into either unilocular sporangia bearing crusts and plurilocular sporangia bearing blades depending on the culture conditions, as revealed for *S. lomentaria,* was not observed.

Wynne (1969), pp. 34–39, fig. 9, pls 18–19.

Scytosiphon lomentaria (Lyngbye) Link (1833), p. 233. Figs 67B, 69; Pl. 11a, b

Chorda lomentaria Lyngbye (1819), p. 74.
Scytosiphon lomentaria var. *typica* Rosenvinge (1893), p. 863.
Scytosiphon pygmaeus Reinke (1888), p. 18.
Scytosiphon lomentaria f. *typicus* Setchell & Gardner (1925), p. 533.

Plants forming erect, single or more commonly gregarious, yellow to dark brown, tubular thalli arising from a small encrusting base and attached by a slightly swollen discoid holdfast; erect thalli simple, fairly flaccid, hollow, cylindrical, either collapsed or

---

Fig. 69 *Scytosiphon lomentaria*
A–B. Surface view of vegetative thallus. C. Surface view of fertile thallus showing plurilocular sporangia and large ascocysts. D. T.S. of young thallus showing outer small cortical cells and inner large medullary cells. E. T.S. of portion of thallus showing outer cortical cells, inner medullary cells and central cavity. F–G. T.S. of fertile thalli showing terminal rows of plurilocular sporangia and large ascocysts. Bar = 20 μm (B–C, G), 50 μm (others).

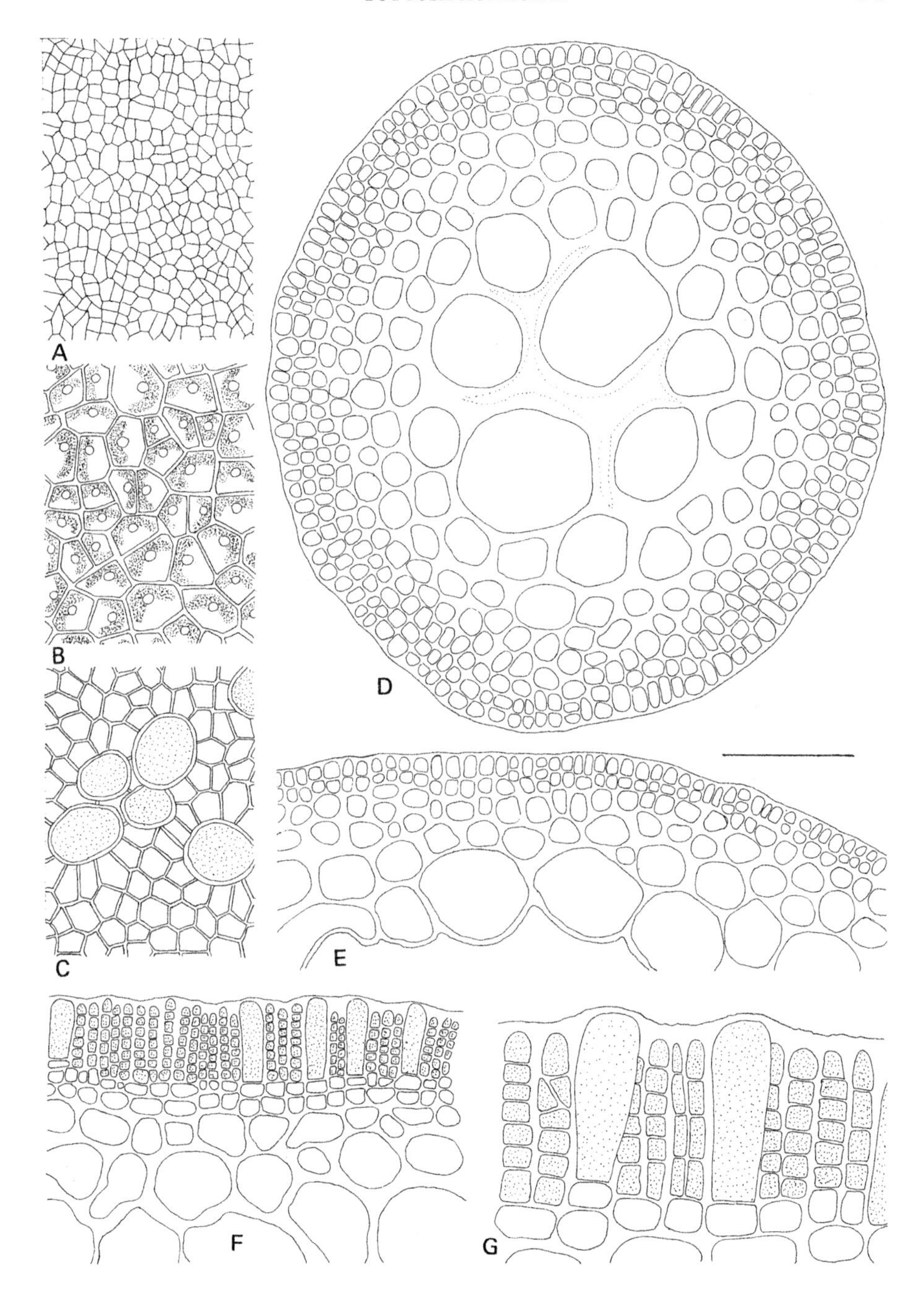
A
B
C
D
E
F
G

inflated, usually characteristically constricted at intervals when mature, to 30 (–40) cm long × 4–6 (–8) mm wide, tapering slowly to base usually more sharply delimited at apex; in surface view cells small, usually irregularly arranged, more rarely in longitudinal rows, 5–13 × 3–9 μm, each with a single plate-like parietal plastid with 1–2 pyrenoids; in section thalli parenchymatous, with a large central cavity enclosed by an inner medulla of 2–5 layers of large, thick-walled, longitudinally elongate, transversely rounded, colourless cells and an outer cortex of 1–3 layers of small, pigmented cells; hairs single, more rarely in tufts, arising from outer cortical cells, sunken in fertile thalli, with basal meristem, lacking obvious sheath, c. 6–8 μm in diameter.

Plurilocular sporangia in dark-coloured continuous sori, extensive in terminal thallus region, in closely packed vertical columns arising by divisions of cortical cells, uni- to multiseriate, to 90 μm (–20 loculi) long, 4–8 μm wide with subquadrate to rectangular loculi, often with accompanying scattered, pyriform, ascocyst-like cells, 18–33 × 11–16 μm; unilocular sporangia unknown on erect thallus but reported on crustose *Microspongium gelatinosum* phase.

Epilithic, in pools/channels and emergent, littoral fringe to lower eulittoral and shallow sublittoral; tolerant of some sand cover.

Common and generally distributed around the British Isles.

Annual; reported throughout the year but most common in spring/early summer; probably persists throughout remainder of year as perennial basal crust.

S. lomentaria has been investigated in laboratory culture throughout the word including the North Atlantic (Reinke, 1878a; Kuckuck, 1898; Sauvageau, 1929; Dammann, 1930; Frye & Phifer, 1930; Kylin, 1933; Dangeard, 1962b, 1963b; Roberts & Ring, 1972; Lüning & Dring, 1973; Rhodes & Connell, 1973; Fletcher, 1974b; Dring & Lüning, 1975a, b; Kristiansen & Pedersen, 1979, Lüning, 1980), Mediterranean (Reinke, 1878a; Berthold, 1881), California (Wynne, 1969), Australia (Clayton, 1976a, b; 1980a, b), and Japan (Abe, 1935; Nakamura, 1965; Tatewaki, 1966; Nakamura & Tatewaki, 1975). In the great majority of these studies a direct asexual life history was revealed with the liberated plurispores from the erect thalli germinating into filamentous, discoid or crustose prostrate systems from which sprouted further generations of thalli. For many isolates environmental conditions appeared to play an important role in determining the relative occurrence of the prostrate and erect systems. For example in general low temperatures and short days favoured erect thallus development whilst warm temperatures and long days favoured prostrate (usually crust) development (Tatewaki, 1966; Wynne, 1969; Rhodes & Connell, 1973; Dring & Lüning, 1975b; 1983). Also red light was shown to enhance erect thallus formation whilst white and blue light favoured crust formation (Lüning & Dring, 1973). Some isolates, however, developed independently of environmental conditions of temperature and daylength (Clayton, 1976a, b; Kristiansen & Pedersen, 1979; Pedersen, 1980). In general, the life history can, therefore, be interpreted as heteromorphic but monophasic with both the prostrate and erect systems possessing the same ploidy level.

Culture studies by Nakamura & Tatewaki (1975) and Clayton (1980a) in the Pacific, however, revealed the occurrence of an additional sexual cycle, with an alternation of heteromorphic generations. The erect thalli represented the sexual haploid generation releasing plurispores which behaved as gametes, whilst the crustose thalli represented the asexual diploid generation bearing unilocular sporangia. Chromosomal studies by Nakamura & Tatewaki further confirmed the occurrence of meiosis in the formation of the unispores and the alternation of nuclear phases in the life cycle. Both Nakamura & Tatewaki and Clayton also reported asexual (parthenogenetic?) development of plurispores released from the erect thalli, the final morphological expression obtained (i.e. crusts or erect thalli) being dependent on environmental conditions; as Nakamura & Tatewaki pointed out, this parthenogenetic development of the 'gametes' is in accordance with the 'direct' life history widely reported for this species.

It is interesting to note that the above reports of sexual reproduction in *S. lomentaria* give some credance to earlier worldwide observations of plurispores behaving as gametes (Berthold, 1881; Kuckuck, 1912; Abe, 1935). The possibility of a sexual life history occurring, albeit of rare occurrence in North Atlantic waters, now needs re-investigation.

A noteworthy feature of the life history of *S. lomentaria* is the wide range of prostrate systems which have been reported and their resemblance to previously described taxa. This has been augmented by field and laboratory studies of the latter which have shown them to be connected in life history with *S. lomentaria*. 'Prostrate' taxa linked in life history include *Streblonema* (Dangeard, 1963b), *Myrionema* (Dangeard, 1963b), *Compsonema* (Loiseaux, 1970b, as *Scytosiphon pygmaeus*; Fletcher, unpublished see p. 222), *Microspongium* (Lund, 1966; McLachlan *et al.*, 1971; Fletcher 1978; Kristiansen & Pedersen, 1979) and *Ralfsia* (Nakamura, 1965; Tatewaki, 1966; Wynne, 1969; Nakamura & Tatewaki, 1975; Clayton, 1980a). To some extent it is likely that *S. lomentaria* represents various genotypes with morphologically different basal expressions. Alternatively the different basal expressions described by the various authors could be due to different degrees of surface contact achieved by the germlings (see Lüning, 1980) or to the use of different culture conditions.

With respect to the crustose expressions implicated in the life history of *S. lomentaria*, these are generally reported to be *Microspongium gelatinosum*-like in the North Atlantic and *Ralfsia*-like in the Pacific supporting the argument that different plants are involved (Lund, 1966). However, this is complicated by reports of *Ralfsia*-like phases also occurring in the life history of Atlantic material of *S. lomentaria* (Rhodes & Connell, 1973) although the evidence is not altogether convincing, as the illustrations provided by these authors (figs 11–12) appear more to resemble *Microspongium* than *'Ralfsia'*. It is further possible that the *'Ralfsia'*-like crusts reported in the life history of Pacific *S. lomentaria* are also based on misidentifications. Illustrations provided by many of the authors (Nakamura, 1965, figs 2–3; Tatewaki, 1966, figs 6–7; Nakamura & Tatewaki, 1975, figs 3B, E, pl. 1D, E; Clayton, 1980a, figs 7–8) certainly quite strongly resemble *'Microspongium'* thalli. Only the illustrations provided by Wynne (1969, pl. 15) of cultured unilocular sporangial crusts of *S. lomentaria* are clearly *Ralfsia*-like. Overall I am inclined to the view that the crustose thallus involved in the life histories of Atlantic and Pacific populations of *S. lomentaria* is *'Microspongium'*-like rather than *'Ralfsia'*-like and can be more specifically identified as Reinke's *M. gelatinosum* bearing unilocular sporangia (see p. 224). Further studies on a worldwide basis are obviously required to determine if the distinction between the *Microspongium* and *Ralfsia*-like phases, based primarily on differences in the cohesion of the erect filaments, can be justifiably maintained.

Reinke (1878a), pp. 267–268, pl. 11, figs 13–14; Sauvageau (1929), pp. 331–333; Frye & Phifer (1930), pp. 234–245, figs 1–39; Newton (1931), pp. 178–179, fig. III; Hamel (1931–39), pp. 194–196, fig. 431; Kylin (1933), pp. 47–49, fig. 18; Rosenvinge & Lund (1947), pp. 27–31. fig. 9; Feldmann (1949), pp. 103–115, figs 1A–B; Taylor (1957), pp. 168–169, pls. 15(2), 16(3);

Lindauer *et al.* (1961), pp. 256–257, fig. 65; Dangeard (1963b), pp. 57–64, pls. IX–XI; Nakamura (1965), pp. 109–110, fig. 5; Gayral (1966), p. 247, pl. XXXIV; Tatewaki (1966), pp. 62–66, figs 1–10; Dangeard (1969), pp. 60–70, fig. 1B; Wynne (1969), pp. 32–34, pls. 14–17; Nakamura (1972), pp. 148–149, fig. 1; Lüning & Dring (1973), pp. 333–338, figs 13–18; Rhodes & Connell (1973), pp. 211–215, figs 1, 11–14; Dring & Lüning (1975a), pp. 107–117 (various figures); Dring & Lüning (1975b), pp. 25–32 (various figures); Nakamura & Tatewaki (1975), pp. 59–65, figs 1–5, pl. 1; Abbott & Hollenberg (1976), p. 198, fig. 162; Clayton (1976a), pp. 187–198, figs 1–9; Clayton (1976b), pp. 201–208, figs 1–7; Kornmann & Sahling (1977), pp. 137–139, fig. 74; Rueness (1977), p. 169, fig. 101, pl. XXII, 4; Fletcher (1978), pp. 371–398, fig. 35; Kristiansen & Pedersen (1979), pp. 31–56, (various figures); Clayton (1980a), pp. 105–118, figs 1–11; Clayton (1980b), pp. 113–118, figs 1–3; Fletcher (1980), pp. 41–42, pl. 14 (4–6); Lüning (1980), pp. 920–930, figs 2–3; Pedersen (1980), pp. 391–398 (various figures); Dring & Lüning (1983), pp. 545–568 (various figures); Stegenga & Mol (1983), p. 105, pl. 37, figs 3–5; Dring (1984), pp. 159–192, figs 1, 4; Kristiansen (1984), pp. 719–724, fig. 1.

STRAGULARIA Strömfelt

STRAGULARIA Strömfelt (1886), p. 173.

Type species: *S. adhaerens* Strömfelt (1886), p. 173 (= *S. clavata* (Harvey in Hooker) Hamel (1931–39), p. xxxi.

Thallus encrusting, discrete and orbicular becoming confluent and irregular, epilithic, epizoic, rarely epiphytic, light yellow brown to dark brown, fairly smooth and firm, sometimes sponge-like, comparatively thin, firmly adherent by undersurface to substratum, usually without rhizoids; structure pseudoparenchymatous, consisting of a monostromatic discoid base of outwardly spreading, frequently branched, quite firmly united filaments, with terminal row of synchronously growing apical cells, giving rise behind, immediately or belatedly, to vertical or slightly curved, simple or little branched, quite well united filaments of cells containing a single, plate-like, parietal plastid with pyrenoid; hairs uncommon, terminal on erect filaments, arising from surface or depressions, with basal meristem, lacking obvious sheath.

Unilocular and plurilocular sporangia borne on the same or different thalli, terminal on erect filaments, grouped in sori; unilocular sporangial sori common, indefinite and extensive over central thallus region, obvious in surface view, slightly raised, sponge-like, light to yellow brown, gelatinous, comprising ovate, obovate-pyriform or elongate-cylindrical sporangia arising laterally at the base of paraphyses, more rarely terminal on erect filaments with or without paraphyses; paraphyses multicellular, simple, linear or clavate, loosely associated and gelatinous; plurilocular sporangial sori present or absent, if present rare, not obvious externally arising as terminal extensions of the erect filaments, cylindrical, uniseriate, rarely biseriate in parts.

The genus *Stragularia* is revived here to include two species previously assigned to the genus *Ralfsia*. *Stragularia* differs from the closely related *Ralfsia* in having vertically directed filaments arising from the basal layer which are not initially almost prostrate, and having expansive, rather than discrete unilocular sporangial sori (see p. 240). Other features which distinguish *Stragularia* from *Ralfsia* include: the crusts are comparatively thin, more closely encrusting with the margin gradually rather than sharply delimited and almost impossible to raise intact; young thalli are light (yellow) brown rather than olive brown, matt rather than glossy; older thalli are fairly soft, smooth and firmly adherent throughout rather than firm, coriaceous, verrucose, bullate with rust-red underside; erect filaments are less firmly coherent and can be squashed apart with relative ease; the single plate-like plastid with pyrenoids is more conspicuous; the plurilocular sporangia are grouped in less conspicuous sori and are not constructed of closely-packed vertical columns each with a terminal sterile cell; plants have a more restricted local (and in most species geographical) distribution, more or less restricted to tidal pools/runnels.

Crustose algae with the above described *Stragularia* features have commanded particular attention over the past two decades in view of their life history connections with macroscopic, erect-bladed members of the Scytosiphonaceae. In most investigations of these heteromorphic life histories the relationship between the crustose and erect fronded expression was shown to be asexual, facultative and determined by environmental conditions. In a few cases a sexual relationship was observed with the erect blades representing the sexual, gametophytic phase and the crustose thalli representing the asexual, sporophytic phase. The so-called '*Stragularia*' stages have been implicated in the life histories of such species as *Petalonia fascia* (Nakamura, 1965; Wynne, 1969; Edelstein *et al.*, 1970; Rhodes & Connell, 1973; Fletcher, 1974b, 1978; Roeleveld *et al.*, 1974), *Petalonia zosterifolia* (Nakamura & Tatewaki, 1975), *Colpomenia bullosa* (Nakamura & Tatewaki, 1975), *Colpomenia peregrina* (Clayton, 1979), *Endarachne binghamiae* J. Agardh (Nakamura & Tatewaki, 1975), and *Scytosiphon lomentaria* (Nakamura, 1965; Tatewaki, 1966; Wynne, 1969; Fletcher, 1974b, 1978; Nakamura & Tatewaki, 1975; Clayton, 1980a). To date only a small number of these '*Stragularia*' taxa have been identified, including *S. clavata* (as *Ralfsia clavata*), *Ralfsia bornetii, R. californica* and *Microspongium gelatinosum*. This is further complicated by species such as *R. californica* reported to be linked in life history with both *S. lomentaria* and *P. fascia* (Wynne, 1969). Also species such as *R. clavata* include geographical isolates which appear to have a direct, monomorphic life history (Kylin, 1934; Loiseaux, 1968). More work, particularly cultural studies, is therefore required to identify and define all the '*Stragularia*'-type crustose algae and the extent of their involvement with the erect, thalloid members of the *Scytosiphonaceae*.

KEY TO SPECIES

Erect filament cells of crust not exceeding 13 μm in diameter; unilocular sporangial sori indefinite and irregularly spreading on crust surface, light brown; paraphyses quite distinct from crust filaments, not exceeding 7 cells, with cells, particularly towards base, markedly longer than wide, 1–7 diameters long . . . . *S. clavata*

Erect filament cells of crust up to 19 μm in diameter; unilocular sporangial sori indefinite and regularly spreading over whole central crust region, yellow brown; paraphyses not markedly different from crust filaments, usually of 8–12 cells, with cells not exceeding 1–2 diameters long . . . . *S. spongiocarpa*

Stragularia clavata (Harv. in Hook.) Hamel (1931–39), p. xxxi. Figs 59C, 70

Myrionema clavata Harvey in Hooker (1833), p. 391.
Ralfsia clavata (Harvey in Hooker) Crouan frat. (1852), no. 56.
Myrionema Henschei Caspary (1871), p. 142.
Stragularia adhaerens Strömfelt (1886), p. 49.
Ralfsia bornetii Kuckuck (1894), p. 245.
Ralfsia tenuis Kylin (1947), p. 45.

Thallus mainly epilithic, encrusting, comparatively thin, discrete and orbicular becoming confluent and irregular to 5 cm or more in extent, frequently with central regions eroded away, light to dark brown/black, firmly adherent to substratum by underside, usually without rhizoids; peripheral region thin, sometimes monostromatic and discoid, light brown, firm, subcoriaceous, closely encrusting and impossible to raise intact; central region dark brown/black, smooth, soft, slightly spongy, comprising rounded or polygonal cells quite closely packed and irregularly displaced in surface view, 5–13 μm in diameter, each with a single, parietal, plate-like plastid with pyrenoid; thallus pseudoparenchymatous in structure, comprising a monostromatic basal layer of outwardly spreading, frequently branched, quite firmly united filaments of mainly rectangular cells, 1–2 diameters long, to 20 × c. 8 μm in diameter; basal layer regularly discoid, with marginal row of distinct, synchronously dividing apical cells, giving rise behind, immediately or belatedly, to vertical or slightly curved, unbranched or little branched compacted filaments; erect filaments quite firmly adjoined, although separable under pressure, to 180 μm (–20 cells) long, comprising quadrate to rectangular cells, 8–12 × 5–13 μm each containing a single, plate-like, parietal, yellow brown plastid with pyrenoid, occupying upper cell region; hairs infrequent, terminal on erect filaments arising superficially or from depressions, with basal meristem but lacking obvious sheath.

Unilocular and plurilocular sporangia borne on the same or different thalli, terminal on erect filaments, grouped in sori; unilocular sporangial sori common, indefinite and irregularly spreading over crust surface, slightly raised and obvious in surface view, light brown, soft, sponge-like, very gelatinous comprising obovate, pyriform, elongate-pyriform or elongate-cylindrical unilocular sporangia, 45–80 (–120) × 16–28 μm, sessile and lateral at the base of paraphyses; paraphyses simple, elongate-clavate to elongate-cylindrical, multicellular, 2–5 (–7) cells, 60–165 μm long, comprising quadrate, more commonly rectangular cells, 1–2 diameters long above, 7–16 × 7–9 μm, markedly elongate below, cylindrical or elongate-clavate, 1–7 diameters long, 24–47 × 4–7 μm, frequently with ruptured outer cell walls; plurilocular sporangial sori rare, terminal on erect filaments, without associated paraphyses, uniseriate or partly biseriate, to 20 μm (–7 loculi) long.

Epilithic especially on soft substrata, epizoic, more rarely epiphytic, upper to lower eulittoral pools/channels, sometimes emergent in damp situations especially under cover of fucoids, in sheltered to moderately exposed areas; tolerant of some sand cover.

Fig. 70 *Stragularia clavata*
A. Surface view of crust cells. Note single plate-like plastid with pyrenoid drawn in some cells. B. V.S. of thallus showing slightly curved erect filaments. C. R.V.S. of thallus margin showing terminal apical cell. D–E. V.S. of fertile thallus showing unilocular sporangial sorus. F. Unilocular sporangia and associated paraphyses. G. V.S. of fertile thallus showing terminal plurilocular sporangia. Bar = 50 μm.

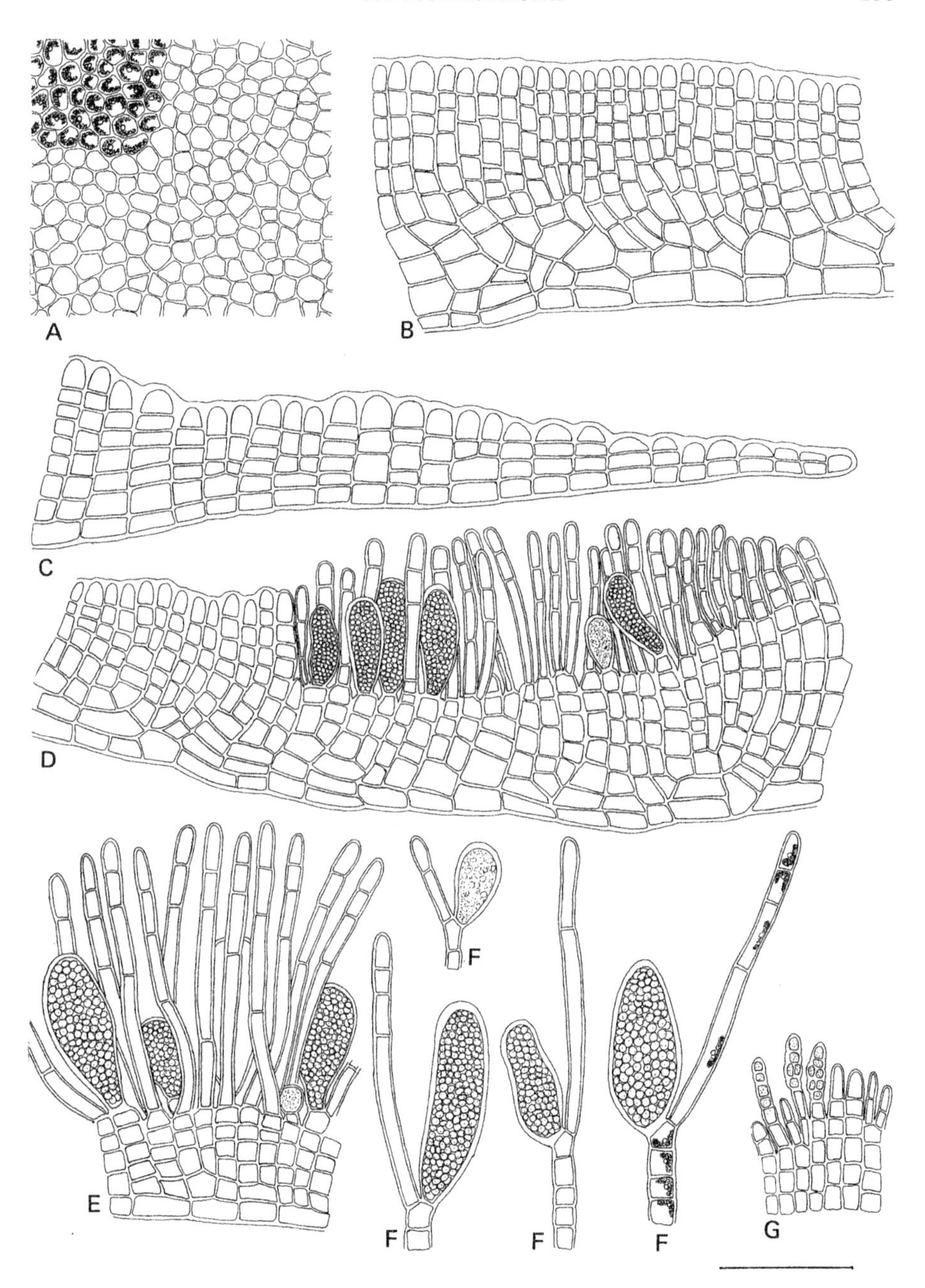
A
B
C
D
E
F
F
F
F
G

Common and widely distributed around the British Isles.

Thalli probably perennial, unilocular sporangia recorded November to May; plurilocular sporangia recorded November (single record). Crusts becoming recognisable on the shore from October onwards, reaching maximum development and fertility in winter/spring and then dying back during the summer or remaining as small pieces of crust edge.

This is the most common and widely distributed species of *Stragularia* present in the British Isles. Although often recorded on the sloping sides of standing shallow pools (rarely in deep pools) it shows a particular preference for runnels and channels associated with running seawater. Associated algae usually include the *Microspongium gelatinosum* phase of *Scytosiphon lomentaria, Ralfsia verrucosa, Petroderma maculiforme* and *Hildenbrandia prototypus* and in particular the erect bladed *Petalonia fascia.* Not uncommonly the crusts are subjected to a periodic cover of sand. It is seldom emergent on rock except under cover of fucoids, in particular *F. serratus*. It has been recorded on a wide variety of natural and man-made substrata although appears to show a preference for softer materials such as chalk, limestone, shale, etc.

S. clavata is based on *Myrionema clavata* described by Harvey in Hooker (1833). Harvey had in turn based his description on *Linckia clavata* originally described for Loch Linnhe on the west coast of Scotland by Captain Carmichael in his manuscripts 'Algae Appinenses' and 'Cryptogam Appinens'. Later the Crouan brothers (1852) removed the species to the genus *Ralfsia* quoting Harvey and thus establishing *R. clavata* (Harvey) Crouan frat. accepted by various authors such as Taylor (1937a) and Parke & Dixon (1976). However, according to Reinke (1889b) and Kylin (1947) this combination is illegitimate as the Crouan's material is identical with *R. verrucosa* (see Batters, 1890, p. 288 footnote). Farlow's (1881) later description of *R. clavata* collected on the coast of New England, U.S.A. appears to agree with our present day concept of this species, and he was thus often associated with the binomial e.g. *R. clavata* Farlow in Newton (1931) and Jaasund (1965) and *R. clavata* (Carmichael) Crouan *sensu* Farlow in Taylor (1957) and South & Hooper (1980). A further development was initiated by Kylin (1947). He cast doubt on the conspecificity of Farlow's material and European material of *R. clavata* as Farlow did state his material was identical to Crouan brothers No. 56 of their 'Algues marines du Finistère; this as stated above, was considered by Reinke and Kylin to be *R. verrucosa*. Kylin also adopted Batters' (1890, p. 288) attitudes that as the description given by Carmichael was very brief and none of the original material has been detected there must always remain some degree of doubt as to whether his plant is identical with the presently recognised species. Because of this doubt Kylin (1947, p. 5) proposed a new name *Ralfsia tenuis* based on *Ralfsia clavata sensu* Reinke (1889b, p. 48) as a replacement for the questionable *R. clavata*. This has been accepted by a number of authors including Lund (1959), Pankow (1971) and Rueness (1977). This binomial is not, however, adopted here as I would prefer to continue to accept, albeit temporarily, Harvey's *Myrionema clavata* as the basis for our present day concept of this taxon, pending the results of further studies. The species is also transferred to the genus *Stragularia* following Hamel (1931–39, p. xxxi).

Included here within the circumscription of *S. clavata* is material resembling *Ralfsia bornetii* originally described by Kuckuck (1894) for Helgoland in the North Sea. The latter species is distinguished from *S. clavata sensu stricto* by the possession of extremely long paraphyses (c. 140–180 μm) with elongate basal cells and associated elongate unilocular sporangia (c. 80–125 μm). Material of *S. clavata sensu stricto* is usually described as having smaller paraphyses (c. 70–100 µm) and smaller, more rounded unilocular sporangia (c. 40–85 µm). Opinion is divided as to whether these represent distinct species. Newton (1931) and Hamel (1931–39) include *R. bornetii* within the synonomy of *R. clavata* whilst it is recognised as a separate species by Sundene (1953), Taylor (1957), Jaasund (1965), Edelstein *et al.* (1970), Rueness (1977), South & Hooper (1980) and Tanaka & Chihara (1980c). Indeed according to Edelstein *et al.* (1970) some of the distinctive features of *R. bornetii* and *R. clavata* were retained in laboratory culture. On the basis of examination of British Isles' material I do not feel that two separate species can be justifiably maintained at present, as there appears to be complete integration with respect to the lengths of paraphyses and sporangia. Therefore, pending further studies, *R. bornetii* is here included within the syonomy of *S. clavata*.

A number of laboratory culture studies have revealed North Atlantic isolates of *S. clavata* (includes *R. bornetii*) to be connected in life history with *Petalonia fascia* (Edelstein *et al.*, 1970; Rhodes & Connell, 1973; Roeleveld *et al.*, 1974; Fletcher, 1974b, 1978) (see. p. 231). The relationship between the crustose and erect bladed expressions appears to be asexual and facultative and determined by environmental conditions rather than differences in ploidy level. There is no evidence of a sexual cycle between an erect-bladed (i.e. *Petalonia*-like) gametophyte and a crustose (i.e. *Stragularia*-like) sporophyte. However, this possibly does require investigation especially in view of reports of sexual life histories in various other members of the Scytosiphonaceae in the North Pacific. It is possible that the crustose and erect bladed expressions are linked by a sexual process in the North Atlantic and this has either merely passed unnoticed or requires precise environmental conditions for its fulfilment. The sexual process might alternatively be confined to a small number of genotypes. Certainly the indications are that different genotypes are present which vary in their life history and developmental patterns. Most isolates investigated in culture revealed a heteromorphic life history; however, there have also been reports of direct, monophasic life histories (Kylin 1934; Loiseaux, 1968). It is also likely that geographically separated ecotypes have developed varying in their response to environmental conditions (as shown for *S. lomentaria* and its crustose expression by Lüning, 1980).

Reinke (1889a), p. 9, pl. 6 (10–19); Batters (1890), p. 288, pl. 8, fig. 22; Kuckuck (1894), pp. 244–246, figs 14–15 (including *R. bornetii*); Newton (1931), p. 154, fig. 95A–B, not C; Hamel (1931–39), p. 108, fig. 26C; Kylin (1934), pp. 17–18, fig. 10; Taylor (1957), pp. 135–136, p. 11, figs 1–2 (includes *R. bornetii*); Lund (1959), pp. 78–79 (as *R. tenuis*); Jaasund (1965), pp. 61–63, fig. 19A (including *R. bornetii*); Loiseaux (1968), pp. 302–308, figs. 3–5; Ardré (1970), pp. 247–248, pl. 36, figs 1–2 (as *R. bornetii*); Edelstein *et al.* (1970), pp. 527–531, figs 2–15; Fletcher (1974b), p. 218; Rueness (1977), pp. 128–130, figs 53, 54B, not C (including *R. bornetii* and *R. tenuis*); Fletcher (1978), pp. 371–398 (various figures): Tanaka & Chihara (1980c), pp. 341–342. fig. 1C (as *R. tenuis*); Stegenga & Mol (1983), p. 82, pl. 28, figs 1–2.

Stragularia spongiocarpa (Batters) Hamel (1931–39), p. 70. Fig 59D, 71–72

Ralfsia spongiocarpa Batters (1888), p. 452.

Thallus encrusting, comparatively thin, discrete and orbicular to 10 mm in diameter, becoming confluent and irregular to 2–3 cm in extent, frequently with central regions eroded away, yellow to dark brown/black, firmly adherent to substratum by underside, usually without rhizoids; in surface view central vegetative cells rounded or polygonal, thick-walled, quite closely packed, 8–16 μm in diameter; thallus pseudoparenchymatous in structure, consisting of a monostromatic basal layer of outwardly spreading, frequently branched, quite firmly adjoined filaments with a marginal row of discrete, synchronously dividing apical cells, giving rise immediately behind to vertical files of simple or little branched, compacted filaments; erect filaments quite firmly adjoined although separable under pressure, to 190 μm high (–20 cells) comprising cells, 1–2 times broader than long below, quadrate or only slightly broader than long above, 7–13 × 10–19 μm each containing a single, plate-like plastid and pyrenoid occupying upper cell region; hairs infrequent, terminal on erect filaments, arising superficially or from depressions, with basal meristem but lacking obvious sheath, c. 6 μm in diameter.

Unilocular sporangia common, borne in slightly raised yellow, spongy, turf-like, gelatinous sori, extensive and occcupying greater part of crust surface, obvious in surface view; unilocular sporangia ovate, obovate, pyriform or elongate-pyriform, 21–70 × 16–35 μm, terminal on erect filament with or without associated paraphysis-like filaments, less frequently lateral at the base of the paraphysis-like filaments, sessile or on 1–2 celled pedicels, rarely reported terminal on paraphyses; paraphysis-like filaments not markedly different from erect vegetative filaments, multicellular, gelatinous, simple, linear, slightly tapering to the apex, 105–120 (–140) μm long, 8–12 cells, comprising quadrate to rectangular cells, 1–2 diameters long throughout, 8–18 × 5–9 μm; plurilocular sporangia unknown.

Epilithic, on the sloping sides of shallow pools, lower eulittoral, frequently covered by sand.

Apparently restricted to southern and eastern shores (Fife, Northumberland, Yorkshire, Sussex, Hampshire, Dorset and South Devon); probably more widely distributed.

Probably perennial, unilocular sporangia recorded December to March.

S. spongiocarpa is not uncommon locally but appears to have a limited geographical distribution. It occurs mainly in the lower eulittoral, on the sides of sand-covered shallow pools, in association with a wide range of other crustose brown and red algae. Fertile plants are particularly common during January and February, after which the central crust areas appear to die away leaving the outer margin portions. In a dried state the plants are readily distinguishable from the morphologically similar *S. clavata* by their extensive and more prominent fructification which is yellow in colour and not unlike the confluent tufts of *Compsonema saxicolum*. *S. spongiocarpa* can further be distinguished from *S. clavata* by the generally wider vegetative cells (10–19 μm compared to 5–13 μm), the greater number of cells in the paraphyses (8–12 compared to 2–7) and the absence of elongated cells in the paraphyses (1–2 compared to 1–7 diameters long).

Fig. 71 *Stragularia spongiocarpa*
A. Surface view of vegetative thallus. B. Surface view of fertile thallus with unilocular sporangia. C. V.S. of crust margin showing single, large, apical cell. D. V.S. of vegetative thalli showing erect filaments. E. V.S. of fertile thalli showing terminal unilocular sporangia. F. Portion of erect filament showing young unilocular sporangia and cells with a single plate-like plastid and pyrenoid. Bar = 20 μm (F). 50 μm (others).

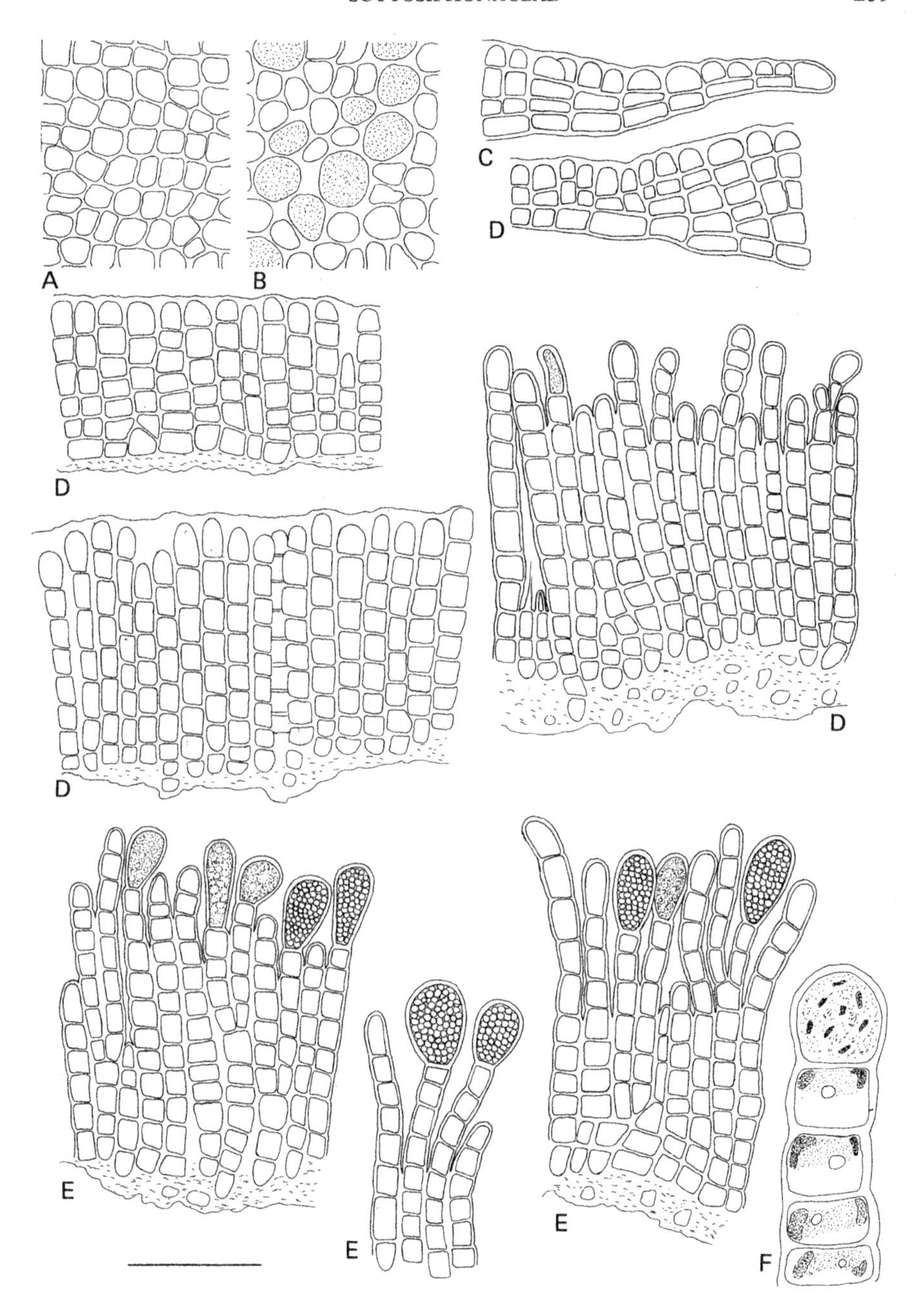
A
B
C
D
D
D
D
E
E
E
F

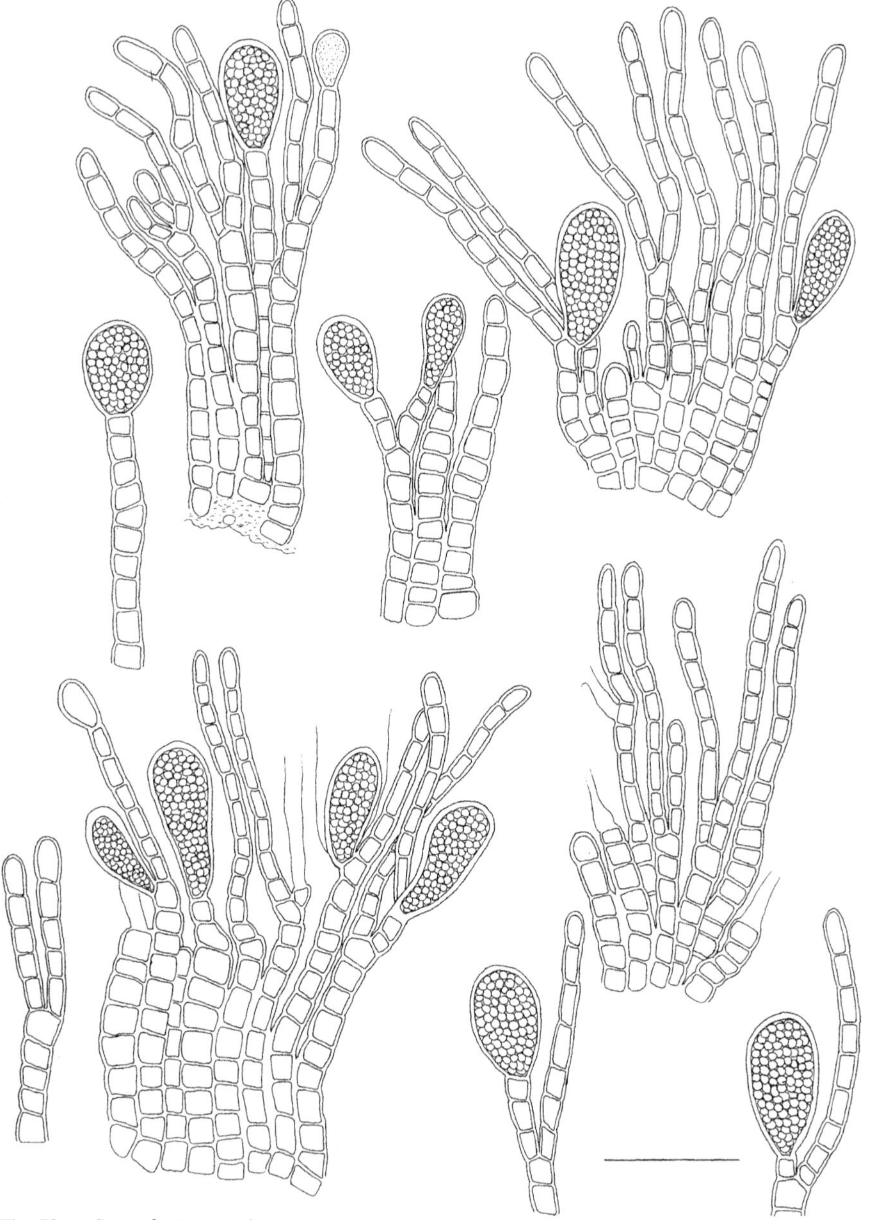

Fig. 72 *Stragularia spongiocarpa*
V.S. and S.P. of fertile thalli showing erect filaments giving rise to paraphyses and unilocular sporangia. Bar = 50 μm.

Initially, unilocular sporangia are formed terminally, in crowded sori on the thallus surface, unaccompanied by the paraphyses. Later, after spore release has occurred, many terminal cells either continue sporangia production within the old sporangial wall husks or continue growth to form the paraphyses. The paraphyses are not clearly distinguishable, as in *S. clavata,* from the erect filaments of the crustose thallus.

Culture studies by Fletcher (1981b) indicate that a direct, monomorphic life history is probably operating in this species. Spores from the unilocular sporangia behaved asexually and developed directly into fertile, crustose thalli similar to the parental thallus. There was no indication that *S. spongiocarpa* is connected in life history with an erect bladed member of the Scytosiphonaceae.

Batters (1888), pp. 452–453, figs 17–21; Batters (1890), p. 289, pl. VIII (17–19); Newton (1931), p. 153; Hamel (1931–39), pp. 109–110; Fletcher (1974b), p. 218; Fletcher (1975), pp. 534–535, figs 1–3; Fletcher (1978), pp. 371–398; Fletcher (1981b), pp. 323–330, figs 1–19.

CUTLERIACEAE Hauck

CUTLERIACEAE Hauck (1883–85), pp. 318, 403

Thallus with fronds erect or procumbent; erect fronds flattened, flaccid, fan-shaped and deeply divided or dichotomously branched and tapering, all branches terminated by a fringe of hair-like, monosiphonous, unbranched, photosynthetic filaments; growth trichothallic, structure parenchymatous, medulla of large colourless thick-walled cells, cortex of small, photosynthetic thin-walled cells; plastids several, discoid without obvious pyrenoids; hairs single or grouped, arising from surface cells, with basal meristem; anisogametes on separate plants, grouped in small, scattered, punctate sori on blade surfaces, terminal or lateral on short, branched or unbranched multicellular filaments; unilocular sporangia unknown on blade surface.

Procumbent thalli macroscopic, orbicular, fan or irregularly shaped, usually deeply divided, lobate, with or without overlapping lobes, loosely attached underneath by rhizoids, with an outer margin of large, apical cells (apical growth) or a fringe of 2 superimposed hair-like, monosiphonous unbranched, pigmented filaments (trichothallic growth); structure parenchymatous with inner medulla of large, colourless cells, lower cortex of small, thick-walled cells, with or without rhizoidal extensions, upper cortex of 1–2 thin walled, photosynthetic cells, or 5–6 cells arranged in vertical, branched, files; upper cortex cells with several discoid plastids without pyrenoids; gametangia (oogonia, antheridia) similar in structure to those on erect thalli, forming mixed sori on thallus surface; unilocular sporangia densely crowded in raised, spreading, mucilaginous sori in separate plants.

Two genera, *Cutleria* and *Zanardinia* occur in the British Isles. They differ in their life history; *Cutleria* has a heteromorphic alternation of phases involving a *'Cutleria'* gametophyte and an *'Aglaozonia'* sporophyte while *Zanardinia* has an isomorphic alternation of phases. The *Aglaozonia* phases in the life histories of species of *Cutleria* were previously described as separate entities. For example, the *'Aglaozonia'* phase of the one British representative of the genus *Cutleria,* i.e. *C. multifida* (Smith) Greville, was described as *Aglaozonia parvula* (Greville) Zanardini. Other reported life history associations include *Cutleria monoica* Ollivier and *Aglaozonia chilosa* Falkenberg in the

Mediterranean and *Cutleria adspersa* (Roth) De Notaris and *Aglaozonia melanoidea* (Schousboe) Sauvageau in both the Mediterranean and North Atlantic.

Cutleria and *Zanardinia* show a number of common features in their structure and reproduction. These include trichothallic growth, a parenchymatous mode of construction with an inner medulla and outer cortex, cells with several discoid plastids, an anisogamous mode of reproduction, and an 'alternation of generations' type of life history. The British representatives of these two genera, *Cutleria multifida* (Smith) Greville and *Zanardinia prototypus* (Nardo) Nardo are easily distinguishable on morphological features, the former producing fairly large, erect thalli, the latter producing fairly small procumbent thalli.

CUTLERIA Greville

CUTLERIA Greville (1830), p. 59.

Type species: *C. multifida* (Smith) Greville (1830), p. 60.

Sexual phase (= *Cutleria*) forming erect blades, usually single, rarely gregarious, arising from a discoid holdfast, bilaterally flattened, flaccid, fan-shaped and deeply divided, becoming dichotomously branched, linear, narrowly tapering, with blunt, dissected apices fringed by a tuft of hair-like, monosiphonous, unbranched, photosynthetic filaments; surface cells irregularly placed, less obviously in longitudinal rows except near apex, with several discoid plastids without pyrenoids; in section, thallus parenchymatous with an inner medulla of 4–6, large, colourless cells and an outer cortex of 1–2 smaller, photosynthetic cells; hairs single or grouped, arising from surface cells, with basal meristem. Anisogametes on separate plants, grouped in punctate sori scattered on both sides of blade surface, terminal on short, branched, multicellular filaments, uni- to triseriate.

Asexual phase (= *Aglaozonia*) forming procumbent thalli, orbicular or irregularly shaped with overlapping lobes, and entire, rounded margins, smooth and membranous becoming subcoriaceous, loosely attached underneath by rhizoids; surface cells in longitudinal rows with several discoid plastids without pyrenoids; in section, thallus parenchymatous, arising from an outer marginal row of large apical cells, comprising an inner medulla of 2–6 large colourless cells and an upper and lower cortex of 1–3 smaller, pigmented cells; hairs single or tufted, arising superficially or from depressions, in upper cortex; unilocular sporangia in extensive, densely crowded, mucilaginous sori on thallus surface, without paraphyses.

One species in the British Isles:

Cutleria multifida (Smith) Greville (1830), p. 60.

Ulva multifida Smith in Smith & Sowerby (1790–1814), pl. 1913.
Dictyota penicillata Lamouroux (1809), p. 13.
Zonaria multifida C. Agardh (1820–28), p. 135.
Zonaria parvula Greville (1828), pl. 360.

Padina parvula (Greville) Greville (1830), p. 63.
Aglaozonia parvula (Greville) Zanardini (1843), p. 38.
Aglaozonia reptans (Crouan frat.) Kützing (1849), p. 566.
Zonaria reptans (Crouan frat.) Crouan frat. (1852), No. 74 (exsiccatum)

Sexual *(Cutleria)* phase Figs 73A, 75

Plants consisting of erect, flattened, solitary blades, light to yellow brown, to 40 cm long, arising from a conical holdfast of fibrous, compacted, colourless, rhizoidal filaments; blades initially fan-shaped, deeply dissected, astipitate, membranous and somewhat striated at base, later extending into dichotomously branched, narrowly tapering, flaccid, linear blades, 1–7 mm (– 12) mm wide; apices blunt, wedge-shaped, finely dissected and fringed terminally by a tuft of hair-like filaments; filaments simple, monosiphonous, 1–2·5 mm long, comprising cells predominantly shorter than wide 2/3–1 diameters long, 8–26 × 13–22 µm, each with numerous discoid plastids lacking pyrenoids; surface cells of thallus irregularly placed, less obviously in longitudinal rows, 10–20 × 7–13 µm, each with numerous disc-shaped plastids without pyrenoids; in section, blades parenchymatous comprising an inner medulla of 4–6 large, colourless, thin-walled cells enclosed by an outer cortex of 1–2 smaller pigmented cells; hairs usually grouped, arising from surface cells, with basal meristem.

Anisogametes occurring on separate plants, forming small, punctate sori sometimes with accompanying hairs, scattered over both sides of the blades; oogonia in small clumps, terminal on short, multicellular (2–4 cells) stalks, 4–5 loculi long, ovoid, usually biseriate, occasionally triseriate, 45–65 × 22–33 µm, comprising loculi 10–16 × 10–13 µm; antheridia more densely clumped, sessile, cylindrical, slightly curved and secundly arranged on short, branched, multicellular filaments, biseriate, more commonly multiseriate, 45–65 × 8–13 µm, 16–19 loculi long, comprising loculi 2–3 × 2–3 µm; unilocular sporangia borne on separate asexual *'Aglaozonia'* plants.

Epilithic, rarely epiphytic or epizoic, sublittoral to 10 m.

Widely distributed around the British Isles, but more common on the south-west coasts and becoming rare further north and on the south-east and east coasts.

Probably annual, young plants appearing in winter and spring and reaching maximum development in the summer; gametangia recorded June–October with oogonial plants more frequently recorded than antheridial plants.

Asexual *(Aglaozonia)* phase Figs 73B, 76

Thallus horizontally expanded and flattened, light olive-brown to black, orbicular and discrete becoming irregular and spreading to several centimetres, with overlapping lobes, loosely attached underneath by rhizoids; surface smooth and membranous, becoming sub-coriaceous, margin entire and rounded; surface cells in longitudinal rows, quadrate or rectangular, 8–16 × 5–12 µm containing densely packed discoid plastids without pyrenoids; in section, thallus parenchymatous, to 195 µm thick arising from a marginal row of large meristematic apical cells, comprising an inner medulla of 2–6 large, thick-walled, elongated, colourless cells with few plastids enclosed by an upper and lower

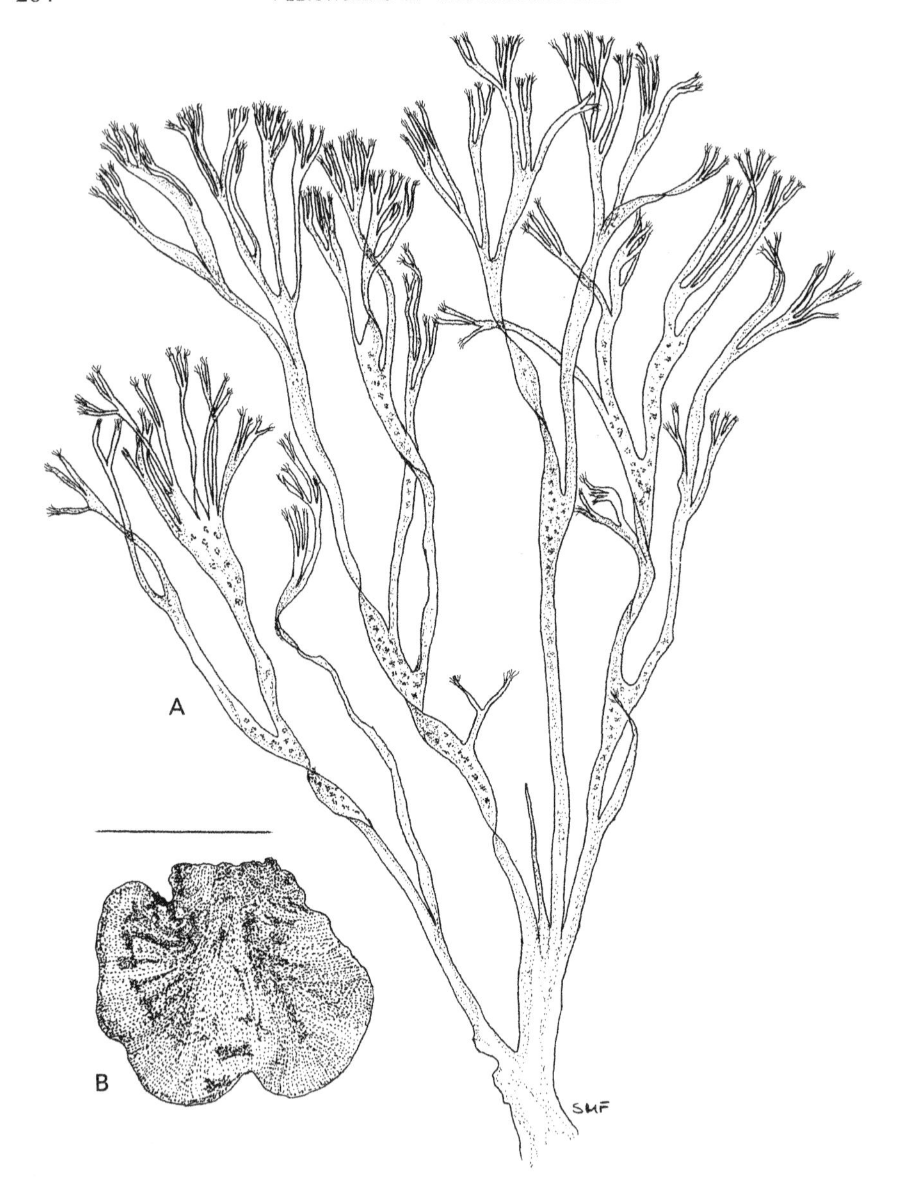

Fig. 73 *Cutleria multifida*
A. Habit of plant (sexual *Cutleria* phase). B. Habit of plant (asexual *Aglaozonia* phase).
Bar = 15 mm.

cortex of smaller, pigmented cells; upper cortex of 2–3 cells enclosed within surface cuticle, cells subquadrate or quadrate, 8–11 high × 7–13 µm wide, lower cortex of 1–2 cells with thick, darkly pigmented walls, cells quadrate, subquadrate or usually much wider than high 14–60 µm wide × 10–16 µm high, commonly producing thick-walled, multicellular branched rhizoids, 9–23 µm wide; hairs uncommon, usually in tufts, superficial or in slight depressions in the upper cortex, with basal meristem.

Unilocular sporangia uncommon, confluent, densely packed, in rounded or irregularly shaped, spreading, slightly raised, mucilaginous sori on thallus surface, 25–37 × 11–15 µm, elongate-pyriform or cylindrical, usually enclosing 8 zoospores; gametangia borne on separate sexual *'Cutleria'* plants.

Epilithic, epiphytic (especially on haptera/stipes of *Laminaria*), less commonly epizoic, lower eulittoral pools, or more commonly sublittoral to 12 m and probably deeper.

Not uncommon and widely distributed all round the British Isles.

Perennial; unilocular sporangia mainly recorded in winter and spring.

Culture studies on European isolates of *C. multifida* by various authors such as Crouan frat. (1855), Derbès & Solier (1856), Reinke (1878b), Falkenberg (1879), Church (1898), Kuckuck (1899b), Yamanouchi (1909, 1912), Kuckuck (1929), Schlösser (1935) and Hartman (1950) have all revealed the occurrence of a heteromorphic life history, with an erect bladed gametophyte (*Cutleria* phase) alternating with a prostrate sporophyte (*Aglaozonia* phase). The sexual plants are dioecious and release anisogametes, which fuse with the aid of a sex attractant (Müller, 1974; Jaenicke, 1977). The resultant zygote develops into the prostrate *Aglaozonia* plant bearing unilocular sporangia in which meiosis occurs. The released unispores then germinate directly to reform the erect bladed *Cutleria* plants. Around the British Isles the gametophyte is mainly a fairly short lived, summer annual whilst the sporophyte is perennial and reproduces during the winter and spring months.

However, evidence from both field and laboratory studies indicates that the life history can abbreviate from such a rigid alternation of sexual phases. For example field studies around the British Isles reveal a disproportionate number of female, compared to male, gametophytes and considerable differences in the relative abundance and geographical distribution of the *Cutleria* and *Aglaozonia* phases. In addition the above mentioned culture studies showed that parthenogenetic development of the female gametes can occur which results in the formation of *Cutleria* and/or (haploid) *Aglaozonia* plants. A similar diversity of development has also been reported for unispores released from the *Aglaozonia* plants. The morphological expression appears, therefore, to be independent of ploidy level. Further, the diversity in the products of germination (e.g. filamentous forms, intermediate growth forms) described by the authors might well be a reflection of the variable culture conditions employed. Such a flexible, heteromorphic life history, markedly influenced by environmental conditions, would be analogous to that revealed for members of the Scytosiphonaceae by Nakamura & Tatewaki (1975).

Sauvageau (1899), pp. 265–362, figs 9, 25–26; Yamanouchi (1909), pp. 380–386; Yamanouchi (1912), pp. 441–502, pls. xxvi–xxxv, figs 1–228; Newton (1931), pp. 197–199, fig. 25; Hamel (1931–39), pp. 321–323, figs 54A–E, G, 55F; Feldmann (1937), pp. 305–308; Fritsch (1945), pp. 157–171, figs 51–55; Kylin (1947), p. 33, pl. 3, fig. 8; Lund (1950), pp. 65–74, figs 13–14; Lindauer *et al.* (1961), pp. 179–182, fig. 26, pl. 3; Gayral (1966), p. 251, pl. XXXVI; Ardré (1970), p. 264; Rueness (1977), p. 170, pl. XXIII, 1a–b.

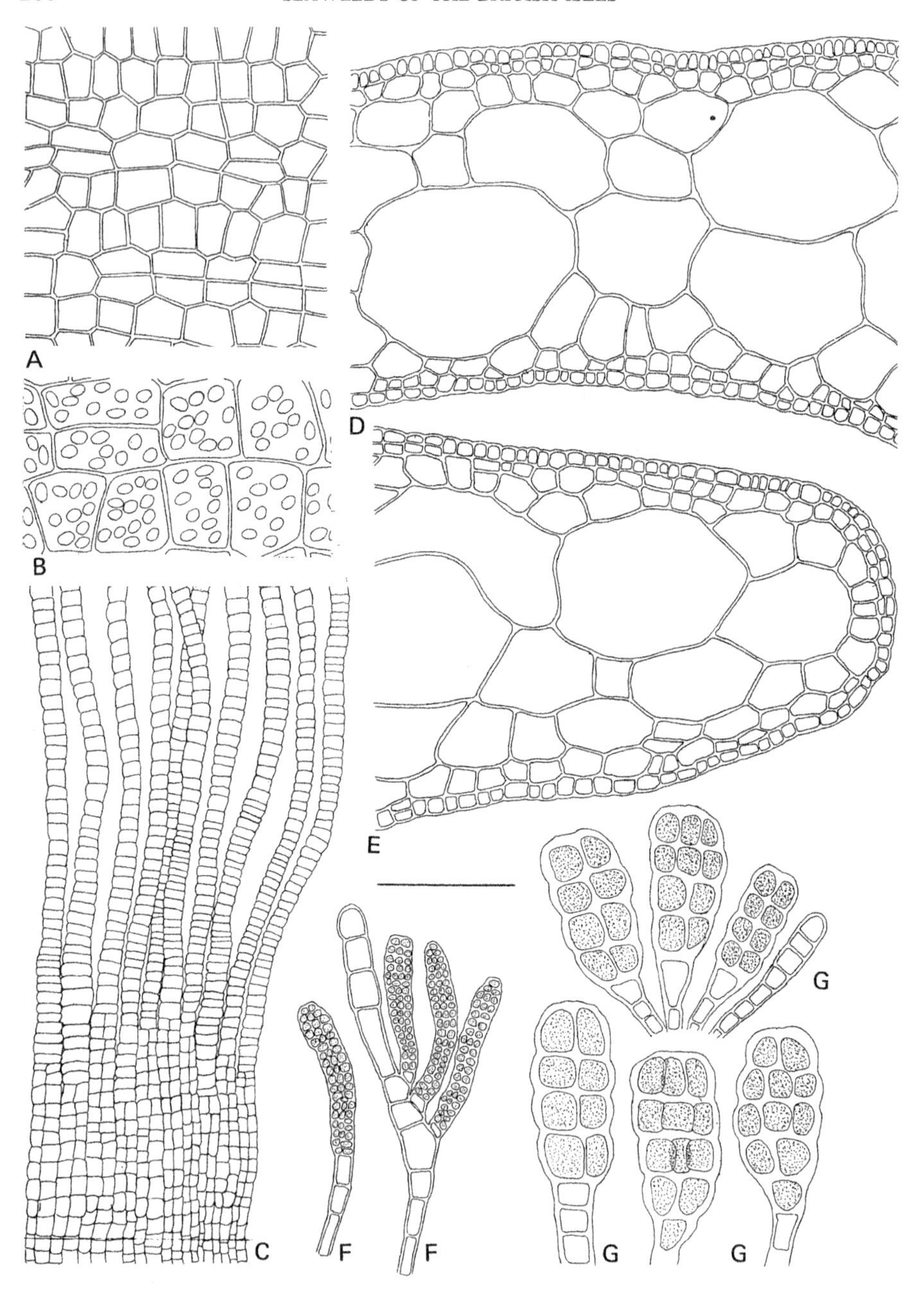
A
B
C
D
E
F
F
G
G
G

ZANARDINIA Nardo ex Crouan frat.

ZANARDINIA Nardo ex Crouan frat. (1857), p. 24.

Type species: *Z. collaris* (C. A. Agardh) Crouan frat. (1857), p. 24 (= *Z. prototypus* (Nardo) Nardo (1841), p. 189).

Thallus procumbent, smooth and membranous, becoming coriaceous, orbicular, fan or sometimes irregularly shaped, lobate, radially ridged, with entire or ruffled margins, with or without a fringe of hair-like filaments, loosely attached underneath by a felt of fibrous branched rhizoids; surface cells longitudinally arranged peripherally, more irregularly arranged centrally, quadrate or rectangular, with several discoid plastids without pyrenoids; in section, thallus parenchymatous, arising from the basal, meristematic cells of a marginal row of two superimposed, monosiphonous, simple, freely extending filaments, comprising an inner medulla of large, thick-walled, colourless, elongated cells and an upper and lower cortex of smaller, pigmented cells; upper cortex photosynthetic, at first of 1–2 cells, later 5–6 cells in vertical, branched files, lower cortex of 1–2 cells, later becoming thick walled and dark coloured, giving rise to multicellular rhizoidal filaments; hairs single or grouped, arising superficially or from depressions in cortex.

Gametangia (oogonia, antheridia) and unilocular sporangia in sori, arising directly from upper cortical cells; gametangia mixed in sori, elongate-cylindrical, terminal, on simple, or little branched, erect filaments; unilocular sporangia on separate plants, crowded, elongate-pyriform enclosing 4–6 zoospores.

One species in the British Isles:

Zanardinia prototypus (Nardo) Nardo (1841), p. 189. Figs 74, 77

Stifftia prototypus Nardo (1831), p. 677.
Zanardinia collaris (Montagne) Crouan frat. (1857), p. 24.
Padina collaris Montagne (1846), p. 33.

Plant consisting of a horizontally expanded, flattened thallus, spherical or fan-like, sometimes irregular in shape, to 20 cm across, olive to dark brown/black, loosely attached on the undersurface by rhizoids; surface smooth, membranous or coriaceous in texture, radially ridged comprising cells at first in longitudinal rows, later becoming irregularly arranged, quadrate or rectangular, 8–17 × 5–11 μm each containing numerous discoid plastids without pyrenoids; margin entire or ruffled with a distinct outer hair-like fringe; in section thallus parenchymatous, 130–400 μm thick, comprising an inner medulla of 5–7 (– 10) large, colourless, thick-walled cells enclosed within an upper and lower cortex of smaller, pigmented, dark coloured cells; outer margin of two superimposed, monosiphonous, multicellular, simple filaments, freely extending 3–4 mm,

---

Fig. 74 *Cutleria multifida* (sexual *Cutleria* phase)
A–B. Surface view of thallus showing cells with numerous discoid plastids. C. Branch apex showing terminal fringe of hair-like filaments. D. T.S. of thallus. E. T.S. of thallus margin. F. Antheridia. G. Oogonia. Bar = 50 μm (A, F, G), 20 μm (B), 100 μm (C–D, E).

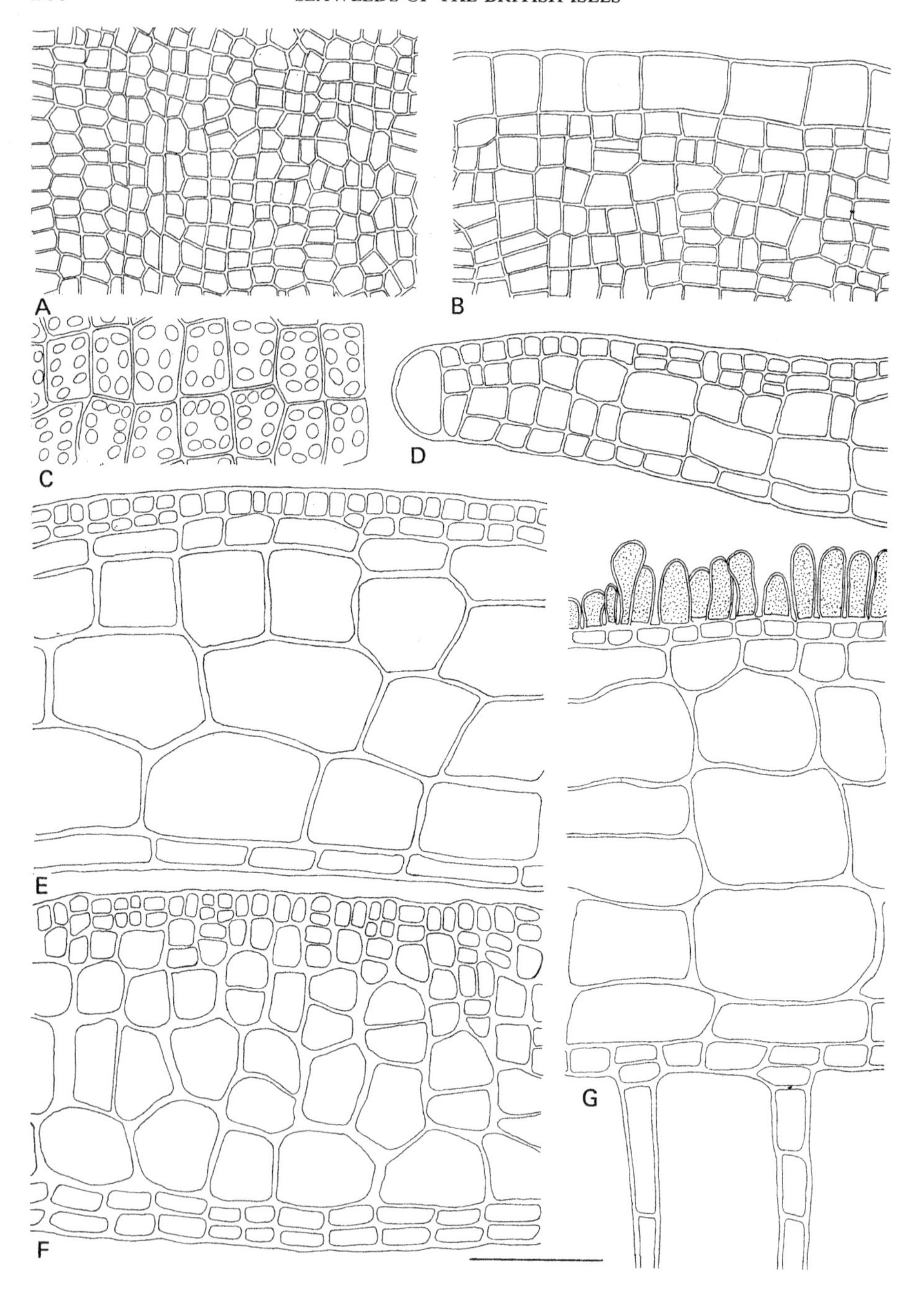
A
B
C
D
E
F
G

comprising cells 1–2 (–2½) diameters long, 13–36 × 14–21 μm, each with numerous discoid plastids without pyrenoids; upper cortex enclosed within a thick cuticle, at first comprising 1–2 pigmented cells, later of 5–6 cells arranged in vertical files of dichotomously branched, attenuating filaments, lower cortex of 1–2 cells, with thick, darkly staining walls, often producing free, multicellular, branched, downwardly extending, thick-walled rhizoids; hairs not observed in the British Isles, reported elsewhere to be single or grouped, arising superficially or from slight depressions in the upper cortex.

Reproductive organs not observed in the British Isles, reported elsewhere to be developed in discrete sori on upper surface of thallus; ansiogametes occurring in the same sori, elongate-cylindrical, terminal on short, simple or little branched, multicellular filaments; oogonia 4–7 cells long, usually biseriate, rarely triseriate; antheridia 14–20 cells long, slightly curved, bi- or triseriate; unilocular sporangia occurring on separate plants, crowded, without accompanying paraphyses, elongate-pyriform, enclosing 4–6 zoospores.

Epilithic, rarely epiphytic, in the sublittoral to 20 m, especially on silty boulders and bedrock at the lower limit of *Laminaria hyperborea*.

Reported from southern and western shores of the British Isles; Channel Isles, eastwards to Hampshire, northwards to Pembroke; recorded in Ireland from Cork.

Probably perennial, although data on seasonal behaviour too inadequate for comment.

Prior to the first report of its occurrence on the south coast of England (Jephson *et al.* 1975) the previous northern limit of *Z. prototypus* was the Channel Isles. It now appears

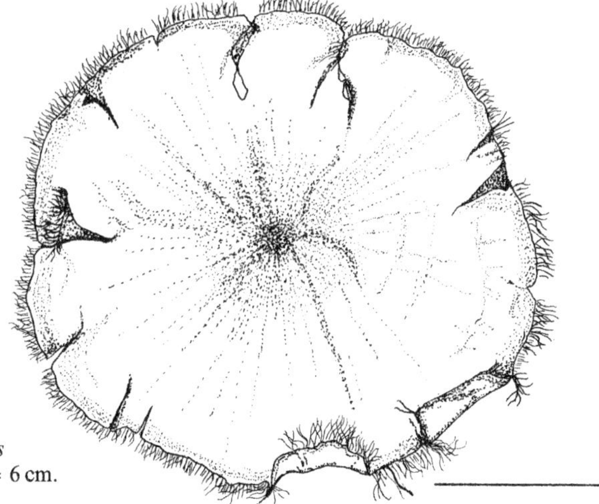

Fig. 76 *Zanardinia prototypus*
Habit of plant. Bar = 6 cm.

Fig. 75 *Cutleria multifida* (asexual *Aglaozonia* phase)
A–B. Surface view of thallus, showing cells with numerous discoid plastids. C. Surface view of thallus margin showing large apical cells. D. V.S. of thallus margin showing large apical cell. E–F. V.S. of vegetative thalli. G. V.S. of fertile thallus showing unilocular sporangia arising from upper surface and rhizoids produced from lower surface. Bar = 20 μm (B), 50 μm (others).

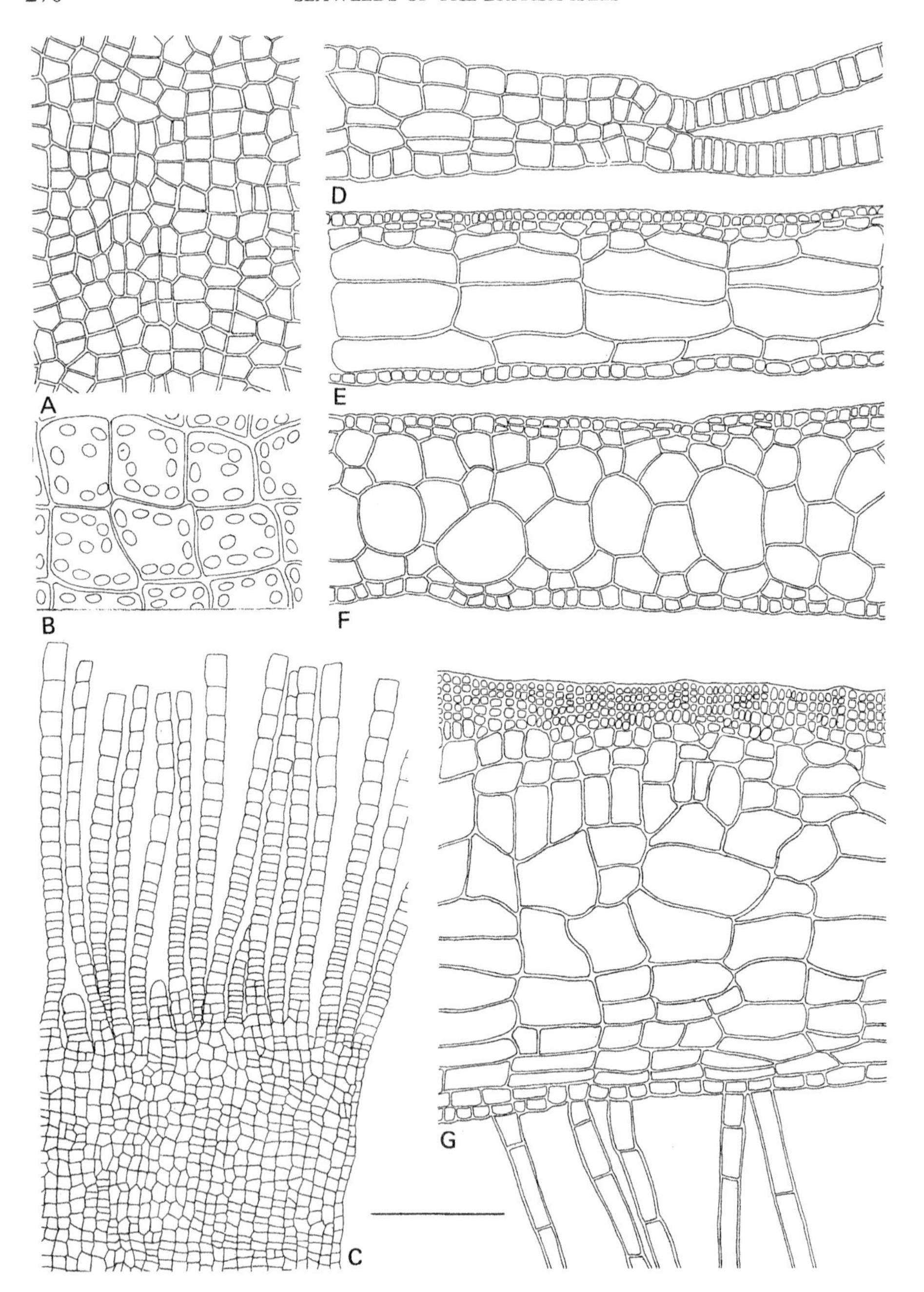
A
B
C
D
E
F
G

to be widespread in south west Britain, and in some localities quite abundant (Hiscock & Maggs, 1982, 1984).

The only reported culture studies on *Z. prototypus* have been by Yamanouchi (1913) who showed the occurrence of a sexual, isomorphic life cycle.

Sauvageau (1899), pp. 286–296, figs 1–4; Yamanouchi (1913), pp. 1–35, pls I–IV, figs 1–108; Newton (1931), p. 199, fig. 128; Hamel (1931–39), pp. 319–321, fig. 53; Feldmann (1937), p. 305; Ardré (1970), pp. 263–264, pl. 37, figs 2–3; Jephson *et al.* (1975), pp. 253–255, figs 1–5; Hiscock & Maggs (1982), pp. 414–416, figs 1–2; (1984), pp. 84–85.

ARTHROCLADIACEAE Chauvin

ARTHROCLADIACEAE Chauvin (1842) p. 66 [as Arthrocladieae].

Thalli erect, terete, much branched, branching opposite, sometimes irregular or alternate, all axes with whorls of 3–4 short, laterally projecting fascicles of hair-like, branched, rarely simple filaments; growth trichothallic, structure pseudoparenchymatous, uniaxial with broad cortex of compacted, colourless cells; unilocular sporangia in chains, fasciculate, borne lateral at the base of the whorled, hair-like filaments; unispores germinating into filamentous, monoecious, microscopic gametophytes bearing antheridia and oogonia.

Only one genus of the Arthrocladiaceae, *Arthrocladia* occurs in the British Isles.

ARTHROCLADIA Duby

ARTHROCLADIA Duby (1830), p. 971.

Type species: *A. villosa* (Hudson) Duby (1830), p. 971.

Plants consisting of erect, much branched, narrow, terete thalli arising from a small, discoid holdfast; thalli with opposite, less frequently with irregular or alternate branches all axes bearing regular whorls of 3–4 fascicles of branched, rarely simple, laterally projecting, hair-like filaments; growth trichothallic, structure pseudoparenchymatous, comprising a large, central, axial cell, surrounded by a broad cortex of longitudinally elongate, transversely rounded or irregular, colourless cells, becoming smaller outwards and enclosed by 1–2 layers of thick-walled, pigmented cells; surface cells in longitudinal rows, much longer than wide, usually with pointed ends; plastids discoid, several per cell, without pyrenoids.

Unilocular sporangia in moniliform, fasciculate chains, lateral at the base of the whorled, hair-like filaments.

Fig. 77 *Zanardinia prototypus*
A–B Surface view of thallus showing cells with numerous plastids. C. Surface view of thallus margin showing outer fringe of hair-like filaments. D. V.S. of thallus margin showing two projecting filaments. E–G. V.S. of vegetative thalli, showing increased anatomical development. Bar = 50 μm (A, D), 20 μm (B), 100 μm (C, E–G).

Germination of unispores gives rise to filamentous, monoecious, microscopic gametophytes bearing oogonia and antheridia.

One species in the British Isles:

Arthrocladia villosa (Hudson) Duby (1830), p. 971. Figs 78, 79; Pl. 12

Conferva villosa Hudson (1778), p. 603.

Plants forming erect, much branched, terete thalli, solitary or more commonly gregarious, light yellow to mid brown, arising from a small discoid holdfast; erect thalli with a distinct main axis, quite firm and cartilaginous throughout becoming soft and flaccid terminally, solid, to 40 (−100) cm long, 0·5 (−0·7) mm wide, branching opposite more rarely irregular, to 2 (−3) orders with branches well spaced apart, divaricate and terminating in a fine hairlike filament; all axes, except in some older basal regions, with regular whorls of 3–4 fascicles of laterally extending, hair-like filaments, up to 2 (−4) mm long, uniseriate, simple or branched, with branching whorled, opposite, irregular or unilateral, comprising cells 13–80 × 8–26 μm, fairly short and broad below, becoming long and narrow above, each with numerous discoid plastids without pyrenoids; surface cells of thalli in longitudinal rows, much longer than wide, 1–5 diameters long, rectangular or more usually with pointed ends, 16–42 × 7–14 μm each with several discoid plastids; in section thalli with a large, central, axial cell, surrounded by a broad cortex of thick-walled, longitudinally elongate, transversely rounded or irregular, colourless cells becoming smaller outwards and enclosed by 1–2 layers of thick-walled, pigmented cells.
Unilocular sporangia borne in place of branches in the lower regions of the whorled, hair-like filaments, usually unilateral, more rarely opposite or irregular, sessile or on 1–3 celled stalks, in moniliform/compressed moniliform, simple, straight or slightly curved, fasciculate chains, to 36 sporangia long (−250 μm), individual sporangia usually much shorter than wide, 4–9 × 11–15 μm, releasing spores via a lateral pore.

Epilithic, particularly on small stones, gravel, shells, etc. more rarely epiphytic, lower eulittoral and sublittoral to at least 8 m.
Generally distributed around the British Isles although more records from south western and western coasts.
Plants annual, June to September, although most records for August and September.

The only reported culture studies of *A. villosa* have been on Mediterranean isolates, by Sauvageau (1931) and Müller & Meel (1982—as *A. villosa* f. *australis*). The life history was only completed by Müller & Meel; they revealed the occurrence of a pronounced heteromorphic life history with the large, macroscopic sporophytes alternating with monoecious, microscopic gametophytes bearing antheridia and oogonia. Fertilisation, however, was not observed and the sporophyte originated apomictically. Culture studies are now required on North Atlantic isolates to determine if a similar apomeiotic and parthenogenetic life history is in operation. In general the culture results gave support to the placement of *Arthrocladia* in the Desmarestiales, but not to Sauvageau's proposal to establish a separate order Arthrocladiales for this genus.

Sauvageau (1931), pp. 95–121, figs 17–22; Newton (1931), p. 167, fig. 104; Hamel (1931–39), pp. 286–287, fig. 49q; Rosenvinge & Lund (1947), pp. 55–57; Taylor (1957), pp. 162–163, pl. 13, fig. 12, pl. 17, figs 7–8; Rueness (1977), p. 171; Müller & Meel (1982), pp. 419–425, figs 1–15.

Fig. 78 A. Habit of plant. B. Portion of thallus showing whorls of branched filaments. Bar = 3 cm (A), 0·65 mm (B).

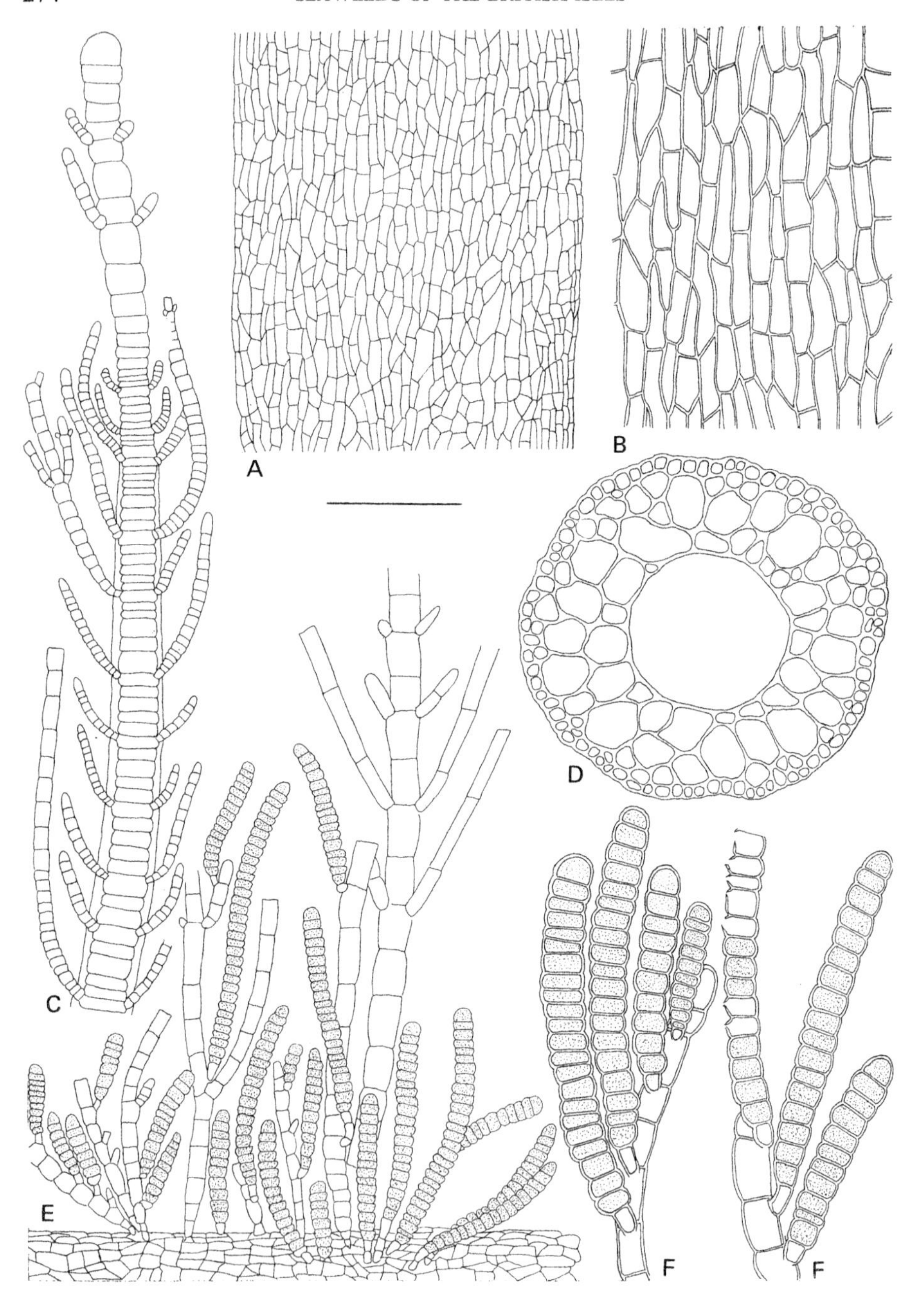
A
B
C
D
E
F
F

DESMARESTIACEAE (Thuret) Kjellman

DESMARESTIACEAE (Thuret) Kjellman (1880), p. 10.

Plants erect, unbranched or branched, terete, slightly compressed to blade-like, branching distichous, alternate or opposite, to several orders, with terminal fascicles of hair-like filaments; growth trichothallic, structure pseudoparenchymatous and solid with central axial filament surrounded by a cortex of large colourless cells, enclosed by 1–3 layers of small pigmented cells; unilocular sporangia scattered over blade surface in little modified surface cells; unispores germinating into microscopic, monoecious or dioecious, branched, filamentous gametophytes bearing oogonia and antheridia.

Only one genus of the Desmarestiaceae, *Desmarestia* occurs in the British Isles.

DESMARESTIA Lamouroux

DESMARESTIA Lamouroux (1813), p. 43 (*nom. cons.*)

Type species: *D. aculeata* (Linnaeus) Lamouroux (1813), p. 25.

Plants erect, much branched, terete, slightly compressed or distinctly flattened and blade-like; branching distichous, alternate or opposite, branches of limited growth, terminal branches at times bearing fascicles of hair-like, pigmented filaments; growth trichothallic, structure pseudoparenchymatous with a central, axial filament surrounded by a solid cortex of large, thick-walled, colourless cells enclosed by 1–2 layers of small, pigmented cells; cells with numerous discoid plastids without pyrenoids.

Unilocular sporangia discrete, scattered, formed in little modified surface cells, slightly immersed; released unispores germinating into separate, microscopic, filamentous, monoecious or dioecious, gametophytes bearing oogonia and antheridia.

Desmarestia is a notoriously polymorphic genus and within the two major filiform and ligulate groups, species delimitation is often considered to be markedly difficult. For example in major revisions of members of both these groups recorded for the Northern Hemisphere, using numerical techniques, Chapman (1972a,b) considerably reduced the number of recognised taxa. In the present work four species of *Desmarestia* are recognised for the British Isles which are easily distinguished on morphological features. These include *D. dresnayi*, retained here as a separate species and not included within the synonomy of *D. ligulata* var. *firma* (C. Agardh) J. Agardh as proposed by Chapman (1972b).

Culture studies on several species of *Desmarestia* around the world (see Abe, 1938; Schreiber, 1932; Kornmann, 1962a; Chapman, 1969; Chapman & Burrows, 1970; Nakahara & Nakamura, 1971; Müller & Lüthe, 1981; Anderson, 1982) have all revealed

---

Fig. 79 *Arthrocladia villosa*
A. Portion of young thallus. B. Surface view of cells. C. Terminal region of branch. D. T.S. of thallus. E. Unilocular sporangia arising from whorls of branched hair-like filaments. F. Unilocular sporangia. Note some sporangia have released contents. Bar = 100 µm (A, D–E), 200 µm (C), 50 µm (B, F).

a life history comprising a heteromorphic alternation between a macroscopic sporophyte and a microscopic gametophyte. As a general rule oppositely branched species were found to be monoecious whilst alternatively branched species were dioecious; the only contradiction to this was Anderson's report of the oppositely branched *D. firma* being dioecious. Notable aspects of these investigatory culture studies on *Desmarestia* species include reports of the influence of environmental conditions such as temperature, light and nutrients on gametophyte maturation (Kornmann, 1962a; Nakahara & Nakamura, 1971; Anderson, 1982; Nakahara, 1984), the involvement of hormonal interactions in gametic discharge and attraction (Müller & Lüthe, 1981) and the occurrence of parthenogenesis in the life history of *D. viridis* (Nakahara, 1984).

Desmarestia plants rapidly deteriorate following collection due to the presence of quite high levels of free sulphuric acid within the cells. The plants quickly turn green, emit a characteristic pungent smell and if not removed, acid released from the thalli will quickly spoil accompanying algae.

Of the four species recorded for the British Isles, *D. dresnayi, D. ligulata* and *D. viridis* are summer annuals which probably overwinter mainly as microscopic gametophytes; *D. aculeata* is perennial.

KEY TO SPECIES

1	Thalli filiform and terete	2
	Thalli flattened and membranous	3
2	Thalli oppositely branched, main axis cartilaginous	*D. viridis*
	Thalli alternately or irregulary branched, main axis coarse and stiff	*D. aculeata*
3	Thalli much branched; axes not more than 7 mm wide	*D. ligulata*
	Thalli unbranched, or rarely with marginal proliferations; axes up to 6 cm wide	*D. dresnayi*

Desmarestia aculeata (Linnaeus) Lamouroux (1813), p. 25. Figs 80, 81; Pl. 13

Fucus aculeatus Linnaeus (1763), p. 1632.

Plants forming erect, solitary, sometimes gregarious, thalli, 0·5–1 (−2) m in length, light brown, cartilaginous when young becoming dark, coarse and stiff later, much branched, main axis fairly distinct, oval, to 3 mm wide at base, slightly more compressed, 1 mm or less above, arising from a small bulbous holdfast; branching regularly or 2–3 times, alternate, more rarely opposite towards base, ultimate branches short and spine-like; in spring/early summer branches bearing opposite fascicles of 1–3 light brown oppositely branched, uniseriate filaments 2–4 mm long, comprising cells 14–120 × 13–28 μm, generally 1 (−2) diameters long below, 3–8 diameters long above, bearing large numbers of discoid plastids without pyrenoids; surface cells of branches irregularly shaped and

Fig. 80 *Desmarestia aculeata*
A. Habit of plant (summer form). B. Part of plant (winter form). Bar = 36 mm (A), 24 mm (B).

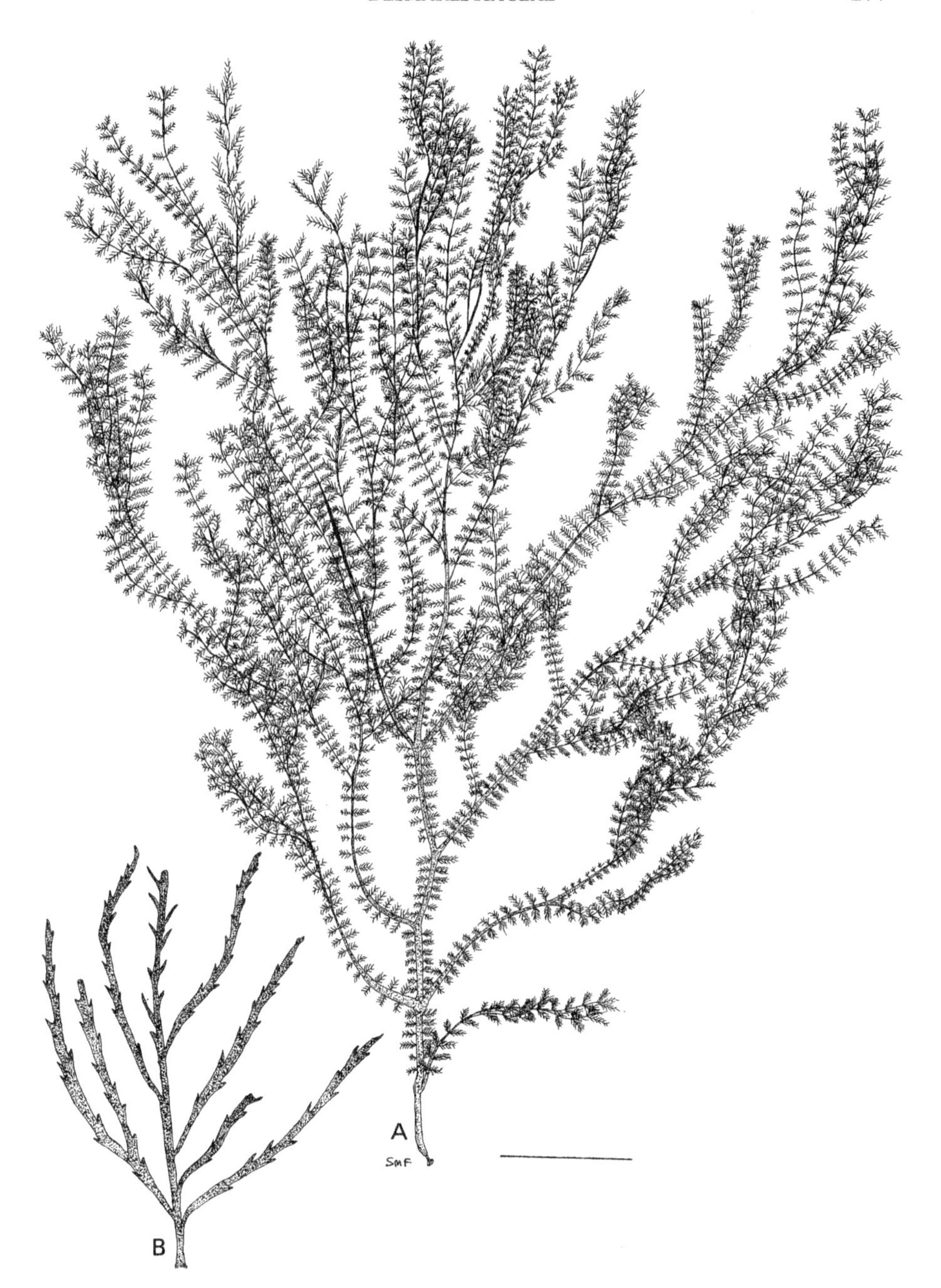
A
SMF
B

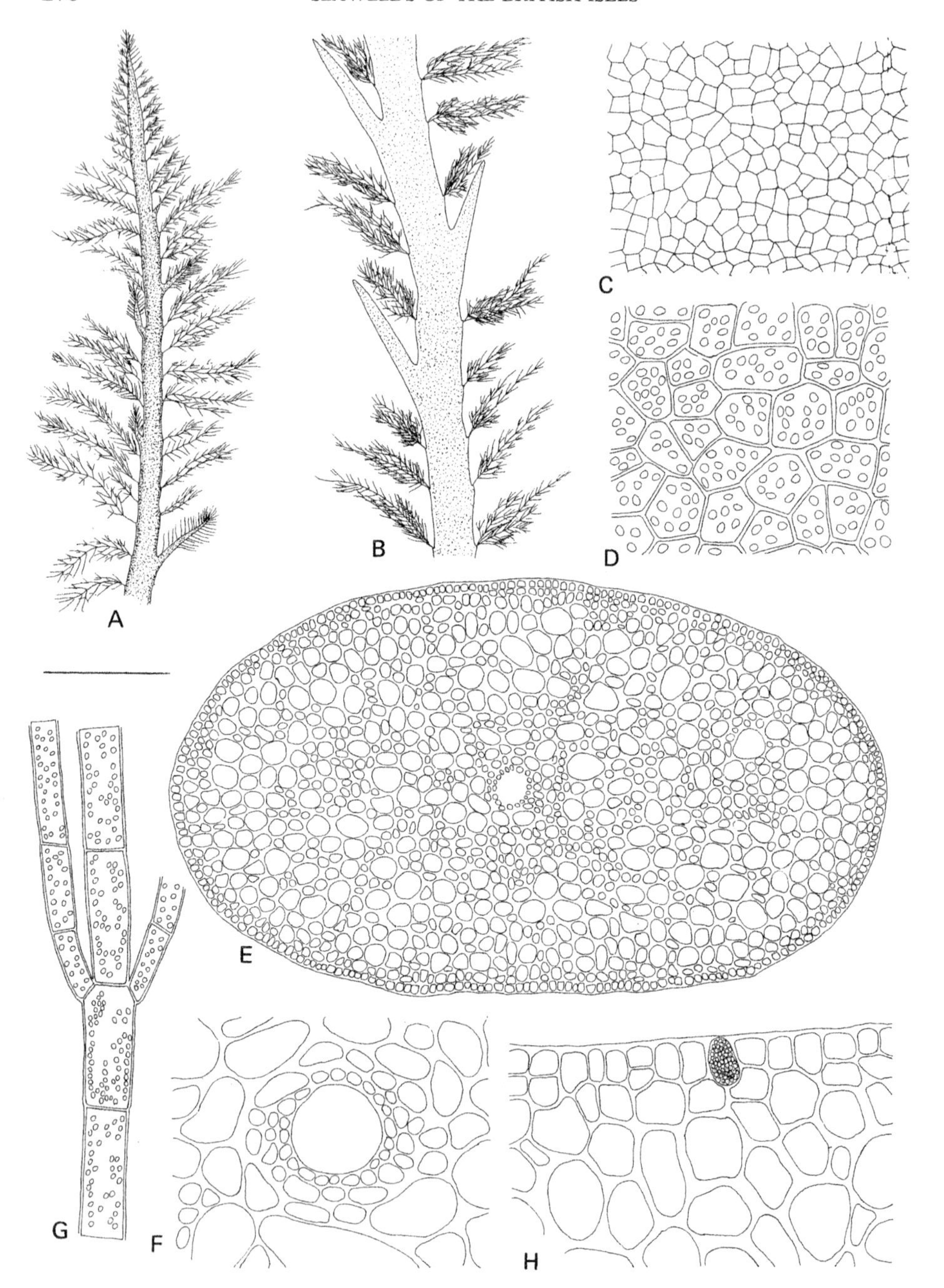
A
B
C
D
E
F
G
H

placed, 8–18 × 5–12 μm, each with numerous, small, discoid plastids; in section, structure pseudoparenchymatous with a large central axial cell, surrounded by a broad cortex of variously sized and irregularly distributed, thick-walled, colourless cells enclosed by an outer layer of 1–2 small, usually taller than wide, pigmented cells.

Unilocular sporangia scattered over both sides of thallus surface, oval in shape and slightly darker than surrounding vegetative cells; in section partially immersed, oval, taller than wide, 15–23 × 9–13 μm.

Epilithic in shaded lower eulittoral pools, or more commonly in the sublittoral to at least 15 m.

Generally distributed around the British Isles.

Plants perennial; spring and early summer plants clothed in opposite fascicles of short, branched, hair-like filaments; these also terminate all branches and are shed by the summer, giving plants a serrated appearance.

Culture studies by Schrieber (1932), Chapman (1969), Chapman & Burrows (1970) and Müller & Lüthe (1981) have revealed North Atlantic isolates of *D. aculeata* to have a heteromorphic life history with the macroscopic sporophyte plant alternating with a microscopic, dioecious gametophyte. Gamete formation in the gametophytic thalli showed a requirement for blue light, whilst discharge and subsequent attraction of the spermatozoids to the egg was controlled by a hormonal interaction (Müller & Lüthe, 1981).

Schreiber (1932), pp. 561–582, figs 1–12; Newton (1931), p. 164, fig. 103; Hamel (1931–39), p. 283, fig. 49m–p; Rosenvinge & Lund (1943), pp. 51–54, fig. 19; Taylor (1957), p. 101, pl. 13, figs 4–5, pl. 14, fig. 7; Gayral (1966), p. 288, pl. LII; Chapman (1969); Chapman & Burrows (1970), pp. 103–108, figs 1–2; Chapman & Burrows (1971), pp. 63–76, figs 1–15; Kornmann & Sahling (1977), p. 140, fig. 75A–D; Rueness (1977), p. 172, pl. 23(3); Fletcher (1980), p. 44, pl. 16, fig. 6; Müller & Lüthe (1981), pp. 351–356, figs 1–14.

Desmarestia dresnayi Lamouroux ex Leman (1819), p. 105. Figs 82, 83

Plants forming erect, solitary, light to yellow brown, becoming olive brown, dorsiventrally flattened blades arising from a distinct, terete stipe and attached by a conical holdfast; erect blades simple, rarely with marginal proliferations, thin, solid, somewhat translucent, membranous, papery, flaccid, delicate, frequently torn, slightly lubricous, linear-lanceolate to strap-shaped, with smooth or scalloped margins, to 35 (−55) cm long, 6 cm wide, narrowing below, stipe 2–4 cm long; blades with distinct midrib extending from stipe, bearing opposite primary veins which branch into secondary veins, primary veins of young thalli only extending to blade margin and terminating in projecting trichothallic, branched, hair-like filaments; in surface view cells irregularly arranged,

Fig. 81 *Desmarestia aculeata*
A. Terminal branch region showing opposite fascicles of filaments. B. Portion of thallus showing fascicles of filaments and short spine-like protuberances. C–D. Surface view of thallus showing cells with numerous discoid plastids. E. T.S. of thallus. F. Central axial filament, in T.S. G. Portion of branched hair-like filament. H. Section of thallus margin showing unilocular sporangium. Bar = 2·5 mm (A), 2 mm (B), 200 μm (E), 50 μm (C, F–H), 20 μm (D).

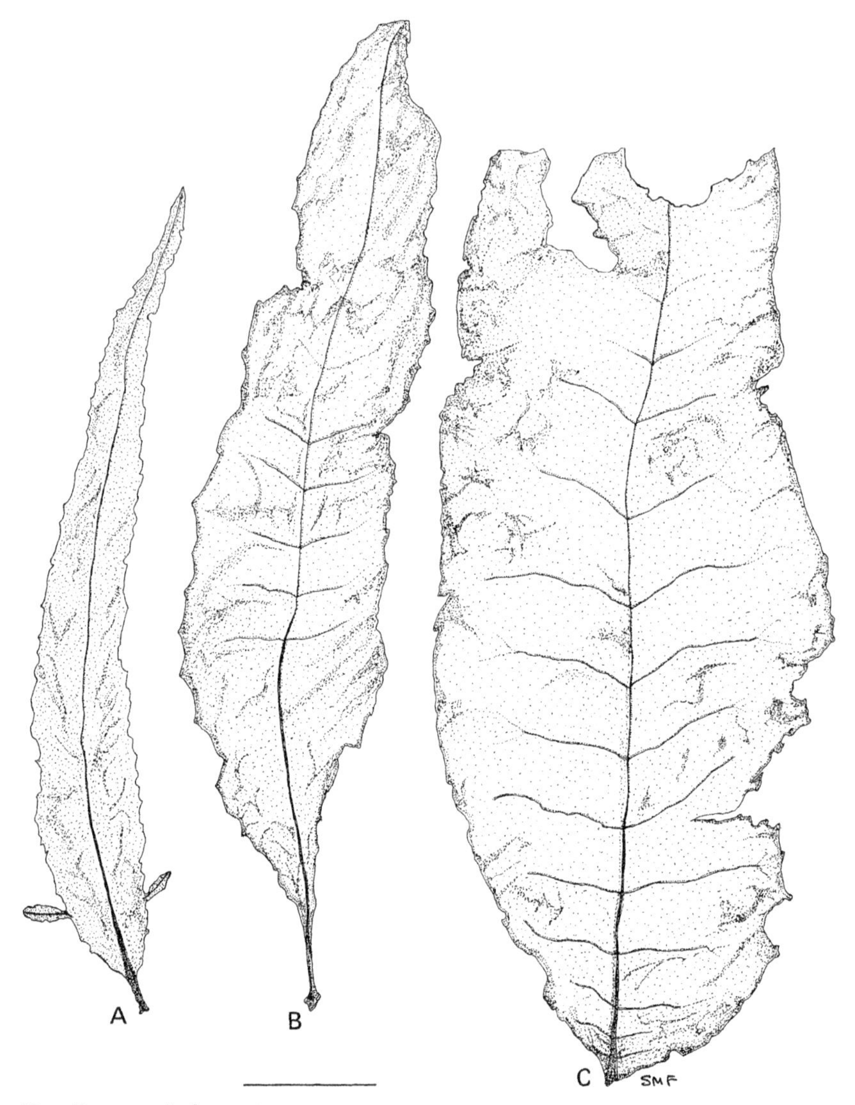

Fig. 82 *Desmarestia dresnayi*
Habit of plants. A–B. Young thalli. C. Portion of older thallus. Bar = 30 mm

Fig. 83 *Desmarestia dresnayi*
A–B. Surface view of thallus showing cells with several discoid plastids. C. T.S. of thallus. Note central axial filament. D–E. Surface view of fertile thallus showing unilocular sporangia. F–G. T.S. of fertile thalli showing unilocular sporangia associated with surface cells. Bar = 50 μm (A, C–D, F), 20 μm (B, E, G).

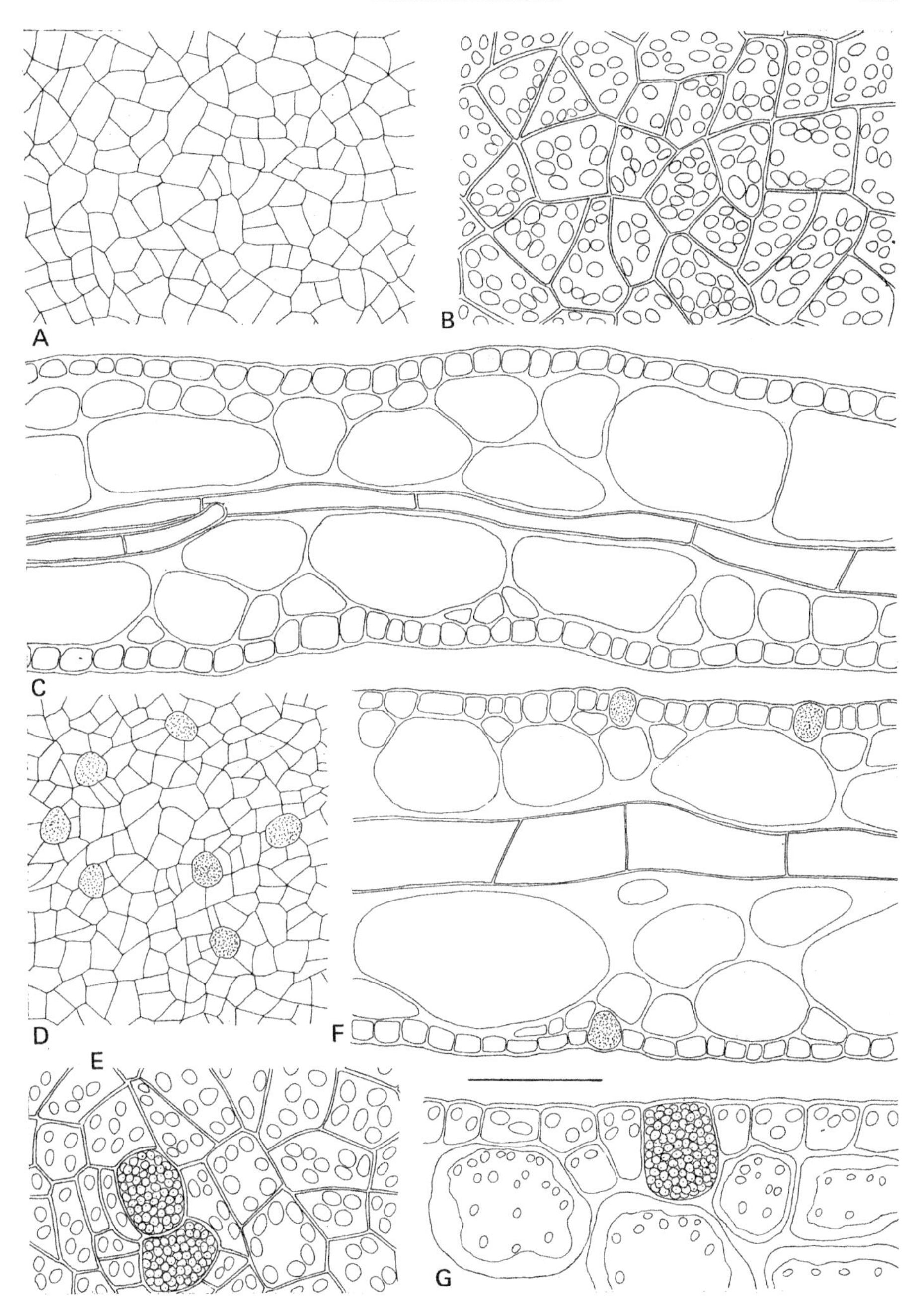
A
B
C
D
E
F
G

variable in shape, although approximately rectangular, 14–32 × 9–17 µm, each with large numbers of discoid plastids without pyrenoids; in section blades to 130 (– 180) µm thick, structure pseudoparenchymatous, comprising a central axial filament, usually intact although not uncommonly disrupted and indiscernible, surrounded by a cortex of 1 (– 2) layers of large, thick-walled, colourless cells, enclosed by an outer layer of small, pigmented cells.

Unilocular sporangia scattered over both sides of blade surface, slighty darker and larger than accompanying vegetative cells, rounded or oval in surface view, 13–21 × 12–16 µm, partly immersed, usually oval, 16–21 × 10–16 µm in vertical section.

Epilithic on small stones and shells embedded in gravel in the sublittoral, in areas of moderate to strong water current, to 18 m.

Only recorded for Devon, Cornwall and Argyll; recorded for Kerry and Donegal in Ireland.

Annual, May to September.

Plants are frequently colonised by the endophytic red alga *Colacodictyon reticulatum* (Batt.) J. Feldmann.

In his taxonomic investigation of the ligulate species of *Desmarestia,* Chapman (1972b) reduced *D. dresnayi* to the synonomy of *D. ligulata* var. *firma.* However, it is retained here as a separate species pending further studies.

Newton (1931), p. 166; Hamel (1931–39), pp. 284–285; Blackler (1961), p. 87; Dizerbo (1965), p. 504; Drew & Robertson (1974), pp. 195–200, figs 1–6.

Desmarestia ligulata (Lightfoot) Lamouroux (1813), p. 25. Figs 84, 85; Pl. 14

Fucus ligulatus Lightfoot (1777), p. 946.

Plants forming erect, solitary, cartilaginous, much branched, dorsiventrally flattened thalli, to 0·5–1 (– 2) m long, 2–3 (– 7) mm wide, light olive-brown in colour, arising from a small terete stipe and lobed holdfast; main axis usually prominent, giving rise to opposite, distichous branches of limited growth similarly branched to several orders, all branches attenuate at base and apex, main branches with distinct midrib; ultimate branches short and spine-like, with or without terminal tufts of hair-like filaments; filaments to 1–2 mm long, uniseriate, oppositely branched, comprising cells 28–80 × 20–28 µm, approximately quadrate below, rectangular to 3 (– 4) diameters long above, containing numerous discoid plastids without pyrenoids; in surface view, thallus cells irregularly shaped and placed, 8–18 × 5–15 µm, each with several discoid plastids; in section thallus pseudoparenchymatous with a large central axial cell surrounded by a broad cortex of large, thin-walled, colourless cells enclosed by 1–2 layers of small, pigmented cells.

Fig. 84 *Desmarestia ligulata*
A. Thallus margin showing tufts of branched filaments. B. Portion of branched filament. C. Surface view of vegetative thallus showing cells with numerous discoid plastids. D. T.S. of vegetative thallus. Note central axial cell. E. Surface view of fertile thallus showing unilocular sporangia. F. Unilocular sporangia associated with surface cells. Bar = 200 µm (A), 50 µm (B), 20 µm (C, E–F), 100 µm (D).

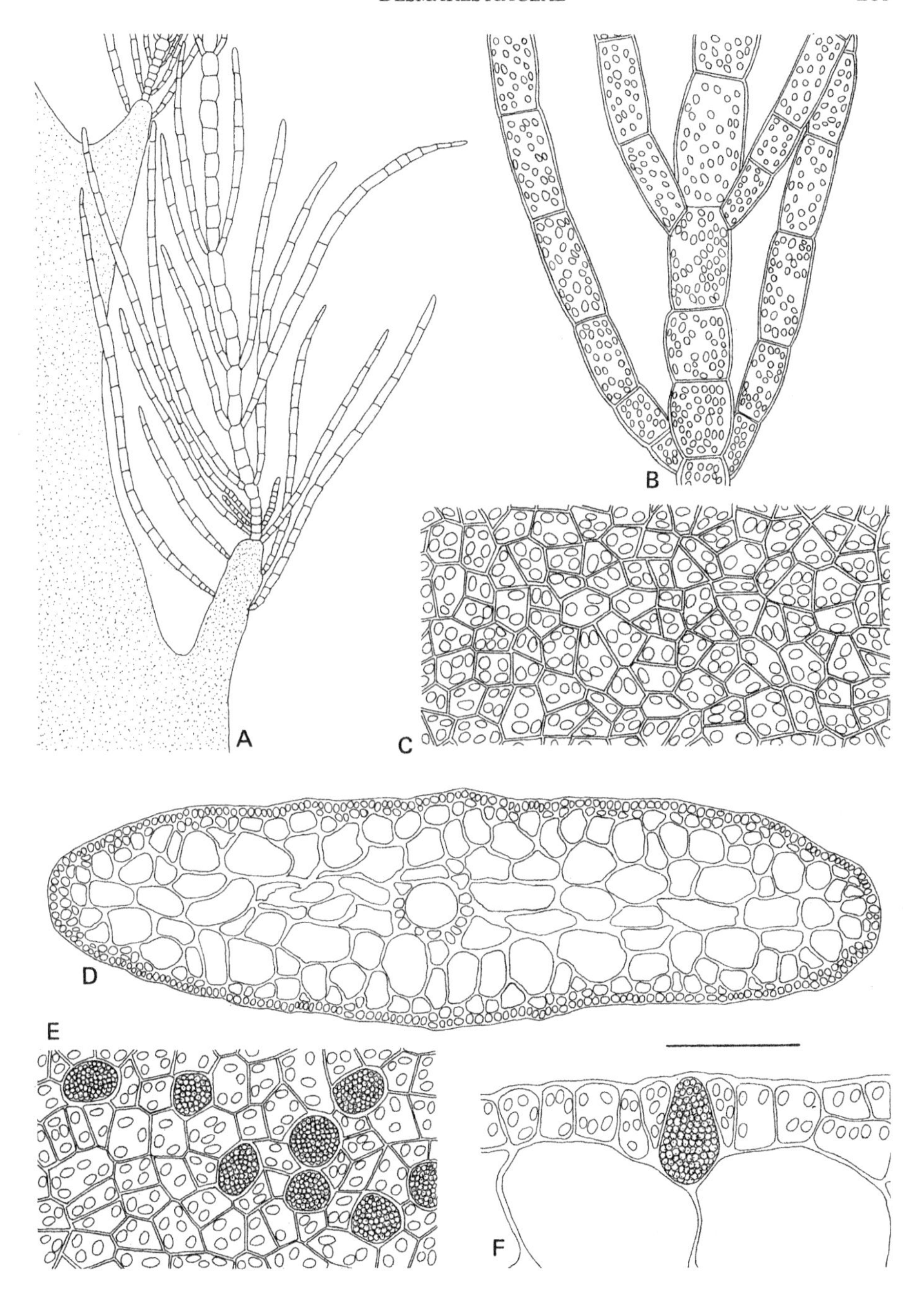
A
B
C
D
E
F

Fig. 85 *Desmarestia viridis*
Habit of plant. Bar = 4 cm.

Unilocular sporangia scattered over both sides of blade surface, rounded or oval in shape, slightly darker than surrounding vegetative cells; in vertical section, oval in shape, taller than wide, partly immersed, 15–21 × 9–12 μm.

Epilithic, lower eulittoral pools, more commonly sublittoral to at least 12 m.

Generally distributed around the British Isles, although appears to be much more common on southern and western shores.

Annual, spring and summer; in spring and early summer ultimate branches bear tufts of hair-like branched filaments; these are shed during the summer. Drift plants commonly washed ashore during the late summer, autumn period.

Culture studies by Nakahara (1984) revealed Japanese isolates of *D. ligulata* to have a heteromorphic life history with the macroscopic sporophyte alternating with a microscopic, monoecious gametophyte.

Newton (1931), pp. 164–165; Hamel (1931–39), pp. 283–284; Gayral (1958), p. 234, pl. 37; Gayral (1966), p. 287, pl. 51; Chapman (1972b), pp. 1–20; Moe & Silva (1977), pp. 159–167; Rueness (1977), p. 172, pl. 23(2); Nakahara (1984), pp. 102–108, figs 14–21, pl. IV, H–L, pl. V, A–H.

Desmarestia viridis O. F. Müller (1771–82), pl. 886. Figs 86, 87; Pl. 15

Sporochnus viridis (O. F. Müller) C. Agardh (1824), p. 259.
Dichloria viridis (O. F. Müller) Greville (1830), p. 39.

Plants forming erect, solitary, light golden brown, much branched, terete thalli, to 0·5 m long, arising from a small bulbous or flattened holdfast; thalli cartilaginous below, becoming soft and flaccid above, slightly lubricous, main axis distinct to 1 (–2) mm wide, slightly compressed, giving rise to regularly opposite, distichous, divaricate, narrowing branches of limited growth, similarly branched to several orders, with or without terminal tufts of branched, hair-like filaments; terminal filaments to 2 (–3) mm long, uniseriate, oppositely branched, comprising cells 18–75 × 15–30 μm, 1 (–1·5) diameters long below, 2–4 diameters long above, each enclosing numerous discoid plastids without pyrenoids; surface cells of axes irregular in shape, irregularly arranged, although in longitudinal rows in terminal branches, 12–20 × 7–15 μm, each enclosing numerous discoid plastids; in section, thallus structure pseudoparenchymatous, with a large, central, axial cell surrounded by a broad cortex of large, colourless cells, enclosed by 1–2 layers of small, pigmented cells.

Unilocular sporangia scattered over thallus surface, slightly larger and darker than accompanying vegetative cells, oval, partly immersed in vertical section, *c.* 10–15 μm in diameter.

Epilithic, occasionally epiphytic on larger algae, in shaded lower eulittoral pools and in the sublittoral to at least 10 m.

Generally distributed around the British Isles.

Annual, spring and summer; in young developing plants terminal fascicles of filaments are present, rendering the thalli more soft and lubricous; these are usually shed by late summer.

Culture studies by Abe (1938), Kornmann (1962a) and Nakahara (1984) on Japanese isolates of *D. viridis* revealed a heteromorphic life history with the macroscopic

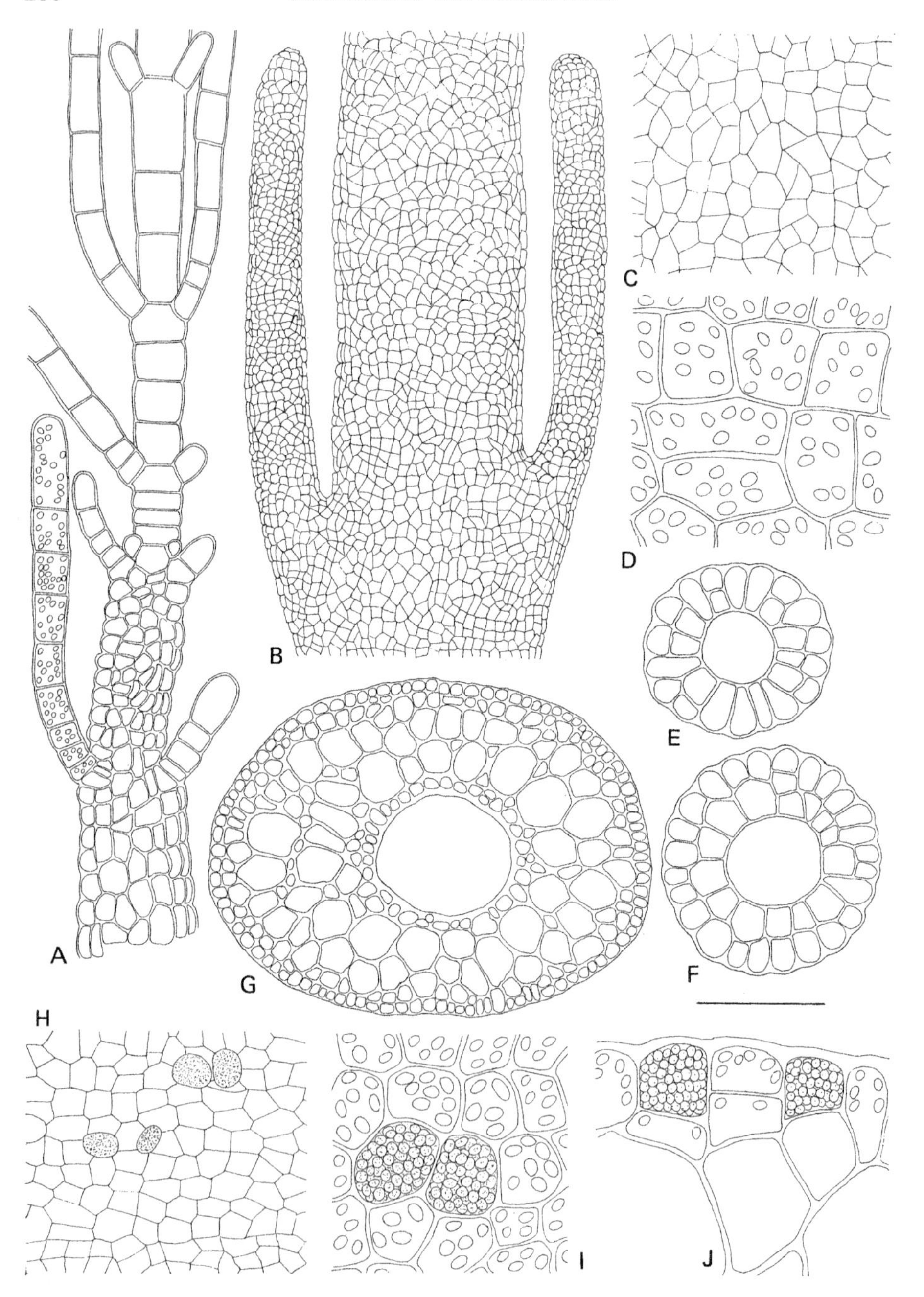
A
B
C
D
E
F
G
H
I
J

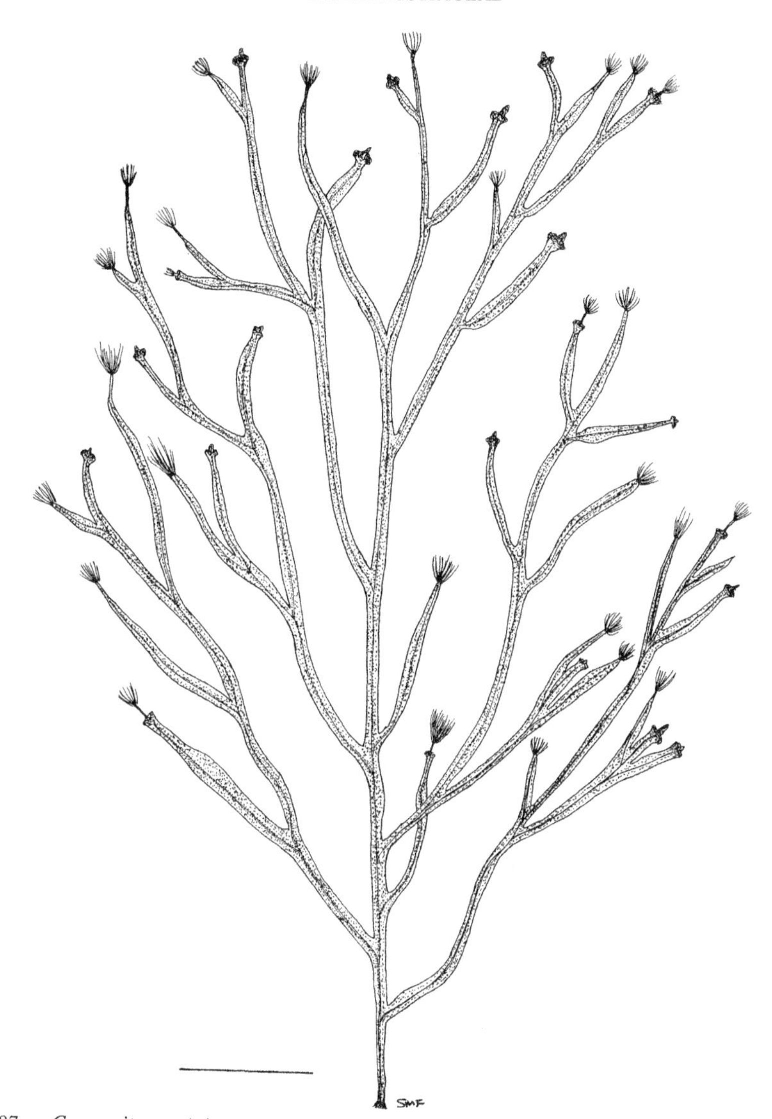

Fig. 87 *Carpomitra costata*
Habit of plant. Bar = 20 mm.

Fig. 86 *Desmarestia viridis*
A. Terminal branch region. B. Portion of oppositely branched thallus. C–D. Surface view of thallus showing cells with numerous discoid plastids. E–G. T.S. of different sized axes. H–I. Surface view of thallus with unilocular sporangia. J. T.S. of fertile thallus showing unilocular sporangia. Bar = 50 μm (A, C, E–F , H), 100 μm (B, G), 20 μm (D, I–J).

sporophyte alternating with a microscopic, monoecious, gametophyte. Nakahara also revealed parthenogenetic development of unfertilised eggs to produce haploid sporophytic plants. Zoospores were produced from the latter without meiosis in the unilocular sporangia.

Newton (1931), p. 164; Hamel (1931–39), p. 282; Abe (1938), pp. 475–482; Rosenvinge & Lund (1947), pp. 54–55; Taylor (1957), pp. 160–161, pl. 13, fig. 3; Kornmann (1962a), p. 287; Chapman (1972a), pp. 225–231, figs 1–4, 10; Kornmann & Sahling (1977), p. 142, pl. 76A–D; Rueness (1977), p. 172, pl. 23(3); Fletcher (1980), p. 44, pl. 16, figs 4–5; Stegenga & Mol (1983), p. 105, pl. 116, fig. 4; Nakahara (1984), pp. 92–102, figs 4–13, pl. II–III, IV, A–G.

SPOROCHNACEAE Greville

SPOROCHNACEAE Greville (1830), p. 36 [as Order IV. Sporochnoideae]

Thallus erect, either terete or slightly ellyptical/dorsiventrally flattened with midrib, much branched, solid, arising from a fibrous holdfast; branching pseudodichotomous, alternate, or triradiate, with or without numerous short lateral branches of limited growth, all branches terminated by a tuft of pigmented, hair-like, unbranched filaments; growth trichothallic, structure pseudoparenchymatous, with a broad central region of large, longitudinally elongate, transversely rounded or irregular, colourless cells enclosed by 1–2 layers of small, pigmented cells; plastids discoid, several in each cell. Unilocular sporangia sessile and clustered, lateral on closely-packed, short, multicellular, branched, paraphyses in slightly raised mucilaginous sori on terminally situated receptacles.

Life history heteromorphic with macroscopic, diploid, sporophytes alternating with microscopic, monoecious or dioecious, haploid gametophytes.

This family is represented in the British Isles by two genera: *Carpomitra* and *Sporochnus,* both easily distinguished on morphological features. Members of this family have a similar life history to those of the Arthrocladiaceae and Desmarestiaceae but differ in having a terminal tuft of unbranched hair-like filaments on all branch-apices and in possessing a multiaxial rather than a uniaxial pseudoparenchymatous mode of construction which is initiated at an early stage by parenchymatous divisions. On the basis of this unique growth mechanism a number of authors, including Fritsch (1945), Lindauer *et al.* (1961), Abbott & Hollenberg (1976), Rueness (1977) and Bold & Wynne (1978) have adopted Sauvageau's (1962a) proposal for ordinal status of the Sporochnaceae.

CARPOMITRA Kützing

CARPOMITRA Kützing (1843), p. 343.

Type species: *C. cabrerae* (Clemente) Kützing (1843) (= *Carpomitra costata* (Stackhouse) Batters (1902), p. 46)

Plants consisting of erect, branched, dorsiventrally flattened to slightly ellyptical, solid thalli arising from a fibrous holdfast; thalli with pseudodichotomous, sometimes triradiate or alternate branches, with a distinct midrib and terminal tuft of pigmented hair-like filaments; young unbranched thalli and terminal young branches often terete; growth trichothallic, structure pseudoparenchymatous comprising a central region of axial cells surrounded by a cortex of longitudinally elongate, transversely rounded or irregular, colourless cells, enclosed by 1–2 layers of small, pigmented, peripheral cells; surface cells of thalli pigmented, in longitudinal rows in midrib region, mainly irregularly arranged elsewhere; cells with numerous discoid plastids without pyrenoids.

Unilocular sporangia borne in continuous sori covering the surface of inverted conical or cylindrical, receptacles which project from branch apices; receptacles usually with basal, surrounding collar and terminal annular swelling from which tufts of pigmented filaments emerge; unilocular sporangia lateral, in clusters, from branched, multicellular, slightly mucilaginous paraphyses.

Germination of unispores gives rise to filamentous, microscopic, monoecious gametophytes bearing oogonia and antheridia.

One species in the British Isles:

Carpomitra costata (Stackhouse) Batters (1902), p. 46. Figs 88, 89

Fucus costatus Stackhouse (1801), p. 110.
Fucus cabrerae Clemente (1807), p. 313.
Carpomitra cabrerae (Clemente) Kützing (1843), p. 343.

Plants forming erect, solitary, much branched, dorsiventrally flattened, slighty ellyptical thalli, 0·1–0·2 m long, 1–3 mm wide, light to olive brown, arising from a small, terete stipe and fibrous holdfast; erect blades quite firm and cartilaginous, solid, with distinct midrib, branching pseudodichotomous, occasionally triradiate or alternate, apex obtuse or truncated, often with young, projecting, cylindrical and terete terminal branches with or without a basal collar and/or terminal tuft of hair-like filaments; surface cells in longitudinal rows in midrib regions, irregularly arranged elsewhere, 8–21 × 6–10 μm, each containing numerous discoid plastids without pyrenoids; terminal filaments 4–6 mm long, simple, attenuate at apex, with short basal meristem, comprising quadrate to rectangular cells 30–160 × 26–34 μm, usually 1–3 diameters long below, 3–6 diameters long above, each densely packed with discoid plastids without pyrenoids; in section thalli with thick-walled, longitudinally elongate, transversely rounded or irregularly shaped cells, distinctly smaller and more thick-walled in the centre, enclosed by a peripheral layer of 1–2, small, pigmented cells.

Unilocular sporangia borne in continuous gelatinous sori covering and encircling the surface of specialised receptacles; receptacles formed from extending, midrib tissue, conical or cylindrical, colourless, to 2 mm long, terminal on branched apices, less frequently in axes of di-trichotomousy branching thalli, usually with basal, surrounding collar and with terminal annular swelling from which tufted filaments extend; unilocular sporangia elongate-clavate or elongate-pyriform, 33–48 × c. 10–13 μm, in large numbers lateral on erect, mucilaginous, densely packed, projecting paraphyses; paraphyses di-trichotomously branched, to 140 μm long, 5 (−7) cells, comprising cells 1–4 diameters long, rectangular or slightly swollen, 10–26 × 5–16 μm with enlarged, bulbous apical cell, 10–23 μm in diameter.

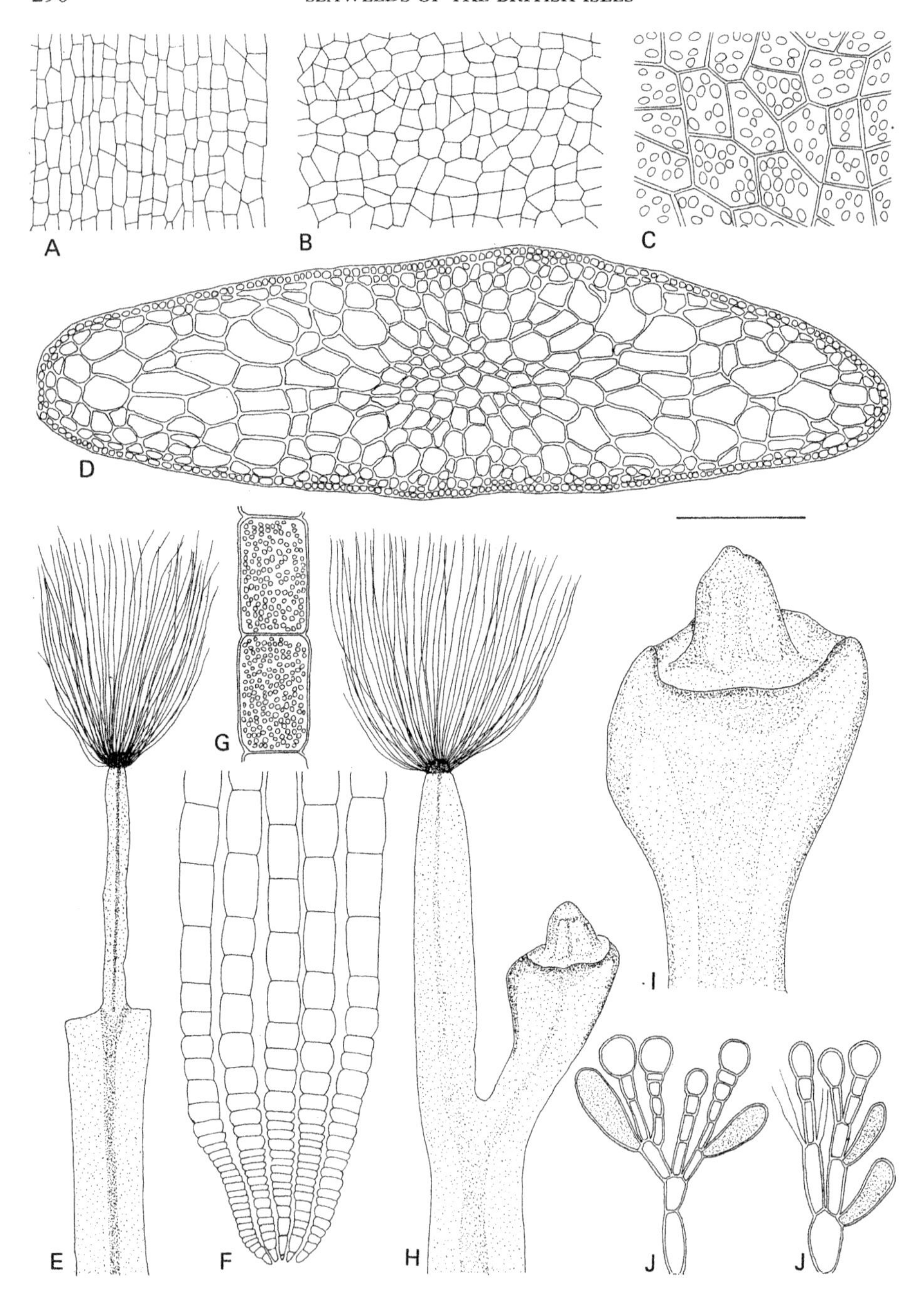
A
B
C
D
E
F
G
H
I
J
J

Epilithic on bedrock and boulders, sublittoral to 37 m; tolerant of sand cover.

Southern and south western shores of the British Isles (Channel Isles, Hampshire, north and south Devon, Cornwall, Pembroke, Isle of Man; recorded for Cork and Donegal in Ireland).

Probably a summer annual; recorded June to September.

C. costata has only been investigated in laboratory culture by Sauvageau (1926) who revealed a heteromorphic life history with the erect, macroscopic sporophyte alternating with a microscopic, filamentous, monoecious gametophyte bearing oogonia and antheridia.

Newton (1931), pp. 137–138, fig. 94; Hamel (1931–39), pp. 274–276, fig. 49d–i; Hai(n)sworth (1976), p. 61; Hiscock & Maggs (1984), p. 85; Sauvageau (1926b), pp. 141–192, figs 1–17; Lindauer *et al.* (1961), pp. 245–246, fig. 61.

SPOROCHNUS C. Agardh

Sporochnus C. Agardh (1817), p. 12.

Type species: *S. pedunculatus* (Hudson) C. Agardh (1817), p. 149.

Plants forming erect, much branched, terete, solid thalli arising from a fibrous holdfast; main axis distinct, branching alternate or irregularly spiralled, to one order only, both primary and secondary axes beset with numerous, short, divaricate branches of limited growth, all branches terminated by a tuft of pigmented, hair-like filaments; growth trichothallic, structure pseudoparenchymatous, with a broad central region of large, fairly thick-walled longitudinally elongate, transversely rounded or irregularly shaped colourless cells, enclosed by 1–2 layers of small, pigmented, peripheral cells; surface cells in longitudinal rows, each containing a few discoid plastids without pyrenoids.

Unilocular sporangia lateral, usually in large numbers, on short, multicellular, branched paraphyses, in slightly raised, mucilaginous sori, developing on the short lateral branches of limited growth.

Life history heteromorphic with unispores germinating into microscopic, either monoecious or dioecious, haploid gametophytes bearing oogonia and antheridia.

Fig. 88 *Carpomitra costata*

A. Surface view of thallus, mid-rib region. B–C. Surface view of thallus, blade region. Note discoid plastids in cells. D. T.S. of thallus. E. Branch apex showing new growth and terminal tuft of filaments. F–G. Portions of filaments. H–I. Branch apices showing receptacles. J. S.P. of unilocular sporangial sori showing unilocular sporangia associated with short paraphyses. Bar = 50 μm (A–B, G–J), 20 μm (C), 100 μm (D, F), 2·5 mm (E, H), 5 mm (I).

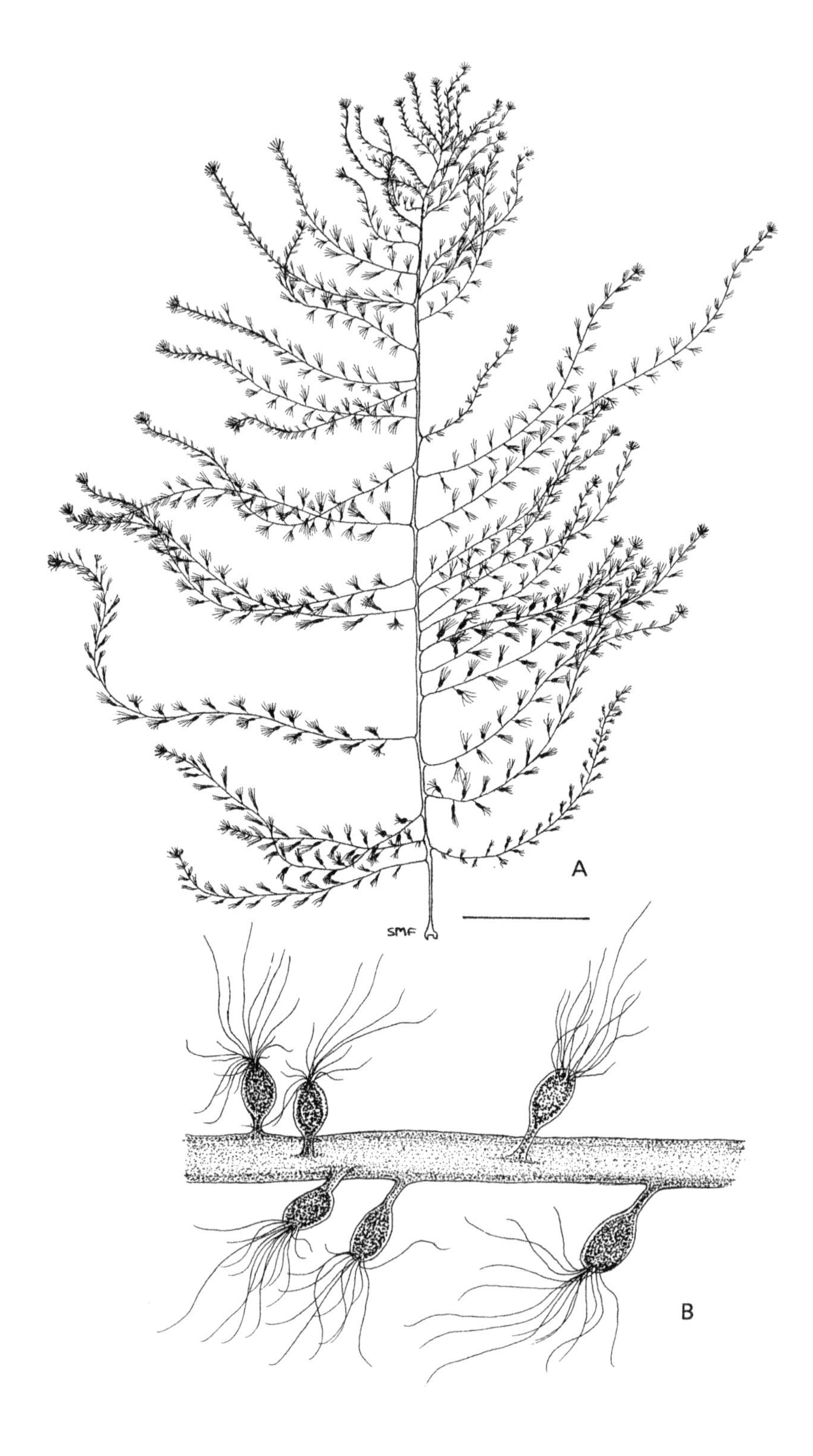
A
SMF
B

Sporochnus pedunculatus (Hudson) C. Agardh (1817), p. 149. Figs 89, 90

Fucus pedunculatus Hudson (1778), p. 587.

Plants forming erect, much branched, terete thalli, usually solitary, occasionally gregarious, light to olive-brown, arising from a small, fibrous holdfast; erect thalli with a distinct main axis, quite firm and cartilaginous becoming soft terminally, solid, to 30 (−50) cm long, 0·5 (−0·65) mm wide, branching alternate, or irregularly spiralled, to one order only, with the branches at well spaced intervals, divaricate, gradually narrowing terminally to a tuft of fine, hair-like filaments; all axes beset with numerous, short, linear, clavate or pyriform, divaricate branches of limited growth, to 3 (−5) mm long, 0·1–0·2 (−0·3) mm wide, with similar terminal tuft of filaments; filaments to 4·5 mm long, uniseriate, simple, with basal meristem, comprising cells 1–3 diameters long below, 28–75 × 16–29 μm, to 10 diameters long above, 40–70 × 5–8 μm, each containing numerous discoid plastids without pyrenoids; surface cells of axes in longitudinal rows, usually longer than wide, 1–6 diameters long, commonly rectangular, occasionally with oblique/pointed end walls, 16–52 × 5–17 μm each with a few discoid plastids without pyrenoids; in section thalli with a broad central region of large, fairly thick-walled, longitudinally elongate, transversely rounded or irregular, colourless cells, enclosed by 1–2 layers of small, slightly pigmented cells.

Unilocular sporangia borne in sori on the short lateral branches of limited growth; sori slightly raised, continuous, gelatinous, completely encircling and swelling the terminal regions of the longer, older branches, appearing pedicellate, to 0·7 mm long × 0·4 mm wide, entirely covering the shorter, younger branches and shortly spreading onto the parental axes, appearing sessile; unilocular sporangia clavate, pyriform, elongate-pyriform or elongate-cylindrical, 26–43 × 8–12 μm, single or grouped, lateral and sessile on short paraphyses; paraphyses produced as lateral extensions of the outer cortical cells, simple, or more commonly much branched, clavate, multicellular, to 3 (−5) cells long, 65 (−90) μm, comprising mainly rectangular, colourless cells, with enlarged bulbous apical cell, 13–16 μm in diameter.

Epilithic, on small stones, gravel, shells, etc., particularly in silty areas and often associated with *Arthrocladia villosa,* in the sublittoral to 18 m.

Rare but generally distributed around the British Isles; more records for southern and western coasts.

Annual, spring and summer, April to September.

Culture studies by Sauvageau (1931) and Caram (1965) revealed a heteromorphic life history with the macroscopic, diploid sporophyte alternating with a microscopic, either monoecious or dioecious, haploid gametophyte. Caram also noted parthenogenetic development of the antherozoids to form microthalli with unilocular sporangia; spores released from the latter recycled the male gametophyte.

Fig. 89 *Sporochnus pedunculatus*
A. Habit of plant. B. Portion of thallus, showing lateral branches of limited growth with terminal tuft of filaments. Bar = 2 cm (A), 2·4 mm (B).

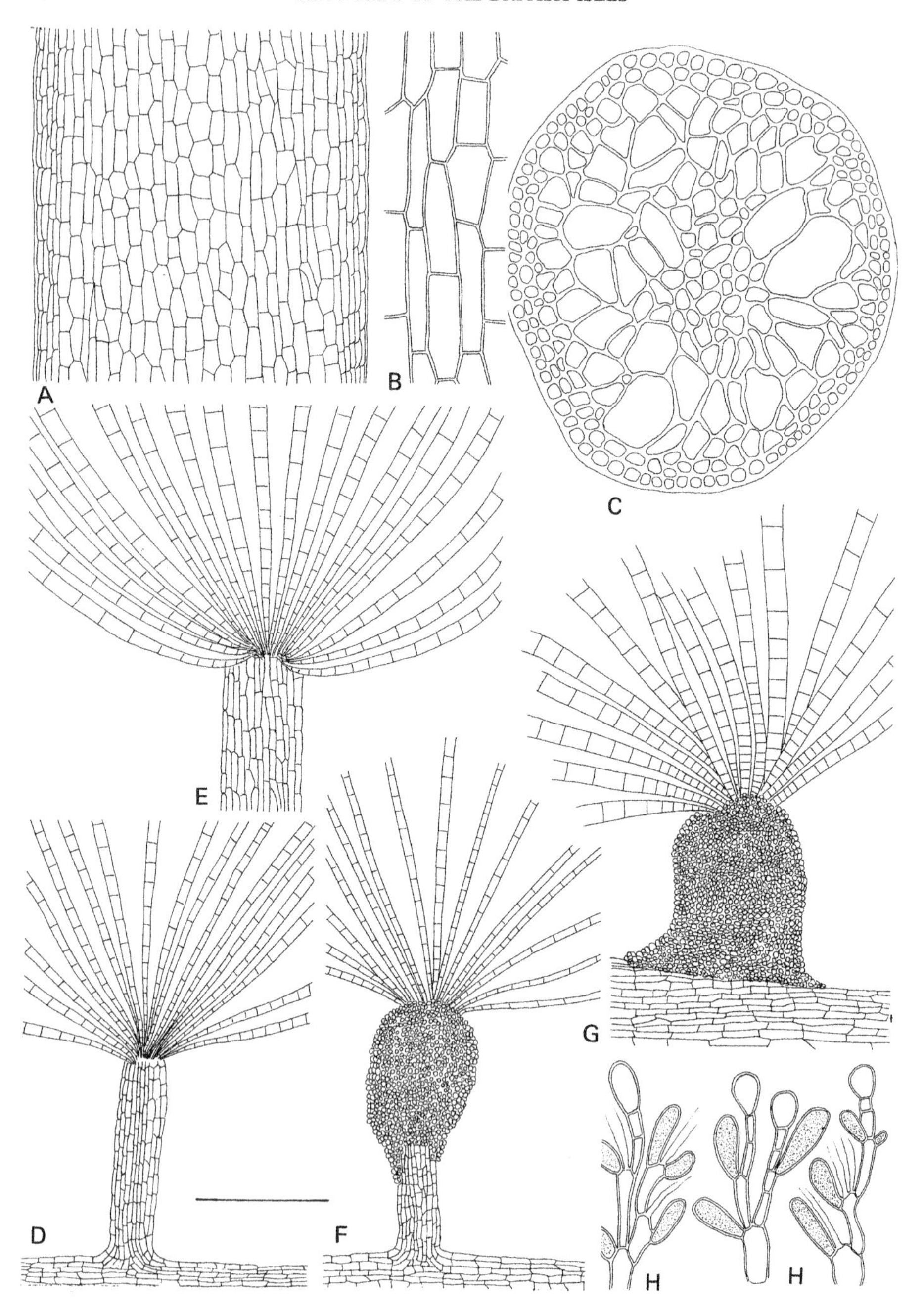
A
B
C
E
D
F
G
H
H

Newton (1931), p. 137, fig. 82; Sauvageau (1931), pp. 122–126, fig. 23; Hamel (1931–39), pp. 276–277, fig. 49j; Kylin (1947), p. 66, fig. 29; Rosenvinge & Lund (1947), pp. 49–50, fig. 18; Lindauer *et al.* (1961), pp. 242–243, fig. 59; Caram (1965), pp. 146–165, figs 1–11, pl. VI; Gayral (1966), p. 285, pl. 50; Abbott & Hollenberg (1976), pp. 184–185), fig. 149; Rueness (1977), p. 171, fig. 102.

Fig. 90 *Sporochnus pedunculatus*
A. Portion of thallus. B. Surface view of thallus. C. T.S. of thallus. D–E. Short lateral branches of limited growth, with terminal tuft of filaments. F–G. Lateral branches clothed in unilocular sporangial sori. H. S.P. of sori showing unilocular sporangia associated with short paraphyses. Bar = 100 µm (A, C), 50 µm (B, H), 200 µm (D–G).

WORLDWIDE DISTRIBUTION OF BROWN ALGAE INCLUDED IN PART 1

	Greenland	Iceland	Faeroes	Norway	Sweden	Baltic	Denmark	Germany (incl. Helgoland)	Netherlands	Belgium	France	Spain	Portugal	Canary Isles	Mediterranean	Black Sea	Red Sea	West Africa	South Africa	East Africa	Indian Ocean	West Atlantic – Canada	West Atlantic – USA	Caribbean	Atlantic – South America	Pacific – Alaska	Pacific – Canada	Pacific – USA	Pacific – South America	Indo-Pacific	Australia	New Zealand
Arthrocladia villosa				+	+		+				+	+	+		+	+							+	+							+	
Asperococcus compressus				+			+				+	+		+	+																	
A. fistulosus			+	+	+	+	+		+	+	+	+		+								+	+									
A. scaber								+							+																	
A. turneri				+	+		+				+	+	+	+	+	+			+											+	+	+
Carpomitra costata											+	+			+																+	+
Chilionema foecundum	+	+	+	+	+	+					+												+									
C. hispanicum											+	+																				
C. ocellatum				+	+	+	+				+																			+		
C. reptans				+	+	+					+	+	+		+							+	+									
Colpomenia peregrina				+			+		+	+	+	+	+	+	+			+				+				+	+	+		+	+	+
Compsonema microspongium				?																												
C. minutum														+	+			?			+											
C. saxicolum 'phase'							+	+																								
Corynophlaea crispa											+			+																		
Cutleria multifida				+	+		+				+	+	+		+			+												+	+	+
Desmarestia aculeata	+	+	+	+	+	+	+	+			+	+	+		+							+	+							+		
D. dresnayi											+	+			+																	
D. ligulata			+	+						+	+	+	+					?							+		+	+	+	+	+	+
D. viridis	+	+	+	+	+	+	+	+	+		+				+							+	+			+	+	+	+	+		
Elachista flaccida									+		+	+	+		+															+		
E. fucicola	+	+	+	+	+	+	+	+	+	+	+	+	+									+	+			+	+	+				
E. scutulata			+	+					+		+	+	+																			
E. stellaris				+	+	+	+		+						+								+									
Halothrix lumbricalis				+		+	+				+											+	+					+		+		
Hecatonema maculans	+	+		+	+	+					+			+								+	+		+				+	+		
Leathesia difformis			+	+	+	+	+		+		+	+	+		+	+			+			+	+		+	+	+	+	+	+		+
Leblondiella densa											+	+											+									

Species																															
Leptonematella fasciculata	+	+	+	+	+	+	+	+	+		+	+			+						+	+					+		+		
Litosiphon laminariae		+	+	+					+		+	+																			
Microcoryne ocellata				+	+		+		+																						
Microspongium globosum	+		+	+	+	+	+				+	+									+	+									
M. gelatinosum 'phase'						+	+														+										
Myriactula areschougii											+																				
M. chordae				+	+		+															+									
M. clandestina	+																														
M. haydenii				+	+		+																								
M. rivulariae							+		+		+	+			+	+															
M. stellulata				+					+		+				+																
Myrionema corunnae		+	+	+	+		+		+		+	+	+								+	+					+		+		
M. liechtensternii															+																
M. magnusii				+	+	+	+		+		+	+		+	+	+		+			+	+					+				
M. papillosum											+																				
M. strangulans	+		+	+	+	+	+		+		+	+	+	+	+	+				+	+	+	+	+	+	+	+	+	+	+	+
Myriotrichia clavaeformis	+	+		+	+	+	+		+		+	+	+		+	+					+	+	+								
Petalonia fascia	+	+	+	+	+	+	+	+	+	+	+	+	+		+			+	+		+	+	+	+	+	+	+	+	+	+	+
P. filiformis		+		?				+																							
P. zosterifolia	+	+	+	+	+	+	+	+	+												+						+		+	+	
Petroderma maculiforme	+	+	+	+	+	+	+	+									+				+	+		+	+		+				
Petrospongium berkeleyi											+	+	+																		
Pogotrichum filiforme	+	+	+	+	+		+	+	+												+	+									
Protectocarpus speciosus			+	+	+		+		+		+	+									+										
Pseudolithoderma extensum	+		+	+	+	+	+		+		+										+	+			+						
P. roscoffense											+																				
Punctaria crispata											+		+								+	+									
P. latifolia			+	+					+		+	+			+	+					+	+	+		+				+	+	+
P. plantaginea	+	+	+	+	+	+	+	+	+			+									+	+		+	?				+		
P. tenuissima			+	+	+	+	+		+			+			+						+	+									
Ralfsia verrucosa	+	+	+	+	+	+	+	+	+	+	+	+	+		+	+	+	+	+	+	+	+							+		+
Scytosiphon dotyi																					+						+				
S. lomentaria	+	+	+	+	+	+	+	+	+	+	+	+	+		+	+	+	+	+		+	+	+	+	+	+	+	+	+	+	+
Sorapion simulans											+																				
Sporochnus pedunculatus				+	+		+				+	+	+		+								+				+				+
Stragularia clavata	+		+	+	+	+	+		+		+	+	+					+			+	+			+				+		
S. spongiocarpa																															
Symphyocarpus strangulans	+			+																											
Ulonema rhizophorum				+	+		+		+												+										
Zanardinia prototypus											+	+	+		+	+															

GLOSSARY

ACUMINATE Tapering gradually to a point.
ACUTE Tapering sharply to a point.
ADAXIAL The side, usually the uppermost, of a lateral adjacent to the axis.
ADHERENT Closely attached.
ADJOINED Joined together, weakly or tightly; said of filaments.
ADVENTITIOUS Arising in an irregular manner or from an abnormal position.
AIR-BLADDER Air-filled vesicle serving for flotation.
ALTERNATE Placed singly, with change of side.
ANISOGAMETES Similar motile gametes of unequal size.
ANISOGAMY The fusion of two similar motile gametes of unequal size.
ANTHERIDIUM Male reproductive organ, producing gametes.
ANTHEROZOID A male gamete.
ANTICLINAL At right angles to the surface.
APICAL CELL Terminal initial cell of a filament or thallus.
APICAL MARGIN Margin of plant entirely comprising apical cells.
APOMEIOSIS The formation of spores without meiosis.
ASCOCYST An abnormal enlarged, hyaline, or darkly staining cell, usually empty when old.
ASEXUAL Not involving the process of sexual fusion.
ASSURGENT Curving upwards; said of some horizontal filaments which turn upwards.
ASTIPITATE Not having a stipe.
ATTENUATE Slender, thin, tapering.
AXIAL FILAMENT The central filament of a thallus or its branch.
AXIS The central portion of the thallus.
BENTHIC Attached to or living on the sea bottom.
BIFLAGELLATE Possessing two flagella, said of gametes or spores.
BIFURCATE Divided into two branches or forks.
BIPHASIC With two ploidy levels (e.g. haploid and diploid).
BISERIATE Consisting, wholly or in part, of two rows of cells in longitudinal series.
BLADE A flattened leaf-like structure.
BULLATE With a bulging or puckering of the surface.
CALCIFIED Encrusted or impregnated with lime.
CARTILAGINOUS Hard, tough and elastic.
CATENATE In the form of a chain.
CLAVATE Club-shaped.
COMPRESSED Flattened from side to side.
CONCEPTACLE A small, usually flask-shaped cavity in which sexual organs develop.
CONFLUENT Becoming united to form one.
CORIACEOUS Leathery.
CORTEX A term loosely used for the outer regions of a thallus, with or without an enclosing epidermis.

CORTICATED Having a cortex.
CRUST A thin, closely adherent outwardly radiating thallus, pseudoparenchymatous in structure with laterally adjoined basal and erect filaments.
CRUSTOSE Resembling a crust; said of thalli.
CRYPTOSTOMATA Conceptacles containing hairs only.
CUNEATE Wedge-shaped.
CUTICLE Thin sheet of tissue on outside of thallus.
CYTOPLASM Living cell material.
DICHOTOMOUS Equally forked; said of branching.
DICHOTOMY Division by means of an apical cell so that forked branching occurs.
DIMORPHIC Occurring in two morphological forms. (also = HETEROMORPHIC).
DIOECIOUS With male and female gametangia on separate thalli.
DIPLOID Having double the haploid number of chromosomes.
DISC A thin, closely adherent, often circular thallus, comprising outwardly radiating, laterally adjoined, usually synchronously extending filaments, monostromatic, or at least partly distromatic, in structure.
DISCOID Like a disc.
DISTICHOUS Arranged in two opposite rows.
DISTROMATIC Composed of a layer, at least in parts, two cells in thickness.
DIVARICATE Spreading at a wide angle; said of branching.
DIVERGENT Spreading at a moderate angle; said of branching.
DORSAL Pertaining to the upper surface of a dorsiventral thallus.
DORSIVENTRAL Said of thalli with the dorsal and ventral surfaces unlike.
ECORTICATE Without a cortex.
ECTOCARPOID Having a structure which resembles that of *Ectocarpus*; said of thalli which comprise erect, branched filaments, and plurilocular sporangia which are siliquose and multiseriate.
EGG A female gamete, without flagella.
ENDOGENOUS Originating from below the surface of a thallus.
ENDOPHYTIC Living within a plant.
ENDOZOIC Living within an animal.
ENTIRE Without divisions, lobes, proliferations or indentations; said of thallus margins.
EPIDERMIS The outermost cell layer of a thallus; usage usually restricted to anatomically complex thalli (e.g. in Laminariales, Fucales).
EPILITHIC Living on rock or stones.
EPIPHYTIC Living on a plant, but attached to the surface only.
EPIZOIC Living on an animal, but attached to the surface only.
EXOGENOUS Originating from the surface of a thallus.
FASCICULATE Forming a dense cluster or bundle; said of filaments or branches.
FERTILE Bearing sporangia or gametangia.
FILAMENT A branched or unbranched row of cells joined end to end.
FILIFORM Thread-like.
FIMBRIATE With a fringed edge or margin.
FLACCID Limp.
FLAGELLATE Bearing one or more flagella.
FLAGELLUM A thread-like outgrowth of the cytoplasm, used for locomotion.
FROND That part of the thallus other than the attachment structure
FOLIACEOUS Leaf-like.

FUCOSAN A substance which accumulates in cells, considered to be a by-product of metabolism which has tannin-like properties.
FUCOXANTHIN A brown pigment in the plastids of the Fucophyceae.
FUSIFORM With the shape of a spindle, tapering at both ends.
GAMETANGIUM A cell producing one or more gametes.
GAMETE A sexual cell capable of uniting with another sexual cell.
GAMETOPHYTE A morphological phase which bears gametangia.
GELATINOUS Slimy and jelly-like
GLOBOSE Spherical or globular.
HAIR A general term for usually unbranched, hyaline deciduous filaments. Two types are distinguished in the Fucophyceae viz. the common *True hairs* which are colourless with a distinct basal meristem and basal collar or sheath and the less common *Pseudo-hairs* which possess plastids, terminate filament tips only and lack a distinct basal meristem and sheath.
HAPLOID Having the number of chromosomes characteristic of the gamete.
HAPTERON (pl. haptera) A specialized multicellular attachment structure; usually cylindrical and branched; present in the Laminariales.
HECATONEMOID Said of a thallus which resembles *Hecatonema* (discoid, distromatic).
HETEROBLASTY The formation at two morphologically different germinating patterns from spores liberated from a single sporangium.
HETEROMORPHIC Having morphologically dissimilar forms; said of life histories.
HETEROTHALLIC With gametes produced from separate thalli.
HETEROTRICHOUS With two parts to the thallus; a prostrate basal portion and an erect one.
HOLDFAST Basal attachment organ; usually discoid or conical, comprising a mass of rhizoids.
HYALINE Colourless; transparent.
IMMERSED Embedded within.
INFLATED Distended with air or gases.
INITIAL The earliest stage of a cell, tissue or structure.
INTERCALARY Occurring in any position in a thallus or filament other than the apex or base.
INTERCELLULAR Lying between cells.
INTERTIDAL Region of shore lying between high and low tide levels.
ISODIAMETRIC Cells of equal diameter; as broad as long.
ISOGAMETES Similar motile gametes of equal size.
ISOGAMY The fusion of two similar motile gametes of equal size.
ISOMORPHIC Having morphologically similar forms; said of life histories.
INTRACELLULAR Within a cell.
LAMINA A thin, plate-like thallus.
LANCEOLATE Narrow and tapering at both ends.
LIGULATE Strap-shaped.
LINEAR Long and narrow with parallel sides.
LITHOPHYTE A plant growing on rock or stones.
LITTORAL Applied to that portion of the shore which is alternately exposed to the air and wetted either by the tide or by splash and spray.
LOCULI The individual compartments of a plurilocular sporangium.
MACROSCOPIC Visible to the unaided eye.

MARGIN The outer edge of a structure or thallus.

MEDULLA A term loosely used for the internal region of a thallus; it may be parenchymatous, pseudoparenchymatous or distinctly filamentous.

MEIOSIS The process of nuclear division in which the chromosome number is halved.

MEMBRANACEOUS Like a membrane.

MERISTEM A group of cells which divide and cause growth.

MICROSCOPIC Not visible to the unaided eye.

MIDLITTORAL Applied to the middle portion of the littoral q.v.

MIDRIB The thickened longitudinal axis of a flattened thallus.

MONILIFORM Bead-like.

MONOECIOUS With male and female gametangia on the same thallus.

MONOMORPHIC Occurring in one morphological form.

MONOPHASIC With a single ploidy level (e.g. haploid or diploid).

MONOSPORANGIUM A sporangium in which a single spore is formed.

MONOSPORE A spore formed in a monosporangium.

MONOSTROMATIC Composed of a single layer of cells.

MUCILAGE A watery sticky substance.

MULTIAXIAL Of a thallus containing several axial filaments.

MULTISERIATE Consisting, wholly or in part, of several rows of cells in longitudinal series.

MYRIONEMOID Said of a thallus which resembles *Myrionema* (discoid, monostromatic).

OOGAMY The fusion of two dissimilar gametes, usually a large non-motile egg and a small motile antherozoid.

OOGONIUM Female reproductive organ in which the egg is formed.

ORGANELLE A specialized structure within a cell e.g. plastid, nucleus.

OSTIOLE A small pore-like opening, usually of a conceptacle.

OVATE Egg-shaped with broader end at base.

PARAPHYSIS (pl. paraphyses) A sterile filament, or cell usually associated with the reproductive organs.

PARENCHYMA A compact tissue formed by cell division in all planes.

PARENCHYMATOUS Composed of parenchyma.

PARIETAL Lying along the wall, peripheral to the cell.

PARTHENOGENETIC Developing without fertilization.

PEDICEL A stalk of a reproductive organ.

PEDICELLATE Borne on a pedicel.

PERICLINAL Parallel to the surface.

PLASTID A specialized cytoplasmic organelle which may contain the photosynthetic pigments.

PLURILOCULAR Containing many loculi or cells, said of reproductive organs.

PLURISERIATE More than two loculi or cells wide.

PLURISPORE A spore produced from a plurilocular sporangium.

POLYMORPHIC Occurring in many morphological forms.

PROCUMBENT Lying flat.

PROLIFERATION Development of new parts of a thallus by vegetative growth.

PSEUDODICHOTOMOUS Forming two unequal dichotomies; said of branching.

PSEUDODISC Having the appearance of a disc but comprising irregularly adjoined and associated filaments.

PSEUDOPARENCHYMA A tissue formed by the aggregation of branched or unbranched filaments and having the appearance of parenchyma.

PYRENOID An organelle occurring within or adjacent to a plastid; often associated with reserve food accumulation.

PYRIFORM Pear-shaped.

PULVINATE Cushion-shaped.

PUNCTATE Characterized by a dot-like appearance.

RALFSIOID Said of a thallus which resembles *Ralfsia* (pseudoparenchymatous and crustose.

RECEPTACLE A specialized structure bearing conceptacles (Fucales) or sporangial sori.

RETICULATE Having the appearance of a network.

RHIZOID A unicellular or multicellular filament formed for attachment, usually arising from the base of a plant.

SACCATE Inflated, or sac-like form.

SAXICOLOUS Growing on rock or stones.

SECUND Arranged on one side only.

SERRATE With a saw-like edge to the thallus.

SESSILE Without a stalk.

SIMPLE Unbranched.

SINUOUS With a wavy outline or margin to the thallus.

SORUS An aggregation of reproductive structures.

SPORANGIUM A structure containing spores.

SPORE An asexual reproductive body which may be motile or non-motile.

SPOROPHYTE A morphological phase which bears sporangia.

STREBLONEMOID Said of a thallus which resembles that of *Streblonema* (prostrate branched filaments).

STIPE The lowermost stalk-like part of an erect frond.

STIPITATE Stalked.

STRIATE Marked with bands, furrows or ridges.

SUBLITTORAL Applied to that portion of the shore which is either totally immersed or only uncovered by the receding tide infrequently and then for very short periods.

SUBSTRATUM The structure on which an alga is growing.

TERETE Circular in transverse section.

TETRASPORANGIUM A sporangium in which four spores are formed.

TETRASPORE One of the four spores formed in a tetrasporangium.

THALLUS General term used for plant body, usually applied to the parenchymatous and pseudoparenchymatous forms of algae.

TRICHOTHALLIC A method of growth, with a meristematic region in an intercalary position, usually near the base of a filament.

TRISERIATE Consisting, wholly or in part, of three rows of cells in longitudinal series.

TUBERCLE A small swelling.

UNIAXIAL Of a thallus containing a single axial filament.

UNILATERAL On one side only.

UNILOCULAR Containing a single loculus or cell, said of reproductive organs.

UNISERIATE Consisting of a single row of cells in longitudinal series.

UNISPORES A spore produced from a unilocular sporangium.

VENTRAL Pertaining to the lower surface of a dorsiventral thallus.

VERRUCOSE Warty.

VERTICILLATE Bearing successive whorls of branches along an axis.

VESICLE An air-bladder.

WHORLED With lateral branches arising in a ring around the main axis.
ZOOSPORE A motile spore.
ZYGOTE Cell formed as a result of the fusion of two gametes.

REFERENCES

ABBOTT, I. A. & HOLLENBERG, G. J. 1976. *Marine algae of California.* Stanford University Press, Stanford.

ABE, K. 1935. Zur Kenntnis der Entwicklungsgeschichte von *Heterochordaria, Scytosiphon* und *Sorocarpus. Sci. Rep. Tohoku Imp. Univ., Biol.* **9:** 329–337.

ABE, K. 1938. Entwicklung der Fortpflanzungsorgane und Keimungsgeschichte von *Desmarestia viridis* (Müll). Lamour. *Sci. Rep. Tohoku Imp. Univ. Biol.* **12:** 475–482.

AGARDH, C. A. 1810–1812. *Dispositio algarum Sueciae.* Lundae.

AGARDH, C. A. 1817. *Synopsis Algarum Scandinaviae.* Lundae.

AGARDH, C. A. 1820–28. *Species Algarum.* **1**(1), **1**(2), **2**(1). Lundae.

AGARDH, C. A. 1824. *Systema Algarum.* Lundae.

AGARDH, C. A. 1827. Aufzahlungen einiger in den österreichischen Ländern gefundenen neuen Gattungen und Arten von Algen. *Flora* **10:** 625–656.

AGARDH, J. G. 1836. *Novitiae florae sveciae ex algarum familia.* Lund.

AGARDH, J. G. 1848–1876. *Species Genera et Ordines Algarum.* Lundae.

AGARDH, J. G. 1896. Analecta algologica Cont. III. *Acta Univ. lund.* Afd. 2 **32**(2): 1–140.

ANAND, P. L. 1937. A taxonomic study of the algae of the British chalk cliffs. *J. Bot., Lond.* **75:** 1–51.

ANDERSON, R. J. 1982. The life history of *Desmarestia firma* (C.Ag.) Skottsb. (Phaeophyceae, Desmarestiales). *Phycologia* **21:** 316–322.

ARDRÉ, F. 1970. Contribution à l'étude des algues marines du Portugal. I–La Flore. *Port. Acta biol.* ser. B10: 137–555 (reprint 1–423).

ARESCHOUG, J. E. 1842. Algarum minus rite cognitarum pugillus primus. *Linnaea* **16:** 225–236.

ARESCHOUG, J. E. 1843. Algarum (phycearum) minus rite cognitarum pugillus secundus. *Linnaea* **17:** 257–269.

ARESCHOUG, J. E. 1847. Enumeratio phycearum in maribus scandinaviae cresentium. Sectio prior. *Nova Acta R. Soc. Scient. upsal.* ser. 2 **13:** 223–382.

ARESCHOUG, J. E. 1875. Observationes phycologicae III. *Nova Acta R. Soc. Scient. upsal.* ser. 3 **10:** 1–36.

ATHANASIADIS, A. 1985. North Aegean marine algae I. New records and observations from the Sithonia Peninsula, Greece. *Botanica mar.* **28:** 453–468.

BAKER, J. R. J. & EVANS, L. V. 1971. A myrionemoid variant of *Ectocarpus fasciculatus* Harv. *Br. phycol. J.* **6:** 73–80.

BATTERS, E. A. L. 1883. Notes on the marine algae of Berwick-upon-Tweed. *Hist. Berwicksh. Nat. Club* **10:** 108–115.

BATTERS, E. A. L. 1885. Notes on the marine algae of Berwick-on-Tweed. *Hist. Berwicksh. Nat. Club* **10:** 535–538.

BATTERS, E. A. L. 1888. A description of three new marine algae. *J. Linn. Soc., Bot.* **24:** 450–453.

BATTERS, E. A. L. 1890. A list of the marine algae of Berwick-on-Tweed. *Hist. Berwicksh. Nat. Club* **12:** 221–392.

BATTERS, E. A. L. 1892a. Additional notes on the marine algae of the Clyde sea area. *J. Bot., Lond.* **30:** 170–177.

BATTERS, E. A. L. 1892b. New or critical British algae. *Grevillea* **21:** 13–23.

BATTERS, E. A. L. 1892c. New or critical British algae. *Grevillea* **21:** 49–53.

BATTERS, E. A. L. 1893a. New or critical British algae. *Grevillea* **22:** 20–24.

BATTERS, E. A. L. 1893b. New or critical British algae. *Grevillea* **22:** 50–52.

BATTERS, E. A. L. 1894. New British marine algae. *Grevillea* **22:** 90–92.

BATTERS, E. A. L. 1895a. Some new British marine algae. *J. Bot., Lond.* **33:** 274–276.

BATTERS, E. A. L. 1895b. On some new British algae. *Ann. Bot.* **9:** 307–321.
BATTERS, E. A. L. 1895c. Some new British algae. *Ann. Bot.* **9:** 168–169.
BATTERS, E. A. L. 1896. Some new British marine algae. *J. Bot., Lond.* **34:** 384–390.
BATTERS, E. A. L. 1897. New or critical British marine algae. *J. Bot., Lond.* **35:** 433–440.
BATTERS, E. A. L. 1900. New or critical British marine algae. *J. Bot., Lond.* **38:** 369–379.
BATTERS, E. A. L. 1902. A catalogue of the British marine algae being a list of all the species of seaweeds known to occur on the shores of the British Islands, with the localities where they are found. *J. Bot., Lond.* **40** (Suppl.): 1–107.
BATTERS, E. A. L. 1906. New or critical British marine algae. *J. Bot., Lond.* **44:** 1–3.
BERKELEY, M. J. 1833. *Gleanings of British Algae.* London.
BERTHOLD, G. 1881. Die geschlechtliche Fortpflanzung der eigentlichen Pheosporeen. *Mitt. zool. Stn Neapel* **11:** 401–413.
BLACKLER, H. 1961. *Desmarestia dudresnayi* Lamouroux in Britain. *Br. phycol. Bull.* **2:** 87.
BLACKLER, [M. C.] H. 1964. Some observations on the genus *Colpomenia* (Endlicher) Derbès et Solier 1851. *Proc. int. Seaweed Symp.* **4:** 50–54.
BLACKLER, H. 1967. The occurrence of *Colpomenia peregrina* (Sauv.) Hamel in the Mediterranean (Phaeophyta, Scytosiphonales). *Blumea* **15:** 5–8.
BLACKLER, H. 1981. Some algal problems with special reference to *Colpomenia peregrina* and other members of the Scytosiphonaceae. *Br. phycol. J.* **16:** 133.
BLACKLER, H. & KATPITIA, A. 1963. Observations of the life history and cytology of *Elachista fucicola. Trans. Proc. bot. Soc. Edinb.* **29:** 392–395.
BOLD, H. C. & WYNNE, M. J. 1978. *Introduction to the algae.* Prentice-Hall, New Jersey.
BØRGESEN, F. 1902. The marine algae of the Faeröes. *Bot. of the Faeröes*, Part II. Copenhagen.
BØRGESEN, F. 1926. Marine algae from the Canary Islands especially from Teneriffe and Gran Canaria. II. Phaeophyceae. *Biol. Meddr* **6:** 1–112.
BORNET, E. & THURET, G. 1876–1880. *Notes Algologiques.* Paris.
BUFFHAM, T. H. 1891. The plurilocular zoosporangia of *Asperococcus bullosus* and *Myriotrichia clavaeformis. J. Bot., Lond.* **29:** 321–323.
CABIOCH, J. 1976. Sur la présence dans la region de Roscoff, de deux Phéophycées non encore mentionnées sur les cotes de France. *Trav. Stn biol. Roscoff N.S.* **23:** 23–26.
CARAM, B. 1957. Sur la sexualité et le developpement d'une Pheophycée: *Cylindrocarpus berkeleyi* (Grev.) Crouan. *C.r. hebd. Séanc. Acad. Sci., Paris* **245:** 440–443.
CARAM, C. 1965. Recherches sur la reproduction et le cycle sexuel de quelques Pheophycées. *Vie Milieu* **16:** 21–221.
CARDINAL, A. 1964. Étude sur les Ectocarpacées de la Manche. *Beih. nov. Hedwigia* **15:** [6]+1–86+[43].
CASPARY, R. 1871. Die Seealgen bei Neukuhren an der samländischen Küste in Preuben nach Hensche's sammlung. *Schr. kgl. physik.-ok. Ges. Königsberg* **12:** 138–146.
CHAPMAN, A. R. O. 1969. An experimental approach to the autecology of *Desmarestia aculeata* (L.) Lamour. *Ph.D. Thesis, University of Liverpool.*
CHAPMAN, A. R. O. 1972a. Species delimitation in the filiform, oppositely branched members of the genus *Desmarestia* Lamour. (Phaeophyceae, Desmarestiales) in the northern Hemisphere. *Phycologia* **11:** 225–231.
CHAPMAN, A. R. O. 1972b. Morphological variation and its taxonomic implications in the ligulate members of the genus *Desmarestia* occurring on the west coast of North America. *Syesis* **5:** 1–20.
CHAPMAN, A. R. O. & BURROWS, E. M. 1970. Experimental investigations into the controlling effects of light conditions on the development and growth of *Desmarestia aculeata* (L.) Lamour. *Phycologia* **9:** 103–108.
CHAPMAN, A. R. O. & BURROWS, E. M. 1971. Field and culture studies of *Desmarestia aculeata* (L.) Lamour. *Phycologia* **10:** 63–76.
CHAUVIN, F. J.-L. 1842. *Recherches sur l'organisation, la fructification et la classification de plusieurs genres d'Algues.* Caen.

CHRISTENSEN, T. 1958. Unilocular sporangia in *Ascocyclus orbicularis*. *Revue algol.* **4:** 129–132.

CHRISTENSEN, T. 1980. *Algae – a taxonomic survey*. Aio Tryk, Odense.

CHURCH, A. H. 1898. The polymorphy of *Cutleria multifida* (Grev.). *Ann. Bot.* **12:** 75–109.

CLAYTON, M. N. 1972. The occurrence of variant forms in cultures of species of *Ectocarpus* and *Giffordia*. *Br. phycol. J.* **7:** 101–108.

CLAYTON, M. N. 1974. Studies on the development, life history and taxonomy of the Ectocarpales (Phaeophyta) in Southern Australia. *Aust. J. Bot.* **22:** 743–813.

CLAYTON, M. N. 1975. A study of variation in Australian species of *Colpomenia* (Phaeophyta, Scytosiphonales). *Phycologia* **14:** 187–195.

CLAYTON, M. N. 1976a. Complanate *Scytosiphon lomentaria* (Lyngbye) J. Agardh (Scytosiphonales: Phaeophyta) from Southern Australia: the effects of season, temperature and daylength on the life history. *J. exp. mar. Biol. Ecol.* **25:** 187–198.

CLAYTON, M. N. 1976b. The morphology, anatomy and life history of a complanate form of *Scytosiphon lomentaria* (Scytosiphonales, Phaeophyta) from Southern Australia. *Mar. Biol.* **28:** 201–208.

CLAYTON, M. N. 1979. The life history and sexual reproduction of *Colpomenia peregrina* (Scytosiphonaceae, Phaeophyta) in Australia. *Br. phycol. J.* **14:** 1–10.

CLAYTON, M. N. 1980a. Sexual reproduction – a rare occurrence in the life history of the complanate form of *Scytosiphon* (Scytosiphonaceae, Phaeophyta) from Southern Australia. *Br. phycol. J.* **15:** 105–118.

CLAYTON, M. N. 1980b. Observations on the factors controlling the reproduction of two common species of brown algae, *Colpomenia peregrina* and *Scytosiphon* sp. (Scytosiphonaceae), in Victoria. *Proc. R. Soc. Vict.* **92:** 113–118.

CLAYTON, M. N. & DUCKER, S. C. 1970. The life history of *Punctaria latifolia* Greville (Phaeophyta) in Southern Australia. *Aust. J. Bot.* **18:** 293–300.

CLEMENTE, Y. RUBIO, S. DE R. 1807. *Ensayo sobre las Variedades de la Vid Comun que Vegetan en Andalucia, con un Indice Etimologico y Tres Listas de Plantas en que se Caracterizan Varias Especias Nuevas*. Madrid.

COLLINS, F. S. 1896. Notes on New England marine algae–VII. *Bull. Torrey bot. Club* **23:** 458–462.

COPPEJANS, E. & DHONDT, F. 1976. Végétation de l'île de Port-Cros (Parc National) XIV. *Myrionema liechtensternii Hauck (Phaeophyta–Chordariales), espèce nouvelle pour la flora algologique de France*. *Biol. Jb. Dobonaea* **44:** 112–117.

COTTON, A. D. 1907. Some British species of Phaeophyceae. *J. Bot., Lond.* **45:** 368–373.

COTTON, A. D. 1908a. *Leathesia crispa* Harv. *J. Bot., Lond.* **46:** 329–331.

COTTON, A. D. 1908b. The appearance of *Colpomenia sinuosa* in Britain. *Bull. misc. Inf. R. bot. Gdns Kew* **1908:** 73–77.

COTTON, A. D. 1908c. *Colpomenia sinuosa* in Britain. *J. Bot., Lond.* **46:** 82–83.

COTTON, A. D. 1911. On the increase of *Colpomenia sinuosa* in Britain. *Bull. misc. Inf. R. bot. Gdns Kew* **1911:** 153–157.

COTTON, A. D. 1912. Marine Algae, *In* Praeger, R. L. A biological survey of Clare Island in the County of Mayo, Ireland and of the adjoining district. *Proc. R. Ir. Acad.* **31** sect. 1(15): 1–178.

CROUAN, P. L. & H. M. 1851. Etudes microscopiques sur quelques algues nouvelles ou peu connues constituant un genre nouveau. *Annls Sci. nat. sér.* 4 Bot. **15:** 359–366.

CROUAN, P. L. & H. M. 1852. *Algues marines due Finistère (Exsiccat)*. Brest.

CROUAN, P. L. & H. M. 1855. Observations microscopiques sur l'organisation, la fructification et la dissémination de plusieurs genres d'Algues appartenant à la famille des Dictyotées. *Bull. Soc. bot. Fr.* **2:** 439–445, 644–652.

CROUAN, P. L. & H. M. 1857. Observations microscopiques sur l'organisation, la fructification et la dissémination de plusierus genres d'Algues appartenant à la famille des Dictyotées. *Bull. Soc. bot. Fr.* **4:** 24–29.

CROUAN, P. L. & H. M. 1867. *Florule du Finistère*. Paris and Brest.

DAMMANN, H. 1930. Entwicklungsgeschichtliche und zytologische Untersuchungen an Helgoländer Meeresalgen. *Wiss. Meeresunters. (Helgol,) N.F.* **18:** 1–36.

DANGEARD, P. 1962a. Sur la réproduction et le développement de *Petalonia zosterifolia* (Reinke) Kuntze. *C.r. hebd. Séanc. Acad. Sci., Paris* **254:** 1895–1896.

DANGEARD, P. 1962b. Sur le développement du *Petalonia fascia* (Müeller) Kuntze et du *Scytosiphon lomentaria* (Lyngbye) Endlicher. *C.r. hebd. Séanc. Acad. Sci., Paris* **254:** 3290–3292.

DANGEARD, P. 1963a. Sur le développement du *Punctaria latifolia* Greville. *Botaniste* **46:** 205–224.

DANGEARD, P. 1963b. Recherches sur le cycle évolutif de quelques Scytosiphonacées. *Botaniste* **46:** 5–129.

DANGEARD, P. 1965a. Sur le cycle évolutif de *Litosiphon pusillus* (Carm.) Harvey. *Botaniste* **49:** 47–62.

DANGEARD, P. 1965b. Une algue nouvelle pour Roscoff: le *Myriactula clandestina* Crouan. *Mém. Soc. bot. Fr.* **11:** 16–18.

DANGEARD, P. 1965c. Recherches sur le cycle évolutif de '*Leathesia difformis*' (L.) Areschoug. *Botaniste* **48:** 5–43.

DANGEARD, P. 1965d. Sur un *Myriotrichia* Harvey recolte à Saint-Vaast-la-Hougue (Cotentin). *Botaniste* **49:** 79–98.

DANGEARD, P. 1966a. Sur le *'Punctaria crouani'* (Thuret) Bornet, récolté à Soulac et sur son développement. *Botaniste* **49:** 157–167.

DANGEARD, P. 1966b. Sur la présence d'un plethysmothalle chez *Leptonematella fasciculatum* (Reinke) Silva. *C.r. hebd. Séanc. Acad. Sci., Paris* **263:** 1692–1694.

DANGEARD, P. 1968a. Sur la presence d'*Elachista stellaris* Areschoug [(*'Areschougia stellaris'* (Aresch.) Menegh.] près d'Erquy (Côtes-du-Nord). *Botaniste* **51:** 87–94.

DANGEARD, P. 1968b Sur la presence à Roscoff du *'Myriactula clandestina'* (Crouan) P. Dangeard. *Botaniste* **51:** 81–86.

DANGEARD, P. 1968c. Recherches sur le cycle évolutif de deux '*Asperococcus*'. *Botaniste* **51:** 59–80.

DANGEARD, P. 1968d. Etude du '*Leptonematella fasciculata* (Reinke) Silva et de son développement en culture. *Botaniste* **51:** 117–130.

DANGEARD, P. 1969. A propos des travaux récents sur le cycle évolutif de quelques Phéophycées, Phéosporées. *Botaniste* **52:** 59–102.

DANGEARD, P. 1970. Sur le *Compsonema minutum* (Ag.) Kuckuck et sur son developpement en culture. *C.r. hebd. Séanc. Acad. Sci., Paris* **280:** 63–65.

DERBÈS, F. & SOLIER, A. J. J. 1850. Sur les organes reproducteurs des algues. *Annls Sci. nat. sér. 3* Bot. **14:** 261–282.

DERBÈS, F. & SOLIER, A. J. J. 1851. *In* J. L. M. Castagne, *Supplément au cataloge des plantes qui croissent naturellement aux environs de Marseille* pp. 93–121. Aix.

DERBÈS, F. & SOLIER, A. J. J. 1856. Memoire sur quelques points de la physiologie des Algues. *C.r. hebd. Séanc. Acad. Sci., Paris* **I:** 1–20.

DE TONI, G. B. 1895. *Sylloge algarum omnium hucusque cognitarum.* III. *Sylloge Fucoidearum . . .* xvi + 638 pp. Patavii

DILLWYN, L. W. 1802–1809. *British Confervae.* London.

DIXON, P. S. & RUSSELL, G. 1964. Miscellaneous notes on algal taxonomy and nomenclature. I. *Bot. Notiser* **117:** 279–284.

DIZERBO, A. H. 1965. *Desmarestia dresnayi* Lamour. ex Leman in France and Spain. *Br. phycol. Bull.* **2:** 504.

DOTY, M. S. 1947. The marine algae of Oregon. Part 1. Chlorophyta and Phaeophyta. *Farlowia* **3:** 1–65.

DREW, E. A. & ROBERTSON, W. A. A. 1974. Direct observations of *Desmarestia dresnayi* Lamour. ex Leman in the British Isles and in the Mediterranean. *Br. phycol. J.* **9:** 195–200.

DRING, M. J. 1984. Photoperiodism and phycology. *In* Round, F. E. & Chapman, D. J. (Eds) *Progress in phycological research* Vol. 3, pp. 159–192. Biopress Ltd., Bristol.

DRING, M. J. & LÜNING, K. 1975a. Induction of two-dimensional growth and hair formation by blue light in the brown alga *Scytosiphon lomentaria.* *Z. Pflanzenphysiol.* **75:** 107–117.

DRING, M. J. & LÜNING, K. 1975b. A photoperiodic response mediated by blue light in the brown alga *Scytosiphon lomentaria.* *Planta* **125:** 25–32.

DRING, M. J. & LÜNING, K. 1983. Photomorphogenesis of marine macro algae. *In* Shropshire, W. & Mohr, H. (Eds) *Encyclopedia of plant physiology. Vol. 16B, Photomorphogenesis,* pp. 545–568. Heidelberg.

DUBY, J. E. 1830 *Botanicon gallicum* ed. 2: 2 Paris.

EDELSTEIN, T., CHEN, L. C.-M. & MCLACHLAN, J. 1970. The life cycle of *Ralfsia clavata* and *R. bornetii*. *Can. J. Bot.* **48:** 527–531.

EDELSTEIN, T., CHEN, L. C.-M. & MCLACHLAN, J. 1971. On the life histories of some brown algae from eastern Canada. *Can. J. Bot.* **49:** 1247–1251.

EDELSTEIN, T. & MCLACHLAN, J. 1969a. Investigations of the marine algae of Nova Scotia. VI. Some species new to North America. *Can. J. Bot.* **47:** 555–560.

EDELSTEIN, T. & MCLACHLAN, J. 1969b. *Petroderma maculiforme* on the coast of Nova Scotia. *Can. J. Bot.* **47:** 561–563.

EDWARDS, P. 1969. Field and cultural studies on the seasonal periodicity of growth and reproduction of selected Texas benthic marine algae. *Contr. mar. Sci. Univ. Tex.* **14:** 59–114.

FALKENBERG, P. 1879. Die Befruchtung und der Generationswechsel von *Culteria*. *Mitt. zool. Stn Neapel* **1:** 420–447.

FARLOW, W. G. 1881. *Marine algae of New England and adjacent coasts.* Washington.

FARNHAM, W. F. 1980. Studies on aliens in the marine flora of southern England. *In* Price, J. H., Irvine, D. E. G. & Farnham, W. F. (Eds) *The shore environment* **2:** *ecosystems* pp. 875–914. Systematics Association Special Volume **17(b).** Academic Press, London and New York.

FELDMANN, J. 1937. Les algues marines de la Côte des Albéres. I-III Cyanophycées, Chlorophycées, Pheophycées. *Rev. algol.* **9:** 141–335.

FELDMANN, J. 1943. Une nouvelle espèce de *Myriactula* parasite du *Gracilaria armata* J. Ag. *Bull. Soc. Hist. nat. Afr. N.* **34:** 222–229.

FELDMANN, J. 1949. L'ordre des Scytosiphonales. *Mém. Hist. Nat. Afr. Nord, Hors-Sér.* **2:** 103–115.

FLETCHER, R. L. 1974a. Studies on the life history and taxonomy of some members of the Phaeophycean families Ralfsiaceae and Scytosiphonaceae. *Ph.D. Thesis, University of London.*

FLETCHER, R. L. 1974b. Studies on the brown algal families Ralfsiaceae and Scytosiphonaceae. *Br. phycol. J.* **9:** 218.

FLETCHER, R. L. 1975. Heteroantagonism observed in mixed algal cultures. *Nature, Lond.* **253:** 534–535.

FLETCHER, R. L. 1978. Studies on the family Ralfsiaceae (Phaeophyta) around the British Isles. *In* Irvine, D. E. G. & Price, J. H. (Eds) *Modern approaches to the taxonomy of red and brown algae* pp. 371–398. Systematics Association Special Volume 10. Academic Press, London and New York.

FLETCHER, R. L. 1980. *Catalogue of main marine fouling organisms.* Vol. 6. *Algae*. O.D.E.M.A. Brussels.

FLETCHER, R. L. 1981a. Studies on the ecology, structure and life history of the brown alga *Petalonia filiformis* (Batt.) Kuntze (Scytosiphonaceae) around the British Isles. *Phycologia* **20:** 103–104.

FLETCHER, R. L. 1981b. Observations on the ecology and life history of *Ralfsia spongiocarpa* Batt. *Proc. int. Seaweed Symp.* **8:** 323–330.

FLETCHER, R. L. 1984. Observations on the life history of the brown alga *Hecatonema maculans* (Coll.) Sauv. (Ectocarpales, Myrionemataceae) in laboratory culture. *Br. phycol. J.* **19:** 193.

FLETCHER, R. L. & MAGGS, C. A. 1985. Two crustose marine brown algae new to Ireland. *Ir. Nat. J.* **21:** 523–526.

FOSLIE, M. 1887. Nye hausalger. *Tromsø Mus. Årsh.* **10:** 175–195.

FOSLIE, M. 1890. Contribution to knowledge of the marine algae of Norway I. East-Finmarken. *Tromsø Mus. Årsh.* **13:** 1–86.

FOSLIE, M. 1894. New or critical Norwegian algae. *K. norske Vidensk. Selsk. Skr.* **1893(b):** 114–144.

FRIES, E. M. 1835. *Corpus Florarum Provincialium Sueciae,* 1 Floram Scanicam. Upsaliae.

FRITSCH, F. E. 1945. *The structure and reproduction of the algae* Vol. 2. Cambridge University Press, Cambridge.

FRYE, T. C. & PHIFER, M. W. 1930. Some questions in the life histories of the Phaeophyceae with particular reference to *Scytosiphon lomentarius*. *In Contributions to Marine Biology* pp. 234–245. Stanford University Press, Stanford.
GATTY, M. 1863. *British seaweeds*. London.
GAYRAL, P. 1958. *Algues de la Côte Atlantique Marocaine*. Rabat.
GAYRAL, P. 1966. *Les Algues des Côtes Françaises (Manche et Atlantique)*. Paris.
GRAN, H. H. 1893. *Algevegetationen i Tönsbergfjorden*. *Förh. VidenskSelsk. Krist.* **7:** 1–38.
GRAN, H. H. 1897. Kristianiafjordens algeflora I. Rhodophyceae og Phaeophyceäe. *Skr. norske Vidensk–Akad.* Mat.–naturv, Kl. **1896**(2), 1–56.
GRAY, S. F. 1821. *A natural arrangement of British plants*. 1. London.
GREVILLE, R. K. 1827. *Scottish cryptogamic flora* **5:** 300. Edinburgh.
GREVILLE, R. K. 1828. *Scottish cryptogamic flora*. **6:** 301–360. Edinburgh.
GREVILLE, R. K. 1830. *Algae britannicae*. Edinburgh and London.
HAI(N)SWORTH, S. 1976. Some interesting additions to the marine fauna of Lundy. *Rep. Lundy Fld Soc.* **26:** 61–62.
HAMEL, G. 1928. Les algues de Vigo. *Revue algol.* **4:** 81–95.
HAMEL, G. 1931–39. *Phéophycées de France*, Paris.
1931. Ectocarpacées pp. 1–80.
1935. Myrionematacées–Spermatochnacées pp. 81–176.
1937. Spermatochnacées–Sphacelariacées pp. 177–240.
1938. Sphacelariacées–Dictyotacées pp. 241–336.
1939. Dictyotacées–Sargassacées pp. 337–432. I–XLVII.
HANNA, H. 1899. The plurilocular sporangia of *Petrospongium berkeleyi*. *Ann. Bot.* **13:** 461–464.
HARTMAN, M. 1950. Beiträge zur Kenntnis der Befruchtung und Sexualitaït mariner Algen. I. Über die Befruchtung von *Cutleria multifida*. *Pubbl. Staz. zool. Napoli* **22:** 120–128.
HARVEY, W. H. 1834. Algological illustrations. No. 1 Remarks on some British algae and descriptions of new species recently added to our flora. *J. Bot., Hooker* **1:** 296–305.
HARVEY, W. H. 1841. *A manual of the British algae*. John van Voorst, London.
HARVEY, W. H. 1846–51. *Phycologia Britannica*.
1846. Vol. 1, pp. l–xv + l–viii, pls (with text) 1–120.
1849. Vol. 2, pp. i–vi, pls (with text) 121–240.
1851. Vol. 3, pp. i–xiv, pls (with text) 241–360.
HARVEY, W. H. 1849. *A manual of the British marine algae*. John van Voorst, London.
HARVEY, W. H. 1857. Short descriptions of some new British algae, with two plates. *Nat. Hist. Rev.* **4:** 201–204.
HAUCK, F. 1877. Beiträge zur Kenntniss der adriatischen Algen. II. *Öst. bot. Z.* **27:** 185–186.
HAUCK, F. 1879. Beiträge zur Kenntniss der adriatischen Algen. X. *Öst. bot. Z.* **29:** 151–154.
HAUCK, F. 1883–85. Die Meeresalgen Deutschlands und Öesterreichs. *In* Rabenhorst, L. *Kryptogamen.-Flora von Deutschland, Oesterreich und der Schweiz*, ed. 2, 2, Leipzig, pp. 1–320. (1883), 321–512 (1884), 513–575 (1885).
HISCOCK, S. & MAGGS, C. A. 1982. Notes on Irish marine algae – 6. *Zanardinia prototypus* (Nardo) Nardo (Phaeophyta). *Ir. Nat. J.* **20:** 414–416
HISCOCK, S. & MAGGS, C. A. 1984. Notes on the distribution and ecology of some new and interesting seaweeds from south-west Britain. *Br. phycol. J.* **19:** 73–87.
HOEK, C. van den, CORTEL-BRÉEMAN, A. M., RIETEMA, H. & WANDERS, J. B. W. 1972. L'interpretation des données obtenues par des cultures unialgales, sur les cycles évolutifs des algues. Quelques examples tirés des recherches conduites au laboratoire de Groningue. *Mém. Soc. bot. Fr.* **1972:** 45–66.
HOLLENBERG, G. J. 1969. An account of the Ralfsiaceae (Phaeophyta) of California. *J. Phycol.* **5:** 290–301.
HOLMES, E. M. & BATTERS, E. A. L. 1891. A revised list of the British marine algae. *Ann. Bot.* **5:** 63–107.
HOOKER, W. J. 1833. Cryptogamia Algae [pp. 264–322] in. Hooker, W. J. *The English Flora of Sir James Edward Smith*. Class XXIV, Cryptogamia. Vol. V, Part 1.

HSIAO, S. I. C. 1969. Life history and iodine nutrition of the marine brown alga, *Petalonia fascia* (O. F. Müll.) Kuntze. *Can. J. Bot.* **47:** 1611–1616.

HSIAO, S. I. C. 1970. Light and temperature effects on the growth, morphology and reproduction of *Petalonia fascia*. *Can. J. Bot.* **48:** 1359–1361.

HUDSON, W. 1778. *Flora Anglica,* ed. 2. London.

JAASUND, E. 1951. Marine algae from northern Norway. I. *Bot. Notiser* **1951:** 128–142.

JAASUND, E. 1957. Marine algae from northern Norway. II. *Bot. Notiser* **110:** 205–231.

JAASUND, E. 1960. *Elachista lubrica* Ruprecht and *Elachista fucicola* (Velley) Areschoug. *Botanica mar.* **1:** 101–107.

JAASUND, E. 1961. A note on *Ectocarpus fasciculatus* (Griff.) Harv. *Bot. Notiser* **114:** 239–241.

JAASUND, E. 1963. Beiträge zur Systematik der norwegischen Braunalgen. *Botanica mar.* **5:** 1–8.

JAASUND, E. 1964. Marine algae from northern Norway. III. *Botanica mar.* **6:** 129–133.

JAASUND, E. 1965. Aspects of the marine algal vegetation of North Norway. *Bot. gothoburg* **4:** 1–174.

JAENICKE, L. 1977. Sex hormones of brown algae. *Naturwissenschaften* **64:** 69–75.

JEPHSON, N. A., FLETCHER, R. L. & BERRYMAN, J. 1975. The occurrence of *Zanardinia prototypus* on the south coast of England. *Br. phycol. J.* **10:** 253–255.

JOHNSON, T. 1892. Seaweeds from the west coast of Ireland. *Ir. Nat.* **1:** 4–6.

JONES, W. E. 1974. Changes in the seaweed flora of the British Isles. *In* Hawksworth, D. L. (Ed) *The changing flora and fauna of Britain* pp. 97–113. Systematics Association Special Volume **6.** Academic Press, London and New York.

JONSSON, H. 1903. The marine algae of Iceland II. Phaeophyceae. *Bot. Tidsskr.* **25:** 141–195.

KARSAKOFF, N. 1892. Quelques remarques sur le genre *Myriotrichia.* *J. Bot. Paris* **6:** 433–444.

KATPITIA, A. D. & BLACKLER, H. 1962. Plurilocular sporangia on *Elachista scutulata* (Sm.) Duby in Britain. *Br. phycol. Bull.* **2:** 173–174.

KJELLMAN, F. R. 1883. The algae of the Arctic Sea. *K. svenska. VetenskAkad. Handl.* **20:** 1–350.

KJELLMAN, F. R. 1890. *Handbok i Skandinaviens hafsalgflora. I. Fucoideae.* Stockholm.

KJELLMAN, F. R. 1893. Encoeliaceae. *In* Engler, A. & Prantl, K. Die naturlichen Pflanzenfamilien. Nachtrage 2 Teil 1, Abt. 2 Leipzig.

KJELLMAN, F. R. & SVEDELIUS, N. L. 1910. Lithodermataceae. *In* Engler, A. & Prantl, K. Die naturlichen Pflanzenfamilien. Nachtrage 2 Tiel 1, Abt. 2. Leipzig.

KJELLMAN, F. R. & SVEDELIUS, N. L. 1910. Lithodermataceae. *In* Engler, A. & Prantl, K. Die naturlichen Pflanzenfamilien. *Nachtrage 2 Teil 1, Abt. 2. Leipzig.*

KNIGHT, M. 1929. Studies in the Ectocarpaceae. II. The life history and cytology of *Ectocarpus siliculosus* Dillw. *Trans. R. Soc. Edinb.* **56:** 307–332.

KNIGHT, M., BLACKLER, M. C. H. & PARKE, M. W. 1935. Notes on the life cycle of species of *Asperococcus.* *Proc. Trans. L'pool biol. Soc.* **48:** 79–97.

KNIGHT, M. & PARKE, M. W. 1931. Manx algae. An algal survey of the south end of the Isle of Man. *Proc. Trans. L'pool biol. Soc.* **45** (Appendix II): 1–155.

KOEMAN, R. P. T. & CORTEL-BREEMAN, A. M. 1976. Observations on the life history of *Elachista fucicola* (Vell.) Aresch. (Phaeophyceae) in culture. *Phycologia* **15:** 107–117.

KORNMANN, P. 1962a. Der Lebenszyklus von *Desmarestia viridis.* *Helgoländer wiss Meeresunters.* **8:** 287–292.

KORNMANN, P. 1962b. Plurilokuläre sporangien bei *Elachista fucicola.* *Helgoländer wiss. Meeresunters.* **8:** 293–297.

KORNMANN, P. & SAHLING, P.-H. 1977. Meeresalgen von Helgoland. Benthische Grün-, Braun- und Rotalgen. *Helgoländer wiss. Meeresunters.* **29:** 1–289.

KORNMANN, P. & SAHLING, P.-H. 1983. Meeresalgen von Helgoland: Ergänzung. *Helgoländer wiss. Meeresunters.* **36:** 1–65.

KRISTIANSEN, AA. 1978. Marine algal vegetation in shallow water around the Danish Island of Saltholm, The Sound. *Bot. Tidsskr.* **72:** 203–226.

KRISTIANSEN, AA. 1984. Experimental field studies on the ecology of *Scytosiphon lomentaria* (Fucophyceae, Scytosiphonales) in Denmark. *Nord. J. Bot* **4:** 719–724.

KRISTIANSEN, A. & PEDERSEN, P. M. 1979. Studies on the life history and seasonal variation of *Scytosiphon lomentaria* (Fucophyceae, Scytosiphonales) in Denmark. *Bot. Tidsskr.* **74:** 31–56.

KUCKUCK, P. 1894. Bemerkungen zur marinen Algenvegetation von Helgoland. I. *Wiss. Meeresunters. (Helgol.) N.F.* **1:** 223–263.

KUCKUCK, P. 1897. Bemerkungen zur marinen Algenvegetation von Helgoland. II. *Wiss. Meeresunters. (Helgol.) N.F.* **2:** 371–400.

KUCKUCK, P. 1898. Ueber die Paarung von Schwärmsporen bei *Scytosiphon*. *Ber. dt. bot. Ges.* **16:** 35–37.

KUCKUCK, P. 1899a. Beiträge zur Kenntnis der Meeresalgen 5–9. *Wiss. Meeresunters. (Helgol.) N.F.* **3:** 13–80.

KUCKUCK, P. 1899b. Ueber den Generationswechsel von *Culteria multifida* (Engl. Bot.) Grev. *Wiss. Meeresunters. (Helgol.) N.F.* **3:** 95–116.

KUCKUCK, P. 1912. Beiträge zur Kenntnis der Meeresalgen. XI. Zur Fortpflanzung der Phaeosporeen. *Wiss. Meeresunters. (Helgol.) N.F.* **5:** 155–188.

KUCKUCK, P. 1917. Über zwerggenerationen bei *Pogotrichum* und über die Fortpflanzung von *Laminaria*. *Ber. dt. bot. Ges.* **35:** 557–578.

KUCKUCK, P. 1929. Fragmente einer Monographie der Phaeosporeen. *Wiss. Meeresunters. (Helgol.) N.F.* **17:** 1–93.

KUCKUCK, P. 1953. Ectocarpaceen – Studien I *Hecatonema, Chilionema, Compsonema*. *Helgoländer wiss. Meeresunters.* **4:** 316–352.

KUCKUCK, P. 1955. Ectocarpaceen – Studien III *Protectocarpus* nov. gen. *Helgoländer wiss. Meeresunters.* **5:** 119–140.

KUNIEDA, H. & ARASAKI, S. 1947. On the life history of *Ilea fascia* and *Punctaria* sp. *Seibutsu* **2:** 185–188.

KUNIEDA, H. & SUTO, S. 1938. The life history of *Colpomenia sinuosa* (Scytosiphonaceae) with special reference to the conjugation of anisogametes. *Bot. Mag., Tokyo* **52:** 539–546.

KUNTZE, O. 1891–98. *Revisio generum plantarum*. Leipzig.
Parts I: I–CLVI + 1–376. 1891.
Parts II: 377–1011. 1891.
Parts III: (1): I–VI + 1–202 + 1–576. 1898.

KÜTZING, F. T. 1843. *Phycologia generalis*. Leipzig.

KÜTZING, F. T. 1845. *Phycologia Germanica*. Nordhausen.

KÜTZING, F. T. 1845–1871. *Tabulae phycologicae*. I–XIX Index Nordhausen. 1845–49 (V.1); 1850–52 (V.2); 1853–69 (V.3–19 respectively); 1871 (index).

KÜTZING, F. T. 1849. *Species Algarum*. Leipzig.

KYLIN, H. 1907. *Studien über die Algenflora der schwedischen Westküste*. Uppsala.

KYLIN, H. 1910. Zur Kenntnis der Algenflora der norwegischen Westküste. *Ark. Bot.* **10** No. 1.

KYLIN, H. 1918. Studien über die Entwicklungsgeschichte der Phaeophyceen. *Svensk bot. Tidskr.* **12:** 1–64.

KYLIN, H. 1933. Über die Engwicklungsgeschichte der Phaeophyceen. *Acta Univ. lund.* **29:** 1–102.

KYLIN, H. 1934. Zur Kenntnis der Entwicklungsgeschichte einiger Phaeophyceen. *Acta Univ. lund.* **30:** 1–18.

KYLIN, H. 1937. Bermerkungen über die Entwicklungsgeschichte einiger Phaeophyceen. *Acta Univ. lund.* **33:** 1–34.

KYLIN, H. 1947. Die Phaeophyceen der schwedischen Westküste. *Acta Univ. lund.* **43:** 1–99.

LAMOUROUX, J. V. 1809. Exposition des caractères du genre *Dictyota* et tableau des especes qu'il renferme. *J. Bot. Paris* **2:** 129–135.

LAMOUROUX, J. V. F. 1813. Essai sur les genres de la famille des thalassiophytes non articulées. *Annls. mus. Hist. nat Paris* **20:** 21–47, 115–139, 267–293. (reprint 1–84).

LE JOLIS, A. 1863. Liste des algues marines de Cherbourg. *Mém. Soc. natn. Sci. nat. math. Cherbourg* **10:** 6–168.

LEMAN, D. S. 1819. *Annales des Sciences Naturelles*. Tom XIII, DEA–DZW. Paris, 104–106.

LEVRING, T. 1937. Zur Kenntnis der Algenflora der norwegischen Westküste. *Acta Univ. lund.* **33:** 1–147.

LEVRING, T. 1940. Studien über die Algenvegetation von Blekinge, Südschweden. Lund.

LIGHTFOOT, J. 1777. *Flora Scotica* 2. London.

LINDAUER, V. W., CHAPMAN, V. J. & AIKEN, M. 1961. The marine algae of New Zealand. Part II. Phaeophyceae. *Nova Hedwigia* **3:** 129–350.

LINK, H. F. 1833. *Handbuch zur Erkennung der nutzbarsten und am häufigsten vorkommenden Gewächse.* Bd 3. Berlin.

LINNEAUS, C. 1755. *Flora Svecica.* Editio Secunda, Stockholmiae.

LINNAEUS, G. 1763. *Species Plantarum,* ed. 2, 2. Holmiae.

LOCKHART, J. C. 1979. Factors determining various forms in *Cladosiphon zosterae* (Phaeophyceae). *Am. J. Bot.* **66:** 836–844.

LOCKHART, J. C. 1982. Influence of light, temperature and nitrogen on morphogenesis of *Desmotrichum undulatum* (J. Agardh) Reinke (Phaeophyta, Punctariaceae). *Phycologia* **21:** 264–272.

LOISEAUX, S. 1964. Sur l'hétéroblastie et le cycle de deux *Ascocyclus* de la région Roscoff. *C.r. hebd. Séanc. Acad. Sci., Paris,* **259:** 2903–2905.

LOISEAUX, S. 1966. Sur le cycle de developpement de *l'Ascocyclus hispanicus* (Phéophycées, Myrionematacées) et la formation en culture de stades coccoides. *C.r. hebd. Séanc. Acad. Sci., Paris,* ser 3, **262:** 68–71.

LOISEAUX, S. 1967a. Morphologie et cytologie des Myrionémacées. Critères taxonomiques. *Rev. gén. Bot.* **74:** 329–347.

LOISEAUX, S. 1967b. Recherches sur les cycles de développement des Myrionématacées (Phéophycées) I–II Hectanonématées et Myrionématées. *Rev. gén Bot.* **74:** 529–576.

LOISEAUX, S. 1968. Recherches sur les cycles de devéloppement des Myrionématacées (Phéophycées) III Tribu des Ralfsiées IV Conclusions générales. *Rev. gén. Bot.* **75:** 295–318.

LOISEAUX, S. 1969. Sur une espece de *Myriotrichia* obtenue en culture à partir zoïdes d'*Hecatonema maculans* Sauv. *Phycologia* **8:** 11–15.

LOISEAUX, S. 1970a. Notes on several Myrionemataceae from California using culture studies. *J. Phycol.* **6:** 248–260.

LOISEAUX, S. 1970b. *Streblonema anomalum* S. et G. and *Compsonema sporangiiferum* S. et G., stages in the life history of a minute *Scytosiphon.* *Phycologia* **9:** 185–191.

LUND, S. 1938. On *Lithoderma fatiscens* Areschoug and *L. fatiscens* Kuckuck. *Meddr Grønland* **116:** 1–18.

LUND, S. 1950. The marine algae of Denmark, contributions to their natural history. Vol. 11, Phaeophyceae, part IV Sphacelariaceae, Cutleriaceae and Dictyotaceae. *Biol. Skr.* **6:** 5–80.

LUND, S. 1959. The marine algae of East Greenland. I. Taxonomical part. *Meddr Grønland* **156:** 1–248.

LUND, S. 1966. On a sporangia-bearing microthallus of *Scytosiphon lomentaria* from nature. *Phycologia* **6:** 67–78.

LÜNING, K. 1980. Control of algal life history by daylength and temperature. *In* Price, J. H., Irvine, D. E. G. & Farnham, W. F. (Eds) *The shore environment* **2:** *ecosystems* 915–945. Systematics Association Special Volume 17(b). Academic Press, London and New York.

LÜNING, K. & DRING, M. J. 1973. The influence of light quality on the development of the brown algae *Petalonia* and *Scytosiphon.* *Br. phycol. J.* **8:** 333–338.

LYNGBYE, H. C. 1819. *Tentamen Hydrophytologiae Danicae.* Hafniae.

MAGNUS, P. 1875. Die botanischen Ergebnisse der Nordseefahrt 1872. *Jhber. Comm. wiss. Unters. Meere* **II:** 60–80. Berlin.

MCLACHLAN, J., CHEN, L. C-M. & EDELSTEIN, T. 1971. The life history of *Microspongium* sp. *Phycologia* **10:** 83–87.

MENEGHINI, G. 1844. Observazioni su alcumi generi della famiglia delle Cordariee. *G. bot. ital.* **1:** 291–295.

MIRANDA, F. 1931. Sobre las algas y cianoficeas del Cantabrico, especialmente de Gijon. *Trab. Mus. nac. Cienc. nat., Madr,* ser. Bot. **25:** 1–106.

MOE, R. L. & SILVA, P. C. 1977. Sporangia in the brown algal genus *Desmarestia* with special reference to Antarctic *D. ligulata.* *Bull. Jap. Soc. Phycol.* **25:** 159–167.

MONTAGNE, J. F. C. 1846. Phyceae *In* Durieu de Maisonneuve, M. C., *Exploration Scientifique de l'Algérie. Sciences Naturelles Botanique* **1:** 1–197.

MÜLLER, D. 1974. Sexual reproduction and isolation of a sex attractant in *Cutleria multifida* (Smith) Grev. (Phaeophyta). *Biochem. Physiol. Pflanzen* **165:** 212–215.

MÜLLER, D. 1975. Experimental evidence against sexual fusions of spores from unilocular sporangia of *Ectocarpus siliculosus* (Phaeophyta). *Br. phycol. J.* **10:** 315–321.

MÜLLER, D. G. & LÜTHE, N. M. 1981. Hormonal interaction in sexual reproduction of *Desmarestia aculeata* (Phaeophyceae). *Br. phycol. J.* **16:** 351–358.

MÜLLER, D. G. & MEEL, H. 1982. Culture studies on the life history of *Arthrocladia villosa* (Desmarestiales, Phaeophyceae). *Br. phycol. J.* **17:** 419–425.

MÜLLER, O. F. 1771–1782. *Flora Danica* **4 & 5.** Havniae.

MUNDA, I. M. 1979. Additions to the check-list of benthic marine algae from Iceland. *Botanica mar.* **22:** 459–463.

NÄGELI, C. 1847. Die neuern Algensysteme. *Neue Denkschr. allg. schweiz. Ges. ges. Naturw.* **9** (unnumbered art. no. 2): 1–275.

NAKAHARA, H. 1984. Alternation of generations of some brown algae in unialgal and axenic cultures. *Scient. Pap. Inst. algol. Res. Hokkaido Univ.* **7:** 77–194.

NAKAHARA, H. & NAKAMURA, Y. 1971. The life history of *Desmarestia tabacoides* Okamura. *Bot. Mag., Tokyo* **84:** 69–75.

NAKAMURA, Y. 1965. Development of zoospores in *Ralfsia*-like thallus, with special reference to the life cycle of the Scytosiphonales. *Bot. Mag., Tokyo* **78:** 109–110.

NAKAMURA, Y. 1972. A proposal on the classification of the Phaeophyta. *In* Abbott, I. A. and Kurogi, M. (Eds) *Contributions to the systematics of marine algae of the North Pacific* pp. 147–156. Japanese Soc. Phycology, Kobe.

NAKAMURA, Y. TATEWAKI, M. 1975. The life history of some species of Scytosiphonales. *Scient. Pap. Inst. algol. Res. Hokkaido Univ.* **6:** 57–93.

NARDO, G. D. 1834. De novo genere *Stifftia* noncupando. *Isis,* Jena. Heft VI & VII, col. 677–678. Leipzig.

NARDO, G. M. 1841. Nuove osservazioni sulla struttura, abituoline e valore dei generi *Stifftia, Hildenbrandtia* et *Agardhina.* Atti 2nd ruin. Sci. ital. in Torino.

NELSON, W. A. 1982. A critical review of the Ralfsiales, Ralfsiaceae and the taxonomic position of *Analipus japonicus* (Harv.) Wynne (Phaeophyta). *Br. phycol. J.* **17:** 311–320.

NEWTON, L. 1931. *A handbook of the British seaweeds.* British Museum (Natural History), London.

NIEUWLAND, J. A. 1917. Critical notes on new and old genera of plants – X. *Am. Midl. Nat.* **5:** 50–52.

NYGREN, S. 1975a. Life history of some Phaeophyceae from Sweden. *Botanica mar.* **18:** 131–141.

NYGREN, S. 1975b. Influence of salinity on the growth and distribution of some Phaeophyceae on the Swedish west coast. *Botanica mar.* **18:** 143–147.

NYGREN, S. 1979. Life histories and chromosome numbers in some Phaeophyceae from Sweden. *Botanica mar.* **22:** 371–373.

OLTMANNS, F. 1922. *Morphologie und Biologie der Algen.* Bd. 1. Jena.

ØRSTED, A.-E. 1844. *De regionibus marinis. Elementa topographiae historiconaturalis freti Oresund.* Havniae.

PANKOW, H. 1971. *Algenflora der Ostsee I. Benthos.* 419 pp. Gustav Fisher Verlag, Stuttgart.

PARKE, M. 1933. A contribution to the knowledge of the Mesogloiaceae and associated families. *Publs Hartley bot. Labs L'pool Univ.* **9:** 1–43.

PARKE, M. 1953. A preliminary check list of British marine algae. *J. mar. biol. Ass. U.K.* **32:** 497–520.

PARKE, M. & DIXON, P. S. 1964. A revised check list of British Marine Algae. *J. mar. biol. Ass. U.K.* **44:** 499–542.

PARKE, M. & DIXON, P. S. 1968. Check list of British marine algae – second revision. *J. mar. biol. Ass. U.K.* **48:** 783–832.

PARKE, M. & DIXON, P. S. 1976. Check list of British marine algae – third revision. *J. mar. biol. Ass. U.K.* **56:** 527–594.

PARSONS, M. J. 1982. *Colpomenia* (Endlicher) Derbés et Solier (Phaeophyta) in New Zealand. *N.S.J. Bot.* **20:** 289–301.

PEDERSEN, P. M. 1976. Marine, benthic algae from southernmost Greenland. *Meddr Gronland* **199:** 1–80.

PEDERSEN, P. M. 1978a. Culture studies on marine algae from West Greenland. III. The life histories and systematic positions of *Pogotrichum filiforme* and *Leptonematella fasciculata* (Phaeophyceae). *Phycologia* **17:** 61–68.

PEDERSEN, P. M. 1978b. Culture studies on the pleomorphic brown alga *Myriotrichia clavaeformis* (Dictyosiphonales, Myriotrichiaceae). *Norw. J. Bot.* **25:** 281–291.

PEDERSEN, P. M. 1979. Culture studies on the brown algae *Halothrix lumbricalis* and *Elachista fucicola* (Elachistaceae). *Bot. Notiser* **132:** 151–159.

PEDERSEN, P. M. 1980. Culture studies on complanate and cylindrical *Scytosiphon* (Fucophyceae, Scytosiphonales) from Greenland. *Br. phycol. J.* **15:** 391–398.

PEDERSEN, P. M. 1981a. The life histories in culture of the brown algae *Gononema alariae* sp. nov. and *G. aecidioides* comb. nov. from Greenland. *Nord. J. Bot.* **1:** 263–270.

PEDERSEN, P. M. 1981b. *Porterinema fluviatile* as a stage in the life history of *Sorapion kjellmanii* (Fucophyceae, Ralfsiaceae). *Proc. int. Seaweed Symp.* **10:** 203–208.

PEDERSEN, P. M. 1981c. Life histories of brown algae. *In* Lobban, C. & Wynne, M. J. (Eds) *The biology of seaweeds* pp. 194–217. Botanical Monographs **17.** Blackwell Scientific Publications, Oxford.

PEDERSEN, P. M. 1983. Notes on marine, benthic algae from Madeira in nature and in culture. *Bocagiana* **70:** 1–8.

PEDERSEN, P. M. 1984. Studies on primitive brown algae (Fucophyceae). *Opera Bot.* **74:** 1–76.

PRICE, J. H., JOHN, D. M. & LAWSON, G. W. 1978. Seaweeds of the Western coast of tropical Africa and adjacent islands: a critical assessment. II. Phaeophyta. *Bull. Br. Mus. nat. Hist.* (Bot.) **6:** 87–182.

PRINTZ, H. 1926. Die Algenvegetation der Trondhjemsfjordes. *Skr. norske Vidensk-Akad.* Mat-naturv. Kl. **5:** 1–274.

PYBUS, C. 1975. A new record of *Corynophlaea crispa* (Harv.) Kuck. *Ir. Nat. J.* **18:** 153–155.

RAUTENBERG, E. 1960. Zur morphologie und ökologie einiger epiphytischer und epiendophytischer Algen. *Botanica mar.* **2:** 133–145.

RIENKE, J. 1978a. Ueber die Entwicklung von *Phyllitis. Scytosiphon* und *Asperococcus. Jb. wiss. Bot.* **11:** 262–273.

REINKE, J. 1878b. Entwicklungsgeschichtliche Untersuchungen über die Culteriaceen des Golfs von Neapel. *Nova Acta Acad. Caesar. Leop.Carol.* **40:** 59–96.

REINKE, J. 1888. Die braunen Algen (Fucaceen und Phaeosporeen) der Kieler Bucht. *Ber. dt. bot Ges.* **6:** 14–20.

REINKE, J. 1889a. *Atlas deutscher Meeresalgen.* Heft I. Paul Parey, Berlin.

REINKE, J. 1889b. Algenflora der westlichen Ostsee, Deutschen Antheils. *Ber. comm. wiss. Untersuch. dt. Meere* **6:** 1–101.

REINKE, J. 1892. *Atlas deutscher Meeresalgen,* Heft II. Paul Parey, Berlin.

RHODES, R. G. 1970. Relations of temperature to development of the macrothallus of *Desmotrichum undulatum. J. phycol.* **6:** 312–314.

RHODES, R. G. & CONNELL, M. V. 1973. The biology of brown algae on the Atlantic coast of Virgina. II. *Petalonia fascia* and *Scytosiphon lomentaria. Chesapeake Sci.* **14:** 211–215.

RIETEMA, H. & HOEK, C. VAN DEN 1981. The life history of *Desmotrichum undulatum* (Phaeophyceae) and its regulation by temperature and light conditions. *Mar. Ecol. Prog. Ser* **4:** 321–335.

ROBERTS, M. & RING, F. M. 1972. Preliminary investigations into conditions affecting the growth of the microscopic phase of *Scytosiphon lomentaria* (Lyngb.) Link. *Mém. Soc. bot. Fr.* **1972:** 117–128.

ROELEVELD, J. G., DUISTERHOF, M. & VROMAN, M. 1974. On the year cycle of *Petalonia fascia* in the Netherlands. *Neth. J. Sea Res.* **8:** 410–426.

ROSENVINGE, L. K. 1893. Grønlands havalger. *Meddr Gronland* **3:** 765–981.

ROSENVINGE, L. K. 1898. Deuxième mémoire sur les algues marines du Groenland. *Meddr Grønland* **20:** 1–128.

ROSENVINGE, L. K. 1935. On some Danish Phaeophyceae. *K. danske Vidensk. Selsk Skr.* **6:** 1–40.

ROSENVINGE, L. K. & LUND, S. 1943. The marine algae of Denmark. Vol. II. Phaeophyceae. II. Corynophlaeceae, Chordariaceae, Acrothrichaceae, Spermatochnaceae, Sporochnaceae, Desmarestiaceae. Arthrocladiaceae with Supplementary comments on Elachistaceae. *Biol. Skr.* **2:** 1–59.

ROSENVINGE, L. K. & LUND, S. 1947. The marine algae of Denmark. Contributions to their natural history. Vol. II. Phaeophyceae. III. Encoeliaceae, Myriotrichiaceae, Giraudiaceae, Striariaceae, Dictyosiphonaceae, Chordaceae and Laminariaceae. *Biol. Skr.* **4:** 1–99.

ROTH, A. W. 1797. *Catalecta Botanica* **1.** Leipzig.

ROTH, A. W. 1800. *Catalecta Botanica* **2.** Leipzig.

ROTH, A. W. 1806. *Catalecta Botanica* **3.** Leipzig.

RUENESS, J. 1977. *Norsk Algeflora.* Scandinavian University Books, Oslo, Bergen and Trondheim.

RUSSELL, G. & FLETCHER, R. L. 1975. A numerical taxonomic study of the British Phaeophyta. *J. mar. biol. Ass. U.K.* **55:** 763–783.

SAUVAGEAU, C. 1892. Sur quelques algues phéosphorées parasites. *J. Bot. Paris* **6:** 1–48.

SAUVAGEAU, C. 1895. Sur les sporanges pluriloculaires de l'*Asperococcus compressus* Griff. *J. Bot. Paris* **16:** 336–338.

SAUVAGEAU, C. 1897a. Sur quelques Myrionematacées. *Annls Sci. nat. sér. 8 Bot.* **5:** 1–130.

SAUVAGEAU, C. 1897b. Note préliminaire sur les algues marine du Golfe du Gascogne. *J. Bot., Paris* **11:** 166–179; 202–214; 252–257; 263–288; 301–311.

SAUVAGEAU, C. 1899. Les Cutlériacées et leur alternance der générations. *Annls. Sci. nat. sér. 8 Bot.* **10:** 265–362.

SAUVAGEAU, C. 1918. Sur le dissémination et la naturalisation de quelques Algues marines. *Bull. Inst. océanogr. Monaco* **342:** 1–28.

SAUVAGEAU, C. 1924. Sur le curieux developpement d'une algue Pheosporée, *Castagnea zosterae* Thuret. *C.r. hebd. Séanc. Acad. Sci., Paris* **179:** 1381–1384.

SAUVAGEAU, C. 1925. Sur le developpement d'une Algue Pheosporée, *Leathesia difformis* Aresch. *C.r. hebd. Séanc. Acad. Sci., Paris* **180:** 1632–1635.

SAUVAGEAU, C. 1926a. Sur un nouveau type d'alterance des generations chez les Algues brunes; les Sporochnales. *C.r. hebd. Séanc. Acad. Sci., Paris* **182:** 361–364.

SAUVAGEAU, C. 1926b. Sur l'alternance des generations chez le *Carpomitra cabrerae* Kütz. *Bull. stn biol. Arcachon* **23:** 141–192.

SAUVAGEAU, C. 1927a. Sur le *Colpomenia sinuosa* Derb. et Sol. *Bull. stn biol. Arcachon* **24:** 309–353.

SAUVAGEAU, C. 1927b. Sur les problèmes du *Giraudya*. *Bull. stn biol. Arcachon* **24:** 1–74.

SAUVAGEAU, C. 1928. Sur les Algues Phéosporées à eclipse ou Eclipsiophycées. *Recl Trav. bot. néerl.* **25:** 260–270.

SAUVAGEAU, C. 1929. Sur le développement de quelques Phéosporées. *Bull. stn biol. Arcachon* **26:** 253–420.

SAUVAGEAU, C. 1931. Sur quelques algues Phéosporées de Guethary (Basses-Pyrenées). *Bull. stn biol. Arcachon* **30:** 1–128.

SAUVAGEAU, C. 1933a. Sur quelques algues phéosporées de Guéthary (Basses-Pyrenées). *Bull. stn biol. Arcachon* **30:** 1–128.

SAUVAGEAU, C. 1933b. Un genre *Symphoriococcus* Reinke est-il justifie? *Bull. stn biol. Arcachon* **30:** 179–188.

SCHIFFNER, V. 1916. Studien über Algen des Adriatischen Meeres. *Wiss. Meeresunters. (Helgol.), N.F.* **11:** 129–198.

SCHLÖSSER, L. A. 1935. Zur Entwicklungsphysiologie des Generationswechsels von *Cutleria*. *Biol. Zbl.* **55:** 198–208.

SCHREIBER, E. 1932. Über die Entwicklungsgeschichte und die systematische stellung der Desmarestiaceen. *Z. Bot.* **25:** 561–582.

SCHUH, R. E. 1900 *Rhabdinocladia*, a new genus of brown algae. *Rhodora* **2:** 111–112.

SEARS, J. R. & WILCE, R. T. 1973. Sublittoral benthic marine algae of southern Cape Cod and adjacent islands: *Pseudolithoderma paradoxum* sp. nov. (Ralfsiaceae, Ectocarpales). *Phycologia* **12:** 75–82.

SETCHELL, W. A. & GARDNER, N. L. 1924. Phycological contributions. VII. *Univ. Calif. Publs. Bot.* **13:** 1–13.

SETCHELL, W. A. & GARDNER, N. L. 1925. The marine algae of the Pacific coast of North America. III Melanophyceae. *Univ. Calif. Publs Bot.* **8:** 383–898.

SHINMURA, I. 1977. Life history of *Cladosiphon okamuranus* Tokida from southern Japan. *Bull. Jap. Soc. Phycol.* **25:** 333–340.

SILVA, P. C. 1959. Remarks on algal nomenclature II. *Taxon* **8:** 60–64.

SKOTTSBERG, C. 1911. Beobachtungen über einige Meeresalgen aus der Gegend von Tvärminne im südwestlichen Finnland. *Acta Soc. Fauna Flora fenn.* **34:** 3–18.

SMITH, G. M. 1944. *Marine algae of the Monterey Peninsula, California.* Stanford University Press, Stanford.

SMITH, G. M. 1955. *Cryptogamic botany.* Vol. 1. McGraw-Hill, New York.

SMITH, J. E. & SOWERBY, J. 1790–1814. *English botany* 1st edn. Vols. 1–36. London.

SMITH, J. E. & SOWERBY, J. 1843. Supplement to the English Botany of the late Sir J. E. Smith and Mr Sowerby. Vol. III [Tabs 2797–2867]. J. Sowerby; Longman & Co. Sherwood & Co. London.

SOUTH, G. R. 1976. A check list of marine algae of Eastern Canada – first revision. *J. mar. biol. Ass. U.K.* **56:** 871–843.

SOUTH, G. R. 1980. Observations on the life histories of *Punctaria plantaginea* (Roth) Greville and *Punctaria orbiculata* Jao (Punctariaceae, Phaeophyta). *Phycologia* **19:** 266–272.

SOUTH, G. R. & HOOPER, R. G. 1980. A catalogue and atlas of the benthic marine algae of the Island of Newfoundland. *Mem. Univ. Nfld. Occas. Pap. Biol.* **3:** 1–136.

STACKHOUSE, J. 1801. *Nereis Britannica.* ed. 1. **3.** Bathoniae and Londini.

STEGENGA, H. & MOL, I. 1983. *Flora van de'Nederlandse zeewieren.* nr 33. Koninklijke Nederlandse Natuurhistorische Vereniging. Bibliotheek. Uitgave.

STRÖMFELT, H. F. G. 1886. Einige für die Wissenschaft neue Meeresalgen aus Island. *Bot. Zbl.* **26:** 172–173.

STRÖMFELT, H. F. G. 1888. Algae novae quas ad litora Scandinaviae indagavit. *Notarisia* **3:** 381–384.

SUNDENE, O. 1953. The algal vegetation of Oslofjord. *Skr. norske Vidensk-Akad.* Mat.-naturv Kl. **2:** 1–244.

SVEDELIUS, N. 1901. Studier öfver Östersjöns Hafsalgflora. 140 pp. *Akad. Afh. Uppsala.*

TANAKA, J. & CHIHARA, M. 1980a. Taxonomic study of the Japanese crustose brown algae (1) General account and the order Ralfsiales. *J. Jap. Bot.* **56:** 193–202.

TANAKA, J. & CHIHARA, M. 1980b. Taxonomic study of the Japanese crustose brown algae (2) *Ralfsia* (Ralfsiaceae, Ralfsiales) (Part 1). *J. Jap. Bot.* **55:** 225–236.

TANAKA, J. & CHIHARA, M. 1980c. Taxonomic study of the Japanese crustose brown algae (3) *Ralfsia* (Ralfsiaceae, Ralfsiales) (Part 2). *J. Jap. Bot.* **55:** 337–342.

TANAKA, J. & CHIHARA, M. 1982. Morphology and taxonomy of *Mesospora schmidtii* Weber van Bosse, Mesosporaceae fam. nov. (Ralfsiales, Phaeophyceae). *Phycologia* **21:** 382–389.

TATEWAKI, M. 1966. Formation of a crustaceous sporophyte with unilocular sporangia in *Scytosiphon lomentaria.* *Phycologia* **6:** 62–66.

TAYLOR, W. R. 1937a. Marine algae of the northeastern coast of North America. *Univ. Mich. Stud. scient. Ser.* **13:** 1–427.

TAYLOR, W. R. 1937b. Notes on north Atlantic marine algae. I. *Pap. Mich. Acad. Sci.* **22:** 225–233.

TAYLOR, W. R. 1957. *Marine algae of the north-eastern coast of North America.* Revised edition. University of Michigan Press, Ann Arbor.

THURET, G. 1850. Recherches sur les zoospores des Algues et les antheridies des Cryptogames. *Annls Sci. nat. sér. 3 Bot.* **14:** 214–260.

Tittley, I. & Shaw, K. M. 1980. Numerical and field methods in the study of the marine flora of chalk cliffs. *In* Price, J. H., Irvine, D. E. G. & Farnham, W. F. (Eds), *The shore environment* **1:** *methods* pp. 213–240. Systematics Association Special Volume 17(a). Academic Press, London and New York.

Tokida, J. 1954. The marine algae of southern Saghalien. *Mem. Fac. Fish. Hokkaido Univ.* **2:** 1–264.

Vandermeulen, H., DeWreede, R. E. & Cole, K. M. 1984. Nomenclatural recommendations for three species of *Colpomenia* (Scytosiphonales, Phaeophyta). *Taxon* **33:** 324–329.

Velley, T. 1795. Coloured figures of marine plants found on the southern coast of England. Bath.

Verlaque, M. & Boudouresque, C. F. 1981. Vegetation marine de la Corse (Mediterranee) V. Documents pour la flore des algues. *Biol. Ecol. Medit.* **8:** 139–156.

Waern, M. 1949. Remarks on Swedish *Lithoderma.* *Svensk bot. Tidskr.* **43:** 633–670.

Waern, M. 1952. Rocky shore algae in the Öregrund Archipelago. *Acta phytogeogr. suec.* **30:** 1–298.

Wanders, J. B. W., Hoek, C. van den & Schillern, E. N. van nes 1972. Observations on the life history of *Elachista stellaris* (Phaeophyceae) in culture. *Neth. J. Sea Res.* **5:** 458–491.

Weber-Van Bosse, A. 1913. Liste des algues du Siboga. I. Myxophyceae, Chlorophyceae, Phaeophyceae. *Siboga Expeditie Monographie* **59a:** 1–186. Leiden.

Wilce, R. T. 1966. *Pleurocladia lacustris* in arctic America. *J. Phycol.* **2:** 57–66.

Wilce, R. T., Webber, E. E. & Sears, J. R. 1970. *Petroderma* and *Porterinema* in the New World. *Mar. Biol.* **5:** 119–135.

Wille, N. & Rosenvinge, L. K. 1885. *Alger fra Novaia-Zemlia og Kara-Havet, samlede paa Dijmphna – Expeditionen 1882–83 af Th. Holm. Dijmphna-Togtets zoologisck-botaniske Udbytte.* Kjøbenhavn.

Wollny, R. 1881. Die Meeresalgen von Helgoland. *Hedwigia* **20:** 1–32.

Wynne, M. J. 1969. Life history and systematic studies of some Pacific North American Phaeophyceae (brown algae). *Univ. Calif. Publs Bot.* **50:** 1–88.

Wynne, M. J. 1972a. Studies on the life forms in nature and in culture of selected brown algae. *In* Abbott, I. A. & Kurogi, M. (Eds) *Contributions to the systematics of benthic marine algae of the North Pacific,* pp. 133–145. Jap. Soc. Phycology, Japan.

Wynne, M. 1972b. Culture studies of Pacific coast Phaeophyceae. *Mém. Soc. bot. Fr.* **1972:** 129–144.

Wynne, M. J. & Loiseaux, S. 1976. Recent advances in the life history studies of the Phaeophyta. *Phycologia* **15:** 435–452.

Yamada, Y. 1935. The marine algae of Urup, Middle Kuriles, especially from the vicinity of Ioma Bay. *Scient. Pap. Inst. algol. Hokkaido.* **1:** 1–26.

Yamanouchi, S. 1909. Cytology of *Cutleria* and *Aglaozonia.* A preliminary paper. *Bot. Gaz.* **48:** 380–386.

Yamanouchi, S. 1912. The life history of *Cutleria.* *Bot. Gaz.* **54:** 441–502.

Yamanouchi, S. 1913. The life history of *Zanardinia. Bot. Gaz.* **56:** 1–35.

Yendo, K. 1919. The germination and development of some marine algae. *Bot. Mag. Tokyo* **33:** 171–184.

Zanardini, G. 1843. *Saggio di Classificazione Naturale delle Ficee.* Venezia.

Plate 1

Myrionema strangulans
a, Habit, on *Ulva*. Bar = 10 mm. b. Habit, on *Enteromorpha*. Bar = 10 mm.

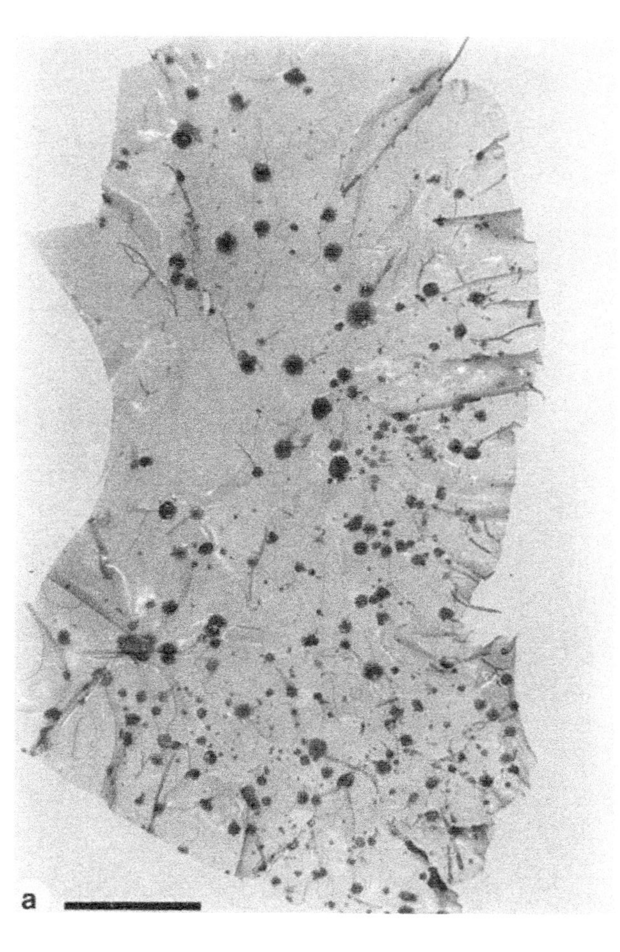
a

b

Plate 2

a. *Myrionema magnusii* Habit, two plants on *Zostera*. Note abundant hyaline ascocysts. Bar = 100 µm. b. *Myrionema strangulans* Squash preparation (S.P.) showing monostromatic basal layer giving rise to paraphyses and a single unilocular sporangium. Bar = 10 µm. c. *Elachista flaccida* Habit of plants on *Cystoseira*. Bar = 5 mm. d. *Elachista scutulata* Vertical section of plant showing large colourless cells of basal cushion giving rise terminally to paraphyses, a single exerted filament (arrowed) and a unilocular sporangium. Bar = 56 µm. e. *Corynophlaea crispa* Vertical section of plant showing colourless cells of basal cushion giving rise terminally to paraphyses, unilocular sporangia and a single hair (arrowed). Bar = 18 µm. f. *Myriactula rivulariae* S.P. of plant showing fusiform paraphyses and basally situated unilocular sporangia. Bar = 50 µm.

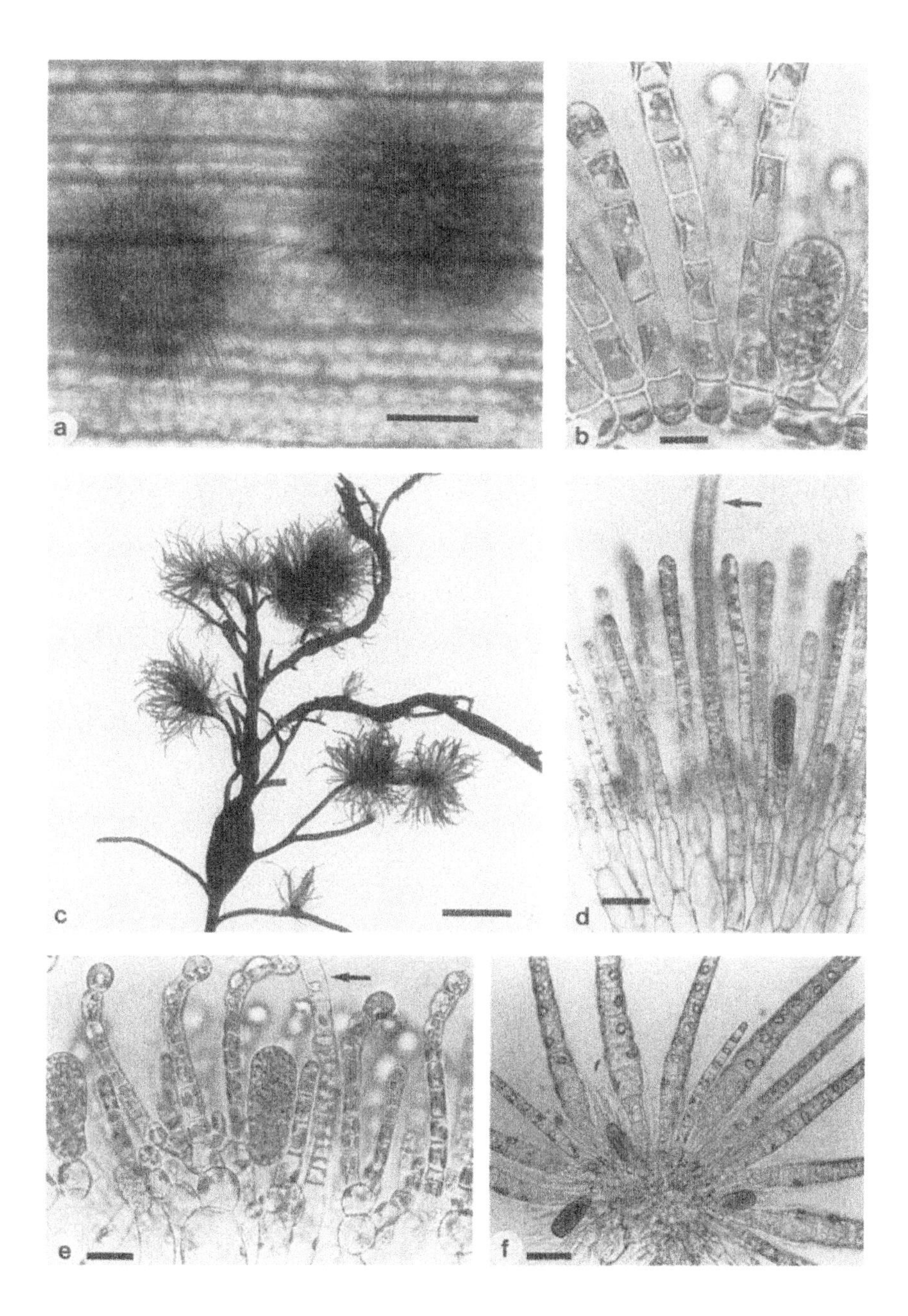
a
b
c
d
e
f

Plate 3

Halothrix lumbricalis

a. Erect filaments of plant with intercalary plurilocular sporangia. Bar = 55 µm. b. Portion of erect filament with intercalary unilocular sporangium. Bar = 27 µm.

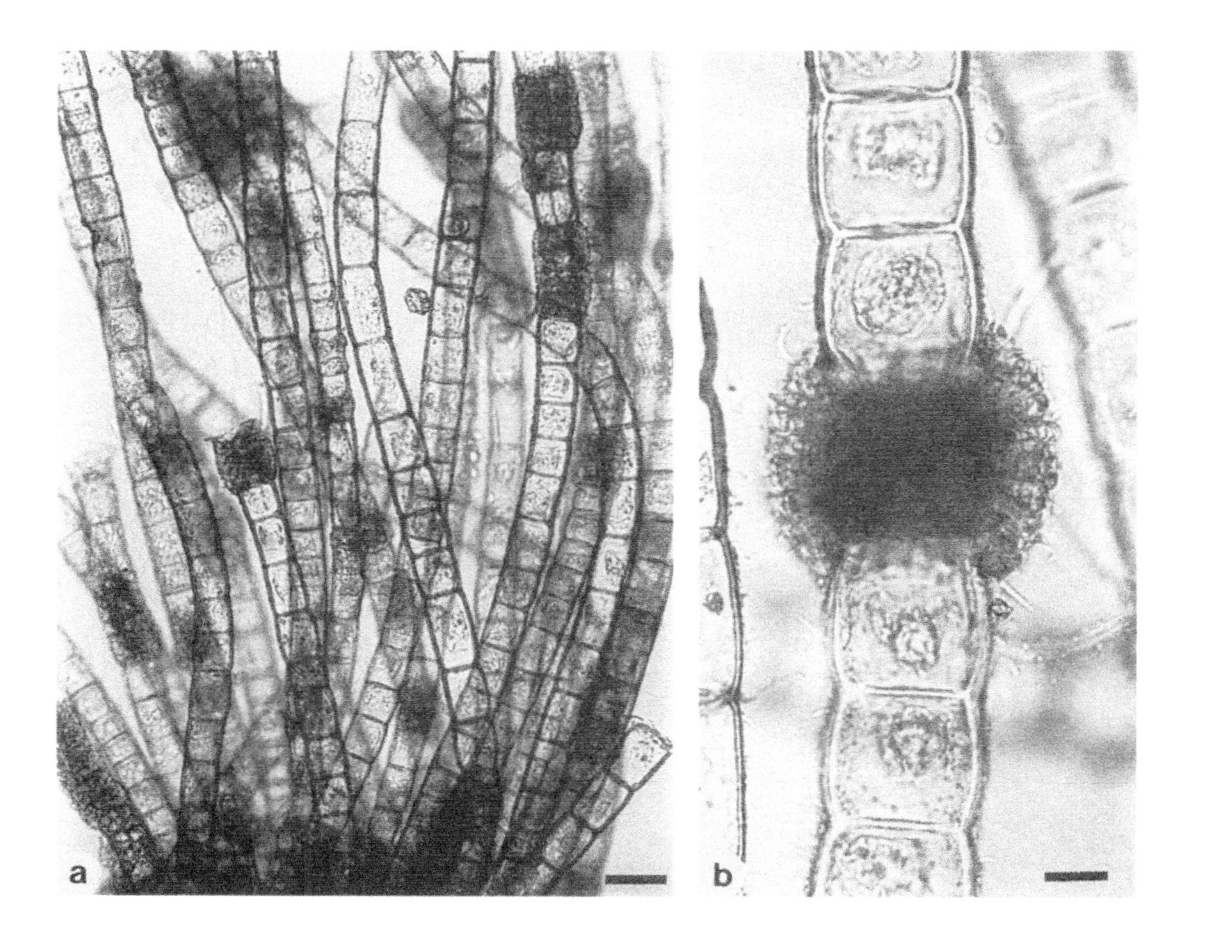
a
b

Plate 4

a. *Microcoryne ocellata* Habit of plants, on *Chorda filum* (NB-herbarium pressed specimens). Bar = 5 mm. b. *Litosiphon laminariae* Habit of plants, on *Chorda filum*. Bar = 5 mm. c *Myriotrichia clavaeformis* Habit of plants, on *Scytosiphon lomentaria*. Bar = 10 mm. d-e. *M. clavaeformis* Portions of thalli at different stages of development. Bar = 80 μm (D), 200 μm (E).

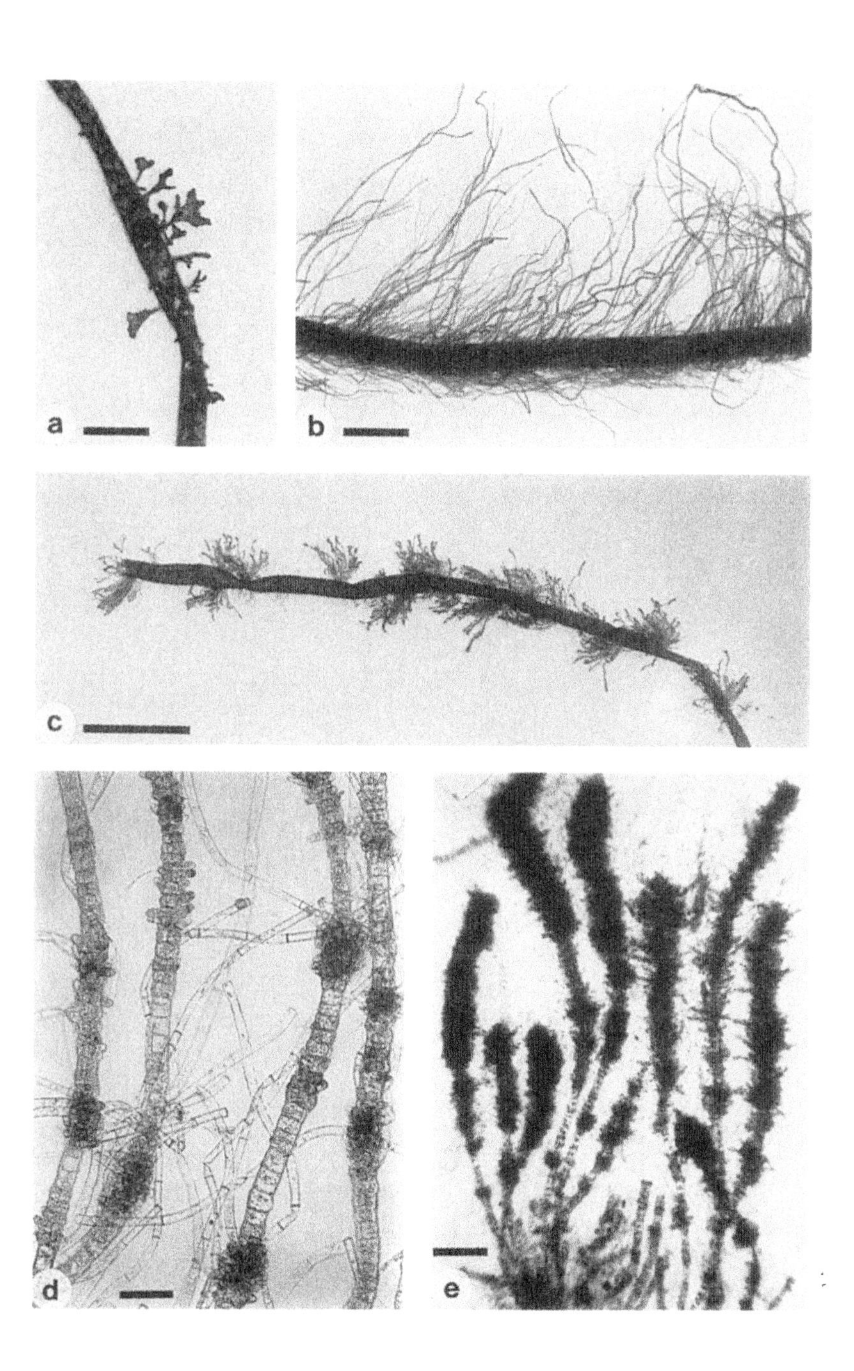
a
b
c
d
e

Plate 5

Asperococcus fistulosus
a. Habit of plant. Bar = 20 mm. b. Portions of erect thalli. Bar = 10 mm.

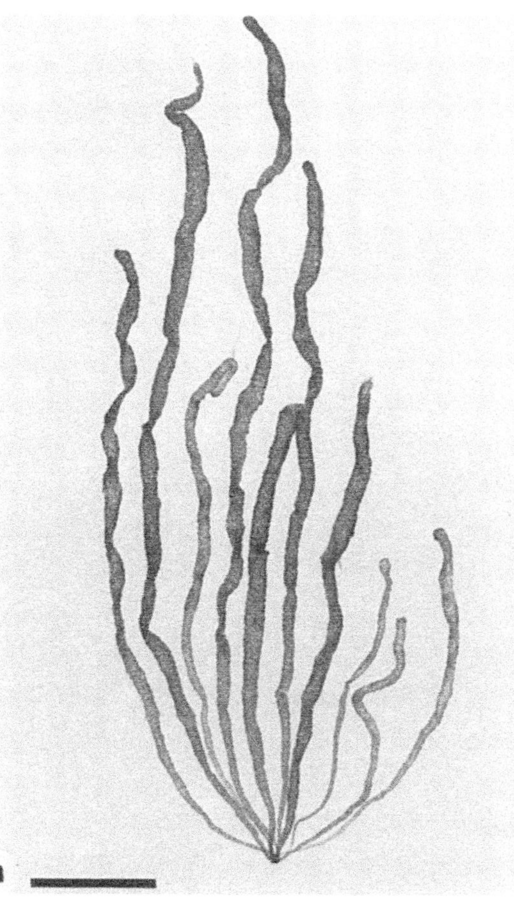
a

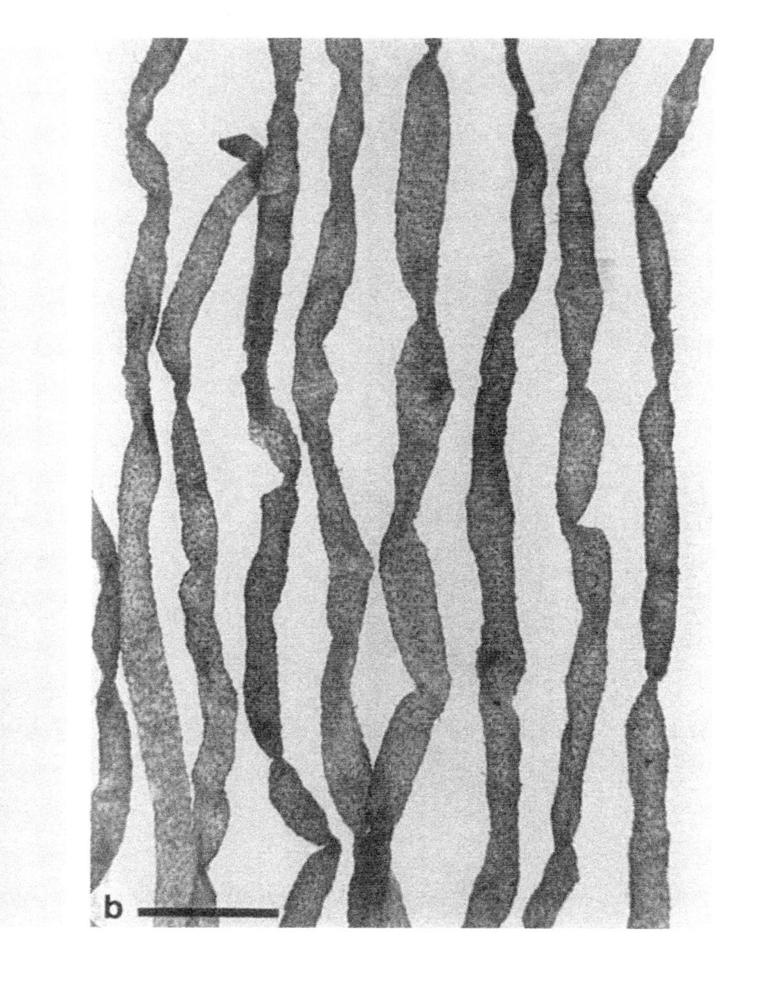
b

Plate 6

a. *Punctaria latifolia* Habit of plant. Bar = 10 mm. b. *P. plantaginea* Habit of plant. Bar = 15 mm.

a

b

Plate 7

Colpomenia peregrina
a. Habit of plants, on *Cystoseria*. Bar = 20 mm. b. Single, large plant. Bar = 12 mm.

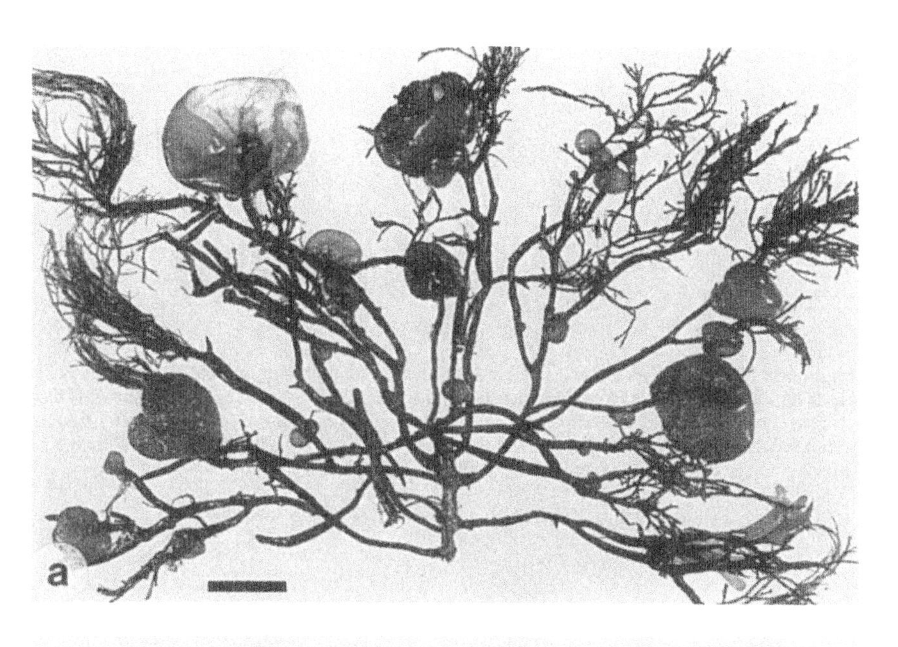
a

b

Plate 8

a. *Compsonema saxicolum* 'phase' S.P. of thallus showing erect filaments and basal unilocular sporangia (arrowed). Bar = 25 μm. b. *Compsonema saxicolum* 'phase' S.P. of thallus portion showing distromatic base giving to erect filaments and unilocular sporangia. Bar = 15 μm. c. *Microspongium gelatinosum* 'phase' Fan-like arrangement of thallus produced by squash preparation, showing basal crust giving rise to erect paraphyses and unilocular sporangia. Bar = 35 μm.

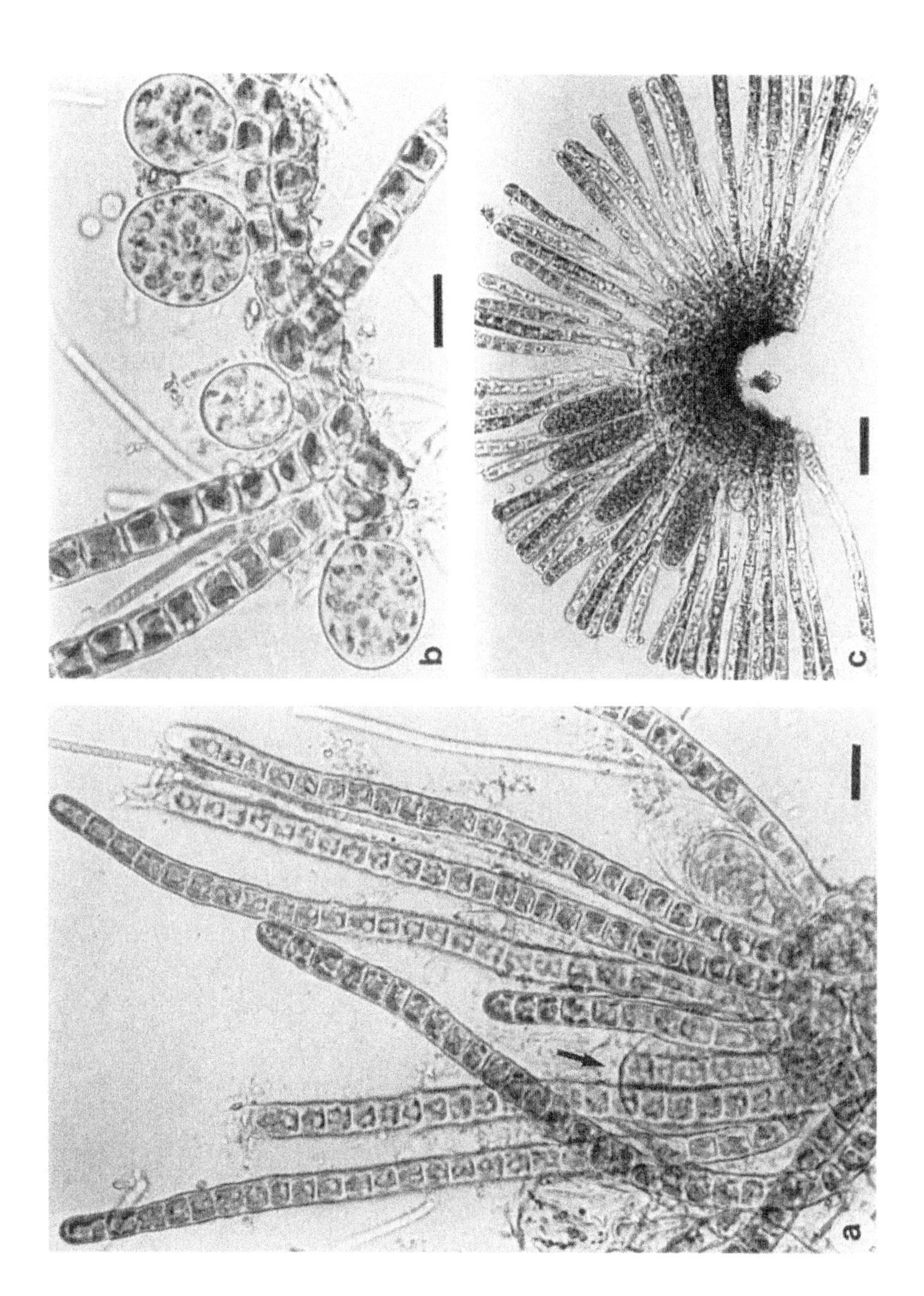

Plate 9

a–c. *Petalonia fascia* Habit of plants showing variations in blade morphology. Bar = 10 mm.

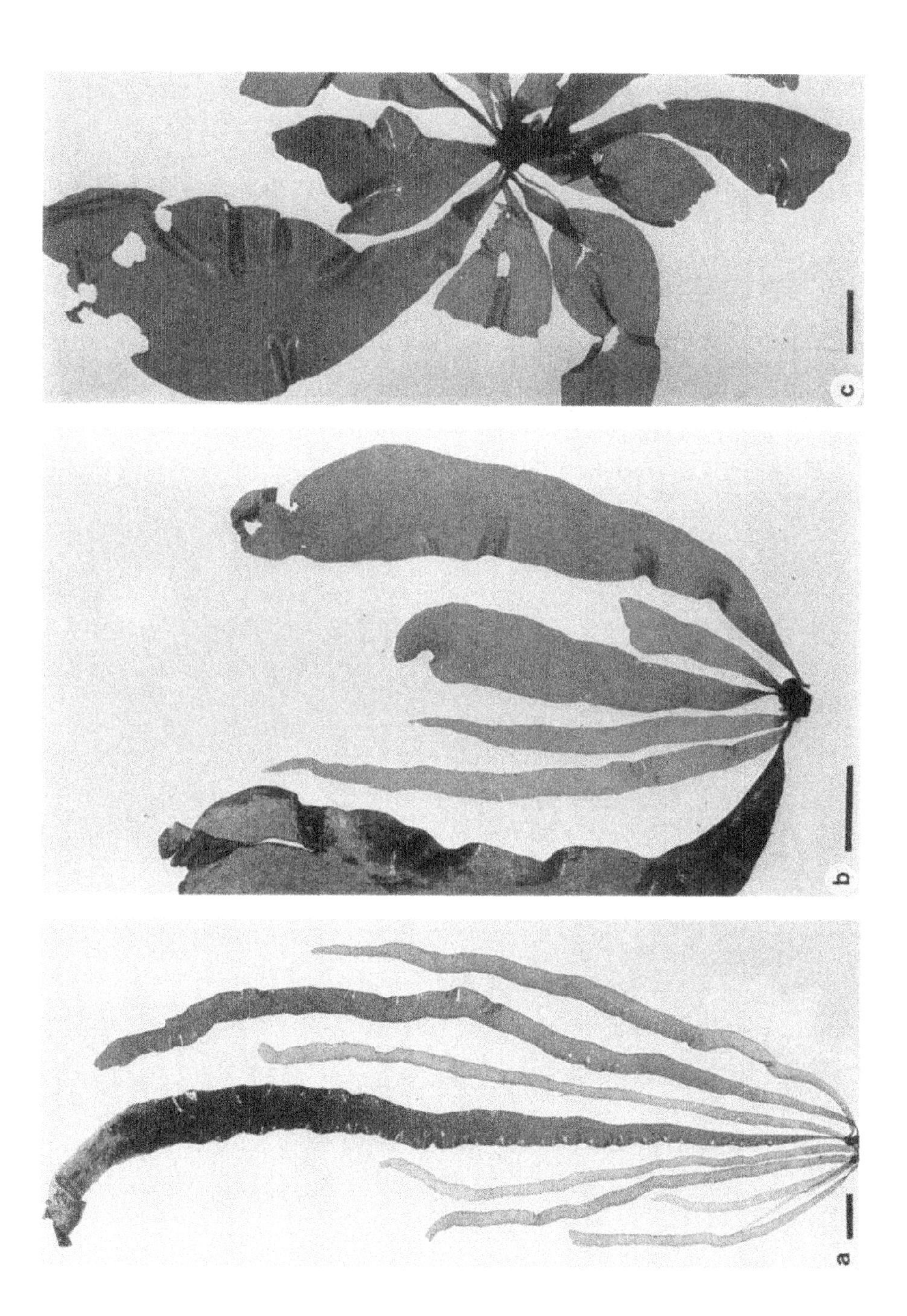

Plate 10

Petalonia filiformis
a. Habit of plants. Bar = 5 mm. b. Portion of erect blade showing spiral twisting. Bar = 180 µm.
Petalonia zosterifolia.
c. Habit of plant. Bar = 10 mm.

a
b
c

Plate 11

Scytosiphon lomentaria
a. Habit of plant; narrow form with few constrictions. Bar = 20 μm. b. Habit of plant; broad form with prominent constrictions. Bar = 12 mm.
Scytosiphon dotyi
c. Habit of plant. Bar = 10 mm.

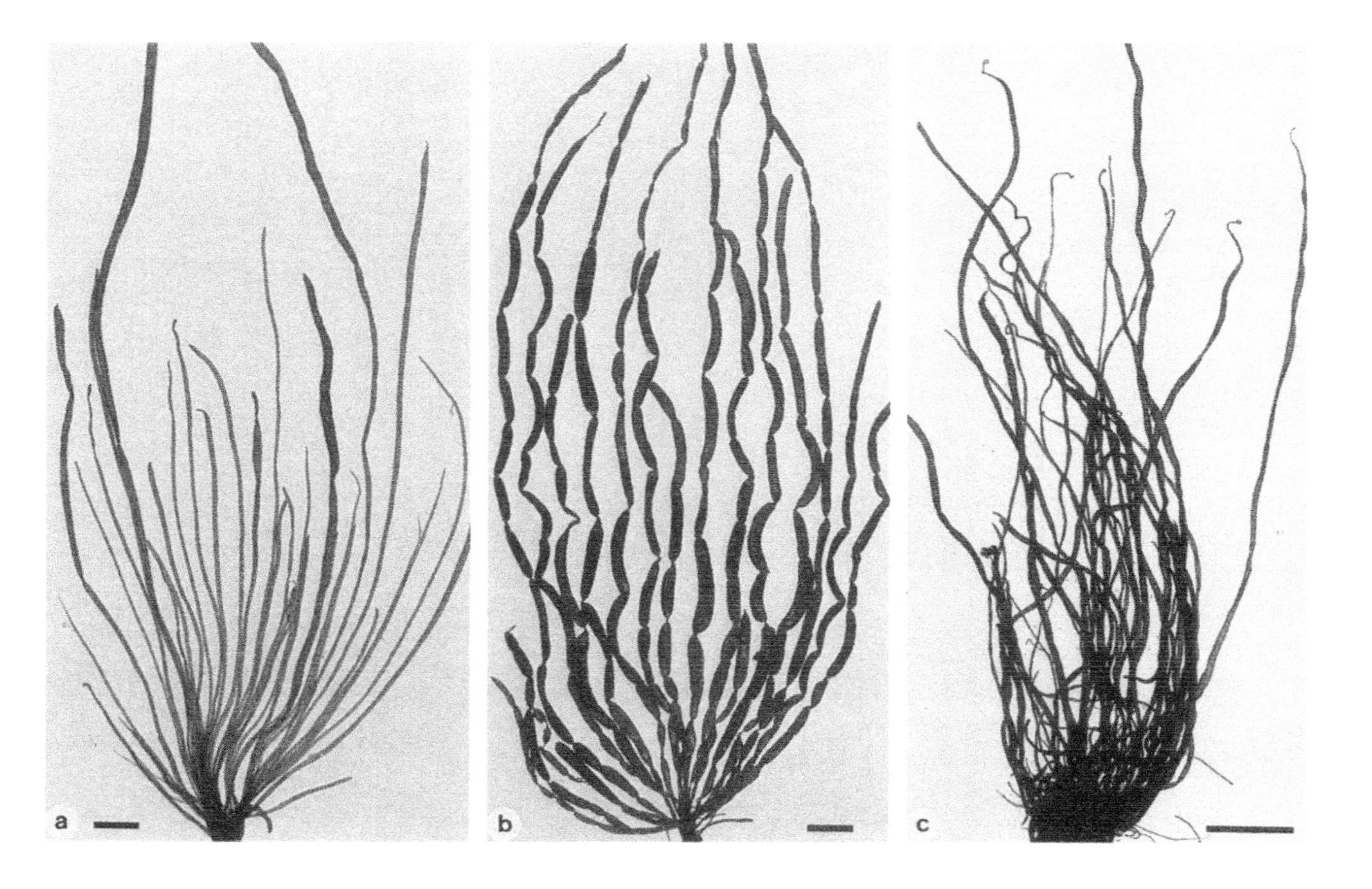
a
b
c

Plate 12

Arthrocladia villosa Habit of plant. Bar = 10 mm.

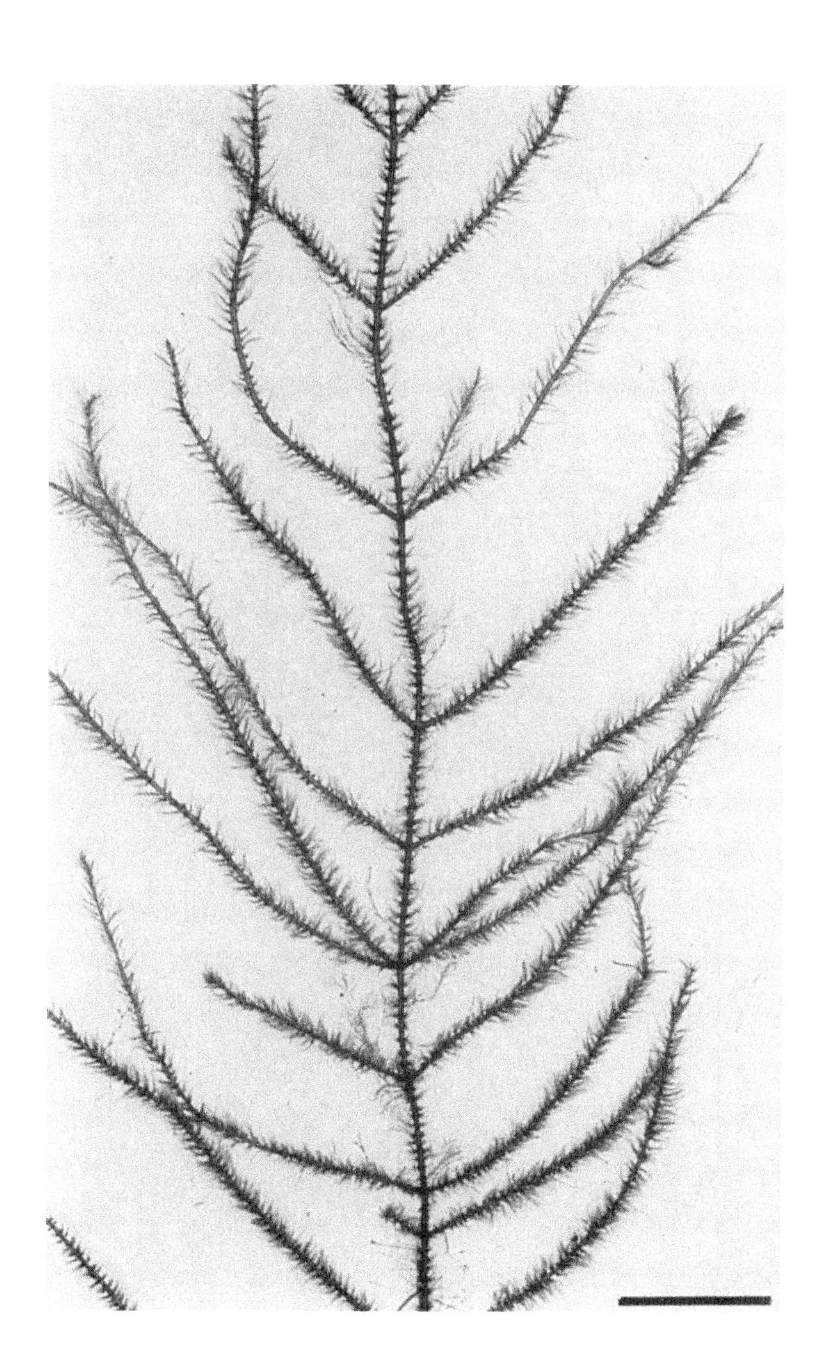

Plate 13

Desmarestia aculeata

a. Portion of plant; summer form. Bar = 14 mm. b. Portion of plant; winter form with prominent serrations. Bar = 10 mm.

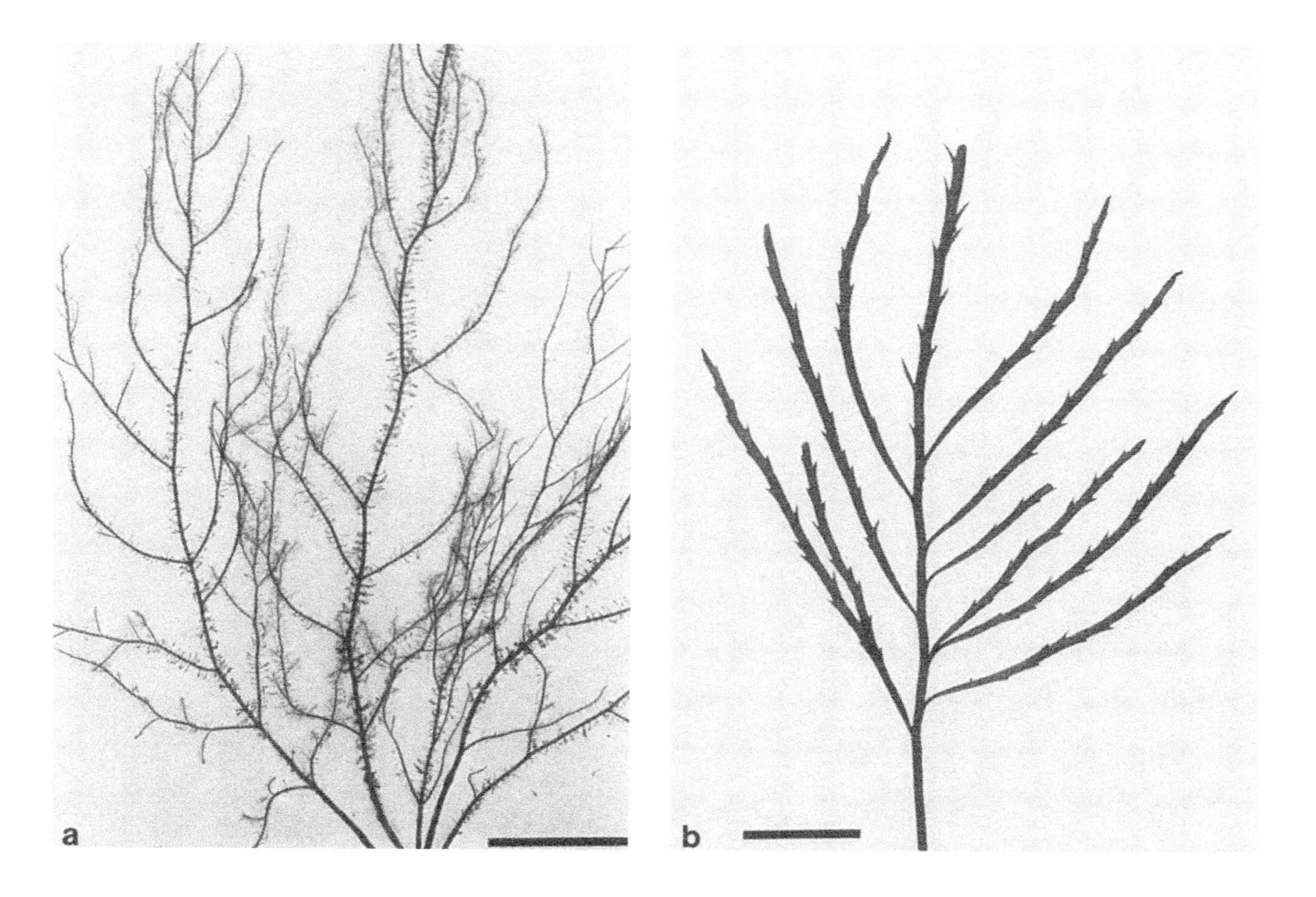
a
b

Plate 14

Desmarestia ligulata

a. Portion of plant. Bar = 20 mm. b. Portion of branch. Bar = 10 mm. c. Edge of branch showing tufts of filaments. Bar = 50 μm.

a
b
c

Plate 15

Desmarestia viridis Habit of plant. Bar = 20 mm.

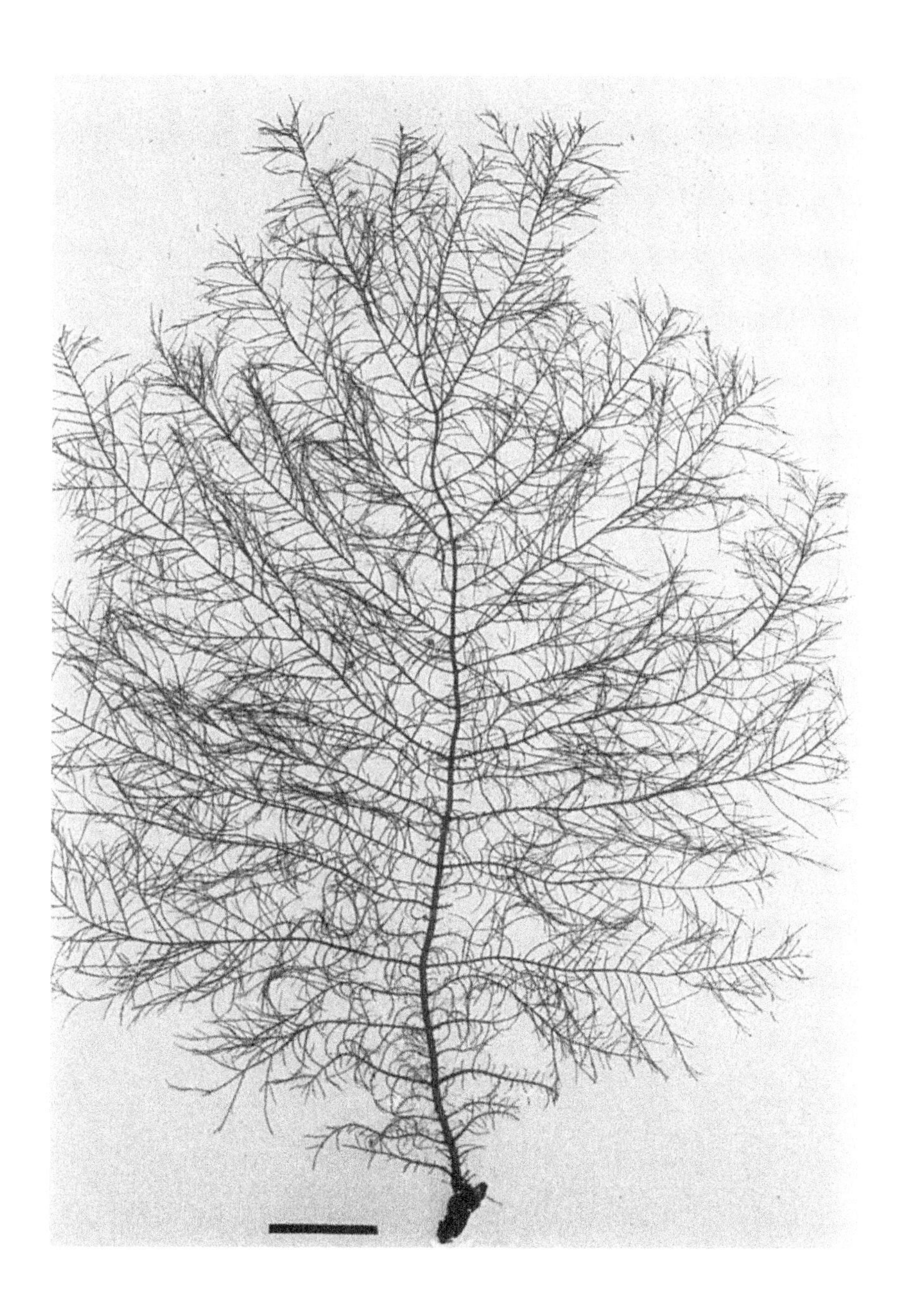

TAXONOMIC INDEX

Orders and families are given in capitals, whilst genera, species and infraspecific taxa are given in roman type. Synonyms are given in *italic* type. All taxa considered in the present treatment and their main page numbers are given in **bold** type–see pp. 52–53 for check-list.

CPSIA information can be obtained at www.ICGtesting.com
Printed in the USA
BVOW08s1714020215

385463BV00006B/53/P

9 781907 807114